Introduction to High-Dimensional Statistics

MONOGRAPHS ON STATISTICS AND APPLIED PROBABILITY

Recently Published Titles

Sufficient Dimension Reduction
Methods and Applications with R
Bing Li 161

Large Covariance and Autocovariance Matrices
Arup Bose and Monika Bhattacharjee 162

The Statistical Analysis of Multivariate Failure Time Data: A Marginal Modeling Approach
Ross L. Prentice and Shanshan Zhao 163

Dynamic Treatment Regimes
Statistical Methods for Precision Medicine
Anastasios A. Tsiatis, Marie Davidian, Shannon T. Holloway, and Eric B. Laber 164

Sequential Change Detection and Hypothesis Testing
General Non-i.i.d. Stochastic Models and Asymptotically Optimal Rules
Alexander Tartakovsky 165

Introduction to Time Series Modeling
Genshiro Kitigawa 166

Replication and Evidence Factors in Observational Studies
Paul R. Rosenbaum 167

Introduction to High-Dimensional Statistics, Second Edition
Christophe Giraud 168

For more information about this series please visit: https://www.crcpress.com/Chapman--HallCRC-Monographs-on-Statistics--Applied-Probability/book-series/CHMONSTAAPP

Introduction to High-Dimensional Statistics

SECOND EDITION

Christophe Giraud

CRC Press is an imprint of the
Taylor & Francis Group, an **informa** business
A CHAPMAN & HALL BOOK

Second edition published 2022

by CRC Press
6000 Broken Sound Parkway NW, Suite 300, Boca Raton, FL 33487-2742

and by CRC Press
2 Park Square, Milton Park, Abingdon, Oxon, OX14 4RN

First edition published by CRC Press 2015

CRC Press is an imprint of Taylor & Francis Group, LLC

Library of Congress Cataloging-in-Publication Data

Names: Giraud, Christophe, author.
Title: Introduction to high-dimensional statistics / Christophe Giraud.
Description: Second edition. | Boca Raton : CRC Press, 2021. | Series:
Chapman & Hall/CRC monographs on statistics and applied probability |
Includes bibliographical references and index.
Identifiers: LCCN 2021009057 (print) | LCCN 2021009058 (ebook) | ISBN
9780367716226 (hardback) | ISBN 9780367746216 (paperback) | ISBN
9781003158745 (ebook)
Subjects: LCSH: Dimensional analysis. | Multivariate analysis. | Big data.
| Statistics.
Classification: LCC QC20.7.D55 G57 2021 (print) | LCC QC20.7.D55 (ebook)
| DDC 519.5/35--dc23
LC record available at https://lccn.loc.gov/2021009057
LC ebook record available at https://lccn.loc.gov/2021009058

ISBN: 978-0-367-71622-6 (hbk)
ISBN: 978-0-367-74621-6 (pbk)
ISBN: 978-1-003-15874-5 (ebk)

DOI: 10.1201/9781003158745

Publisher's note: This book has been prepared from camera-ready copy provided by the authors.

Contents

Preface, second edition **xiii**

Preface **xv**

Acknowledgments **xvii**

1 Introduction **1**
1.1 High-Dimensional Data 1
1.2 Curse of Dimensionality 3
1.2.1 Lost in the Immensity of High-Dimensional Spaces 3
1.2.2 Fluctuations Cumulate 6
1.2.3 Accumulation of Rare Events May Not Be Rare 10
1.2.4 Computational Complexity 13
1.3 High-Dimensional Statistics 13
1.3.1 Circumventing the Curse of Dimensionality 13
1.3.2 A Paradigm Shift 16
1.3.3 Mathematics of High-Dimensional Statistics 17
1.4 About This Book 18
1.4.1 Statistics and Data Analysis 18
1.4.2 Purpose of This Book 19
1.4.3 Overview 19
1.5 Discussion and References 20
1.5.1 Take-Home Message 20
1.5.2 References 21
1.6 Exercises 21
1.6.1 Strange Geometry of High-Dimensional Spaces 21
1.6.2 Volume of a p-Dimensional Ball 21
1.6.3 Tails of a Standard Gaussian Distribution 22
1.6.4 Principal Component Analysis 22
1.6.5 Basics of Linear Regression 23
1.6.6 Concentration of Square Norm of Gaussian Random Variable 24
1.6.7 A Simple Proof of Gaussian Concentration 25

2 Model Selection **27**
2.1 Statistical Setting 27
2.2 Selecting among a Collection of Models 30
2.2.1 Models and Oracle 30
2.2.2 Model Selection Procedures 33
2.3 Risk Bound for Model Selection 37
2.3.1 Oracle Risk Bound 37
2.4 Computational Issues 41
2.5 Illustration 43
2.6 Alternative Point of View on Model Selection 44
2.7 Discussion and References 45
2.7.1 Take-Home Message 45
2.7.2 References 46
2.8 Exercises 46
2.8.1 Orthogonal Design 46
2.8.2 Risk Bounds for the Different Sparsity Settings 48
2.8.3 Collections of Nested Models 50
2.8.4 Segmentation with Dynamic Programming 51
2.8.5 Goldenshluger–Lepski Method 52
2.8.6 Estimation under Convex Constraints 53

3 Minimax Lower Bounds **55**
3.1 Minimax Risk 55
3.2 Recipe for Proving Lower Bounds 56
3.2.1 Fano's Lemma 57
3.2.2 From Fano's Lemma to a Lower Bound over a Finite Set 60
3.2.3 Back to the Original Problem: Finding a Good Discretization 62
3.3 Illustration 62
3.4 Minimax Risk for Coordinate-Sparse Regression 63
3.4.1 Lower Bound on the Minimax Risk 64
3.4.2 Minimax Optimality of the Model Selection Estimator 66
3.4.3 Frontier of Estimation in High Dimensions 67
3.5 Discussion and References 68
3.5.1 Take-Home Message 68
3.5.2 References 68
3.6 Exercises 68
3.6.1 Kullback–Leibler Divergence between Gaussian Distribution 68
3.6.2 Spreading Points in a Sparse Hypercube 69
3.6.3 Some Other Minimax Lower Bounds 69
3.6.4 Non-Parametric Lower Bounds 70
3.6.5 Data Processing Inequality and Generalized Fano Inequality 71

4 Aggregation of Estimators **75**
4.1 Introduction 75
4.2 Gibbs Mixing of Estimators 75
4.3 Oracle Risk Bound 76
4.4 Numerical Approximation by Metropolis–Hastings 79
4.5 Numerical Illustration 82
4.6 Discussion and References 83
4.6.1 Take-Home Message 83
4.6.2 References 83
4.7 Exercises 83
4.7.1 Gibbs Distribution 83
4.7.2 Orthonormal Setting with Power Law Prior 84
4.7.3 Group-Sparse Setting 84
4.7.4 Gain of Combining 84
4.7.5 Online Aggregation 85
4.7.6 Aggregation with Laplace Prior 86

5 Convex Criteria **89**
5.1 Reminder on Convex Multivariate Functions 89
5.1.1 Subdifferentials 89
5.1.2 Two Useful Properties 90
5.2 Lasso Estimator 90
5.2.1 Geometric Insights 92
5.2.2 Analytic Insights 92
5.2.3 Oracle Risk Bound 95
5.2.4 Computing the Lasso Estimator 98
5.2.5 Removing the Bias of the Lasso Estimator 102
5.3 Convex Criteria for Various Sparsity Patterns 103
5.3.1 Group-Lasso for Group Sparsity 103
5.3.2 Sparse-Group Lasso for Sparse-Group Sparsity 106
5.3.3 Fused-Lasso for Variation Sparsity 107
5.4 Discussion and References 107
5.4.1 Take-Home Message 107
5.4.2 References 107
5.5 Exercises 108
5.5.1 When Is the Lasso Solution Unique? 108
5.5.2 Support Recovery via the Witness Approach 109
5.5.3 Lower Bound on the Compatibility Constant 111
5.5.4 On the Group-Lasso 112
5.5.5 Dantzig Selector 113
5.5.6 Projection on the ℓ^1-Ball 114
5.5.7 Ridge and Elastic-Net 115
5.5.8 Approximately Sparse Linear Regression 117
5.5.9 Slope Estimator 118
5.5.10 Compress Sensing 120

6 Iterative Algorithms 123
6.1 Iterative Hard Thresholding 123
6.1.1 Reminder on the Proximal Method 124
6.1.2 Iterative Hard-Thresholding Algorithm 124
6.1.3 Reconstruction Error 125
6.2 Iterative Group Thresholding 131
6.3 Discussion and References 133
6.3.1 Take-Home Message 133
6.3.2 References 134
6.4 Exercices 134
6.4.1 Linear versus Non-Linear Iterations 134
6.4.2 Group Thresholding 135
6.4.3 Risk Bound for Iterative Group Thresholding 136

7 Estimator Selection 139
7.1 Estimator Selection 140
7.2 Cross-Validation Techniques 141
7.3 Complexity Selection Techniques 143
7.3.1 Coordinate-Sparse Regression 144
7.3.2 Group-Sparse Regression 145
7.3.3 Multiple Structures 146
7.4 Scale-Invariant Criteria 146
7.5 Discussion and References 151
7.5.1 Take-Home Message 151
7.5.2 References 153
7.6 Exercises 154
7.6.1 Expected V-Fold CV ℓ^2-Risk 154
7.6.2 Proof of Corollary 7.5 155
7.6.3 Some Properties of Penalty (7.4) 155
7.6.4 Selecting the Number of Steps for the Forward Algorithm 157

8 Multivariate Regression 159
8.1 Statistical Setting 159
8.2 Reminder on Singular Values 160
8.3 Low-Rank Estimation 161
8.3.1 When the Rank of A^* is Known 161
8.3.2 When the Rank of A^* Is Unknown 165
8.4 Low Rank and Sparsity 167
8.4.1 Row-Sparse Matrices 167
8.4.2 Criterion for Row-Sparse and Low-Rank Matrices 168
8.4.3 Convex Criterion for Low-Rank Matrices 172
8.4.4 Computationally Efficient Algorithm for Row-Sparse and Low-Rank Matrices 175
8.5 Discussion and References 176
8.5.1 Take-Home Message 176

8.5.2 References 176
8.6 Exercises 176
8.6.1 Hard Thresholding of the Singular Values 176
8.6.2 Exact Rank Recovery 177
8.6.3 Rank Selection with Unknown Variance 177

9 Graphical Models 179
9.1 Reminder on Conditional Independence 180
9.2 Graphical Models 180
9.2.1 Directed Acyclic Graphical Models 180
9.2.2 Non-Directed Models 182
9.3 Gaussian Graphical Models 185
9.3.1 Connection with the Precision Matrix and the Linear Regression 185
9.3.2 Estimating g by Multiple Testing 186
9.3.3 Sparse Estimation of the Precision Matrix 187
9.3.4 Estimation by Regression 188
9.4 Practical Issues 192
9.5 Discussion and References 193
9.5.1 Take-Home Message 193
9.5.2 References 193
9.6 Exercises 194
9.6.1 Factorization in Directed Models 194
9.6.2 Moralization of a Directed Graph 194
9.6.3 Convexity of $-\log(\det(K))$ 195
9.6.4 Block Gradient Descent with the ℓ^1/ℓ^2 Penalty 195
9.6.5 Gaussian Graphical Models with Hidden Variables 195
9.6.6 Dantzig Estimation of Sparse Gaussian Graphical Models 196
9.6.7 Gaussian Copula Graphical Models 198
9.6.8 Restricted Isometry Constant for Gaussian Matrices 200

10 Multiple Testing 203
10.1 Introductory Example 203
10.1.1 Differential Expression of a Single Gene 203
10.1.2 Differential Expression of Multiple Genes 204
10.2 Statistical Setting 205
10.2.1 p-Values 205
10.2.2 Multiple-Testing Setting 207
10.2.3 Bonferroni Correction 207
10.3 Controlling the False Discovery Rate 207
10.3.1 Heuristics 208
10.3.2 Step-Up Procedures 208
10.3.3 FDR Control under the WPRD Property 211
10.4 Illustration 214
10.5 Discussion and References 215

10.5.1 Take-Home Message 215
10.5.2 References 216
10.6 Exercises 216
10.6.1 FDR versus FWER 216
10.6.2 WPRD Property 217
10.6.3 Positively Correlated Normal Test Statistics 217

11 Supervised Classification 219
11.1 Statistical Modeling 219
11.1.1 Bayes Classifier 219
11.1.2 Parametric Modeling 220
11.1.3 Semi-Parametric Modeling 221
11.1.4 Non-Parametric Modeling 223
11.2 Empirical Risk Minimization 223
11.2.1 Misclassification Probability of the Empirical Risk Minimizer 224
11.2.2 Vapnik–Chervonenkis Dimension 228
11.2.3 Dictionary Selection 231
11.3 From Theoretical to Practical Classifiers 233
11.3.1 Empirical Risk Convexification 233
11.3.2 Statistical Properties 236
11.3.3 Support Vector Machines 240
11.3.4 AdaBoost 244
11.3.5 Classifier Selection 245
11.4 Discussion and References 246
11.4.1 Take-Home Message 246
11.4.2 References 246
11.5 Exercises 246
11.5.1 Linear Discriminant Analysis 246
11.5.2 VC Dimension of Linear Classifiers in $\mathbb{R}^d$ 247
11.5.3 Linear Classifiers with Margin Constraints 248
11.5.4 Spectral Kernel 248
11.5.5 Computation of the SVM Classifier 248
11.5.6 Kernel Principal Component Analysis 249

12 Clustering 251
12.1 Proximity-Separation-Based Clustering 252
12.1.1 Clustering According to Proximity and Separation 252
12.1.2 Kmeans Paradigm 252
12.1.3 Hierarchical Clustering Algorithms 253
12.2 Model-Based Clustering 255
12.2.1 Gaussian Sub-Population Model 255
12.2.2 MLE Estimator 256
12.3 Kmeans 257
12.3.1 Kmeans Algorithm 257

12.3.2 Bias of Kmeans 258
12.3.3 Debiasing Kmeans 259
12.4 Semi-Definite Programming Relaxation 261
12.5 Lloyd Algorithm 264
12.5.1 Bias of Lloyd Algorithm 264
12.6 Spectral Algorithm 267
12.7 Recovery Bounds 269
12.7.1 Benchmark: Supervised Case 270
12.7.2 Analysis of Spectral Clustering 271
12.7.3 Analysis of Lloyd Algorithm 276
12.8 Discussion and References 283
12.8.1 Take-Home Message 283
12.8.2 References 284
12.9 Exercises 284
12.9.1 Exact Recovery with Hierarchical Clustering 284
12.9.2 Bias of Kmeans 285
12.9.3 Debiasing Kmeans 285
12.9.4 Bayes Classifier for the Supervised Case 287
12.9.5 Cardinality of an ε-Net 287
12.9.6 Operator Norm of a Random Matrix 288
12.9.7 Large Deviation for the Binomial Distribution 288
12.9.8 Sterling Numbers of the Second Kind 289
12.9.9 Gaussian Mixture Model and EM Algorithm 290

Appendix A Gaussian Distribution 293
A.1 Gaussian Random Vectors 293
A.2 Chi-Square Distribution 294
A.3 Gaussian Conditioning 295

Appendix B Probabilistic Inequalities 297
B.1 Basic Inequalities 297
B.2 Concentration Inequalities 299
B.2.1 McDiarmid Inequality 299
B.2.2 Gaussian Concentration Inequality 301
B.2.3 Concentration of Quadratic Forms of Gaussian Vectors 303
B.3 Symmetrization and Contraction Lemmas 305
B.3.1 Symmetrization Lemma 305
B.3.2 Contraction Principle 306
B.4 Birgé's Inequality 309

Appendix C Linear Algebra 311
C.1 Singular Value Decomposition 311
C.2 Moore–Penrose Pseudo-Inverse 312
C.3 Matrix Norms 313
C.4 Matrix Analysis 314

C.4.1 Characterization of the Singular Values 314
C.4.2 Best Low-Rank Approximation 315
C.5 Perturbation Bounds 316
C.5.1 Weyl Inequality 316
C.5.2 Eigenspaces Localization 317

Appendix D Subdifferentials of Convex Functions 321
D.1 Subdifferentials and Subgradients 321
D.2 Examples of Subdifferentials 323

Appendix E Reproducing Kernel Hilbert Spaces 325

Notations 329

Bibliography 331

Index 343

Preface, second edition

It has now been five years since the publication of the first edition of *Introduction to High-Dimensional Statistics*. High-dimensional statistics is a fast-evolving field and much progress has been made on a large variety of topics, providing new insights and methods. I felt it was time to share a selection of them, and also to add or complement some important topics that were either absent from, or insufficiently covered in the first edition.

This second edition of *Introduction to High-Dimensional Statistics* preserves the philosophy of the first edition: to be a concise guide for students and researchers discovering the area, and interested in the mathematics involved. The main concepts and ideas are presented in simple settings, avoiding thereby unessential technicalities. As I am convinced of the effectiveness of *learning by doing*, for each chapter, extensions and more advanced results are exposed via (fully) detailed exercises. The interested reader can then discover these topics by proving the results by himself. As in the first edition, everyone is welcome to share his own solutions to the exercises on the wiki-site
`http://high-dimensional-statistics.wikidot.com`
which has been updated.

Convex relaxation methods have been ubiquitous in the last twenty years. In the recent years, there has been also a renewed interest in iterative algorithms, which are computationally efficient competitors for solving large-scale problems. The theory developed for analyzing iterative algorithms is now quite robust, and the mathematical statistics community has been able to handle a large variety of problems. It was time to include a new Chapter 6 on this topic, in the simple setting of sparse linear regression. A distinctive feature of this theory, compared to classical statistical theory, is that the analysis must handle together the statistical and the optimization aspects, which is a recent trend in statistics and machine-learning theory. Iterative methods also show up in several other chapters, in particular a theoretical analysis of the Lloyd algorithm can be found in the last new chapter on clustering.

Unsupervised classification is an important topic in the area of big data gathering data from inhomogeneous subpopulations. I could not spare writing a chapter on this essential topic. So, I have added a new Chapter 12 on clustering, building on the impressive recent progress on the theory of clustering (point clustering, or clustering in graphs) both for convex and iterative algorithms. The theory for this last chapter is somewhat more involved than for the other chapters of the book, so it is a good transition towards some more advanced books.

Years after years, I felt that a simple exposition of minimax lower bounds was missing in the first version of the book. Minimax lower bounds can be found in a large fraction of PhD theses, so I believe that it was useful to add a simple and transparent introduction to this topic in a new Chapter 3. The presentation is made as little "magical" as possible, following a pedestrian and simple proof of the main results issued from information theory. Some more principled extensions are then given as (detailed) exercises.

The chapters from the first edition have also been revised, with the inclusion of many additional materials on some important topics, including estimation with convex constraints, aggregation of a continuous set of estimators, simultaneously low-rank and row sparse linear regression, the slope estimator or compress sensing. The Appendices have also been enriched, mainly with the addition of the Davis-Kahan perturbation bound and of two simple versions of the Hanson-Wright concentration inequality.

Despite my sustained efforts to remove typos, I am sure that some of them have escaped my vigilance. Samy Clementz and Etienne Peyrot have already spotted several of them, I warmly thank them for their feedback. The reader is welcome to point out any remaining typos to
`high.dimensional.statistics@gmail.com`
for errata that will be published on the book's website,
`http://sites.google.com/site/highdimensionalstatistics`.

Enjoy your reading!

Christophe Giraud
Orsay, France

Preface

Over the last twenty years (or so), the dramatic development of data acquisition technologies has enabled devices able to take thousands (up to millions) of measurements simultaneously. The data produced by such wide-scale devices are said to be *high-dimensional.* They can be met in almost any branch of human activities, including medicine (biotech data, medical imaging, etc.), basic sciences (astrophysics, environmental sciences, etc.), e-commerce (tracking, loyalty programs, etc.), finance, cooperative activities (crowdsourcing data), etc. Having access to such massive data sounds like a blessing. Unfortunately, the analysis of high-dimensional data is extremely challenging. Indeed, separating the useful information from the noise is *generally* almost impossible in high-dimensional settings. This issue is often referred to as the *curse of dimensionality.*

Most of the *classical* statistics developed during the twentieth century focused on data where the number n of *experimental units* (number of individuals in a medical cohort, number of experiments in biology or physics, etc.) was large compared to the number p of *unknown features.* Accordingly, most of the *classical* statistical theory provides results for the asymptotic setting where p is fixed and n goes to infinity. This theory is very insightful for analyzing data where "n is large" and "p is small," but it can be seriously misleading for modern high-dimensional data. Analyzing "large p" data then requires some new statistics. It has given rise to a huge effort from the statistical and data analyst community for developing new tools able to circumvent the curse of dimensionality. In particular, building on the concept of *sparsity* has shown to be successful in this setting.

This book is an introduction to the mathematical foundations of high-dimensional statistics. It is intended to be a concise guide for students and researchers unfamiliar with the area, and interested in the mathematics involved. In particular, this book is not conceived as a comprehensive catalog of statistical methods for high-dimensional data. It is based on lectures given in the Master programs "Mathematics for Life Sciences" and "Data Sciences," from Paris Sud University (Orsay), Ecole Polytechnique (Palaiseau), Ecole Normale Supérieure de Cachan (Cachan), and Telecom ParisTech (Paris). The primary goal is to explain, as simply as possible, the main concepts and ideas on some selected topics of high-dimensional statistics. The focus is mainly on some simple settings, avoiding, thereby, unessential technicalities that could blur the main arguments. To achieve this goal, the book includes significantly streamlined proofs issued from the recent research literature.

Each chapter of the book ends with some exercises, and the reader is invited to share his solutions on the wiki-site
`http://high-dimensional-statistics.wikidot.com`.

Finally, I apologize for any remaining typos and errors (despite my efforts to remove them). The reader is invited to point them out to
`high.dimensional.statistics@gmail.com`
for errata that will be published on the book's website,
`http://sites.google.com/site/highdimensionalstatistics`.

Enjoy your reading!

Christophe Giraud
Orsay, France

Acknowledgments

This book has strongly benefited from the contributions of many people.

My first thanks go to Sylvain Arlot, Marc Hoffmann, Sylvie Huet, Guillaume Lecué, Pascal Massart, Vincent Rivoirard, Etienne Roquain, François Roueff, and Nicolas Verzelen, who very kindly agreed to review one or several chapters of this book. Thank you very much to all of you for your suggestions, and corrections, which have significantly improved the quality of the manuscript. I also very warmly thank Marten Wegkamp and Irina Gaynanova for their feedback, suggestions and corrections on an early version of my lecture notes. My warm thanks go also to all my students, who have pointed out some typos and errors in the first draft of my lecture notes.

I have been lucky to receive some advice from famous statisticians on the writing of a book. Thank you, Florentina Bunea, Pascal Massart, and Sacha Tsybakov for your invaluable advice from which this book has strongly benefited. Thank you, also Elisabeth Gassiat, for your cheerful encouragement, especially during my final months of writing.

This book would not exist without the invitation of David Grubbs to convert my initial lecture notes into a book. I am very grateful to you and the team at Chapman & Hall for your support, enthusiasm, and kindness. I am also deeply grateful to the anonymous reviewers for their insightful suggestions, comments, and feedback.

I have learned statistics mostly from my collaborators, so this book is theirs in some sense (of course, typos and errors are mine). Thank you all so much! In particular, I am happy to thank Yannick Baraud and Sylvie Huet for introducing me to statistics years ago. My deepest gratitude goes to Nicolas Verzelen. Thanks for all I have learned from you, for the many invaluable scientific discussions we had, for the uncountable references you have pointed out to me, and for your major contribution to the reviewing of this book.

Last but not least, I am extremely grateful to you, Dominique, for your patience during these last months, your beautiful illustrations, and all the nice times we have together.

I am really indebted to all of you.
Merci beaucoup!

Acknowledgments

Chapter 1

Introduction

1.1 High-Dimensional Data

The sustained development of technologies, data storage resources, and computing resources give rise to the production, storage, and processing of an exponentially growing volume of data. Data are ubiquitous and have a dramatic impact on almost every branch of human activities, including science, medicine, business, finance and administration. For example, wide-scale data enable to better understand the regulation mechanisms of living organisms, to create new therapies, to monitor climate and biodiversity changes, to optimize the resources in the industry and in administrations, to personalize the marketing for each individual consumer, etc.

A major characteristic of modern data is that they often record simultaneously thousands up to millions of *features* on each *object* or *individual*. Such data are said to be *high-dimensional*. Let us illustrate this characteristic with a few examples. These examples are relevant at the time of writing and may become outdated in a few years, yet we emphasize that the mathematical ideas conveyed in this book are independent of these examples and will remain relevant.

- **Biotech data:** Recent biotechnologies enable to acquire high-dimensional data on single individuals. For example, DNA microarrays measure the transcription level[1] of tens of thousands of genes simultaneously; see Figure 1.1. Next generation sequencing (NGS) devices improve on these microarrays by allowing to sense the "transcription level" of virtually any part of the genome. Similarly, in proteomics some technologies can gauge the abundance of thousands of proteins simultaneously. These data are crucial for investigating biological regulation mechanisms and creating new drugs. In such biotech data, the number p of "variables" that are sensed scales in thousands and is most of the time much larger than the number n of "individuals" involved in the experiment (number of repetitions, rarely exceeding a few hundreds).
- **Images (and videos):** Large databases of images are continuously collected all around the world. They include medical images, massive astrophysic images, video surveillance images, etc. Each image is made of thousands to millions of

[1]The transcription level of a gene in a cell at a given time corresponds to the quantity of ARNm associated to this gene present at this time in the cell.

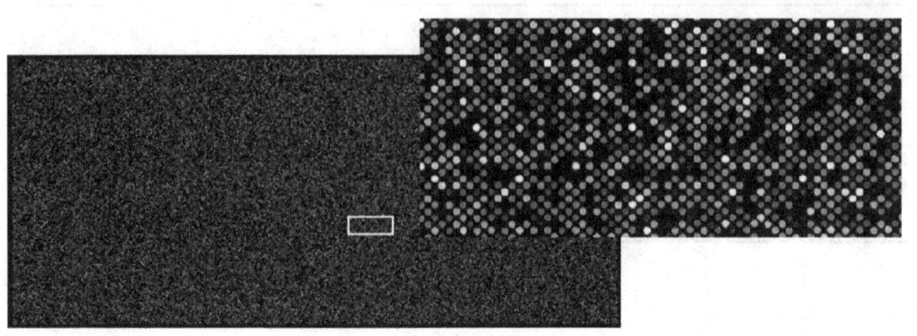

Figure 1.1 *Whole human genome microarray covering more than 41,000 human genes and transcripts on a standard* $1'' \times 3''$ *glass slide format. ©Agilent Technologies, Inc. 2004. Reproduced with permission, courtesy of Agilent Technologies, Inc.*

pixels or voxels. For medical images, as for biotech data, the number p of pixels can be much larger than the number n of patients in the cohort under study.

- **Consumers preferences data:** Websites and loyalty programs collect huge amounts of information on the preferences and the behaviors of customers. These data are processed for marketing purposes, for recommendations, and for fixing personalized prizes. For example, recommendation systems (for movies, books, music, etc.) gather the customers' ratings on various products, together with some personal data (age, sex, location), and guess from them which products could be of interest for a given consumer.
- **Business data:** Every major company has its own chief data officer who supervises the optimal exploitation of internal and external data. For example, logistic and transportation companies intensively process internal and geo-economic data in order to optimize the allocation of their resources and to try to forecast precisely the future demand. Insurance companies widely rely on various sources of data in order to control their risk and allocate at best their financial resources. Many profitable activities of the financial industry are based on the intensive processing of transaction data from all over the world. Again, the dimensionality of the data processed in these examples can scale in thousands.
- **Crowdsourcing data:** The launch of websites dedicated to participative data recording together with the spreading of smartphones enable volunteers to record online massive data sets. For example, the Cornell Lab of Ornithology and the National Audubon Society have jointly launched a crowdsourcing program, "eBird" `http://ebird.org`, inviting all bird-watchers from North America to record via an online checklist all the birds they have seen and heard during their last birding session. The purpose of this program is to monitor birds abundances and their evolutions across North America. In 2014, eBird involved tens of thousands of participants, which had already recorded millions of observations.

Blessing?

Being able to sense simultaneously thousands of variables on each "individual" sounds like good news: Potentially we will be able to scan every variable that may influence the phenomenon under study. The statistical reality unfortunately clashes with this optimistic statement: Separating the signal from the noise is *in general* almost impossible in high-dimensional data. This phenomenon described in the next section is often called the "curse of dimensionality."

1.2 Curse of Dimensionality

The impact of high dimensionality on statistics is multiple. First, high-dimensional spaces are vast and data points are isolated in their immensity. Second, the accumulation of small fluctuations in many different directions can produce a large global fluctuation. Third, an event that is an accumulation of rare events may not be rare. Finally, numerical computations and optimizations in high-dimensional spaces can be overly intensive.

1.2.1 Lost in the Immensity of High-Dimensional Spaces

Let us illustrate this issue with an example. We consider a situation where we want to explain a response variable $Y \in \mathbb{R}$ by p variables $X_1, \ldots, X_p \in [0,1]$. Assume, for example, that each variable X_k follows a uniform distribution on $[0,1]$. If these variables are independent, then the variable $X = (X_1, \ldots, X_p) \in [0,1]^p$ follows a uniform distribution on the hypercube $[0,1]^p$. Our data consist of n independent and identically distributed (i.i.d.) observations $(Y_i, X^{(i)})_{i=1,\ldots,n}$ of the variables Y and X. We model them with the classical regression equation

$$Y_i = f(X^{(i)}) + \varepsilon_i, \quad i = 1, \ldots, n,$$
$$\text{with } f : [0,1]^p \to \mathbb{R} \text{ and } \varepsilon_1, \ldots, \varepsilon_n \text{ independent and centered.}$$

Assuming that the function f is smooth, it is natural to estimate $f(x)$ by some average of the Y_i associated to the $X^{(i)}$ in the vicinity of x. The most simple version of this idea is the k-Nearest Neighbors estimator, where $f(x)$ is estimated by the mean of the Y_i associated to the k points $X^{(i)}$, which are the nearest from x. Some more sophisticated versions of this idea use a weighted average of the Y_i with weights that are a decreasing function of the distance $\|X^{(i)} - x\|$ (like kernel smoothing). The basic idea being in all cases to use a local average of the data. This idea makes perfect sense in low-dimensional settings, as illustrated in Figure 1.2.

Unfortunately, when the dimension p increases, the notion of "nearest points" vanishes. This phenomenon is illustrated in Figure 1.3, where we have plotted the histograms of the distribution of the pairwise-distances $\{\|X^{(i)} - X^{(j)}\| : 1 \leq i < j \leq n\}$ for $n = 100$ and dimensions $p = 2, 10, 100$, and 1000. When the dimension p increases, we observe in Figure 1.3 that

- the minimal distance between two points increases,

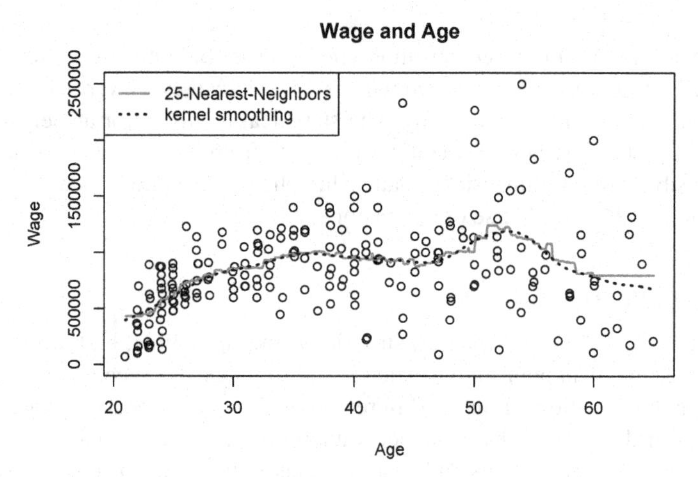

Figure 1.2 *Canadian high school graduate earnings from the 1971 Canadian Census Public Use Tapes. Black dots: records. Gray line: 25-Nearest Neighbors estimator. Dashed line: local averaging by kernel smoothing.*

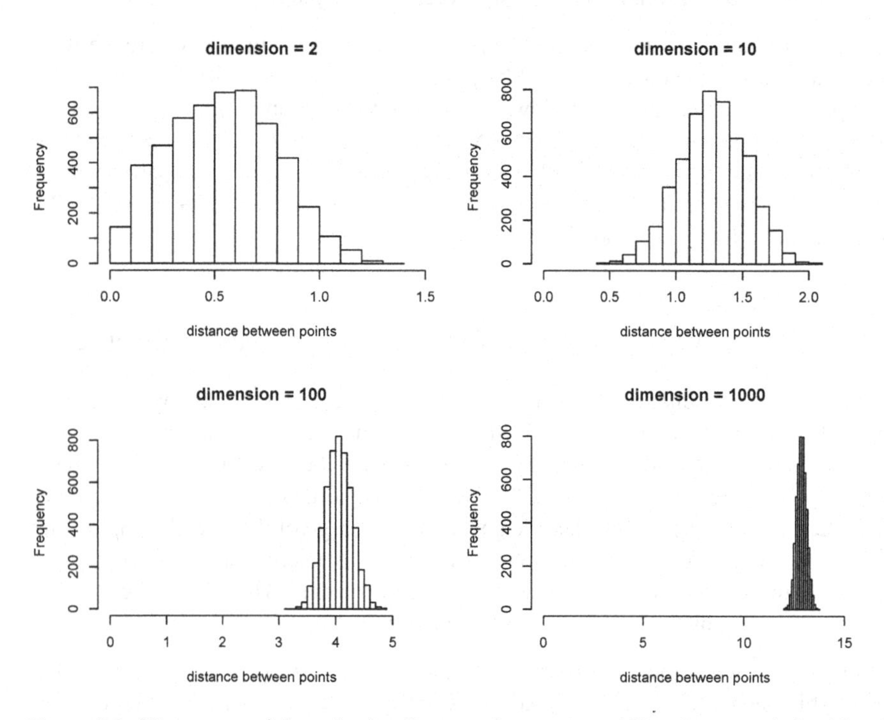

Figure 1.3 *Histograms of the pairwise-distances between $n = 100$ points sampled uniformly in the hypercube $[0,1]^p$, for $p = 2, 10, 100,$ and 1000.*

- all the points are at a similar distance from the others, so the notion of "nearest points" vanishes.

In particular, any estimator based on a local averaging will fail with such data.

Let us quantify roughly the above observations. Writing U and U' for two independent random variables with uniform distribution on $[0,1]$, the mean square distance between $X^{(i)}$ and $X^{(j)}$ is

$$\mathbb{E}\left[\|X^{(i)}-X^{(j)}\|^2\right]=\sum_{k=1}^{p}\mathbb{E}\left[\left(X_k^{(i)}-X_k^{(j)}\right)^2\right]=p\mathbb{E}\left[(U-U')^2\right]=p/6,$$

and the standard deviation of this square distance is

$$\text{sdev}\left[\|X^{(i)}-X^{(j)}\|^2\right]=\sqrt{\sum_{k=1}^{p}\text{var}\left[\left(X_k^{(i)}-X_k^{(j)}\right)^2\right]}=\sqrt{p\text{var}\left[(U'-U)^2\right]}\approx 0.2\sqrt{p}.$$

In particular, we observe that the typical square distance between two points sampled uniformly in $[0,1]^p$ grows linearly with p, while the scaled deviation $\text{sdev}\left[\|X^{(i)}-X^{(j)}\|^2\right]/\mathbb{E}\left[\|X^{(i)}-X^{(j)}\|^2\right]$ shrinks like $p^{-1/2}$.

How Many Observations Do We Need?

Figure 1.3 shows that if the number n of observations remains fixed while the dimension p of the observations increases, the observations $X^{(1)},\ldots,X^{(n)}$ get rapidly very isolated and local methods cannot work. If for any $x\in[0,1]^p$ we want to have at least one observation $X^{(i)}$ at distance less than one from x, then we must increase the number n of observations. How should this number n increase with the dimension p? We investigate below this issue by computing a lower bound on the number n of points needed in order to fill the hypercube $[0,1]^p$ in such a way that at any $x\in[0,1]^p$ there exists at least one point at distance less than 1 from x.

The volume $V_p(r)$ of a p-dimensional ball of radius $r>0$ is equal to (see Exercise 1.6.2)

$$V_p(r)=\frac{\pi^{p/2}}{\Gamma(p/2+1)}r^p \overset{p\to\infty}{\sim} \left(\frac{2\pi e r^2}{p}\right)^{p/2}(p\pi)^{-1/2}, \qquad (1.1)$$

where Γ represents the Gamma function $\Gamma(x)=\int_0^\infty t^{x-1}e^{-t}\,dt$ for $x>0$.

If $x^{(1)},\ldots,x^{(n)}$ are such that for any $x\in[0,1]^p$ there exists a point $x^{(i)}$ fulfilling $\|x^{(i)}-x\|\leq 1$, then the hypercube is covered by the family of unit balls centered in $x^{(1)},\ldots,x^{(n)}$,

$$[0,1]^p\subset\bigcup_{i=1}^{n}B_p(x^{(i)},1).$$

As a consequence, the volume of the union of the n unit balls is larger than the volume of the hypercube, so $1\leq nV_p(1)$. According to Equation (1.1), we then need at least

$$n\geq\frac{\Gamma(p/2+1)}{\pi^{p/2}} \overset{p\to\infty}{\sim} \left(\frac{p}{2\pi e}\right)^{p/2}\sqrt{p\pi} \qquad (1.2)$$

The Strange Geometry of High-Dimensional Spaces (I)

High-dimensional balls have a vanishing volume!

From Formula (1.1), we observe that for any $r > 0$, the volume $V_p(r)$ of a ball of radius r goes to zero more than exponentially fast with the dimension p. We illustrate this phenomenon by plotting $p \to V_p(1)$. We observe that for $p = 20$ the volume of the unit ball is already almost 0.

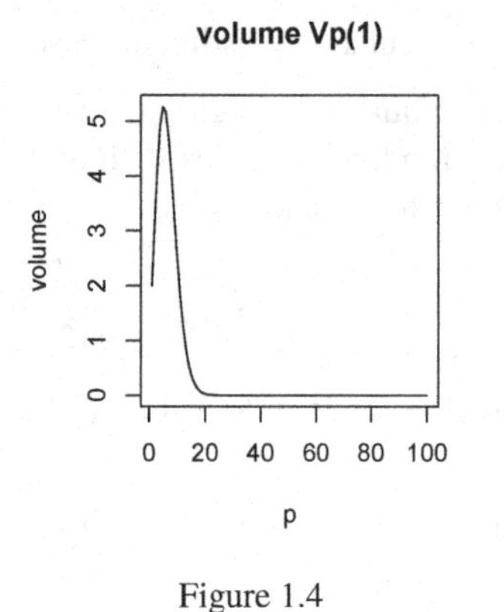

Figure 1.4

points in order to fill the hypercube $[0,1]^p$. This number of points then grows more than exponentially fast with p. If we come back to our above example in the regression setting, it means that if we want a local average estimator to work with observations uniformly distributed in $[0,1]^p$ with p larger than a few tens, then we would need a number n of observations, which is completely unrealistic (see Table 1.1).

p	20	30	50	100	150	200
n	39	45630	$5.7\,10^{12}$	$42\,10^{39}$	$1.28\,10^{72}$	larger than the estimated number of particles in the observable universe

Table 1.1 *Lower Bound (1.2) on the required number of points for filling the hypercube* $[0,1]^p$.

The moral of this example is that we have to be very careful with our geometric intuitions in high-dimensional spaces. These spaces have some counterintuitive geometric properties, as illustrated in Figures 1.4 and 1.5.

1.2.2 Fluctuations Cumulate

Assume that you want to evaluate some function $F(\theta_1)$ of some scalar value $\theta_1 \in \mathbb{R}$. Assume that you have only access to a noisy observation of θ_1, denoted by $X_1 = \theta_1 + \varepsilon_1$, with $\mathbb{E}[\varepsilon_1] = 0$ and $\text{var}(\varepsilon_1) = \sigma^2$. If the function F is 1-Lipschitz, then the mean square error is

$$\mathbb{E}\left[\|F(X_1) - F(\theta_1)\|^2\right] \leq \mathbb{E}\left[|\varepsilon_1|^2\right] = \sigma^2.$$

In particular, if the variance σ^2 of the noise is small, then this error is small.

Assume now that you need to evaluate a function $F(\theta_1, \ldots, \theta_p)$ from noisy observations $X_j = \theta_j + \varepsilon_j$ of the θ_j. Assume that the noise variables $\varepsilon_1, \ldots, \varepsilon_p$ are all centered

The Strange Geometry of High-Dimensional Spaces (II)

The volume of a high-dimensional ball is concentrated in its crust!

Let us write $B_p(0,r)$ for the p-dimensional ball centered at 0 with radius $r > 0$, and $C_p(r)$ for the "crust" obtained by removing from $B_p(0,r)$ the sub-ball $B_p(0,0.99r)$. In other words, the "crust" gathers the points in $B_p(0,r)$, which are at a distance less than $0.01r$ from its surface.

We plot as a function of p the ratio of the volume of $C_p(r)$ to the volume of $B_p(0,r)$

$$\frac{\text{volume}(C_p(r))}{\text{volume}(B_p(0,r))} = 1 - 0.99^p,$$

which goes exponentially fast to 1.

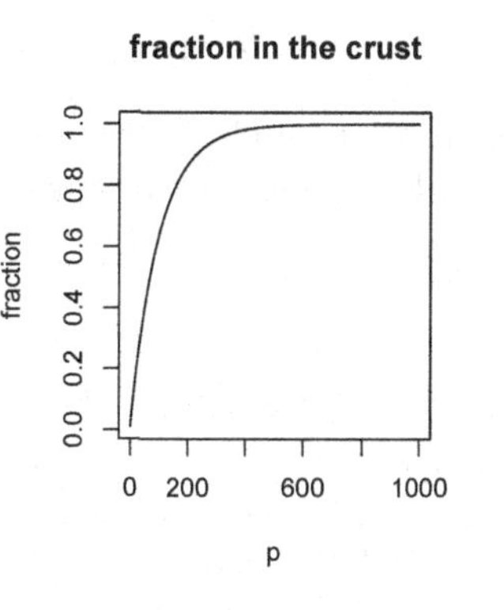

Figure 1.5

with variance σ^2. If as before F is 1-Lipschitz, we have

$$\mathbb{E}\left[\|F(X_1,\ldots,X_p) - F(\theta_1,\ldots,\theta_p)\|^2\right] \leq \mathbb{E}\left[\|(\varepsilon_1,\ldots,\varepsilon_p)\|^2\right] = \sum_{j=1}^{p} \mathbb{E}\left[\varepsilon_j^2\right] = p\sigma^2.$$

Furthermore, if F fulfills $\|F(x+h) - F(x)\| \geq c\|h\|$ for some $c > 0$, then the mean square error $\mathbb{E}\left[\|F(X_1,\ldots,X_p) - F(\theta_1,\ldots,\theta_p)\|^2\right]$ scales like $p\sigma^2$. This error can be very large in high-dimensional settings, even if σ^2 is small. A central example where such a situation arises is in the linear regression model with high-dimensional covariates.

High-Dimensional Linear Regression

Assume that we have n observations $Y_i = \langle x^{(i)}, \beta^* \rangle + \varepsilon_i$ for $i = 1,\ldots,n$, with the response Y_i in $\mathbb{R}$ and the covariates $x^{(i)}$ in $\mathbb{R}^p$. We want to estimate $\beta^* \in \mathbb{R}^p$, and we assume that $\varepsilon_1,\ldots,\varepsilon_n$ are i.i.d. centered, with variance σ^2. Writing

$$Y = \begin{pmatrix} Y_1 \\ \vdots \\ Y_n \end{pmatrix}, \quad \mathbf{X} = \begin{pmatrix} (x^{(1)})^T \\ \vdots \\ (x^{(n)})^T \end{pmatrix} \quad \text{and} \quad \varepsilon = \begin{pmatrix} \varepsilon_1 \\ \vdots \\ \varepsilon_n \end{pmatrix},$$

we have $Y = \mathbf{X}\beta^* + \varepsilon$. A classical estimator of β^* is the least-squares (LS) estimator

$$\widehat{\beta} \in \underset{\beta \in \mathbb{R}^p}{\operatorname{argmin}} \|Y - \mathbf{X}\beta\|^2,$$

which is uniquely defined when the rank of $\mathbf{X}$ is p. Let us focus on this case. The solution of this minimization problem is $\widehat{\beta} = (\mathbf{X}^T\mathbf{X})^{-1}\mathbf{X}^T Y$, which fulfills (see Exercise 1.6.5)

$$\mathbb{E}\left[\|\widehat{\beta}-\beta^*\|^2\right] = \mathbb{E}\left[\|(\mathbf{X}^T\mathbf{X})^{-1}\mathbf{X}^T\varepsilon\|^2\right] = \mathrm{Tr}\left((\mathbf{X}^T\mathbf{X})^{-1}\right)\sigma^2.$$

Assume for simplicity that the columns of $\mathbf{X}$ are orthonormal (i.e., orthogonal with norm 1). Then, the mean square error is

$$\mathbb{E}\left[\|\widehat{\beta}-\beta^*\|^2\right] = p\sigma^2.$$

So the more high-dimensional is the covariate $x^{(i)}$, the larger is this estimation error.

We cannot give a direct picture of a linear regression in dimension p, with p larger than 2. Yet, we can illustrate the above phenomenon with the following example. Assume that the covariates $x^{(i)}$ are given by $x^{(i)} = \phi(i/n)$, where $\phi : [0,1] \to \mathbb{R}^p$ is defined by $\phi(t) = [\cos(\pi jt)]_{j=1,\dots,p}$. Then we observe

$$Y_i = \sum_{j=1}^{p} \beta_j^* \cos(\pi ji/n) + \varepsilon_i = f_{\beta^*}(i/n) + \varepsilon_i, \quad \text{for } i = 1,\dots,n, \tag{1.3}$$

with $f_\beta(t) = \sum_{j=1}^{p} \beta_j \cos(\pi jt)$. We can illustrate the increase of the error $\|\widehat{\beta}-\beta^*\|^2$ with p by plotting the function f_{β^*} and $f_{\widehat{\beta}}$ for increasing values of p. In Figure 1.6, the noise $\varepsilon_1,\dots,\varepsilon_n$ is i.i.d., with $\mathcal{N}(0,1)$ distribution and the function f_{β^*} has been generated by sampling the β_j^* independently with $\mathcal{N}(0, j^{-4})$ distribution. We choose $n = 100$ and the four figures correspond to $p = 10$, 20, 50 and 100, respectively. We observe that when p increases, the estimated function $f_{\widehat{\beta}}(t)$ becomes more and more wavy. These increasing oscillations are the direct consequence of the increasing error $\|\widehat{\beta}-\beta^*\|^2 \approx p\sigma^2$.

Tails of High-Dimensional Gaussian Distributions Are Thin but Concentrate the Mass

Gaussian distributions are known to have very thin tails. Actually, the density $g_p(x) = (2\pi)^{-p/2}\exp(-\|x\|^2/2)$ of a standard Gaussian distribution $\mathcal{N}(0, I_p)$ in $\mathbb{R}^p$ decreases exponentially fast with the square norm of x. Yet, when p is large, most of the mass of the standard Gaussian distribution lies in its tails!

First, we observe that the maximal value of $g_p(x)$ is $g_p(0) = (2\pi)^{-p/2}$, which decreases exponentially fast toward 0 when p increases, so the Gaussian distribution in high dimensions is much more flat than in dimension one or two. Let us compute the mass in its "bell" (central part of the distribution where the density is the largest). Let $\delta > 0$ be a small positive real number and write

$$B_{p,\delta} = \left\{x \in \mathbb{R}^p : g_p(x) \geq \delta g_p(0)\right\} = \left\{x \in \mathbb{R}^p : \|x\|^2 \leq 2\log(\delta^{-1})\right\}$$

for the ball gathering all the points $x \in \mathbb{R}^p$, such that the density $g_p(x)$ is larger or

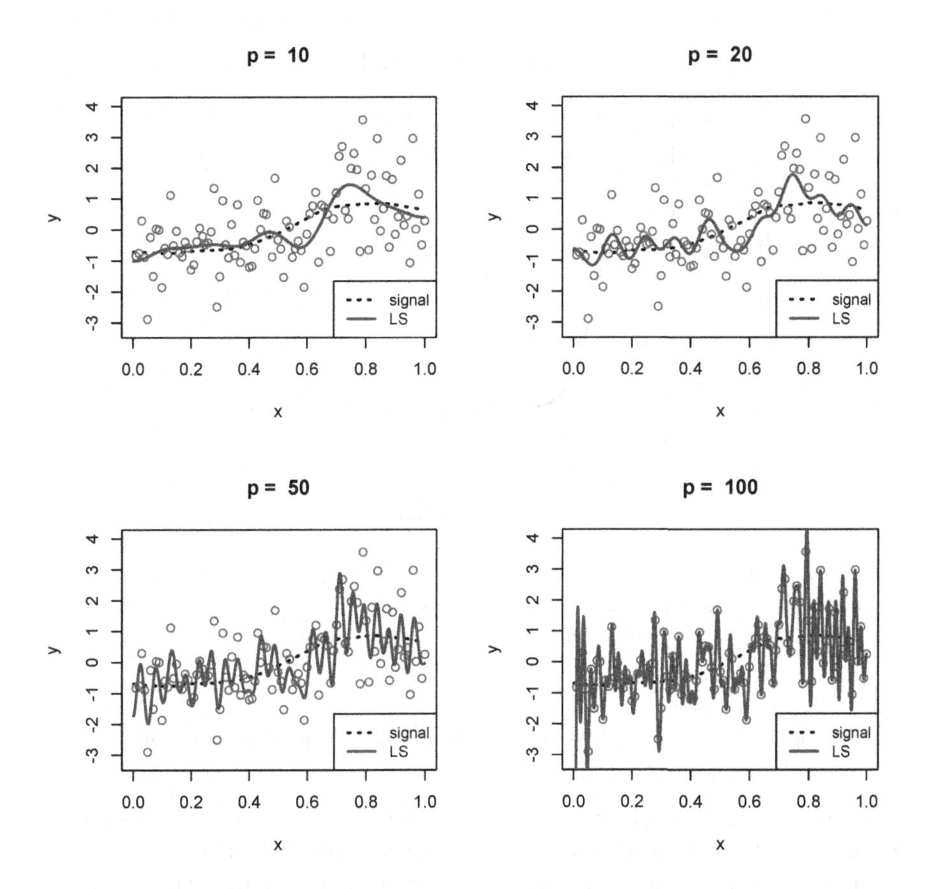

Figure 1.6 *Least-squares (LS) estimator* $f_{\widehat{\beta}}$ *in the setting (1.3), with* $n = 100$ *and for* $p =$ 10, 20, 50, *and* 100. *Gray dots: observations. Dashed line: function* f_{β^*}. *Gray line: LS estimator* $f_{\widehat{\beta}}$.

equal to δ times the density at 0. Our intuition from the one-dimensional case is that for δ small, say $\delta = 0.001$, the probability for a $\mathcal{N}(0, I_p)$ Gaussian random variable X to be in the "bell" $B_{p,\delta}$ is close to one. Yet, as illustrated in Figure 1.7, it is the opposite!

Actually, from the Markov inequality (Lemma B.1 in Appendix B), we have

$$\begin{aligned}\mathbb{P}\left(X \in B_{p,\delta}\right) &= \mathbb{P}\left(e^{-\|X\|^2/2} \geq \delta\right) \\ &\leq \frac{1}{\delta}\mathbb{E}\left[e^{-\|X\|^2/2}\right] = \frac{1}{\delta}\int_{x\in\mathbb{R}^p} e^{-\|x\|^2}\,\frac{dx}{(2\pi)^{p/2}} = \frac{1}{\delta\, 2^{p/2}}.\end{aligned}$$

So most of the mass of the standard Gaussian distribution is in the tail

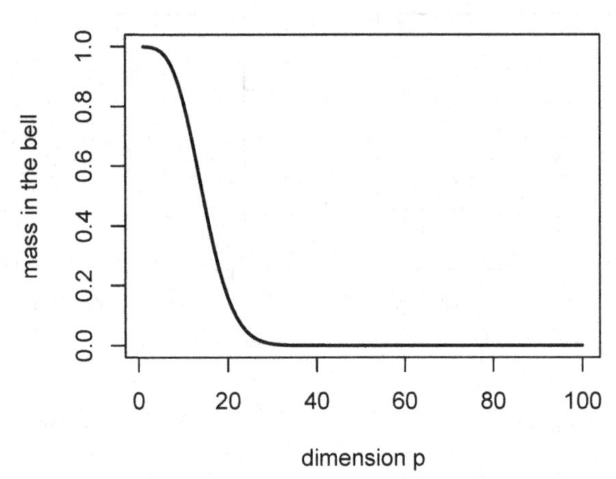

Figure 1.7 *Mass of the standard Gaussian distribution* $g_p(x)\,dx$ *in the "bell"* $B_{p,0.001} = \{x \in \mathbb{R}^p : g_p(x) \geq 0.001 g_p(0)\}$ *for increasing values of* p.

$\{x \in \mathbb{R}^p : g_p(x) < \delta g_p(0)\}$. If we want to have $\mathbb{P}\left(X \in B_{p,\delta}\right) \geq 1/2$, we must choose $\delta \leq 2^{-p/2+1}$ which is exponentially small.

How can we explain this counterintuitive phenomenon? It is related to the geometric properties of the high-dimensional spaces described above. We have seen that the volume of a ball of radius r grows like r^p. So when r increases, the density $g_p(x)$ on the ring $\{x : r < \|x\| < r + dr\}$ shrinks like $e^{-r^2/2}$, but at the same time the volume $V_p(r+dr) - V_p(r)$ of the thin ring $\{x : r < \|x\| < r + dr\}$ grows with r like r^{p-1}, so the probability for X to be in $\{x : r < \|x\| < r + dr\}$ evolves as $r^{p-1} e^{-r^2/2}$ with $r > 0$. In particular, this probability is maximal for $r^2 = p - 1$, and the Gaussian density at a point x with norm $\|x\|^2 = p - 1$ fulfills $g_p(x) = e^{-(p-1)^2/2} g_p(0) \ll g_p(0)$: most of the mass of a Gaussian distribution is located in areas where the density is extremely small compared to its maximum value.

1.2.3 Accumulation of Rare Events May Not Be Rare

Assume that we have an observation Z_1 of a single quantity θ_1 blurred by some $\mathcal{N}(0,1)$ Gaussian noise ε_1. From Lemma B.4 in Appendix B we have $\mathbb{P}(|\varepsilon_1| \geq x) \leq e^{-x^2/2}$ for $x > 0$, so with probability at least $1 - \alpha$, the noise ε_1 has an absolute value smaller than $\left(2\log(1/\alpha)\right)^{1/2}$.

Assume that we observe now p quantities $\theta_1, \ldots, \theta_p$ blurred by $\varepsilon_1, \ldots, \varepsilon_p$ i.i.d. with

$\mathcal{N}(0,1)$ Gaussian distribution. We have

$$\mathbb{P}\left(\max_{j=1,\ldots,p}|\varepsilon_j| \geq x\right) = 1-(1-\mathbb{P}(|\varepsilon_1|\geq x))^p \overset{x\to\infty}{\sim} p\mathbb{P}(|\varepsilon_1|\geq x).$$

If we want to bound simultaneously the absolute values $|\varepsilon_1|,\ldots,|\varepsilon_p|$ with probability $1-\alpha$, then we can only guarantee that $\max_{j=1,\ldots,p}|\varepsilon_j|$ is smaller than $(2\log(p/\alpha))^{1/2}$. This extra $\log(p)$ factor can be a serious issue in practice as illustrated below.

False Discoveries

Assume that for a given individual i, we can measure for p genes simultaneously the ratios between their expression levels in two different environments (e.g., with microarrays). For the individual i, let us denote by $Z_j^{(i)}$ the log-ratio of the expression levels of the gene j between the two environments. Assume that (after some normalization) these log-ratios can be modeled by

$$Z_j^{(i)} = \theta_j + \varepsilon_j^{(i)}, \quad j=1,\ldots,p, \; i=1,\ldots,n,$$

with the $\varepsilon_j^{(i)}$ i.i.d. with $\mathcal{N}(0,1)$ Gaussian distribution. Our goal is to detect the genes j, such that $\theta_j \neq 0$, which means that they are involved in the response to the change of environment (such a gene is said to be "positive").

We write $X_j = n^{-1}(Z_j^{(1)}+\ldots+Z_j^{(n)})$ for the mean of the observed log-ratios for the gene j. The random variables $n^{1/2}X_1,\ldots,n^{1/2}X_p$ are independent, and $n^{1/2}X_j$ follows a $\mathcal{N}(\sqrt{n}\theta_j,1)$ Gaussian distribution. For W with $\mathcal{N}(0,1)$ Gaussian distribution, we have $\mathbb{P}[|W|>1.96]\approx 5\%$, so a natural idea is to declare "positive" all the genes j, such that $n^{1/2}|X_j|$ is larger than 1.96. Nevertheless, this procedure would produce many false positives (genes declared "positive" while they are not) and thereby many false discoveries of genes responding to the change of environment. Let us illustrate this point. Assume that $p=5000$, and among them 200 genes are positive. Then, the average number of false positive genes is

$$\text{card}\left\{j:\theta_j=0\right\}\times 0.05 = 4800\times 0.05 = 240 \text{ false positive genes},$$

which is larger than the number of positive genes (200). It means that, on average, more than half of the discoveries will be false discoveries. If we want to avoid false positives, we must choose a threshold larger than 1.96. From Exercise 1.6.3, for $W_1,\ldots,W_p$ i.i.d. with $\mathcal{N}(0,1)$ Gaussian distribution, we have

$$\mathbb{P}\left(\max_{j=1,\ldots,p}|W_j| \geq \sqrt{\alpha\log(p)}\right) = 1-\exp\left(-\sqrt{\frac{2}{\alpha\pi}}\frac{p^{1-\alpha/2}}{(\log p)^{1/2}} + O\left(\frac{p^{1-\alpha/2}}{(\log p)^{3/2}}\right)\right)$$

$$\overset{p\to\infty}{\longrightarrow} \begin{cases} 0 & \text{if } \alpha\geq 2 \\ 1 & \text{if } \alpha<2. \end{cases}$$

Therefore, in order to avoid false positives, it seems sensible to declare positive the genes j, such that $n^{1/2}|X_j| \geq \sqrt{2\log(p)}$. With this choice, we will roughly be able to detect the θ_j whose absolute value is larger than $\sqrt{2\log(p)/n}$. We then observe that the larger p, the less we are able to detect the nonzero θ_j. This can be a severe issue when p scales in thousands and n is only a few units. We refer to Chapter 10 for techniques suited to this setting.

Empirical Covariance Is Not Reliable in High-Dimensional Settings

Another important issue with high-dimensional data is that the empirical covariance matrix of a p-dimensional vector is not reliable when p scales like the sample size n. Let us illustrate briefly this point. Assume that we observe some i.i.d. random vectors $X^{(1)},\ldots,X^{(n)}$ in $\mathbb{R}^p$, which are centered with covariance matrix $\mathrm{cov}(X^{(i)}) = I_p$. The empirical covariance matrix $\widehat{\Sigma}$ associated to the observations $X^{(1)},\ldots,X^{(n)}$ is given by

$$\widehat{\Sigma}_{ab} = \frac{1}{n}\sum_{i=1}^{n} X_a^{(i)} X_b^{(i)}, \quad \text{for } a,b = 1,\ldots,p.$$

We have $\mathbb{E}\left[X_a^{(i)} X_b^{(i)}\right] = \mathbf{1}_{\{a=b\}}$, so by the strong law of large numbers, $\widehat{\Sigma}_{a,b}$ tends to $\mathbf{1}_{\{a=b\}}$ almost surely when n goes to infinity. Therefore, the empirical covariance matrix $\widehat{\Sigma}$ converges almost surely to the identity matrix when n goes to infinity with p fixed. In particular, the spectrum of $\widehat{\Sigma}$ is concentrated around 1 when n is large and p small. This property is lost when p increases proportionally to n. Figure 1.8 displays three histograms of the spectral values of $\widehat{\Sigma}$ when $X^{(1)},\ldots,X^{(n)}$ are i.i.d. with standard Gaussian $\mathcal{N}(0,I_p)$ distribution, with $n = 1000$ and $p = n/2$, $p = n$, and $p = 2n$, respectively.

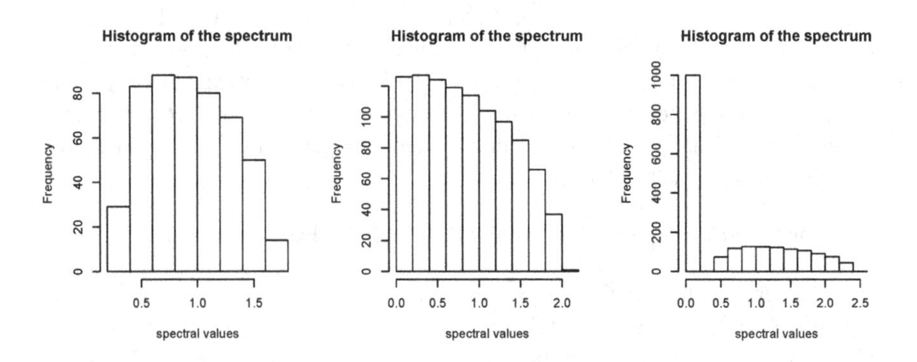

Figure 1.8 *Histogram of the spectral values of the empirical covariance matrix $\widehat{\Sigma}$, with $n = 1000$ and $p = n/2$ (left), $p = n$ (center), $p = 2n$ (right).*

We observe in the three cases that the spectrum of the empirical covariance matrix $\widehat{\Sigma}$ is very different from the spectrum of the identity, so the empirical covariance $\widehat{\Sigma}$ is a very poor approximation of the covariance I_p in this setting. From the theory of random matrices, we know the limit distribution of the spectral values of $\widehat{\Sigma}$ when n goes to infinity and $p \sim \alpha n$ with $\alpha > 0$ (see Section 1.5 for references). The support of this limit distribution (known as the Marchenko–Pastur distribution) is actually equal to $[(1-\sqrt{\alpha})^2, (1+\sqrt{\alpha})^2]$ up to a singleton at 0 when $\alpha > 1$. This means that we cannot trust the empirical covariance matrix $\widehat{\Sigma}$ when n and p have a similar size.

1.2.4 Computational Complexity

Another burden arises in high-dimensional settings: numerical computations can become very intensive and largely exceed the available computational (and memory) resources. For example, basic operations with $p \times p$ matrices (like multiplication, inversion, etc.) require at least p^{α} operations with $\alpha > 2$. When p scales in thousands, iterating such operations can become quite intensive.

The computational complexity appears to be really problematic in more involved problems. For example, we have seen above that the mean square error $\|\widehat{\beta} - \beta^*\|^2$ in the linear regression model

$$y = \sum_{j=1}^{p} \beta_j^* x_j + \varepsilon$$

typically scales linearly with p. Yet, it is unlikely that all the covariates x_j are influential on the response y. So we may wish to compare the outcomes of the family of regression problems

$$y = \sum_{j \in m} \beta_j^* x_j + \varepsilon \quad \text{for each } m \subset \{1, \ldots, p\}. \tag{1.4}$$

Unfortunately, the cardinality of $\{m : m \subset \{1, \ldots, p\}\}$ is 2^p, which grows exponentially with p. So, when p is larger than a few tens, it is impossible to compute the 2^p estimators $\widehat{\beta}_m$ associated to the model (1.4). This issue is detailed in Chapters 2 and 5.

1.3 High-Dimensional Statistics

1.3.1 Circumventing the Curse of Dimensionality

As explained in the previous section, the high dimensionality of the data, which seems at first to be a blessing, is actually a major issue for statistical analyzes. In light of the few examples described above, the situation may appear hopeless. Fortunately, high-dimensional data are often much more low-dimensional than they seem to be. Usually, they are not "uniformly" spread in $\mathbb{R}^p$, but rather concentrated around some low-dimensional structures. These structures are due to the relatively small complexity of the systems producing the data. For example,

- pixel intensities in an image are not purely random since there exist many geometrical structures in an image;
- biological data are the outcome of a "biological system", which is strongly regulated and whose regulation network has a relatively small complexity;

PCA in action

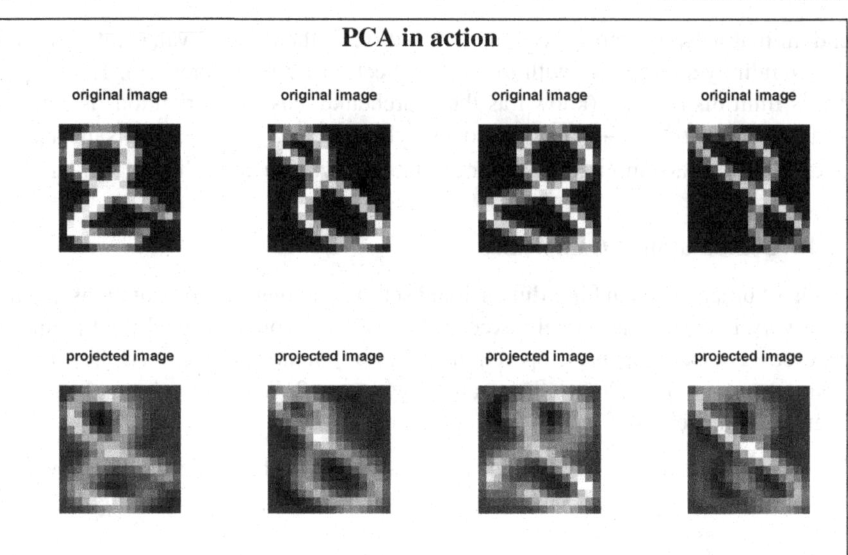

Figure 1.9 *The Modified National Institute of Standards and Technology (MNIST) data set [104] gathers 1100 scans of each digit. Each scan is a* 16×16 *image that is encoded by a vector in* $\mathbb{R}^{256}$*. The above pictures represent the projections of four images onto the space* V_{10}*, computed according to (1.5) from the 1100 scans of the digit 8 in the MNIST database. The original images are displayed in the first row, with their projection onto* V_{10} *in the second row.*

- marketing data strongly reflects some social structures in a population, and these structures are relatively simple; and
- technical data are the outcome of human technologies, whose complexity remains limited.

So in many cases, the data have an intrinsic low complexity, and we can hope to extract useful information from them. Actually, when the low-dimensional structures are known, we are back to some more classical "low-dimensional statistics." The major issue with high-dimensional data is that these structures are usually unknown. The main task will then be to identify at least approximately theses structures. This issue can be seen as the central issue of high-dimensional statistics.

A first approach is to try to find directly the low-dimensional structures in the data. The simplest and the most widely used technique for this purpose is the principal component analysis (PCA). For any data points $X^{(1)}, \ldots, X^{(n)} \in \mathbb{R}^p$ and a given dimension $d \leq p$, the PCA computes the linear span V_d fulfilling

$$V_d \in \underset{\dim(V) \leq d}{\operatorname{argmin}} \sum_{i=1}^{n} \|X^{(i)} - \operatorname{Proj}_V(X^{(i)})\|^2, \tag{1.5}$$

where the minimum is taken over all the subspaces V of $\mathbb{R}^p$, with dimension not larger than d, and $\text{Proj}_V : \mathbb{R}^p \to \mathbb{R}^p$ is the orthogonal projector onto V. We refer to Figure 1.9 for an illustration of the PCA in action and to Exercise 1.6.4 for the mathematical details.

Another approach, developed throughout this book, is to perform an "estimation-oriented" search of these low-dimensional structures. With this approach, we only seek the low-dimensional structures that are useful for our estimation problem. In principle, this approach allows for more precise results. We illustrate this feature in Figure 1.10, which is based on a data set of Sabarly *et al.* [138] gathering the measurement of 55 chemical compounds for 162 strains of the bacteria *E. coli*. Some of these bacteria are pathogens for humans, others are commensal. Our goal is to find a classification rule that enables us to separate from the chemical measurements the strains that are pathogens from those that are commensal. In Figure 1.10, the pathogen bacteria are denoted by "IPE" (internal pathogens) and "EPE" (external pathogens) and the commensal bacteria are denoted by "Com." The left-hand-side figure displays the data projected on the two-dimensional space V_2 given by a PCA. The right-hand-side figure displays the data projected on a two-dimensional space S_2, which aims to separate at best the different classes of bacteria (more precisely, S_2 is spanned by the directions orthogonal to the separating hyperplanes of a linear discriminant analysis (LDA) ; see Exercise 11.5.1 for details on the LDA). Clearly, the plane S_2 is more useful than V_2 for classifying strains of bacteria according to their pathogenicity. This better result is simply due to the fact that V_2 has been computed

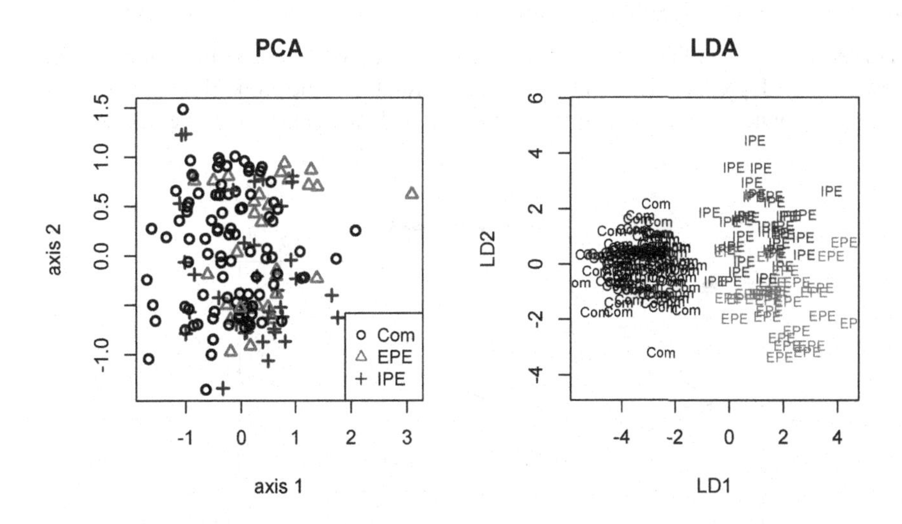

Figure 1.10 *Dimension reduction for a data set gathering 55 chemical measurements of 162 strains of* E. coli. *Commensal strains are labelled "Com," pathogen strains are labelled "EPE" (external pathogens) and "IPE" (internal pathogens). Left: The data is projected on the plane given by a PCA. Right: The data is projected on the plane given by an LDA.*

independently of the classification purpose, whereas S_2 has been computed in order to solve at best this classification problem.

1.3.2 A Paradigm Shift

Classical statistics provide a very rich theory for analyzing data with the following characteristics:

- a small number p of parameters
- a large number n of observations

This setting is typically illustrated by the linear regression $y = ax + b + \varepsilon$ plotted in Figure 1.11, where we have to estimate 2 parameters a and b, with $n = 100$ observations $(X_i, Y_i)_{i=1,\dots,n}$. Classical results carefully describe the asymptotic behavior of estimators when n goes to infinity (with p fixed), which makes sense in such a setting.

As explained in Section 1.1, in many fields, current data have very different characteristics:

- a huge number p of parameters
- a sample size n, which is either roughly of the same size as p, or sometimes much smaller than p

The asymptotic analysis with p fixed and n goes to the infinity does not make sense anymore. What is worse is that it can lead to very misleading conclusions. We must change our point of view on statistics!

In order to provide a theory adapted to the 21st century data, two different points of view can be adopted. A first point of view is to investigate the properties of estimators in a setting where both n and p go to infinity, with $p \sim f(n)$ for some function f;

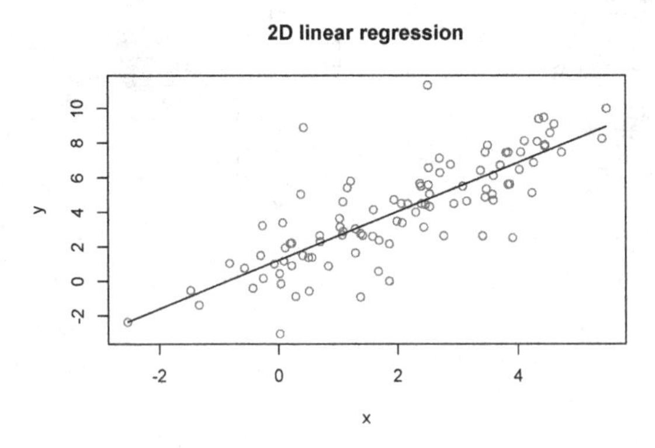

Figure 1.11 *Iconic example of classical statistics: $n = 100$ observations (gray dots) for estimating the $p = 2$ parameters of the regression line (in black).*

for example, $f(n) = \alpha n$, or $f(n) = n^2$, or $f(n) = e^{\alpha n}$, etc. Such a point of view is definitely more suited to modern data than the classical point of view. Yet, it is sensitive to the choice of the function f. For example, asymptotic results for $f(n) = n^2$ and $f(n) = e^{\alpha n}$ can be very different. If $p = 1000$ and $n = 33$, are you in the setting $f(n) = n^2$ or $f(n) = e^{n/5}$?

An alternative point of view is to treat n and p as they are and provide a non-asymptotic analysis of the estimators, which is valid for any value of n and p. Such an analysis avoids the above caveat of the asymptotic analysis. The main drawback is that non-asymptotic analyzes are much more involved than asymptotic analyzes. They usually require much more elaborate arguments in order to provide precise enough results.

1.3.3 Mathematics of High-Dimensional Statistics

In order to quantify non-asymptotically the performances of an estimator, we need some tools to replace the classical convergence theorems used in classical statistics. A typical example of convergence theorem is the central limit theorem, which describes the asymptotic convergence of an empirical mean toward its expected value: for $f : \mathbb{R} \to \mathbb{R}$ and $X_1, \ldots, X_n$ i.i.d. such that $\operatorname{var}(f(X_1)) < +\infty$, we have when $n \to +\infty$

$$\sqrt{\frac{n}{\operatorname{var}(f(X_1))}} \left(\frac{1}{n} \sum_{i=1}^{n} f(X_i) - \mathbb{E}[f(X_1)] \right) \xrightarrow{\text{d}} Z, \quad \text{with } Z \sim \mathcal{N}(0,1).$$

Loosely speaking, the difference between the empirical mean and the statistical mean $\mathbb{E}[f(X_1)]$ behaves roughly as $\sqrt{n^{-1}\operatorname{var}(f(X_1))}\, Z$ when n is large. Let us assume that f is L-Lipschitz, and let X_1, X_2 be i.i.d. with finite variance σ^2. We have

$$\operatorname{var}(f(X_1)) = \frac{1}{2}\mathbb{E}\left[(f(X_1) - f(X_2))^2\right] \leq \frac{L^2}{2}\mathbb{E}\left[(X_1 - X_2)^2\right] = L^2\sigma^2,$$

so, the central limit theorem ensures that for $X_1, \ldots, X_n$ i.i.d. with finite variance σ^2 and a L-Lipschitz function f,

$$\lim_{n\to\infty} \mathbb{P}\left(\frac{1}{n} \sum_{i=1}^{n} f(X_i) - \mathbb{E}[f(X_1)] \geq \frac{L\sigma}{\sqrt{n}} x \right) \leq \mathbb{P}(Z \geq x) \leq e^{-x^2/2}, \quad \text{for } x > 0 \quad (1.6)$$

(the last inequality follows from Lemma B.4, page 298 in Appendix B).

Concentration inequalities provide some non-asymptotic versions of such results. Assume, for example, that $X_1, \ldots, X_n$ are i.i.d., with $\mathcal{N}(0, \sigma^2)$ Gaussian distribution. The Gaussian concentration inequality (Theorem B.7, page 301 in Appendix B, see also Exercise 1.6.7) claims that for any L-Lipschitz function $F : \mathbb{R}^n \to \mathbb{R}$ we have

$$F(X_1, \ldots, X_n) - \mathbb{E}[F(X_1, \ldots, X_n)] \leq L\sigma\sqrt{2\xi}, \quad \text{where } \mathbb{P}(\xi \geq t) \leq e^{-t} \text{ for } t > 0.$$

When $f : \mathbb{R} \to \mathbb{R}$ is L-Lipschitz, the Cauchy–Schwartz inequality gives

$$\left| \frac{1}{n} \sum_{i=1}^{n} f(X_i) - \frac{1}{n} \sum_{i=1}^{n} f(Y_i) \right| \leq \frac{L}{n} \sum_{i=1}^{n} |X_i - Y_i| \leq \frac{L}{\sqrt{n}} \sqrt{\sum_{i=1}^{n} (X_i - Y_i)^2}\,,$$

so the function $F(X_1,\dots,X_n) = n^{-1}\sum_{i=1}^n f(X_i)$ is $(n^{-1/2}L)$-Lipschitz. According to the Gaussian concentration inequality, we then have for $x > 0$ and $n \geq 1$

$$\mathbb{P}\left(\frac{1}{n}\sum_{i=1}^n f(X_i) - \mathbb{E}[f(X_1)] \geq \frac{L\sigma}{\sqrt{n}}x\right) \leq \mathbb{P}\left(\sqrt{2\xi} \geq x\right) = e^{-x^2/2},$$

which can be viewed as a non-asymptotic version of (1.6). Concentration inequalities are central tools for the non-asymptotic analysis of estimators, and we will meet them in every major proof of this book. Appendix B gathers a few classical concentration inequalities (with proofs); see also Exercises 1.6.6 and 1.6.7 at the end of this chapter.

1.4 About This Book

1.4.1 Statistics and Data Analysis

Data science is an ever-expanding field. The world is awash in data, and there is a sustained effort by the data analyst community for developing statistical procedures and algorithms able to process these data. Most of this effort is carried out according to the following track:

1. identification of an issue (new or not) related to some kind of data;
2. proposition of a statistical procedure based on some heuristics; and
3. implementation of this procedure on a couple of data sets (real or simulated), with comparison to some existing procedures, when available.

This line of research feeds hundreds of scientific conferences every year covering a wide range of topics, ranging from biology to economy via humanities, communication systems and astrophysics. It has led to some dazzling success, at least in the technological area.

In the above process, mathematical formalism is used all along the way, but there are (almost) no mathematics there, in the sense that there is no mathematical analysis of the implemented procedures. In particular, even if we restrict to the academic community, the statistician community, as part of the mathematical community, is a minority. Let us call "mathematical statistics" the field of research that

1. formalizes precisely a statistical issue (new or not),
2. formalizes precisely a statistical procedure for handling this issue (new or not),
3. provides a mathematical analysis (theorem) of the performance of this statistical procedure, with a special attention to its optimality and its limitations.

Since the mathematical analysis is the limiting step (due to its difficulty), the models and statistical procedures investigated in mathematical statistics are usually quite simpler than those implemented by "mainstream" data analysts. Some may directly conclude that mathematical statistics are useless and have no interest. We want to argue that mathematical statistics are important and actually have their own interest:

- Mathematical statistics provide some strong guarantees on statistical procedures and they identify to what extent we can trust these procedures.

- Mathematical statistics enable us to identify precisely the frontier between the problems where estimation is possible and those where it is hopeless.
- Mathematical statistics provide mathematical foundations to data analysis.
- Mathematical statistics can be one of the remedies against the lack of reproducibility observed in many data-based scientific fields.

Again, most of the works in mathematical statistics concern some simple models, but they provide some useful intuitions for the more involved settings met in practice.

1.4.2 Purpose of This Book

This book focuses on the mathematical foundations of high-dimensional statistics, which is clearly in the "mathematical statistics" stream. Its goal is to present the main concepts of high-dimensional statistics and to delineate in some fundamental cases the frontier between what is achievable and what is impossible.

The book concentrates on state-of-the-art techniques for handling high-dimensional data. They do not strive for gathering a comprehensive catalog of statistical methods; they try instead to explain for some selected topics the key fundamental concepts and ideas. These concepts and ideas are exposed in simple settings, which allow concentration on the main arguments by avoiding unessential technicalities. The proofs issued from the recent research literature have been intensively streamlined in order to enlighten the main arguments. The reader is invited to adapt these ideas to more complex settings in the detailed exercises at the end of each chapter. He is also welcome to share his solutions to the exercises on the dedicated wiki-site `http://high-dimensional-statistics.wikidot.com`.

1.4.3 Overview

As explained in Section 1.3, for analyzing high-dimensional data, we must circumvent two major issues:

- the intrinsic statistical difficulty related to the curse of dimensionality, and
- the computational difficulty: procedures must have a low computational complexity in order to fit the computational resources.

To bypass the statistical curse of dimensionality, the statistician must build on the low-dimensional structures hidden in the data. Model selection, mainly developed in the late -'90s, is a powerful theory for tackling this problem. It provides very clear insights on the frontier between the problems where estimation is possible and those where it is hopeless. A smooth version of this theory is presented in Chapter 2. An alternative to model selection is model aggregation introduced in Chapter 4. While the underlying principles of model selection and model aggregation are essentially the same, they lead to different estimation schemes in practice. Model selection and model aggregation provide powerful tools able to break the statistical curse of dimensionality, yet they both suffer from a very high computational complexity, which is prohibitive in practice.

A powerful strategy to bypass this second curse of dimensionality is to "convexify" in some way the model selection schemes. This strategy has been intensively developed in the last decade with bright success. Chapter 5 is a concise introduction to this topic. It is mainly illustrated with the celebrated Lasso estimator, for which a thorough mathematical analysis is provided. The statistical procedures presented in Chapter 5 succeed to circumvent both the statistical and computational curses of high dimensionality. Yet, they suffer from two caveats. First, for each specific class of low-dimensional structures (which is often unknown) corresponds a specific estimator. Second, all these estimators depend on a "tuning parameter" that needs to be chosen according to the variance of the noise, usually unknown. Some estimator selection is then required in order to handle these two issues. A selection of such procedures are sketched in Chapter 7.

Chapters 8 and 9 go one step beyond in the statistical complexity. They present recent theories that have been developed in order to take advantage of handling simultaneously several statistical problems. Chapter 8 explores how we can exploit via some rank constraints the correlations between these statistical problems, and improve thereby the accuracy of the statistical procedures. Chapter 9 focuses on the simultaneous estimation of the conditional dependencies among a large set of variables via the theory of graphical models.

The last two chapters deal with issues arising in high-dimensional data classification. Chapter 10 details the mathematical foundations of multiple testing, with a special focus on False Discovering Rate (FDR) control. Chapter 11 gives a concise presentation of Vapnik's theory for supervised classification.

The theories presented in this book heavily rely on some more or less classical mathematical tools, including some probabilistic inequalities (concentration inequalities, contraction principle, Birgé's Lemma, etc.), some matrix analysis (singular value decomposition, Ky–Fan norms, etc.), and some classical analysis (subdifferential of convex functions, reproducing Hilbert spaces, etc.). The five appendices provide self-contained and compact introductions to these mathematical tools. The notations used throughout the book are gathered at the end of the volume.

1.5 Discussion and References

1.5.1 Take-Home Message

Being able to sense simultaneously thousands of parameters sounds like a blessing, since we collect huge amounts of data. Yet, the information is awash in the noise in high-dimensional settings. Fortunately, the useful information usually concentrates around low-dimensional structures, and building on this feature allows us to circumvent this curse of the dimensionality.

This book is an introduction to the main concepts and ideas involved in the analysis of the high-dimensional data. Its focus is on the mathematical side, with the choice to concentrate on simple settings in order to avoid unessential technical details that could blur the main arguments.

1.5.2 References

The book by Hastie, Tibshirani, and Friedman [91] is an authoritative reference for the data scientist looking for a pedagogical and (almost) comprehensive catalog of statistical procedures. The associated website
`http://statweb.stanford.edu/~tibs/ElemStatLearn/`
provides many interesting materials both for learners and academics.

We refer to the books by Ledoux [105] and by Boucheron, Lugosi, and Massart [38] for thorough developments on concentration inequalities. The asymptotic distribution of the spectral values of a random matrix has been described first by Marchenko and Pastur [117]. We refer to Vershynin [158] for a non-asymptotic analysis. We also recommend the books by Vershynin [159] and Wainwright [163] for learning further on the topic of high-dimensional probability and high-dimensional statistics.

The lecture notes [65] by Donoho give an interesting point of view on high-dimensional statistics and their mathematical challenges. Finally, the papers by Jin [96], Verzelen [160], Donoho, Johnstone, and Montanari [66] or Berthet and Rigollet [28] are examples of papers providing insightful information on the frontier of successful estimation in high-dimensional settings.

1.6 Exercises

1.6.1 Strange Geometry of High-Dimensional Spaces

Figures 1.4 and 1.5 give examples of counter-intuitive results in high-dimensional spaces. Let us give another one. Let $e_1,\dots,e_p$ be the canonical basis in $\mathbb{R}^p$, and let us denote by v the diagonal of the hypercube $[0,1]^p$. Check that the angle θ_i between the vector e_i and v fulfills

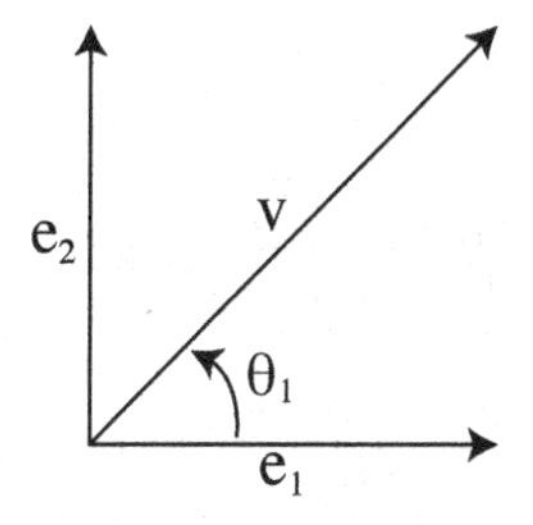

$$\cos(\theta_i) = \frac{\langle e_i, v\rangle}{\|e_i\|\,\|v\|} = \frac{1}{\sqrt{p}} \stackrel{p\to\infty}{\to} 0.$$

So the diagonal of a hypercube tends to be orthogonal to all the edges of the hypercube in high-dimensional spaces!

1.6.2 Volume of a p-Dimensional Ball

We will prove the Formula (1.1) for the volume $V_p(r)$ of a p-dimensional ball of radius $r > 0$.

1. Prove that the gamma function $\Gamma(\alpha) = \int_0^\infty x^{\alpha-1}e^{-x}\,dx$, $\alpha > 0$ fulfills the equalities $\Gamma(1) = 1$, $\Gamma(1/2) = \sqrt{\pi}$, and $\Gamma(\alpha+1) = \alpha\Gamma(\alpha)$ for any $\alpha > 0$. Deduce the two formulas

$$\Gamma(p+1) = p! \quad \text{and} \quad \Gamma(p+3/2) = \frac{(2p+1)(2p-1)\dots 1}{2^{p+1}}\sqrt{\pi} \quad \text{for } p \in \mathbb{N}.$$

2. Prove that $V_p(r) = r^p V_p(1)$ for any $p \geq 1$ and check that $V_1(1) = 2$ and $V_2(1) = \pi$.
3. For $p \geq 3$, prove that

$$\begin{aligned} V_p(1) &= \int_{x_1^2+x_2^2 \leq 1} V_{p-2}\left(\sqrt{1-x_1^2-x_2^2}\right) dx_1\, dx_2 \\ &= V_{p-2}(1) \int_{r=0}^{1} \int_{\theta=0}^{2\pi} (1-r^2)^{p/2-1} r\, dr\, d\theta = \frac{2\pi}{p} V_{p-2}(1). \end{aligned}$$

4. Conclude that

$$V_{2p}(1) = \frac{\pi^p}{p!} \quad \text{and} \quad V_{2p+1}(1) = \frac{2^{p+1}\pi^p}{(2p+1)(2p-1)\ldots 3}.$$

5. With the Stirling expansion $\Gamma(\alpha) = \alpha^{\alpha-1/2} e^{-\alpha} \sqrt{2\pi}\left(1 + O(\alpha^{-1})\right)$ for $\alpha \to +\infty$, prove (1.1).

1.6.3 Tails of a Standard Gaussian Distribution

Let Z be a $\mathcal{N}(0,1)$ standard Gaussian random variable.

1. For $z > 0$, prove (with an integration by parts) that

$$\begin{aligned} \mathbb{P}(|Z| \geq z) &= \sqrt{\frac{2}{\pi}} \frac{e^{-z^2/2}}{z} - \sqrt{\frac{2}{\pi}} \int_z^{\infty} x^{-2} e^{-x^2/2}\, dx \\ &= \sqrt{\frac{2}{\pi}} \frac{e^{-z^2/2}}{z} \left(1 + O\left(\frac{1}{z^2}\right)\right). \end{aligned}$$

2. For $Z_1, \ldots, Z_p$ i.i.d. with $\mathcal{N}(0,1)$ standard Gaussian distribution and $\alpha > 0$, prove that when $p \to \infty$

$$\begin{aligned} &\mathbb{P}\left(\max_{j=1,\ldots,p} |Z_j| \geq \sqrt{\alpha \log(p)}\right) \\ &= 1 - \left(1 - \mathbb{P}\left(|Z_1| \geq \sqrt{\alpha \log(p)}\right)\right)^p \\ &= 1 - \exp\left(-\sqrt{\frac{2}{\alpha\pi}} \frac{p^{1-\alpha/2}}{(\log p)^{1/2}} + O\left(\frac{p^{1-\alpha/2}}{(\log p)^{3/2}}\right)\right). \end{aligned}$$

1.6.4 Principal Component Analysis

The Principal Component Analysis (PCA) is tightly linked to the Singular Value Decomposition (SVD). We refer to Appendix C for a reminder on the SVD.

For any data points $X^{(1)},\dots,X^{(n)} \in \mathbb{R}^p$ and any dimension $d \le p$, the PCA computes the linear span in $\mathbb{R}^p$

$$V_d \in \underset{\dim(V)\le d}{\text{argmin}} \sum_{i=1}^{n} \|X^{(i)} - \text{Proj}_V X^{(i)}\|^2,$$

where Proj_V is the orthogonal projection matrix onto V. Let us denote by $\mathbf{X} = \sum_{k=1}^{r} \sigma_k u_k v_k^T$ a SVD of the $n \times p$ matrix

$$\mathbf{X} = \begin{pmatrix} (X^{(1)})^T \\ \vdots \\ (X^{(n)})^T \end{pmatrix}$$

with $\sigma_1 \ge \sigma_2 \ge \ldots \ge \sigma_r > 0$.

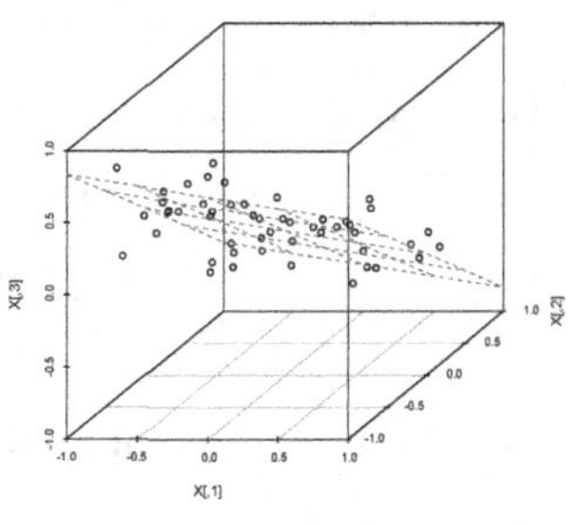

V_2 *in dimension* $p = 3$.

1. With the Theorem C.5 in Appendix C, page 315, prove that for any $d \le r$

$$\sum_{i=1}^{n} \|X^{(i)} - \text{Proj}_V X^{(i)}\|^2 = \|\mathbf{X} - \mathbf{X}\text{Proj}_V\|_F^2 \ge \sum_{k=d+1}^{r} \sigma_k^2,$$

   where $\|\cdot\|_F$ denotes the Frobenius norm $\|A\|_F^2 = \sum_{i,j} A_{ij}^2$.

2. Write V_d for the linear space spanned by $\{v_1,\dots,v_d\}$. Prove that for any $d \le r$ we have

$$\|\mathbf{X} - \mathbf{X}\text{Proj}_{V_d}\|^2 = \sum_{k=d+1}^{r} \sigma_k^2.$$

3. Conclude that V_d minimizes (1.5).
4. Prove that the coordinates of $\text{Proj}_{V_d} X^{(i)}$ in the orthonormal basis $(v_1,\dots,v_d)$ of V_d are given by $(\sigma_1\langle e_i,u_1\rangle,\dots,\sigma_d\langle e_i,u_d\rangle)$.

Terminology: The right-singular vectors $v_1,\dots,v_r$ are called the principal axes. The vectors $c_k = \mathbf{X}v_k = \sigma_k u_k$ for $k = 1,\dots,r$ are called the principal components. The principal component c_k gathers the coordinates of $X^{(1)},\dots,X^{(n)}$ on v_k.

Remark: Since V_d is a linear span and not an affine span, it is highly recommended to first center the data points

$$\widetilde{X}^{(i)} = X^{(i)} - \frac{1}{n}\sum_{i=1}^{n} X^{(i)}$$

and then proceed with a PCA on the $\widetilde{X}^{(1)},\dots,\widetilde{X}^{(n)}$.

1.6.5 Basics of Linear Regression

We consider the linear regression model $Y = \mathbf{X}\beta^* + \varepsilon$, with $Y, \varepsilon \in \mathbb{R}^n$, $\beta^* \in \mathbb{R}^p$ and $\mathbf{X}$ an $n \times p$-matrix with rank p. We assume that the components $\varepsilon_1,\dots,\varepsilon_n$ of ε are i.i.d. centered, with variance σ^2. We set $F(\beta) = \|Y - \mathbf{X}\beta\|^2$.

1. Prove that F is convex and has a gradient $\nabla F(\beta) = 2\mathbf{X}^T(\mathbf{X}\beta - Y)$.
2. Prove that the least-squares estimator $\widehat{\beta} \in \operatorname{argmin}_{\beta \in \mathbb{R}^p} F(\beta)$ solves $\mathbf{X}^T\mathbf{X}\widehat{\beta} = \mathbf{X}^T Y$, so $\widehat{\beta} = (\mathbf{X}^T\mathbf{X})^{-1}\mathbf{X}^T Y$ when the rank of $\mathbf{X}$ is p.
3. Let A be a $p \times n$ matrix. Prove that $\mathbb{E}\left[\|A\varepsilon\|^2\right] = \sigma^2 \mathrm{Tr}(A^T A)$.
4. Conclude that mean square error of the least-squares estimator $\widehat{\beta}$ is

$$\mathbb{E}\left[\|\widehat{\beta} - \beta^*\|^2\right] = \mathrm{Tr}\left((\mathbf{X}^T\mathbf{X})^{-1}\right)\sigma^2.$$

1.6.6 Concentration of Square Norm of Gaussian Random Variable

In the regression setting, we often need a bound on the square norm of a $\mathcal{N}(0, I_n)$ Gaussian random variable ε; see, for example, the proof of Theorem 2.2 in Chapter 2. The expectation of $\|\varepsilon\|^2$ is $\mathbb{E}\left[\|\varepsilon\|^2\right] = n$, so all we need is to get a probabilistic bound on the deviations of $\|\varepsilon\|^2$ above its expectation. Since the map $\varepsilon \to \|\varepsilon\|$ is 1-Lipschitz, the Gaussian concentration inequality (Theorem B.7, page 301, in Appendix B) ensures that

$$\mathbb{P}\left(\|\varepsilon\| \leq \mathbb{E}\left[\|\varepsilon\|\right] + \sqrt{2x}\right) \geq 1 - e^{-x}, \quad \text{for any } x > 0.$$

From Jensen inequality, we have $\mathbb{E}\left[\|\varepsilon\|\right]^2 \leq \mathbb{E}\left[\|\varepsilon\|^2\right] = n$, so we have the concentration inequality

$$\mathbb{P}\left(\|\varepsilon\|^2 \leq n + 2\sqrt{2nx} + 2x\right) \geq 1 - e^{-x}, \quad \text{for any } x > 0. \tag{1.7}$$

A) Concentration from Above

In the following, we give a direct proof of (1.7) based on the simple Markov inequality (Lemma B.1, page 297, in Appendix B).

1. Check that for $0 < s < 1/2$, we have $\mathbb{E}\left[\exp\left(s\|\varepsilon\|^2\right)\right] = (1-2s)^{-n/2}$.
2. With the Markov inequality, prove that $\mathbb{P}\left(\|\varepsilon\|^2 > n+t\right) \leq e^{-s(n+t)}(1-2s)^{-n/2}$ for any $t > 0$ and $0 < s < 1/2$.
3. Check that the above bound is minimal for $s = t/(2(n+t))$, and hence

$$\mathbb{P}\left(\|\varepsilon\|^2 > n+t\right) \leq e^{-t/2}(1+t/n)^{n/2}, \quad \text{for any } t > 0.$$

4. Check that $\log(1+u) \leq u - u^2/(2+2u)$ for any $u \geq 0$, and hence

$$\mathbb{P}\left(\|\varepsilon\|^2 > n+t\right) \leq \exp\left(-\frac{t^2}{4(n+t)}\right), \quad \text{for any } t > 0.$$

5. Prove Bound (1.7) for $0 < x \leq n$.
6. From the bound $\log(1+u) \leq \sqrt{u}$ for $u \geq 0$, check that

$$\frac{n}{2}\log\left(1 + \frac{2\sqrt{2nx} + 2x}{n}\right) \leq \sqrt{2nx}, \quad \text{for } x \geq n.$$

7. Prove (1.7) for $x \geq n$.

B) Concentration from Below

1. Prove that for $0 \leq t < n$ and $s \geq 0$

$$\mathbb{P}\left[n - \|\varepsilon\|^2 \geq t\right] \leq \frac{e^{s(n-t)}}{(1+2s)^{n/2}}.$$

2. Check that the above bound is minimal for $s = t/(2(n-t))$.
3. With the upper bound $\log(1-u) \leq -u + u^2/2$ for $0 \leq u < 1$, prove that for $0 \leq t < n$

$$\mathbb{P}\left[\|\varepsilon\|^2 \leq n - t\right] \leq e^{-t^2/(4n)}.$$

4. Conclude that for any $x \geq 0$

$$\mathbb{P}\left[\|\varepsilon\|^2 \geq n - 2\sqrt{nx}\right] \geq 1 - e^{-x}. \tag{1.8}$$

1.6.7 A Simple Proof of Gaussian Concentration

This exercise provides a simple proof of a weak version of the Gaussian concentration inequality (Theorem B.7, page 301 in Appendix B). The price to pay for the simplicity of the proof is a non-tight constant and a differentiability assumption (that can be dropped however). The main argument of this proof is due to Pisier [129] and Maurey.

We will prove here that, for any $F : \mathbb{R}^d \to \mathbb{R}$ which is differentiable with $\|\nabla F(x)\| \leq L$ for all $x \in \mathbb{R}^d$, and for Z with Gaussian $\mathcal{N}(0, I_d)$ distribution, we have

$$\mathbb{P}\left(F(Z) - \mathbb{E}\left[F(Z)\right] \geq L\sigma t\right) \leq \exp\left(-2t^2/\pi^2\right). \tag{1.9}$$

We observe that $F(Z) = L\sigma\tilde{F}(Z/\sigma)$ with Z/σ following a Gaussian $\mathcal{N}(0, I_d)$ distribution and $\tilde{F}(x) = F(\sigma x)/(L\sigma)$ fulfilling $\|\nabla\tilde{F}(x)\| \leq 1$ for all $x \in \mathbb{R}^d$. So, with no loss of generality, we assume henceforth that $L = 1$ and $\sigma = 1$.

The core of the Pisier-Maurey argument is the following inequality proved in part (B) of the exercise. For any convex function $\varphi : \mathbb{R} \to \mathbb{R}$, we have

$$\mathbb{E}\left[\varphi(F(Z) - \mathbb{E}\left[F(Z)\right])\right] \leq \mathbb{E}\left[\varphi\left(\frac{\pi}{2}\langle\nabla F(Z_1), Z_2\rangle\right)\right], \tag{1.10}$$

where Z_1, Z_2 are two independent Gaussian $\mathcal{N}(0, I_d)$ random variables.

A) Deriving the Concentration Bound (1.9) from Pisier-Maurey inequality (1.10)

Before proving Pisier-Maurey inequality (1.10), we explain how the concentration bound (1.9) follows from it.

1. Check that the distribution of the random variable $\langle\nabla F(Z_1), Z_2\rangle$ conditionally on Z_1 is the Gaussian $\mathcal{N}(0, \|\nabla F(Z_1)\|^2)$ distribution.
2. For $s \geq 0$, prove with (1.10) that

$$\mathbb{E}\left[\exp\left(s(F(Z) - \mathbb{E}\left[F(Z)\right])\right)\right] \leq \mathbb{E}\left[\exp\left((s\pi\|\nabla F(Z_1)\|)^2/8\right)\right] \leq e^{(s\pi)^2/8}.$$

3. Conclude the proof of (1.9) by applying Chernoff lemma B.2, on page 297.

B) Proof of Pisier-Maurey Inequality (1.10)

Let Z_1, Z_2 be two independent Gaussian $\mathcal{N}(0, I_d)$ random variables. Let us define the random function $W : [0, \pi/2] \to \mathbb{R}$, by $W(\theta) = Z_1 \sin(\theta) + Z_2 \cos(\theta)$.

1. Check that

$$F(Z_1) - F(Z_2) = \int_0^{\pi/2} \langle \nabla F(W(\theta)), W'(\theta) \rangle \, d\theta.$$

2. Applying twice Jensen inequality, prove that

$$\begin{aligned}\mathbb{E}_{Z_1}\left[\varphi\left(F(Z_1) - \mathbb{E}_{Z_2}\left[F(Z_2)\right]\right)\right] &\leq \mathbb{E}_{Z_1,Z_2}\left[\varphi\left(F(Z_1) - F(Z_2)\right)\right] \\ &\leq \frac{2}{\pi}\int_0^{\pi/2} \mathbb{E}_{Z_1,Z_2}\left[\varphi\left(\frac{\pi}{2}\langle \nabla F(W(\theta)), W'(\theta)\rangle\right)\right] d\theta.\end{aligned}$$

3. Check that $(W(\theta), W'(\theta))$ has the same distribution as (Z_1, Z_2) and conclude the proof of (1.10).

Chapter 2

Model Selection

Model selection is a key conceptual tool for performing dimension reduction and exploiting hidden structures in the data. The general idea is to compare different statistical models corresponding to different possible hidden structures and then select among them the one that is more suited for estimation. Model selection is a very powerful theory, but it suffers in many cases from a very high computational complexity that can be prohibitive. When model selection cannot be implemented due to its computational cost, it remains a good guideline for developing computationally tractable procedures, as we will see in Chapters 5 and 6.

In this chapter, we present the theory of model selection in a simple — yet useful — setting: the Gaussian regression model. References for a broader theory are given in Section 2.7.

2.1 Statistical Setting

We consider in this chapter the regression model

$$y_i = f(x^{(i)}) + \varepsilon_i, \quad i = 1, \ldots, n,$$

which links a quantity of interest $y \in \mathbb{R}$ to p variables whose values are stored in a p-dimensional vector $x \in \mathbb{R}^p$. We give several examples below corresponding to the regression model.

Example 1: Sparse piecewise constant regression (variation sparsity)

It corresponds to the case where $x \in \mathbb{R}$ and f is piecewise constant with a small number of jumps. This situation appears, for example, in biology in the analysis of the copy number variations along a DNA chain.

Assume that the jumps of f are located in the set $\{z_j : j \in \mathcal{J}\}$. Then a right-continuous piecewise constant function f can be written as

$$f(x) = \sum_{j \in \mathcal{J}} c_j \mathbf{1}_{x \geq z_j}.$$

When the function f only has a few jumps, then only a small fraction of the $\{c_j : j \in \mathcal{J}\}$ are nonzero.

Example 2: Sparse basis/frame expansion

It corresponds to the case where $f : \mathbb{R} \to \mathbb{R}$, and we estimate f by expanding it on a basis or frame $\{\varphi_j\}_{j \in \mathscr{J}}$

$$f(x) = \sum_{j \in \mathscr{J}} c_j \varphi_j(x),$$

with a small number of nonzero c_j. This situation arises, for example, for denoising, for representing cortex signals, etc. Typical examples of basis are Fourier basis, splines, wavelets, etc. The most simple example is the piecewise linear decomposition

$$f(x) = \sum_{j \in \mathscr{J}} c_j (x - z_j)_+, \tag{2.1}$$

where $z_1 < z_2 < \ldots$ and $(x)_+ = \max(x, 0)$.

In these first two examples, x is low dimensional, but the function to be estimated can be complex and requires to estimate a large number of coefficients. At the opposite, we can have x in a high-dimensional space $\mathbb{R}^p$ and f very simple. The most popular case is the linear regression.

Example 3: Sparse linear regression

It corresponds to the case where f is linear: $f(x) = \langle \beta, x \rangle$ with $\beta \in \mathbb{R}^p$. We say that the linear regression is sparse when only a few coordinates of β are nonzero.

This model can be too rough to model the data. Assume, for example, that we want to model the relationship between some phenotypes and some gene expression levels. We expect from biology that only a small number of genes influence a given phenotype, but the relationship between these genes and the phenotype is unlikely to be linear. We may consider in this case a slightly more complex model.

Example 4: Sparse additive model and group-sparse regression

In the sparse additive model, we expect that $f(x) = \sum_k f_k(x_k)$ with most of the f_k equal to 0.

If we expand each function f_k on a frame or basis $\{\varphi_j\}_{j \in \mathscr{J}_k}$ we obtain the decomposition

$$f(x) = \sum_{k=1}^{p} \sum_{j \in \mathscr{J}_k} c_{j,k} \varphi_j(x_k),$$

where most of the vectors $\{c_{j,k}\}_{j \in J_k}$ are zero.

Such a model can be hard to fit from a small sample, since it requires to estimate a relatively large number of nonzero $c_{j,k}$. Nevertheless, in some cases, the basis expansion of f_k can be sparse itself, as in Example 2, producing a more complex pattern of sparsity. An interesting example is the following model.

Example 5: Sparse additive piecewise linear regression

The sparse additive piecewise linear model is a sparse additive model $f(x) = \sum_k f_k(x_k)$, with sparse piecewise linear functions f_k. We then have two levels of sparsity:

1. Most of the f_k are equal to 0;
2. The nonzero f_k have a sparse expansion in the following representation

$$f_k(x_k) = \sum_{j \in \mathcal{J}_k} c_{j,k} (x_k - z_{j,k})_+$$

In other words, the matrix $c = [c_{j,k}]$ of the sparse additive model has only a few nonzero columns, and these nonzero columns are sparse.

It turns out that all the above models correspond to a sparse linear model, in the sense that we have a representation of $f^* = \left[f(x^{(i)})\right]_{i=1,\dots,n}$ of the form $f_i^* = \langle \alpha, \psi_i \rangle$ for $i = 1, \dots, n$, with α sparse in some sense. Let us identify this representation in the five above examples.

Examples 1, 2, 3, 4, 5 (continued): Representation $f_i^* = \langle \alpha, \psi_i \rangle$

- Sparse piecewise constant regression: $\psi_i = e_i$ with $\{e_1, \dots, e_n\}$ the canonical basis of $\mathbb{R}^n$ and $\alpha = f^*$ is piecewise constant.
- Sparse basis expansion: $\psi_i = [\varphi_j(x^{(i)})]_{j \in \mathcal{J}}$ and $\alpha = c$.
- Sparse linear regression: $\psi_i = x^{(i)}$ and $\alpha = \beta$.
- Sparse additive models: $\psi_i = [\varphi_j([x_k^{(i)}])]_{\substack{k=1,\dots,p \\ j \in \mathcal{J}_k}}$ and $\alpha = [c_{j,k}]_{\substack{k=1,\dots,p \\ j \in \mathcal{J}_k}}$.

Since the five above examples can be recast in a linear regression framework, we will consider the linear regression model as the canonical example and use the following notations.

Linear model

We have $f_i^* = \langle \beta^*, x^{(i)} \rangle$. We define the $n \times p$ matrix $\mathbf{X} = [x_j^{(i)}]_{i=1,\dots,n,\ j=1,\dots,p}$ and the n-dimensional vector $\varepsilon = [\varepsilon_i]_{i=1,\dots,n}$. With these notations, we have the synthetic formula

$$Y = f^* + \varepsilon = \mathbf{X}\beta^* + \varepsilon. \tag{2.2}$$

The above examples correspond to different sparsity patterns for β^*. We use throughout the book the following terminology.

Sparsity patterns

- **Coordinate sparsity:** Only a few coordinates of β^* are nonzero. This situation arises in Examples 2 and 3.
- **Group sparsity:** The coordinates of β^* are clustered into groups, and only a few groups are nonzero. More precisely, we have a partition $\{1,\ldots,p\} = \bigcup_{k=1}^{M} G_k$, and only a few vectors $\beta^*_{G_k} = (\beta^*_j)_{j\in G_k}$ are nonzero. This situation arises in Example 4.
- **Sparse-group sparsity:** In the same notation as the group-sparse setting, only a few vectors $\beta^*_{G_k}$ are nonzero, and in addition they are sparse. This situation arises in Example 5.
- **Variation sparsity (or fused sparsity):** The vector $\Delta\beta^* = [\beta^*_j - \beta^*_{j-1}]_{j=2,\ldots,p}$ is coordinate-sparse. This situation arises in Example 1.

In the remainder of this chapter, we will focus on the Gaussian regression setting

$$Y_i = f^*_i + \varepsilon_i,\ i = 1,\ldots,n, \quad \text{with the } \varepsilon_i \text{ i.i.d. with } \mathcal{N}(0,\sigma^2) \text{ distribution.} \tag{2.3}$$

2.2 Selecting among a Collection of Models

2.2.1 Models and Oracle

Let us consider the problem of linear regression in the coordinate-sparse setting. If we knew that only the coordinates β^*_j with $j \in m^*$ are nonzero, then we would remove all the other variables $\{x_j : j \notin m^*\}$ and consider the simpler linear regression model

$$y_i = \langle \beta^*_{m^*}, x^{(i)}_{m^*} \rangle + \varepsilon_i = \sum_{j\in m^*} \beta^*_j x^{(i)}_j + \varepsilon_i.$$

In the other sparse settings described above, we would work similarly if we knew exactly the sparsity pattern (which groups are nonzero for group sparsity, where the jumps are located for variation sparsity, etc.).

More generally, if we knew that f^* belongs to some linear subspace S of $\mathbb{R}^n$, instead of estimating f^* by simply maximizing the likelihood, we would rather estimate f^* by maximizing the likelihood under the constraint that the estimator belongs to S. For example, in the coordinate-sparse setting where we know that the nonzero coordinates are the $\{\beta^*_j : j \in m^*\}$, we would take $S = \text{span}\{x_j,\ j \in m^*\}$. In our Gaussian setting (2.3), the log-likelihood is given by

$$-\frac{n}{2}\log(2\pi\sigma^2) - \frac{1}{2\sigma^2}\|Y - f\|^2,$$

so the estimator maximizing the likelihood under the constraint that it belongs to S

is simply $\widehat{f} = \mathrm{Proj}_S Y$, where $\mathrm{Proj}_S : \mathbb{R}^n \to \mathbb{R}^n$ is the orthogonal projection operator onto S.

If we do not know *a priori* that f^* belongs to a *known* linear subspace S of $\mathbb{R}^n$, then we may wish to

1. consider a collection $\{S_m,\ m \in \mathcal{M}\}$ of linear subspaces of $\mathbb{R}^n$, called *models*;
2. associate to each subspace S_m the constrained maximum likelihood estimators $\widehat{f}_m = \mathrm{Proj}_{S_m} Y$; and
3. finally estimate f by the *best* estimator among the collection $\{\widehat{f}_m,\ m \in \mathcal{M}\}$.

To give a meaning to *best*, we need a criterion to quantify the quality of an estimator. In the following, we will measure the quality of an estimator $\widehat{f}$ of f^* by its ℓ^2-risk

$$R(\widehat{f}) = \mathbb{E}\left[\|\widehat{f} - f^*\|^2\right]. \tag{2.4}$$

Each estimator $\widehat{f}_m$ has a ℓ^2-risk $r_m = R(\widehat{f}_m)$, and the best estimator in terms of the ℓ^2-risk is the so-called *oracle estimator*

$$\widehat{f}_{m_o} \quad \text{with} \quad m_o \in \underset{m \in \mathcal{M}}{\operatorname{argmin}} \{r_m\}. \tag{2.5}$$

We emphasize that the signal f^* may not belong to the oracle model S_{m_o}. It may even not belong to any of the models $\{S_m,\ m \in \mathcal{M}\}$.

The oracle estimator is the best estimator in terms of the ℓ^2-risk, so we would like to use this estimator to estimate f^*. Unfortunately, we cannot use it in practice, since it cannot be computed from the data only. Actually, the index m_o depends on the collection of risks $\{r_m,\ m \in \mathcal{M}\}$, which is unknown to the statisticians since it depends on the unknown signal f^*.

A natural idea to circumvent this issue is to replace the risks r_m in (2.5) by some estimators $\widehat{r}_m$ of the risk and therefore estimate f^* by

$$\widehat{f}_{\widehat{m}} \quad \text{with} \quad \widehat{m} \in \underset{m \in \mathcal{M}}{\operatorname{argmin}} \{\widehat{r}_m\}. \tag{2.6}$$

The estimator $\widehat{f}_{\widehat{m}}$ can be computed from the data only, but we have no guarantee that it performs well. The main object of this chapter is to provide some suitable $\widehat{r}_m$ for which we can guarantee that the selected estimator $\widehat{f}_{\widehat{m}}$ performs almost as well as the oracle $\widehat{f}_{m_o}$.

Examples of collections of models

Let us describe the collections of models suited to the different sparsity patterns described in Section 2.1. In the following, we write $\mathbf{X}_j$ for the j-th column of $\mathbf{X}$.

Coordinate sparsity

Let us define $\mathcal{M} = \mathcal{P}(\{1,\ldots,p\})$, where $\mathcal{P}(\{1,\ldots,p\})$ denotes the set of all the subsets of $\{1,\ldots,p\}$. In this setting, we will consider the collection of models $\{S_m,\ m \in \mathcal{M}\}$, with the models S_m defined by $S_m = \text{span}\left\{\mathbf{X}_j,\ j \in m\right\}$ for $m \in \mathcal{M}$. The model S_m then gathers all the vectors $\mathbf{X}\beta$, with β fulfilling $\beta_j = 0$ if $j \notin m$.

Group sparsity

In this setting, we will consider a collection of models $\{S_m,\ m \in \mathcal{M}\}$ indexed by the set $\mathcal{M} = \mathcal{P}(\{1,\ldots,M\})$ of all the subsets of $\{1,\ldots,M\}$. For $m \in \mathcal{M}$, the model S_m is defined by $S_m = \text{span}\left\{\mathbf{X}_j\ :\ j \in \bigcup_{k\in m} G_k\right\}$. The model S_m then gathers all the vectors $\mathbf{X}\beta$, with β fulfilling $\beta_{G_k} = 0$ for $k \notin m$.

Sparse-group sparsity

The description of the collection of models $\{S_m,\ m \in \mathcal{M}\}$ in this setting is slightly more involved. To a subset $g \subset \{1,\ldots,M\})$, we associate the set of indices

$$\mathcal{M}_g = \left\{\{J_k\}_{k\in g}\ :\ \text{where } J_k \subset G_k \text{ for all } k \in g\right\}.$$

We defined the set of indices by

$$\mathcal{M} = \bigcup_{g\subset\{1,\ldots,M\}}\ \bigcup_{\{J_k\}_{k\in g}\in\mathcal{M}_g} \left\{(g, \{J_k\}_{k\in g})\right\},$$

and to an index $m = (g, \{J_k\}_{k\in g})$ in $\mathcal{M}$, we associate the model $S_m = \text{span}\left\{\mathbf{X}_j\ :\ j \in \bigcup_{k\in g} J_k\right\}$. The model $S_{(g,\{J_k\}_{k\in g})}$ then gathers all the vectors $\mathbf{X}\beta$, with β fulfilling $\beta_j = 0$ if $j \notin \bigcup_{k\in g} J_k$.

Variation sparsity

Writing $\Delta\beta = [\beta_k - \beta_{k-1}]_{k=1,\ldots,p}$ (with the convention $\beta_0 = 0$), we have $\beta_j = \sum_{k\le j} [\Delta\beta]_k$ and

$$\mathbf{X}\beta = \sum_{j=1}^{p} \sum_{k\le j} [\Delta\beta]_k \mathbf{X}_j = \sum_{k=1}^{p} [\Delta\beta]_k \sum_{j\ge k} \mathbf{X}_j\,.$$

In view of this formula, we will consider for the variation sparse setting the collection of models $\{S_m,\ m \in \mathcal{M}\}$ indexed by the set $\mathcal{M} = \mathcal{P}(\{1,\ldots,p\})$ of all the subsets of $\{1,\ldots,p\}$, with models S_m defined by $S_m = \text{span}\left\{\sum_{j\ge k} \mathbf{X}_j,\ k \in m\right\}$. The model S_m then gathers all the vectors $\mathbf{X}\beta$, with β fulfilling $\beta_j = \beta_{j-1}$ if $j \notin m$.

Models are (usually) wrong

We emphasize that it is unlikely that our data exactly corresponds to one of the settings described in the five examples of Section 2.1. For example, it is unlikely that a signal $f(x)$ is exactly sparse linear, yet a good sparse linear approximation of it may exist. In particular, we do not have in mind that one of the models is exact and we seek it. We rather try to select the best model in the collection in order to estimate f^*, keeping in mind that all these models can be wrong. The best model for estimation corresponds to the oracle model S_{m_o}, which is our benchmark for comparison.

2.2.2 Model Selection Procedures

Risk r_m of $\widehat{f}_m$

Let us compute the risk $r_m = R(\widehat{f}_m)$ of the estimator $\widehat{f}_m$. Starting from $Y = f^* + \varepsilon$, we obtain the decomposition $f^* - \widehat{f}_m = (I - \mathrm{Proj}_{S_m})f^* - \mathrm{Proj}_{S_m}\varepsilon$. Then, the Pythagorean formula and Lemma A.3, page 294 in Appendix A, give

$$r_m = \mathbb{E}\left[\|f^* - \widehat{f}_m\|^2\right] = \mathbb{E}\left[\|(I - \mathrm{Proj}_{S_m})f^*\|^2 + \|\mathrm{Proj}_{S_m}\varepsilon\|^2\right]$$
$$= \|(I - \mathrm{Proj}_{S_m})f^*\|^2 + d_m\sigma^2,$$

where $d_m = \dim(S_m)$. The risk r_m involves two terms. The first term $\|(I - \mathrm{Proj}_{S_m})f^*\|^2$ is a bias term that reflects the quality of S_m for approximating f^*. The second term $d_m\sigma^2$ is a variance term that increases linearly with the dimension of S_m. In particular, we notice that enlarging S_m reduces the first term but increases the second term. The oracle model S_{m_o} is then the model in the collection $\{S_m,\ m \in \mathcal{M}\}$, which achieves the best trade-off between the bias and the variance.

Unbiased estimator of the risk

A natural idea is to use an unbiased estimator $\widehat{r}_m$ of the risk r_m in Criterion (2.6). It follows from the decomposition $Y - \widehat{f}_m = (I - \mathrm{Proj}_{S_m})(f^* + \varepsilon)$ that

$$\mathbb{E}\left[\|Y - \widehat{f}_m\|^2\right] = \mathbb{E}\left[\|(I - \mathrm{Proj}_{S_m})f^*\|^2 + 2\langle(I - \mathrm{Proj}_{S_m})f^*, \varepsilon\rangle + \|(I - \mathrm{Proj}_{S_m})\varepsilon\|^2\right]$$
$$= \|(I - \mathrm{Proj}_{S_m})f^*\|^2 + (n - d_m)\sigma^2$$
$$= r_m + (n - 2d_m)\sigma^2.$$

As a consequence,

$$\widehat{r}_m = \|Y - \widehat{f}_m\|^2 + (2d_m - n)\sigma^2 \tag{2.7}$$

is an unbiased estimator of the risk. Note that dropping the $-n\sigma^2$ term in $\widehat{r}_m$ does not change the choice of $\widehat{m}$ in (2.6). This choice gives the Akaike Information Criterion (AIC)

$$\widehat{m}_{\mathrm{AIC}} \in \underset{m \in \mathcal{M}}{\mathrm{argmin}} \left\{\|Y - \widehat{f}_m\|^2 + 2d_m\sigma^2\right\}. \tag{2.8}$$

This criterion is very natural and popular. Nevertheless, it can produce very poor results in some cases because it does not take into account the variability of the estimated risks $\widehat{r}_m$ around their mean r_m. This is, for example, the case when the number of models S_m with dimension d grows exponentially with d; see Exercise 2.8.1. It is

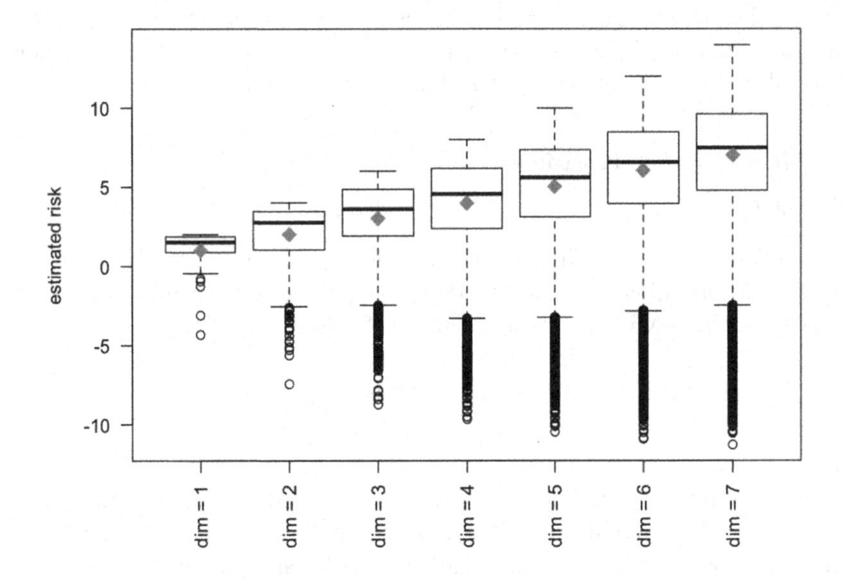

Figure 2.1 *We consider the pure noise case ($f^* = 0$) in the coordinate-sparse setting with orthogonal design, as in Exercise 2.8.1, with $p = 40$. For each dimension $d = 1,\dots,7$, the boxplot of $\{\widehat{r}_m : d_m = d\}$ is represented, with $\widehat{r}_m$ defined by (2.7). The true risks $r_m = d_m\sigma^2$ are represented by gray diamonds. We observe that when dimension d grows, there are more and more outliers, and* $\min\{\widehat{r}_m : d_m = d\}$ *tends to decrease with d. The AIC model $S_{\widehat{m}_{AIC}}$ then has a large dimension.*

due to the fact that in this setting, for large dimensions d, we have a huge number of models S_m with dimension d. Therefore, we have a huge number of estimators $\widehat{r}_m$, and, due to the randomness, some of them deviate seriously from their expected value r_m. In particular, some $\widehat{r}_m$ are very small, much smaller than $\widehat{r}_{m_o}$ associated to the oracle m_o. This leads the AIC criterion to select a model $S_{\widehat{m}}$ much bigger than S_{m_o} with very high probability. We illustrate this issue in Figure 2.1.

Penalized estimator of the risk

How can we avoid the undesirable phenomenon described above? The AIC criterion (2.8) involves two terms. The first term $\|Y - \widehat{f}_m\|^2 = \|Y - \mathrm{Proj}_{S_m} Y\|^2$ is decreasing with the size of S_m, whereas the second term $2d_m\sigma^2$ is increasing with the dimension of S_m. In order to avoid the selection of a model $S_{\widehat{m}}$ with an overly large dimension $d_{\widehat{m}}$ (as in Exercise 2.8.1), we shall replace the second term $2d_m\sigma^2$ by a term taking into account the number of models per dimension.

Following this idea, we focus henceforth on a selection criterion of the form

$$\widehat{m} \in \underset{m \in \mathcal{M}}{\operatorname{argmin}} \left\{ \|Y - \widehat{f}_m\|^2 + \sigma^2 \mathrm{pen}(m) \right\},$$

where the function pen : $\mathcal{M} \to \mathbb{R}^+$ is called the *penalty* function. A proper strategy to build a good penalty function is to perform a non-asymptotic analysis of the risk $R(\widehat{f}_{\widehat{m}})$ and then choose the penalty $\mathrm{pen}(m)$ in order to have a risk $R(\widehat{f}_{\widehat{m}})$ as close as possible to the oracle risk $R(\widehat{f}_{m_o})$. Such an analysis, leading to the choice of the selection criterion presented below, is detailed in the proof of the forthcoming Theorem 2.2.

To start with, we associate to the collection of models $\{S_m,\ m \in \mathcal{M}\}$ a probability distribution $\pi = \{\pi_m,\ m \in \mathcal{M}\}$ on $\mathcal{M}$. For a given probability π and a given $K > 1$, we select $\widehat{m}$ according to the criterion

$$\widehat{m} \in \underset{m \in \mathcal{M}}{\operatorname{argmin}} \left\{ \|Y - \widehat{f}_m\|^2 + \sigma^2 \mathrm{pen}(m) \right\},$$
$$\text{with} \quad \mathrm{pen}(m) = K \left(\sqrt{d_m} + \sqrt{2 \log(1/\pi_m)} \right)^2. \tag{2.9}$$

The resulting estimator is $\widehat{f} = \widehat{f}_{\widehat{m}}$.

At first sight, it is not obvious why we should use a penalty $\mathrm{pen}(m)$ with such a shape. We refer to Remark 2, page 40, for explanations on the shape of Criterion (2.9) in the light of the proof of Theorem 2.2.

Criterion (2.9) depends heavily on the probability π. It is then crucial to choose this probability π properly. This probability distribution can reflect our knowledge on the likelihood of being in one model rather than in another one, but most of the time it is chosen in a completely *ad-hoc* way. Actually, as we will see in Theorem 2.2, the ℓ^2-risk of the estimator $\widehat{f} = \widehat{f}_{\widehat{m}}$, with $\widehat{m}$ selected by (2.9), roughly behaves like

$$\min_{m \in \mathcal{M}} \left(r_m + \sigma^2 \log \frac{1}{\pi_m} \right). \tag{2.10}$$

Therefore, the choice of π_m will be driven by an attempt to make this term as close as possible to the oracle risk r_{m_o}. It will be the case, for example, if $\log(\pi_m^{-1})\sigma^2$ behaves like the variance $d_m\sigma^2$ of $\widehat{f}_m$: since $r_m \geq d_m\sigma^2$, the minimum (2.10) is upper-bounded by $2r_{m_o}$ when $\log(\pi_m^{-1}) \leq d_m$. Unfortunately, we cannot choose $\log(\pi_m^{-1}) \leq \alpha d_m$ with α a numerical constant for any collection of models. Actually, increasing the size of $\mathcal{M}$ requires to decrease the size of π_m (since π is a probability distribution on $\mathcal{M}$), and when $\mathcal{M}$ is very large, as in the coordinate-sparse setting, the sum $\sum_{m \in \mathcal{M}} e^{-\alpha d_m}$ cannot be bounded independently of p for a fixed $\alpha > 0$.

Below, we give examples of probabilities π that produce good results. They are built by slicing the collection of models $\{S_m,\ m \in \mathcal{M}\}$ according to some notion of complexity, most of the time connected to the variance $d_m\sigma^2$ of $\widehat{f}_m$.

Examples of probability π_m

The risk r_m equals $\|(I-\text{Proj}_{S_m})f^*\|^2 + d_m\sigma^2$, where $d_m = \dim(S_m)$. Since we want the bound $\min_{m\in\mathcal{M}}\left\{r_m + \sigma^2\log\left(1/\pi_m\right)\right\}$ to be as close as possible to r_{m_o}, we build below probabilities π, such that $\log\left(1/\pi_m\right)$ remains of a size comparable to d_m. We refer to Section 2.2.1 for the definition of the collections of models $\{S_m, m\in\mathcal{M}\}$ in the different sparsity settings.

We will write henceforth C_p^d for $p!/(d!(p-d)!)$. It fulfills the following upper bound.

Lemma 2.1

For $0\le d\le p$, we have the upper bound $\log\left(C_p^d\right)\le d(1+\log(p/d))$, with the convention $0\log 0 = 0$.

Proof. The result is obvious for $d=0$ and $d=1$. Assume that the bound holds true for C_p^{d-1}. Since $(1+1/k)^k\le e$ for any integer k, we have

$$C_p^d = C_p^{d-1}\frac{p-d}{d}\le\left(\frac{ep}{d-1}\right)^{d-1}\frac{p}{d}\le\left(\frac{ep}{d}\right)^{d-1}\left(1+\frac{1}{d-1}\right)^{d-1}\frac{p}{d}\le\left(\frac{ep}{d}\right)^{d}.$$

We conclude by induction on d. □

Coordinate sparsity

In this setting, we slice the collection of models according to the cardinality of m, and we give the same probability to the C_p^d models with m of cardinality d. We consider two choices of probability π on $\mathcal{M}=\mathcal{P}(\{1,\ldots,p\})$.

1. A simple choice is $\pi_m = (1+1/p)^{-p}p^{-|m|}$, for which we have the upper bound $\log(1/\pi_m)\le 1+|m|\log(p)$ since $(1+1/p)^p\le e$.
2. Another choice is $\pi_m = \left(C_p^{|m|}\right)^{-1}e^{-|m|}(e-1)/(e-e^{-p})$. According to Lemma 2.1, we have

$$\log\left(\frac{1}{\pi_m}\right)\le\log(e/(e-1)) + |m|(2+\log(p/|m|)).$$

We notice that for the two above choices, $\log(\pi_m^{-1})$ roughly behaves like $2|m|\log(p)$ for $|m|$ small compared to p. There is then an additional $\log(p)$ factor compared to the AIC penalty $\text{pen}_{\text{AIC}}(m) = 2|m|\sigma^2$. This $\log(p)$ factor is due to the fact that the number of models with dimension d roughly grows with d, like $d\log(p)$ in the logarithmic scale, so this term shall appear in the definition of π_m. Exercise 2.8.1 and Section 3.4.1 show that this $\log(p)$ factor is actually unavoidable. Since the upper Bound (2.10) involves $\log(\pi_m^{-1})$, we then lose a $\log(p)$ factor between the upper Bound (2.10) and the oracle risk r_{m_o}.

Group sparsity

Here, we slice again the collection of models according to the cardinality of m, which corresponds to the number of groups G_k for which $\beta_{G_k} \neq 0$. As before, we consider two choices.

1. A simple choice $\pi_m = (1+1/M)^{-M} M^{-|m|}$, for which we have the upper bound $\log(1/\pi_m) \leq 1 + |m| \log(M)$.
2. Another choice is $\pi_m = \left(C_M^{|m|}\right)^{-1} e^{-|m|}(e-1)/(e-e^{-M})$, for which we have the upper bound $\log(1/\pi_m) \leq \log(e/(e-1)) + |m|(2+\log(M/|m|))$.

Sparse-group sparsity

In this case, we slice $\mathcal{M}$ according to both the number of groups and the cardinality of the nonzero elements in each groups. More precisely, for $m = (g, \{J_k\}_{k\in g})$, we set

$$\pi_m = \frac{1}{Z} \times \frac{e^{-|g|}}{C_M^{|g|}} \prod_{k\in g} \frac{e^{-|J_k|}}{C_{|G_k|}^{|J_k|}} \tag{2.11}$$

with Z such that $\sum_m \pi_m = 1$. In the case where all the G_k have the same size p/M, we have

$$Z = \frac{\alpha - \alpha^{M+1}}{1-\alpha} \quad \text{where } \alpha = \frac{1-e^{-p/M}}{e(e-1)}.$$

We have the upper bound

$$\log(1/\pi_m) \leq |g|(2+\log(M/|g|) + \sum_{k\in g} |J_k|(2+\log(|G_k|/|J_k|)).$$

Variation sparsity

We can choose the same π_m as in the coordinate-sparse setting.

2.3 Risk Bound for Model Selection

2.3.1 Oracle Risk Bound

We consider a collection of models $\{S_m,\ m \in \mathcal{M}\}$, a probability distribution $\pi = \{\pi_m,\ m \in \mathcal{M}\}$ on $\mathcal{M}$, a constant $K > 1$, and the estimator $\widehat{f} = \widehat{f}_{\widehat{m}}$ given by (2.9). We have the following risk bound on $R(\widehat{f})$.

Theorem 2.2 Oracle risk bound for model selection

For the above choice of $\widehat{f}$, there exists a constant $C_K > 1$ depending only on $K > 1$, such that

$$\mathbb{E}\left[\|\widehat{f} - f^*\|^2\right] \leq C_K \min_{m\in\mathcal{M}} \left\{ \mathbb{E}\left[\|\widehat{f}_m - f^*\|^2\right] + \sigma^2 \log\left(\frac{1}{\pi_m}\right) + \sigma^2 \right\}. \tag{2.12}$$

Proof. The analysis of the risk of $\widehat{f}$ relies strongly on the Gaussian concentration Inequality (B.2) (see Appendix B, page 301). The strategy for deriving the risk Bound (2.12) is to start from Definition (2.9) of $\widehat{m}$, which ensures that

$$\|Y - \widehat{f}_{\widehat{m}}\|^2 + \sigma^2 \text{pen}(\widehat{m}) \leq \|Y - \widehat{f}_m\|^2 + \sigma^2 \text{pen}(m), \quad \text{for all } m \in \mathcal{M}, \tag{2.13}$$

and then to control the fluctuations of these quantities around their mean with the Gaussian concentration Inequality (B.2). The first step is to bound $\|\widehat{f} - f^*\|^2$ in terms of $\|\widehat{f}_m - f^*\|^2$ up to an additional random term. The second step is to upper bound the expectation of this additional random term.

Basic inequalities

Let us fix some $m \in \mathcal{M}$. Since $Y = f^* + \varepsilon$, expanding the square $\|\varepsilon + (f^* - \widehat{f})\|^2$ in the Inequality (2.13) gives

$$\|f^* - \widehat{f}\|^2 \leq \|f^* - \widehat{f}_m\|^2 + 2\langle \varepsilon, f^* - \widehat{f}_m \rangle + \text{pen}(m)\sigma^2 + 2\langle \varepsilon, \widehat{f} - f^* \rangle - \text{pen}(\widehat{m})\sigma^2. \tag{2.14}$$

From Lemma A.3 in Appendix A, page 294, the random variable $\|\text{Proj}_{S_m} \varepsilon\|^2$ follows a chi-square distribution of d_m degrees of freedom, so

$$\mathbb{E}\left[\langle \varepsilon, f^* - \widehat{f}_m \rangle\right] = \mathbb{E}\left[\langle \varepsilon, f^* - \text{Proj}_{S_m} f^* \rangle\right] - \mathbb{E}\left[\|\text{Proj}_{S_m} \varepsilon\|^2\right] = 0 - d_m \sigma^2, \tag{2.15}$$

which is non-positive. Since

$$\begin{aligned} \text{pen}(m)\sigma^2 = K\left(\sqrt{d_m} + \sqrt{2\log(1/\pi_m)}\right)^2 \sigma^2 &\leq 2K\left(d_m + 2\log(1/\pi_m)\right)\sigma^2 \\ &\leq 2K\mathbb{E}\left[\|\widehat{f}_m - f^*\|^2\right] + 4K\log(1/\pi_m)\sigma^2, \end{aligned} \tag{2.16}$$

in order to prove Bound (2.12), we only need to prove that for some constants $a > 1$ and $c \geq 0$ there exists a random variable Z, such that

$$2\langle \varepsilon, \widehat{f} - f^* \rangle - \text{pen}(\widehat{m}) \leq a^{-1}\|\widehat{f} - f^*\|^2 + Z \tag{2.17}$$

and $\mathbb{E}[Z] \leq c\sigma^2$. Actually, combining (2.14), (2.15), (2.16), and (2.17) we have

$$\frac{a-1}{a}\,\mathbb{E}\left[\|f^* - \widehat{f}\|^2\right] \leq (1+2K)\mathbb{E}\left[\|f^* - \widehat{f}_m\|^2\right] + 4K\log(1/\pi_m)\sigma^2 + c\sigma^2,$$

for any $m \in \mathcal{M}$, and Bound (2.12) follows.

Let us prove (2.17). We denote by $< f^* >$ the line spanned by f^*, by $\bar{S}_m$ the space $\bar{S}_m = S_m + < f^* >$, and by $\tilde{S}_m$ the orthogonal of $< f^* >$ in $\bar{S}_m$. In particular, $\bar{S}_m$ is the orthogonal sum of $< f^* >$ and $\tilde{S}_m$, which is denoted by $\bar{S}_m = < f^* > \bigoplus \tilde{S}_m$. Applying the inequality $2\langle x, y \rangle \leq a\|x\|^2 + \|y\|^2/a$ for $a > 1$, we obtain

$$\begin{aligned} 2\langle \varepsilon, \widehat{f} - f^* \rangle &= 2\langle \text{Proj}_{\bar{S}_{\widehat{m}}} \varepsilon, \widehat{f} - f^* \rangle \\ &\leq a\|\text{Proj}_{\bar{S}_{\widehat{m}}} \varepsilon\|^2 + a^{-1}\|\widehat{f} - f^*\|^2 \\ &\leq aN^2\sigma^2 + aU_{\widehat{m}}\sigma^2 + a^{-1}\|\widehat{f} - f^*\|^2, \end{aligned}$$

where $N^2 = \|\text{Proj}_{<f^*>}\varepsilon\|^2/\sigma^2$ and $U_m = \|\text{Proj}_{\bar{S}_m}\varepsilon\|^2/\sigma^2$. According to Lemma A.3, N^2 follows a χ^2 distribution of dimension 1, and U_m follows a χ^2 distribution of dimension d_m if $f^* \notin S_m$ and $d_m - 1$ if $f^* \in S_m$. We then have (2.17), with $Z = aN^2\sigma^2 + aU_{\widehat{m}}\sigma^2 - \text{pen}(\widehat{m})\sigma^2$. All we need is to prove that there exists a constant $c \geq 0$, such that

$$\mathbb{E}[Z] = a\sigma^2 + \sigma^2\mathbb{E}[aU_{\widehat{m}} - \text{pen}(\widehat{m})] \leq c\sigma^2. \tag{2.18}$$

We emphasize that the index $\widehat{m}$ depends on ε, so it is not independent of the sequence $\{U_m,\ m \in \mathcal{M}\}$. As a consequence, even if each U_m follows a χ^2 distribution, the variable $U_{\widehat{m}}$ does not follow a χ^2 distribution.

Stochastic control of $\mathbb{E}[aU_{\widehat{m}} - \text{pen}(\widehat{m})]$

We remind the reader that $\text{pen}(m) = K\left(\sqrt{d_m} + \sqrt{2\log(1/\pi_m)}\right)^2$. In the following, we choose $a = (K+1)/2 > 1$, and we start from the basic inequality

$$\begin{aligned}
&\mathbb{E}\left[\frac{K+1}{2}U_{\widehat{m}} - \text{pen}(\widehat{m})\right] \\
&\leq \frac{K+1}{2}\mathbb{E}\left[\max_{m\in\mathcal{M}}\left(U_m - \frac{2}{K+1}\text{pen}(m)\right)\right] \\
&\leq \frac{K+1}{2}\sum_{m\in\mathcal{M}}\mathbb{E}\left[\left(U_m - \frac{2K}{K+1}\left(\sqrt{d_m} + \sqrt{2\log(1/\pi_m)}\right)^2\right)_+\right].
\end{aligned} \tag{2.19}$$

The map $\varepsilon \to \|\text{Proj}_{\bar{S}_m}\varepsilon\|$ is 1-Lipschitz, so the Gaussian concentration Inequality (B.2), page 301, ensures that for each m there exists ξ_m with exponential distribution, such that

$$\|\text{Proj}_{\bar{S}_m}\varepsilon\| \leq \mathbb{E}\left[\|\text{Proj}_{\bar{S}_m}\varepsilon\|\right] + \sigma\sqrt{2\xi_m}.$$

Since $U_m = \|\text{Proj}_{\bar{S}_m}\varepsilon\|^2/\sigma^2$ follows a χ^2 distribution with d_m or $d_m - 1$ degrees of freedom, we have $\mathbb{E}\left[\|\text{Proj}_{\bar{S}_m}\varepsilon\|\right] \leq \mathbb{E}\left[\|\text{Proj}_{\bar{S}_m}\varepsilon\|^2\right]^{1/2} \leq \sigma\sqrt{d_m}$. As a consequence, we have the upper bound

$$\begin{aligned}
U_m &\leq \left(\sqrt{d_m} + \sqrt{2\xi_m}\right)^2 \\
&\leq \left(\sqrt{d_m} + \sqrt{2\log(1/\pi_m)} + \sqrt{2(\xi_m - \log(1/\pi_m))_+}\right)^2 \\
&\leq \frac{2K}{K+1}\left(\sqrt{d_m} + \sqrt{2\log(1/\pi_m)}\right)^2 + \frac{4K}{K-1}(\xi_m - \log(1/\pi_m))_+,
\end{aligned}$$

where we used for the last line the inequality $(x+y)^2 \leq (1+\alpha)x^2 + (1+\alpha^{-1})y^2$, with $\alpha = (K-1)/(K+1)$. Since $\mathbb{E}[(\xi_m - \log(1/\pi_m))_+] = \exp(-\log(1/\pi_m)) = \pi_m$, we have

$$\mathbb{E}\left[\left(U_m - \frac{2K}{K+1}\left(\sqrt{d_m} + \sqrt{2\log(1/\pi_m)}\right)^2\right)_+\right] \leq \frac{4K}{K-1}\,\pi_m.$$

Combining this bound with (2.19), we finally end with

$$\mathbb{E}\left[\frac{K+1}{2}U_{\widehat{m}} - \mathrm{pen}(\widehat{m})\right] \leq \frac{2K(K+1)}{K-1}\sigma^2 \sum_{m\in\mathcal{M}} \pi_m = \frac{2K(K+1)}{K-1}\sigma^2, \tag{2.20}$$

since $\sum_{m\in\mathcal{M}} \pi_m = 1$. This proves (2.18) and thus (2.17). The proof of (2.16) is complete. □

Remark 1. We observe from the last inequality in the above proof that Theorem 2.2 remains valid when $\{\pi_m : m \in \mathcal{M}\}$ do not sum to one, as long as the sum $\sum_{m\in\mathcal{M}} \pi_m$ remains small (since the constant C_K is proportional to $\sum_{m\in\mathcal{M}} \pi_m$). Asking for $\{\pi_m : m \in \mathcal{M}\}$ to sum to one is therefore not mandatory, it is simply a convenient convention.

Remark 2. Let us explain the shape of Penalty (2.9) in light of the above proof. From Inequality (2.17) we observe that we need a penalty, such that there exist some $a > 1$, $c \geq 0$, and some random variables Z_m fulfilling

$$2\langle \varepsilon, \widehat{f}_m - f^* \rangle - \mathrm{pen}(m) \leq a^{-1}\|\widehat{f}_m - f^*\|^2 + Z_m, \quad \text{for all } m \in \mathcal{M},$$

and $\mathbb{E}[\sup_{m\in\mathcal{M}} Z_m] \leq c\sigma^2$. Since

$$\begin{aligned}
2\langle \varepsilon, \widehat{f}_m - f^* \rangle &\leq a \left\langle \varepsilon, \frac{\widehat{f}_m - f^*}{\|\widehat{f}_m - f^*\|} \right\rangle^2 + a^{-1}\|\widehat{f}_m - f^*\|^2 \\
&\leq a \sup_{f\in\bar{S}_m} \left\langle \varepsilon, \frac{f}{\|f\|} \right\rangle^2 + a^{-1}\|\widehat{f}_m - f^*\|^2 \\
&\leq a\|\mathrm{Proj}_{\bar{S}_m}\varepsilon\|^2 + a^{-1}\|\widehat{f}_m - f^*\|^2,
\end{aligned}$$

we need a penalty fulfilling $\mathbb{E}\left[\sup_{m\in\mathcal{M}} \left(a\|\mathrm{Proj}_{\bar{S}_m}\varepsilon\|^2 - \mathrm{pen}(m)\sigma^2\right)\right] \leq c\sigma^2$. As explained in the second part of the proof of Theorem 2.2, Penalty (2.9) fulfills such a bound; see Bound (2.20).

Remark 3. We mention that there are some alternatives to Criterion (2.9) for selecting a model, for example the Goldenshluger–Lepski method. We refer to Exercise 2.8.5 for a simple version of this method and to Section 2.7 for references.

Discussion of Theorem 2.2

The risk Bound (2.12) shows that the estimator (2.9) almost performs the best trade-off between the risk r_m and the complexity term $\log(\pi_m^{-1})\sigma^2$. For collections of sufficiently small complexity, we can choose a probability π, such that $\log(\pi_m^{-1})$ is of the same order as d_m. In this case, r_m is larger (up to a constant) than the complexity term $\log(\pi_m^{-1})\sigma^2$, and the risk of the estimator $\widehat{f}$ is bounded by a constant times the risk of the oracle $\widehat{f}_{m_o}$; see Exercise 2.8.3 for an example. In particular, the AIC can produce some good results in this case. For some very large collections, such as the one for coordinate sparsity, we cannot choose $\log(\pi_m^{-1})$ of the same order as d_m, and Exercise 2.8.1 shows that the AIC fails. In this case, the Bound (2.12) cannot be directly compared to the oracle risk $r_{m_o} = R(\widehat{f}_{m_o})$. We refer to Exercise 2.8.2 for a discussion of Bound (2.12) in the coordinate-sparse, the group-sparse, and the sparse group-sparse settings. In these cases, where (2.12) cannot be directly compared to the oracle risk r_{m_o}, two natural questions arise:

1. Can an estimator perform significantly better than the estimator (2.9)?
2. Can we choose a smaller penalty in (2.9)?

Addressing these two questions in full generality is beyond the scope of this book. Rather, we will focus on the classical coordinate-sparse setting.

To answer the first question on the optimality of the estimator (2.9), we need to introduce a suitable notion of optimality and to develop a theory to assess the best performance that an estimator can achieve. This is the topic of the next chapter. We will see there, that the estimator (2.9) is optimal in some sense, made precise in Chapter 3.

Let us now consider the second question. Does it make sense to choose a penalty smaller than $\text{pen}(m) = K\left(\sqrt{d_m} + \sqrt{2\log(1/\pi_m)}\right)^2$, with $K > 1$ and π_m described in Section 2.2.2 ? In the coordinate-sparse and group-sparse settings the answer is no, in the sense that we can prove that when $f^* = 0$, the selected $\widehat{m}$ can have a very large size when $K < 1$; see Exercise 2.8.1, parts B and C. In particular, the AIC criterion can lead to a strong overfitting in these cases.

2.4 Computational Issues

We have seen in the previous section that the estimator (2.9) has some very nice statistical properties. Unfortunately, in many cases, it cannot be implemented in practice due to its prohibitive computational complexity.

Actually, computing (2.9) requires to screen the whole family $\mathcal{M}$ except in a few cases listed below. The size of the families $\mathcal{M}$ introduced in Section 2.2.1 becomes huge for moderate or large p. For example, the cardinality of the family $\mathcal{M}$ corresponding to the coordinate-sparse setting is 2^p, for the group-sparse setting it is 2^M, etc. As a consequence, the complexity of the selection procedure (2.9) becomes prohibitive for p or M larger than a few tens; see the table below.

number of variables (or groups)	p	10	20	40	80	270
cardinality of $\mathcal{M}$	2^p	1 024	$1.1\,10^6$	$1.1\,10^{12}$	$1.2\,10^{24}$	number of particles in the universe

Model selection is therefore a very powerful conceptual tool for understanding the nature of high-dimensional statistical problems, but it cannot be directly implemented except in the few cases listed below or when the signal is very sparse [92]. Yet, it is a useful baseline for deriving computationally feasible procedures, as we will see in Chapters 5, 6 and 7.

Cases where model selection can be implemented

There are two cases where we can avoid to screen the whole family $\mathcal{M}$ for minimizing (2.9).

1. The first case is when the columns of $\mathbf{X}$ are orthogonal. Minimizing (2.9) then essentially amounts to threshold the values of $\mathbf{X}_j^T Y$, for $j = 1, \dots, p$, which has a very low complexity; see Exercise 2.8.1, part A, for details.
2. The second case is for the variation sparse setting, where minimizing (2.9) can be performed via dynamic programming with an overall complexity of the order of n^3. We refer to Exercise 2.8.4 for the details.

Finally, we emphasize that for small families $\{S_m, \ m \in \mathcal{M}\}$ of models (for example, given by a PCA of $\mathbf{X}$), the algorithmic complexity remains small and the estimator (2.9) can be implemented; see Exercise 2.8.3 for examples.

Approximate computation

As mentioned above, some computationally efficient procedures can be derived from the model selection Criterion (2.9). We refer to Chapters 5, 6 and 7 for details. An alternative point of view is to perform an approximate minimization of Criterion (2.9). Many algorithms, of a deterministic or stochastic nature, have been proposed. We briefly described one of the most popular ones: the forward–backward minimization.

For simplicity, we restrict to the coordinate-sparse setting and only describe one of the multiple versions of the forward–backward algorithm. The principle of forward–backward minimization is to approximately minimize Criterion (2.9) by alternatively trying to add and remove variables from the model. More precisely, it starts from the null model and then builds a sequence of models $(m_t)_{t \in \mathbb{N}}$, where two consecutive models differ by only one variable. At each step t, the algorithm alternatively tries to add or remove a variable to m_t in order to decrease Criterion (2.9). When the criterion cannot be improved by the addition or deletion of one variable, the algorithm stops and returns the current value. Below $\text{crit}(m)$ refers to Criterion (2.9).

Forward–backward algorithm

Initialization: start from $m_0 = \emptyset$ and $t = 0$.

Iterate: until convergence

- **forward step:**
  - search $j_{t+1} \in \text{argmin}_{j \notin m_t} \text{crit}(m_t \cup \{j\})$
  - if $\text{crit}(m_t \cup \{j_{t+1}\}) \leq \text{crit}(m_t)$, then $m_{t+1} = m_t \cup \{j_{t+1}\}$ else $m_{t+1} = m_t$
  - increase t of one unit
- **backward step:**
  - search $j_{t+1} \in \text{argmin}_{j \in m_t} \text{crit}(m_t \setminus \{j\})$
  - if $\text{crit}(m_t \setminus \{j_{t+1}\}) < \text{crit}(m_t)$, then $m_{t+1} = m_t \setminus \{j_{t+1}\}$ else $m_{t+1} = m_t$
  - increase t of one unit

Output: $\widehat{f}_{m_t}$

In practice, the forward–backward algorithm usually converges quickly. So the final estimator $\widehat{f}_{m_t}$ can be computed efficiently. Yet, we emphasize that there is no guarantee at all, that the final estimator $\widehat{f}_{m_t}$ corresponds to the estimator $\widehat{f}_{\widehat{m}}$ minimizing the Criterion (2.9). Yet, some strong results are given by Zhang [169] for a variant of this algorithm.

2.5 Illustration

We briefly illustrate the theory of model selection on a simulated example. The graphics are reported in Figure 2.2. We have $n = 60$ noisy observation $Y_i = f(i/n) + \varepsilon_i$ (gray dots in Figure 2.2) of an unknown undulatory signal $f : [0,1] \to \mathbb{R}$ (dotted curve) at times $i/60$ for $i = 1, \ldots, 60$, with $\varepsilon_1, \ldots, \varepsilon_{60}$ i.i.d. with $\mathcal{N}(0,1)$ Gaussian distribution. We want to estimate this signal by expanding the observations on the Fourier basis. More precisely, we consider the setting of Example 2, page 28, with the φ_j given by

$$\begin{cases} \varphi_1(x) = 1 & \\ \varphi_{2k}(x) = \cos(2\pi k x) & \text{for } k = 1, \ldots, 25 \\ \varphi_{2k+1}(x) = \sin(2\pi k x) & \text{for } k = 1, \ldots, 25. \end{cases}$$

The top-right plot in Figure 2.2 represents the classical estimator $\widehat{f}^{\text{OLS}}$ (in gray) obtained by simple likelihood maximization

$$\widehat{f}^{\text{OLS}}(x) = \sum_{j=1}^{p} \widehat{\beta}_j^{\text{OLS}} \varphi_j(x) \quad \text{with} \quad \widehat{\beta}^{\text{OLS}} \in \underset{\beta \in \mathbb{R}^p}{\text{argmin}} \sum_{i=1}^{n} \Big(Y_i - \sum_{j=1}^{p} \beta_j \varphi_j(i/n) \Big)^2.$$

We observe that the estimator $\widehat{f}^{\text{OLS}}$ is overly wavy and provides a poor estimation of f. If we knew the oracle model m_o, then we could estimate f with $\widehat{f}_{m_o}$. This

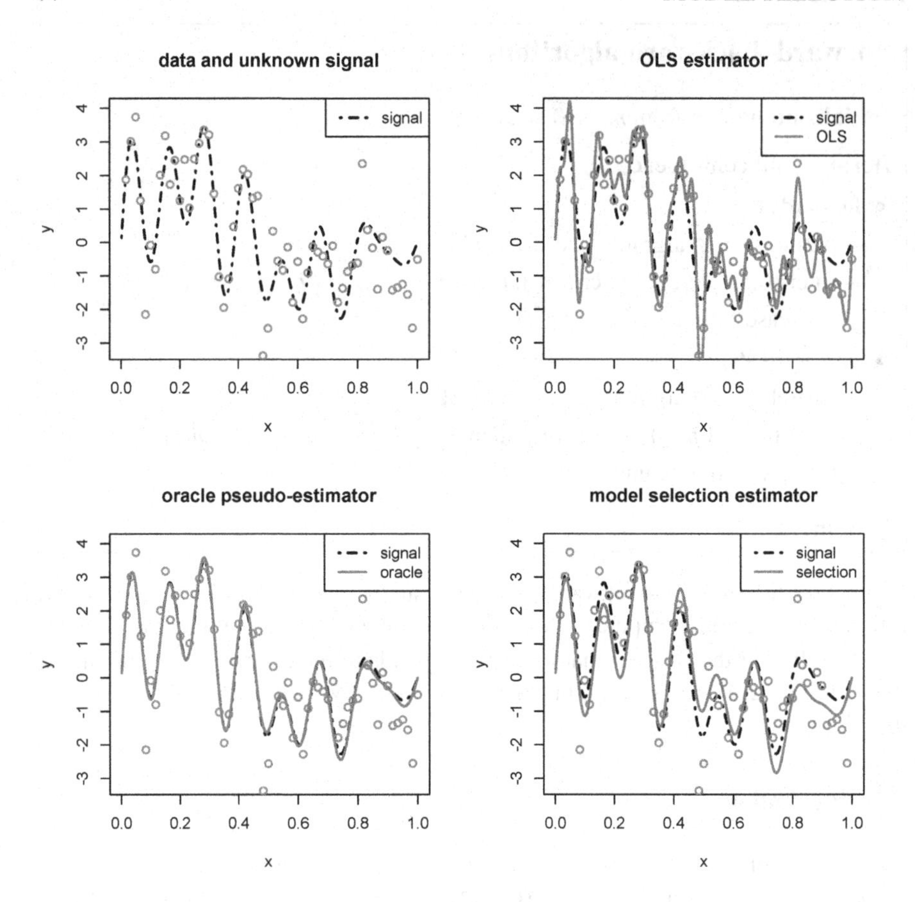

Figure 2.2 *Top-left: The observed data (gray points) and the unknown signal f (dotted line) that we want to estimate. Top-right: OLS estimator (in gray). Bottom-left: Oracle pseudo-estimator (gray). Bottom-right: The Estimator (2.9) in gray.*

pseudo-estimator is plotted in gray at the bottom-left of Figure 2.2. We observe that it is very close to f. In practice, we do not know the oracle model m_o, but we can select a model $\widehat{m}$ according to the model selection procedure (2.9), with a collection of models and a probability distribution π corresponding to the coordinate-sparse setting. The constant K in the penalty term is chosen to be equal to 1.1. The resulting estimator $\widehat{f}_{\widehat{m}}$ is plotted in gray at the bottom-right of Figure 2.2. We observe that this estimator is much better than $\widehat{f}^{\text{OLS}}$, and it is almost as good as $\widehat{f}_{m_o}$.

2.6 Alternative Point of View on Model Selection

Before concluding this chapter, we point out an alternative strategy for deriving a selection criterion based on the Bayesian paradigm. In this perspective, the mean f^*

of Y is assumed to be the outcome of the following sampling scheme: A model S_{m^*} is sampled according to a distribution $(\pi_m)_{m\in\mathcal{M}}$, and then f^* is sampled according to a distribution $d\Pi(f|m^*)$ on S_{m^*}. The mean f^* of Y is then sampled according to the distribution

$$d\Pi(f) = \sum_{m\in\mathcal{M}} \pi_m \, d\Pi(f|m),$$

with $(\pi_m)_{m\in\mathcal{M}}$ a distribution on $\mathcal{M}$ and $d\Pi(f|m)$ a distribution on S_m. In such a probabilistic setting, where m^*, f^*, and Y are random variables, it makes sense to consider the probability $\Pi(m|Y)$ of a model S_m given the observations Y. Computing this conditional probability, we obtain

$$\Pi(m|Y) = \frac{\pi_m d\pi(Y|m)}{d\pi(Y)} = \frac{\pi_m \int_{f\in S_m} e^{-\|Y-f\|^2/(2\sigma^2)}\, d\Pi(f|m)}{\sum_{m'} \pi_{m'} \int_{f\in S_{m'}} e^{-\|Y-f\|^2/(2\sigma^2)}\, d\Pi(f|m')}.$$

Under some technical conditions on the distribution $d\Pi(f|m)$, an asymptotic expansion when $n\to\infty$ gives for $m, m' \in \mathcal{M}$

$$\log\left(\frac{\Pi(m|Y)}{\Pi(m'|Y)}\right) \overset{n\to\infty}{\approx} \frac{\|Y-\widehat{f}_{m'}\|^2 - \|Y-\widehat{f}_m\|^2}{2\sigma^2} + \frac{d_{m'}-d_m}{2}\log(n) + \log\left(\frac{\pi_m}{\pi_{m'}}\right) + O(1); \quad (2.21)$$

see [152]. This asymptotic expansion suggests to choose m by minimizing the criterion

$$\operatorname{crit}(m) = \|Y-\widehat{f}_m\|^2 + \sigma^2 d_m \log(n) + 2\sigma^2 \log\left(\pi_m^{-1}\right). \quad (2.22)$$

Assuming a uniform distribution π_m on $\mathcal{M}$, we obtain the popular Bayesian Information Criterion (BIC)

$$\widehat{m}_{\mathrm{BIC}} \in \operatorname*{argmin}_{m\in\mathcal{M}} \left\{ \|Y-\widehat{f}_m\|^2 + \sigma^2 d_m \log(n) \right\}. \quad (2.23)$$

The term $\sigma^2 d_m \log(n)$ appearing in the BIC increases more rapidly than the term $2d_m\sigma^2$ appearing in the AIC. Yet, the model $\widehat{m}_{\mathrm{BIC}}$ can still be overly large when there is an exponential number of models per dimension; see Exercise 2.8.1, part B. This failure has two origins. First, the asymptotic expansion (2.21) fails to be valid in a high-dimensional setting. Second, the choice of a uniform distribution π on $\mathcal{M}$ does not enable the BIC estimator $\widehat{f}_{\widehat{m}_{\mathrm{BIC}}}$ to adapt to the complexity of $\mathcal{M}$ by taking into account the stochastic variability in (2.21). Actually, Criterion (2.22) is quite similar to Criterion (2.9), and choosing π_m as in Section 2.2.2 would avoid the overfitting described in Exercise 2.8.1.

2.7 Discussion and References

2.7.1 Take-Home Message

Model selection is a powerful theory for conceptualizing estimation in high-dimensional settings. Except in the orthogonal setting of Exercise 2.8.1, the variation sparse setting of Exercise 2.8.4 and the nested setting of Exercise 2.8.3, its computational complexity is prohibitive for implementation in high-dimensional settings. Yet, it remains a good baseline for designing computationally efficient procedures and for building criteria for selecting among different estimators. These two issues are developed in Chapters 5, 6 and 7.

2.7.2 References

Model selection ideas based on penalized criteria date back to Mallows [116] and Akaike [2] who have introduced the AIC. The BIC has been later derived by Schwarz [141]. The non-asymptotic analysis of model selection can be found in Barron, Birgé, and Massart [22] and Birgé and Massart [32]. In particular, Theorem 2.2 is adapted from Birgé and Massart [32].

The concept of model selection can be applied in many different settings, including density estimation, classification, non-parametric regression, etc. In particular, model selection can handle models more general than linear spans. We refer to Massart's Saint-Flour lecture notes [118] for a comprehensive overview on model selection based on penalized empirical risk criteria. An alternative to these penalized empirical risk criteria is the Goldenshluger–Lepski method; see Goldenschluger and Lepski [85], Chagny [53], and the references therein. A simplified version of this method is presented in Exercise 2.8.5.

The results presented above can be extended to the setting where the variance σ^2 of the noise is unknown; see Baraud *et al.* [19]. The rough idea is to plug for each m an estimator $\widehat{\sigma}_m^2$ of σ^2 in Criterion (2.9) and to adapt the shape of the penalty $\text{pen}(m)$ accordingly. Another strategy is the slope heuristic developed by Birgé and Massart [33] (see also Arlot and Massart [9] and Baudry *et al.* [23] for a review). The slope heuristic is based on the observation that the dimension $d_{\widehat{m}}$ is small for $K > 1$ and large for $K < 1$. The rough idea is then to replace $K\sigma^2$ by a constant $\kappa > 0$ in (2.9) and search for a constant $\widehat{\kappa}_c$, such that $d_{\widehat{m}}$ is large for $\kappa < \widehat{\kappa}_c$ and small for $\kappa > \widehat{\kappa}_c$. The Criterion (2.9) is finally applied with $K\sigma^2$ replaced by $2\widehat{\kappa}_c$.

2.8 Exercises

2.8.1 Orthogonal Design

We consider the linear model (2.2) in the coordinate-sparse setting. We assume in this exercise that the columns of $\mathbf{X}$ are orthogonal. We consider the family $\mathcal{M}$ and the models S_m of the coordinate-sparse setting and take in Criterion (2.9) the penalty $\text{pen}(m) = \lambda|m|$ for some $\lambda > 0$. Note that with $\lambda = K\big(1+\sqrt{2\log(p)}\big)^2$, this penalty is almost equal to the penalty $\text{pen}(m) = K\big(\sqrt{d_m}+\sqrt{2\log(1/\pi_m)}\big)^2$, with the probability $\pi_m = (1+1/p)^{-p}p^{-|m|}$.

For $\lambda > 0$, we define $\widehat{m}_\lambda$ as a minimizer of (2.9) with $\text{pen}(m) = \lambda|m|$.

A) Hard thresholding

1. Check that, in this setting, Criterion (2.9) is equal to

$$\|Y - \text{Proj}_{S_m} Y\|^2 + \text{pen}(m)\sigma^2 = \|Y\|^2 + \sum_{j \in m} \left(\lambda \sigma^2 - \left(\frac{\mathbf{X}_j^T Y}{\|\mathbf{X}_j\|} \right)^2 \right),$$

where $\mathbf{X}_j$ is the j-th column of $\mathbf{X}$.

2. Prove that a minimizer $\widehat{m}_\lambda$ of (2.9) is given by $\widehat{m}_\lambda = \left\{ j : \; (\mathbf{X}_j^T Y)^2 > \lambda \|\mathbf{X}_j\|^2 \sigma^2 \right\}$.
3. What is the consequence in terms of computational complexity?

B) Minimal penalties

We next check that the penalty $\text{pen}(m) = 2|m|\log(p)$ is minimal in some sense. We assume in this part that $f^* = 0$ so that the oracle model m_o is the void set. An accurate value of λ must then be such that $\mathbb{E}[|\widehat{m}_\lambda|]$ is small.

1. For $j = 1, \ldots, p$, we set $Z_j = \mathbf{X}_j^T \varepsilon / (\|\mathbf{X}_j\| \sigma)$. Prove that the Z_j are i.i.d. with $\mathcal{N}(0,1)$ distribution.
2. Prove that $|\widehat{m}_\lambda|$ follows a binomial distribution with parameters p and $\eta_\lambda = \mathbb{P}(Z^2 > \lambda)$, with Z a standard Gaussian random variable.
3. We recall that the AIC is defined by (2.8). Check that $\mathbb{E}[|\widehat{m}_{\text{AIC}}|] \approx 0.16p$. In particular, the AIC is not suited for this context.
4. We recall that for a standard Gaussian random variable Z, we have

$$\mathbb{P}(Z^2 > x) \overset{x \to \infty}{\sim} \sqrt{\frac{2}{\pi x}} \, e^{-x/2}.$$

Check that for $K > 0$ we have

$$\mathbb{E}\left[|\widehat{m}_{2K\log(p)}|\right] \overset{p \to \infty}{\sim} \frac{p^{1-K}}{\sqrt{\pi K \log(p)}}.$$

Conclude that for $K < 1$ the mean size of the selected model $\widehat{m}_{2K\log(p)}$ grows like a fractional power of p. Choosing $\text{pen}(m) = 2K|m|\log(p)$ with $K < 1$ can then produce poor results in the coordinate-sparse setting. In this sense the penalty $\text{pen}(m) = 2|m|\log(p)$ is minimal in this setting.

5. We recall that the BIC is defined by (2.23). Check that for $p \sim n$ we have

$$\mathbb{E}[|\widehat{m}_{\text{BIC}}|] \overset{p \to \infty}{\sim} \sqrt{\frac{2p}{\pi \log(p)}}.$$

In particular, the BIC is not suited for this context.

C) Overfitting with $K < 1$

We do not assume anymore that $f^* = 0$. We consider again the penalty $\text{pen}(m) = \lambda|m|$. We prove here that the choice $\lambda = 2K\log(p)$ with $K < 1$ produces an estimator with a risk much larger than the minimax risk.

1. As before, we set $Z_j = \mathbf{X}_j^T \varepsilon/(\|\mathbf{X}_j\|\sigma)$. We assume henceforth that $f^* \in S_{m^*}$ with $|m^*| = D^*$. Writing P_j for the projection on the line spanned by $\mathbf{X}_j$, prove that

$$\|\widehat{f}_{\widehat{m}_\lambda} - f^*\|^2 = \|f^* - \sum_{j\in\widehat{m}_\lambda} P_j f^*\|^2 + \sum_{j\in\widehat{m}_\lambda} Z_j^2\sigma^2$$
$$\geq \sum_{j\in\widehat{m}_\lambda\setminus m^*} Z_j^2\sigma^2 \geq (|\widehat{m}_\lambda| - D^*)\lambda\sigma^2.$$

2. Prove that for $a, x \in \mathbb{R}^+$, we have

$$\int_{x-a}^{x} e^{-z^2/2}\,dz \geq \int_{x}^{x+a} e^{-z^2/2}\,dz.$$

   As a consequence, for a standard Gaussian random variable Z and for any $a, x \in \mathbb{R}$, prove that $\mathbb{P}\left((Z+a)^2 > x^2\right) \geq \mathbb{P}\left(Z^2 > x^2\right)$.
3. Check that $\mathbb{E}\left[|\widehat{m}_\lambda|\right] \geq p\,\mathbb{P}\left(Z^2 > \lambda\right)$, with Z a standard Gaussian random variable.
4. For $K < 1$ and $D^* \ll p^{1-K}(\log(p))^{-1/2}$, prove that

$$\mathbb{E}\left[\|\widehat{f}_{\widehat{m}_{2K\log(p)}} - f^*\|^2\right] \geq \left(\mathbb{E}\left[|\widehat{m}_{2K\log(p)}|\right] - D^*\right) 2K\log(p)\sigma^2$$
$$\geq \left(p\,\mathbb{P}\left(Z^2 > 2K\log(p)\right) - D^*\right) 2K\log(p)\sigma^2 \overset{p\to\infty}{\sim} p^{1-K}\sigma^2\sqrt{\frac{4K\log(p)}{\pi}}.$$

5. We use the notations $V_D(\mathbf{X})$ and $R[X, D]$ defined in Section 3.4.1. Conclude that for $K < 1$ and $D^* \ll p^{1-K}(\log(p))^{-1/2}$, we have for any $f^* \in V_{D^*}(\mathbf{X})$

$$\mathbb{E}\left[\|\widehat{f}_{\widehat{m}_{2K\log(p)}} - f^*\|^2\right] \gg R[X, D^*], \quad \text{as } p \to +\infty.$$

   In particular, the choice $\lambda = 2K\log(p)$ with $K < 1$ produces an estimator with a risk overly large on $V_{D^*}(\mathbf{X})$ for $D^* \ll p^{1-K}(\log(p))^{-1/2}$.

This last part emphasizes that the AIC estimator overfits in this setting, as well as the BIC estimator when $p \sim n^\alpha$ with $\alpha > 1/2$. They should then be avoided.

2.8.2 Risk Bounds for the Different Sparsity Settings

We consider the linear regression model (2.2).

A) Coordinate-sparse setting

We focus here on the coordinate-sparse setting. We use the models defined in Section 2.2.1 for this setting and the weights $\pi_m = \left(C_p^{|m|}\right)^{-1} e^{-|m|}(e-1)/(e-e^{-p})$.

1. We define $\text{supp}(\beta) = \{j : \beta_j \neq 0\}$ and $|\beta|_0 = \text{card}(\text{supp}(\beta))$. Prove that for any $m \in \mathcal{M}$ we have

$$\mathbb{E}\left[\|\widehat{f}_m - f^*\|^2\right] = \inf_{\beta:\text{supp}(\beta)=m} \left\{\|\mathbf{X}\beta - \mathbf{X}\beta^*\|^2 + |\beta|_0\sigma^2\right\}.$$

2. With Theorem 2.2, prove that there exists a constant $C_K > 1$ depending only on $K > 1$, such that the estimator $\widehat{f} = \mathbf{X}\widehat{\beta}$ defined by (2.9) fulfills

$$\begin{aligned}
&\mathbb{E}\left[\|\mathbf{X}\widehat{\beta} - \mathbf{X}\beta^*\|^2\right] \\
&\leq C_K \min_{m\in\mathcal{M}\setminus\emptyset} \inf_{\beta:\text{supp}(\beta)=m} \left\{\|\mathbf{X}\beta - \mathbf{X}\beta^*\|^2 + |\beta|_0\left[1 + \log\left(\frac{p}{|\beta|_0}\right)\right]\sigma^2\right\} \\
&\leq C_K \inf_{\beta\neq 0} \left\{\|\mathbf{X}\beta - \mathbf{X}\beta^*\|^2 + |\beta|_0\left[1 + \log\left(\frac{p}{|\beta|_0}\right)\right]\sigma^2\right\}. \qquad (2.24)
\end{aligned}$$

B) Group-sparse setting

We focus here on the group-sparse setting, and we assume that the M groups have the same cardinality p/M. We use the models defined in Section 2.2.1 for this setting and the weights $\pi_m = \left(C_M^{|m|}\right)^{-1} e^{-|m|}(e-1)/(e-e^{-M})$.

1. We write $\mathcal{K}(\beta) = \{k : \beta_{G_k} \neq 0\}$. Prove that

$$\mathbb{E}\left[\|\widehat{f}_m - f^*\|^2\right] = \inf_{\beta:\mathcal{K}(\beta)=m} \left\{\|\mathbf{X}\beta - \mathbf{X}\beta^*\|^2 + \frac{|\mathcal{K}(\beta)|p}{M}\sigma^2\right\}.$$

2. With Theorem 2.2, prove that there exists a constant $C_K > 1$ depending only on $K > 1$, such that the estimator $\widehat{f} = \mathbf{X}\widehat{\beta}$ defined by (2.9) fulfills

$$\begin{aligned}
&\mathbb{E}\left[\|\mathbf{X}\widehat{\beta} - \mathbf{X}\beta^*\|^2\right] \\
&\leq C_K \min_{m\in\mathcal{M}\setminus\emptyset} \left\{\mathbb{E}\left[\|\widehat{f}_m - f^*\|^2\right] + |m|\left(1 + \log(M/|m|)\right)\sigma^2\right\} \\
&\leq C_K \inf_{\beta\neq 0} \left\{\|\mathbf{X}\beta - \mathbf{X}\beta^*\|^2 + |\mathcal{K}(\beta)|\left[1 + \frac{p}{M} + \log\left(\frac{M}{|\mathcal{K}(\beta)|}\right)\right]\sigma^2\right\}.
\end{aligned}$$

3. Observe that for some β we have $|\mathcal{K}(\beta)| = |\beta|_0 M/p$ and

$$|\mathcal{K}(\beta)|\left[1 + \frac{p}{M} + \log\left(\frac{M}{|\mathcal{K}(\beta)|}\right)\right] = |\beta|_0 + \frac{M}{p}|\beta|_0\left[1 + \log\left(\frac{p}{|\beta|_0}\right)\right].$$

   This inequality enlightens the gain of using the group-sparse models instead of the coordinate-sparse model when β^* is group-sparse and $M \ll p$.

C) Sparse group-sparse setting

We focus here on the sparse group-sparse setting, and we assume again that the M groups have the same cardinality p/M. We use the models and the probability π defined in Sections 2.2.1 and 2.2.2 for this setting.

1. We write $\mathscr{J}_k(\beta) = \{j \in G_k : \beta_j \neq 0\}$. Prove that

$$\mathbb{E}\left[\|\widehat{f}_{(g,\{J_k\}_{k\in g})} - f^*\|^2\right] = \inf_{\beta:\mathscr{K}(\beta)=g,\, \mathscr{J}_k(\beta)=J_k} \left\{\|\mathbf{X}\beta - \mathbf{X}\beta^*\|^2 + \sum_{k\in g} |\mathscr{J}_k(\beta)|\, \sigma^2\right\}.$$

2. With Theorem 2.2, prove that there exists a constant $C_K > 1$ depending only on $K > 1$, such that the estimator $\widehat{f} = \mathbf{X}\widehat{\beta}$ defined by (2.9) fulfills

$$\mathbb{E}\left[\|\mathbf{X}\widehat{\beta} - \mathbf{X}\beta^*\|^2\right] \leq C_K \inf_{\beta\neq 0} \left\{\|\mathbf{X}\beta - \mathbf{X}\beta^*\|^2 + |\mathscr{K}(\beta)| \left[1 + \log\left(\frac{M}{|\mathscr{K}(\beta)|}\right)\right]\sigma^2 \right.$$
$$\left. + \sum_{k\in g} |\mathscr{J}_k(\beta)| \left[1 + \log\left(\frac{p}{M|\mathscr{J}_k(\beta)|}\right)\right]\sigma^2\right\}.$$

Compared with the group-sparse setting, we observe that the term $|\mathscr{K}(\beta)|p/M$ is replaced by

$$\sum_{k\in g} |\mathscr{J}_k(\beta)| \left[1 + \log\left(\frac{p}{M|\mathscr{J}_k(\beta)|}\right)\right],$$

which can be much smaller when $|\beta|_0 \ll |\mathscr{K}(\beta)|p/M$.

2.8.3 Collections of Nested Models

We consider a collection of models $\{S_m,\ m \in \mathscr{M}\}$ indexed by $\mathscr{M} = \{1,\ldots,M\}$ and fulfilling $\dim(S_m) = m$. Such a collection of models arises naturally, when expanding a signal (as in Example 2) on the m first coefficients of the Fourier basis $\{\varphi_j : j \in \mathbb{N}\}$. In this case we have $S_m = \operatorname{span}\{\varphi_j : j = 1,\ldots,m\}$. Another example is when S_m is spanned by the m first principal axes given by a Principal Component Analysis of $\mathbf{X}$.

For any $\alpha > 0$, we choose the probability distribution on $\mathscr{M}$ defined by $\pi_m = e^{-\alpha m}(e^{\alpha} - 1)/(1 - e^{-\alpha M})$ for $m \in \mathscr{M}$. In this case, we can compare directly the risk of the estimator (2.9) to the risk of the oracle $\widehat{f}_{m_o}$.

1. Check that the estimator defined by (2.9) fulfills the oracle inequality

$$\mathbb{E}\left[\|\widehat{f}_{\widehat{m}} - f^*\|^2\right] \leq C_{K,\alpha} \inf_{m=1,\ldots,M} \mathbb{E}\left[\|\widehat{f}_m - f^*\|^2\right],$$

   for some constant $C_{K,\alpha} > 1$ depending only on K and α. In particular, the performance of the estimator (2.9) is almost as good as the performance of the oracle $\widehat{f}_{m_o}$ up to a universal (multiplicative) constant $C_{K,\alpha}$.

2. For $K \in]1,2[$, we set $\alpha = (\sqrt{2/K} - 1)^2$ and $\pi_m = \exp(-\alpha m)$.
   (a) Compute the penalty $\operatorname{pen}(m) = K\left(\sqrt{m} + \sqrt{\log(1/\pi_m)}\right)^2$ for this choice of π_m. What do you recognize?
   (b) Compute the sum $\sum_{m\in\mathscr{M}} \pi_m$. This sum is not equal to one, but we know from the remark below the proof of Theorem 2.2 that the risk Bound (2.12) remains valid when the sum $\sum_{m\in\mathscr{M}} \pi_m$ is not equal to one, as long as the sum remains of a moderate size. What it the size of this sum for $K = 1.1$?

The conclusion is that the AIC fulfills an oracle inequality like (2.12) when the collection of models is nested. Using the AIC in this setting thus perfectly makes sense.

2.8.4 Segmentation with Dynamic Programming

We consider here the setting where the coordinates f_i^* are piecewise constant as described in Example 1, page 27. It corresponds to the variation sparse setting with $p = n$ and with $\mathbf{X}_1, \ldots, \mathbf{X}_n$ the canonical basis of $\mathbb{R}^n$.

We want to estimate f^* by model selection. We then consider the family of models introduced in Section 2.2.2 for the variation sparse setting. This family is indexed by the 2^n subsets of $\{1, \ldots, n\}$. A naive minimization of Criterion (2.9) would then require 2^n evaluations of the criterion, which is prohibitive in practice. Yet, in this setting, Criterion (2.9) can be minimized much more efficiently by dynamic programing. This exercise is adapted from Lebarbier [102].

We write $N^2(k) = Y_1^2 + \ldots + Y_{k-1}^2$ for $2 \leq k \leq n$ and $N^2(1) = 0$. For $1 \leq k < j \leq n+1$, we set

$$\bar{y}_{(k,j)} = \frac{1}{j-k} \sum_{i=k}^{j-1} Y_i \quad \text{and} \quad R^2(k,j) = \sum_{i=k}^{j-1} \left(Y_i - \bar{y}_{(k,j)}\right)^2.$$

1. For $m = \{i_1, \ldots, i_D\} \subset \{1, \ldots, n\}$, check that $\widehat{f}_m$ is given by

$$\widehat{f}_m = \sum_{q=1}^{D} \bar{y}_{i_q, i_{q+1}} \mathbf{1}_{i_q}^{i_{q+1}},$$

   where $i_{D+1} = n+1$ and where $\mathbf{1}_k^j = \mathbf{X}_k + \ldots + \mathbf{X}_{j-1}$ is the vector with i^{th} coordinate equal to 1 if $k \leq i < j$ and equal to 0 else.

2. We assume in the following that the probability π_m depends on m only through its cardinality $|m|$, as in the examples given in Section 2.2.2 for the variation sparse setting. We will write $\text{pen}(|m|)$ instead of $\text{pen}(m)$ in order to emphasize this dependence. We also define for $0 \leq D \leq n$

$$\widehat{m}_D = \underset{m : |m| = D}{\text{argmin}} \|Y - \widehat{f}_m\|^2 \quad \text{and} \quad C_n(D) = \|Y - \widehat{f}_{\widehat{m}_D}\|^2.$$

   Prove that the minimizer $\widehat{m}$ of (2.9) is given by $\widehat{m} = \widehat{m}_{\widehat{D}}$, where

$$\widehat{D} = \underset{D=0,\ldots,n}{\text{argmin}} \left\{C_n(D) + \sigma^2 \text{pen}(D)\right\}.$$

3. For $1 \leq D \leq n$, prove that $C_n(D)$ and $\widehat{m}_D$ are solutions of

$$C_n(D) = \min_{1 \leq i_1 < \ldots < i_D \leq n} \left\{N^2(i_1) + \sum_{q=1}^{D} R^2(i_q, i_{q+1})\right\}$$

$$\text{and} \quad \widehat{m}_D = \underset{1 \leq i_1 < \ldots < i_D \leq n}{\text{argmin}} \left\{N^2(i_1) + \sum_{q=1}^{D} R^2(i_q, i_{q+1})\right\},$$

   with the convention $i_{D+1} = n+1$.

4. Check that for $2 \leq D \leq n$, we have the recursive formula

$$C_n(D) = \min_{i=D,\ldots,n} \left\{C_{i-1}(D-1) + R^2(i,n+1)\right\}. \tag{2.25}$$

5. Check that computing all the $N^2(k)$ and $R^2(k,j)$ for $1 \leq k < j \leq n+1$ requires $O(n^3)$ operations. Building on the recursive Formula (2.25), propose an algorithm that computes $C_n(D)$ and $\widehat{m}_D$ for $D = 1,\ldots,n$ with $O(n^3)$ operations. Conclude that $\widehat{m}$ can be computed with only $O(n^3)$ operations.

2.8.5 Goldenshluger–Lepski Method

The Goldenshluger–Lepski method [107, 85] is an alternative to the penalized empirical risk Criterion (2.9). It is initially designed for selecting the bandwidth of kernel estimators, but the method has been recently extended to model selection by Chagny [53]. The following exercise is an adaptation of this work.

We consider a set $\mathcal{M} = \{1,\ldots,M\}$ and a collection of models $\{S_m,\ m \in \mathcal{M}\}$, such that $S_m \subset S_{m'}$ for $m \leq m'$. The main idea underlying Goldenshluger–Lepski method is to estimate directly the bias term $\|f^* - \text{Proj}_{S_m} f^*\|^2$ by a term $B(m)$ involving $\|\widehat{f}_{m'} - \widehat{f}_{m\wedge m'}\|^2$ with $m' \neq m$. More precisely, let us consider a probability distribution $\pi = \{\pi_m,\ m \in \mathcal{M}\}$ on $\mathcal{M}$ and define $\text{pen}(m)$ as in (2.9). In the following, we consider the model selection procedure

$$\widehat{m} \in \operatorname*{argmin}_{m\in\mathcal{M}} \left\{B(m) + \text{pen}(m)\sigma^2\right\},$$
$$\text{where} \quad B(m) = \max_{m'\in\mathcal{M}} \left[\|\widehat{f}_{m'} - \widehat{f}_{m'\wedge m}\|^2 - \text{pen}(m')\sigma^2\right]_+. \tag{2.26}$$

1. Considering apart the cases $m \leq \widehat{m}$ and $m > \widehat{m}$, prove that

$$\begin{aligned}
\|\widehat{f}_{\widehat{m}} - f^*\|^2 &\leq 2\max\left(\|\widehat{f}_{\widehat{m}} - \widehat{f}_{\widehat{m}\wedge m}\|^2, \|\widehat{f}_m - \widehat{f}_{\widehat{m}\wedge m}\|^2\right) + 2\|\widehat{f}_m - f^*\|^2 \\
&\leq 2(B(\widehat{m}) + \text{pen}(m)\sigma^2 + B(m) + \text{pen}(\widehat{m})\sigma^2) + 2\|\widehat{f}_m - f^*\|^2 \\
&\leq 4(B(m) + \text{pen}(m)\sigma^2) + 2\|\widehat{f}_m - f^*\|^2.
\end{aligned}$$

2. Prove that for any $m \in \mathcal{M}$ and $\eta > 0$ we have

$$\begin{aligned}
B(m) \leq & \frac{\eta+1}{\eta} \max_{m'\geq m} \|\text{Proj}_{S_{m'}} f^* - \text{Proj}_{S_m} f^*\|^2 \\
& + \sum_{m'\geq m} \left[(1+\eta)\|\text{Proj}_{S_{m'}} \varepsilon - \text{Proj}_{S_m} \varepsilon\|^2 - \text{pen}(m')\sigma^2\right]_+ \\
\leq & \frac{\eta+1}{\eta} \|f^* - \text{Proj}_{S_m} f^*\|^2 + \sum_{m'\in\mathcal{M}} \left[(1+\eta)\|\text{Proj}_{S_{m'}} \varepsilon\|^2 - \text{pen}(m')\sigma^2\right]_+.
\end{aligned}$$

We set in the following $\eta = (K-1)/2$, where $K > 1$ is the constant involved in the definition of $\text{pen}(m)$. We extract from the proof of Theorem 2.2 that the expectation of the above sum is upper-bounded by $2K(K+1)\sigma^2/(K-1)$.

3. Prove the following risk bound for the estimator $\widehat{f}_{\widehat{m}}$ with $\widehat{m}$ selected according to (2.26)

$$\mathbb{E}\left[\|\widehat{f}_{\widehat{m}} - f^*\|^2\right] \leq \inf_{m\in\mathcal{M}} \left\{\frac{6K+2}{K-1}\mathbb{E}\left[\|\widehat{f}_m - f^*\|^2\right] + 4\text{pen}(m)\sigma^2\right\} + \frac{8K(K+1)}{K-1}\sigma^2.$$

Compare this bound with the bound in Theorem 2.2.

4. Let us assume that $d_m = m$, for $m = 1,\ldots,M$. Check that with the choice $\pi_m = e^{-m}(e-1)/(1-e^{-M})$, we have

$$\mathbb{E}\left[\|\widehat{f}_{\widehat{m}} - f^*\|^2\right] \leq C_K \inf_{m=1,\ldots,M} \mathbb{E}\left[\|\widehat{f}_m - f^*\|^2\right]$$

for some constant C_K depending only on K.

Remark: Note that the positive part in the definition of $B(m)$ did not play a role in the above analysis. Yet, this positive part is natural, since $B(m)$ is an estimator of the bias term $\|f^* - \text{Proj}_{S_m} f^*\|^2$, which is always non-negative. Similarly, we can consider maximizing only over the $m' \geq m$ in the definition of $B(m)$. As an exercise, you can check that if we remove the positive part in the definition of $B(m)$ and maximize only on $m' \geq m$, then the selection criteria (2.9) and (2.26) are completely equivalent.

2.8.6 Estimation under Convex Constraints

Let us consider the estimation problem $Y = f + \sigma\varepsilon \in \mathbb{R}^n$ with $\sigma > 0$ and $\varepsilon \sim \mathcal{N}(0, I_n)$. Throughout this chapter, for simplicity, we have only considered linear models S_m for estimating f. In the analysis, we have heavily used that the orthogonal projection onto a linear space is a linear operator. When moving to more general models, this feature is lost and the analysis is more involved. In this exercise, we investigate the theoretical properties of the estimators constrained to be in a closed convex model. We focus in particular on their deviations, which are the key to design sensible model selection criterions. The exercise is mainly adapted from [55].

Let $\mathscr{C}$ be a closed convex set of $\mathbb{R}^n$ and let us define the projection

$$\pi_{\mathscr{C}} y = \underset{u\in\mathscr{C}}{\text{argmin}} \|y - u\|^2$$

where $\|.\|$ represents the usual Euclidean norm in $\mathbb{R}^n$. Our goal is to investigate the properties of the estimator $\widehat{f}_{\mathscr{C}} = \pi_{\mathscr{C}} y$.

A) Basic facts

1. Let us fix $u \in \mathscr{C}$ and $0 < t < 1$. Why do we have $\|y - (tu + (1-t)\pi_{\mathscr{C}} y)\|^2 \geq \|y - \pi_{\mathscr{C}} y\|^2$?
2. Investigating this inequality for t small, prove that

$$\langle u - \pi_{\mathscr{C}} y, y - \pi_{\mathscr{C}} y\rangle \leq 0 \quad \text{and} \quad \|\pi_{\mathscr{C}} y - y\|^2 \leq \|u - y\|^2 - \|u - \pi_{\mathscr{C}} y\|^2.$$

3. Let K be a closed convex set (if $u \in K$, then $tu \in K$ for all $t \geq 0$). Prove that

$$\langle \Pi_K y, y \rangle = \|\Pi_K y\|^2 = \left(\sup_{u \in K,\ \|u\|=1} \langle u, y \rangle \right)^2 .$$

B) Global width

Let $u \in \mathbb{R}^n$. We define $K_{u,\mathscr{C}}$ as the closure of the cone $\{t(c-u) : t \geq 0,\ c \in \mathscr{C}\}$ and we set

$$\delta(\mathscr{C}, u) = \mathbb{E}\left[\|\Pi_{K_{u,\mathscr{C}}} \varepsilon\|^2 \right].$$

1. Compute $\delta(\mathscr{C}, u)$ when $\mathscr{C}$ is a linear span of dimension d and $u \in \mathscr{C}$.
2. Prove that for all $u \in \mathscr{C}$ we have

$$\begin{aligned} \|\widehat{f}_{\mathscr{C}} - f\|^2 &\leq \|u - f\|^2 + 2\sigma \langle \varepsilon, \widehat{f}_{\mathscr{C}} - u \rangle - \|u - \widehat{f}_{\mathscr{C}}\|^2 \\ &\leq \|u - f\|^2 + \sigma^2 \|\Pi_{K_{u,\mathscr{C}}} \varepsilon\|^2. \end{aligned}$$

3. Prove that with probability at least $1 - e^{-L}$, we have

$$\|\widehat{f}_{\mathscr{C}} - f\|^2 \leq \inf_{u \in \mathscr{C}} \left\{ \|u - f\|^2 + \sigma^2 \left(\sqrt{\delta(\mathscr{C}, u)} + \sqrt{2L} \right)^2 \right\}. \tag{2.27}$$

C) Local width

Let $t > 0$ and $u \in \mathbb{R}^n$. We define

$$F_{u,\mathscr{C}}(t) = \mathbb{E}\left[\sup_{c \in \mathscr{C},\ \|u-c\| \leq t} \langle \varepsilon, c - u \rangle \right].$$

1. Prove that when $\|\widehat{f}_{\mathscr{C}} - u\|^2 \leq t$, then

$$\|\widehat{f}_{\mathscr{C}} - f\|^2 \leq \|u - f\|^2 + 2\sigma Z_t, \quad \text{where} \quad Z_t = \sup_{c \in \mathscr{C},\ \|u-c\| \leq t} \langle \varepsilon, c - u \rangle.$$

2. Prove that when $\|\widehat{f}_{\mathscr{C}} - u\| > t$, then

$$\|\widehat{f}_{\mathscr{C}} - f\|^2 \leq \|u - f\|^2 + \left(\frac{\sigma Z_t}{t} \right)^2 .$$

3. Let $t(u) > 0$ be such that $F_{u,\mathscr{C}}(t(u)) \leq (t(u))^2/(2\sigma)$. Prove that with probability at least $1 - e^{-L}$ we have

$$\|\widehat{f}_{\mathscr{C}} - f\|^2 \leq \inf_{u \in \mathscr{C}} \left\{ \|u - f\|^2 + \left(t(u) + \sigma\sqrt{2L} \right)^2 \right\}. \tag{2.28}$$

Comparing (2.28) with (2.27), we observe that $\sigma\sqrt{\delta(\mathscr{C}, u)}$ has been replaced by $t(u)$. The main difference between these two measures of complexity is that $\delta(\mathscr{C}, u)$ measures the complexity of the whole set $\mathscr{C}$, while $t(u)$ measures the local complexity of $\mathscr{C}$ around u, which can be much smaller at some points $u \in \mathscr{C}$.

Chapter 3

Minimax Lower Bounds

The goal of the statistician is to infer information as accurately as possible from data. From a theoretical perspective, when investigating a given statistical model, her goal is to propose an estimator with the smallest possible risk, ideally with a low computational complexity. In particular, when analyzing a given estimator, not only we must derive an upper bound on the risk as in the previous chapter, but also, we must derive a lower bound on the risk achievable by the best possible estimator. Then, we can compare if the upper and lower bounds match. If so, we have the guarantee that the proposed estimator is optimal (in terms of the chosen risk).

Deriving lower bounds is then a common task in mathematical statistics. We present in this chapter the most common technique for deriving lower bounds, based on some inequality taken from information theory. This technique is then applied to show some optimality results on the model selection estimator (2.9) of Chapter 2, in the coordinate-sparse setting.

3.1 Minimax Risk

Let us consider a set $(\mathbb{P}_f)_{f \in \mathscr{F}}$ of distributions on a measurable space $(\mathscr{Y}, \mathscr{A})$. Let d be a distance on $\mathscr{F}$. We assume that we only have access to an observation $Y \in \mathscr{Y}$ distributed as $\mathbb{P}_f$ and our goal is to recover f from Y. Hence, we want to design an estimator $\hat{f} : \mathscr{Y} \to \mathscr{F}$ such that $d(\hat{f}(Y), f)$ is as small as possible. For example, we seek for $\hat{f}$ such that, for some $q > 0$ the expected error[1] $\mathbb{E}_f\left[d(\hat{f}(Y), f)^q\right]$ is as small as possible.

It turns out that seeking for $\hat{f}$ such that $\mathbb{E}_f\left[d(\hat{f}(Y), f)^q\right]$ is as small as possible is a degenerate problem. Indeed, we have for $f \in \mathscr{F}$

$$\min_{\hat{f}:\mathscr{Y} \xrightarrow{\text{measurable}} \mathscr{F}} \mathbb{E}_f\left[d(\hat{f}(Y), f)^q\right] = 0,$$

(the minimum is taken over all the measurable applications $\hat{f} : \mathscr{Y} \to \mathscr{F}$), with the minimum reached for the constant application $\hat{f}(y) = f$. Hence, we will not consider pointwise optimality (i.e., for a single f) but optimality on the class $\mathscr{F}$. A popular

[1] Throughout this chapter, we use the notation $\mathbb{E}_f[\phi(Y)] = \int_{\mathscr{Y}} \phi(y)\, d\mathbb{P}_f(y)$ and $\mathbb{E}_{\mathbb{Q}}[\phi(Y)] = \int_{\mathscr{Y}} \phi(y)\, d\mathbb{Q}(y)$.

notion of risk is the minimax risk which corresponds to best possible error uniformly over the class $\mathscr{F}$

$$\mathscr{R}^*(\mathscr{F}) := \min_{\hat{f}:\mathscr{Y} \xrightarrow{\text{measurable}} \mathscr{F}} \max_{f \in \mathscr{F}} \mathbb{E}_f \left[d(\hat{f}(Y), f)^q \right], \tag{3.1}$$

where, again, the minimum is taken over all the measurable applications $\hat{f} : \mathscr{Y} \to \mathscr{F}$.

Our goal in this chapter is to derive a lower bound on $\mathscr{R}^*(\mathscr{F})$. Such a lower bound is useful in statistics, as, if we find an estimator $\hat{f}$ with a max-risk over $\mathscr{F}$ similar to the lower bound

$$\max_{f \in \mathscr{F}} \mathbb{E}_f \left[d(\hat{f}(Y), f)^q \right] \approx \text{lower bound}, \quad \text{where} \quad \text{lower bound} \leq \mathscr{R}^*(\mathscr{F}),$$

then it means that the estimator $\hat{f}$ performs almost as well as the best possible estimator in terms on the max-risk over $\mathscr{F}$.

3.2 Recipe for Proving Lower Bounds

In probability theory, it is often delicate to handle suprema over an infinite, possibly uncountable, space $\mathscr{F}$. When the objective function, here $f \to \mathbb{E}_f \left[d(\hat{f}(Y), f)^q \right]$, is smooth, a standard recipe is to replace the maximum over $\mathscr{F}$ by a maximum over a finite set $\{f_1, \ldots, f_N\}$. Indeed, if any point $f \in \mathscr{F}$ can be well approximated (in terms of the distance d) by one of the $f_1, \ldots, f_N$, then the maximum over $\mathscr{F}$ and the maximum over $\{f_1, \ldots, f_N\}$ will be close.
Once we have discretized the problem, then it is possible to use lower bounds lifted from information theory, in order to get a lower bound on the minimax risk $\mathscr{R}^*(\mathscr{F})$.

Kullback–Leibler Divergence

A useful "metric" for deriving lower-bounds is the Kullback–Leibler (KL) divergence between $\mathbb{P}$ and $\mathbb{Q}$, defined by

$$KL(\mathbb{P}, \mathbb{Q}) = \begin{cases} \int \log\left(\frac{d\mathbb{P}}{d\mathbb{Q}}\right) d\mathbb{P} & \text{when } \mathbb{P} \ll \mathbb{Q} \\ +\infty & \text{else.} \end{cases}$$

Important properties:

1. **Non-negativity.** For $\mathbb{P} \ll \mathbb{Q}$, we have the alternative formula for the Kullback–Leibler divergence

$$KL(\mathbb{P}, \mathbb{Q}) = \mathbb{E}_{\mathbb{Q}} \left[\phi\left(\frac{d\mathbb{P}}{d\mathbb{Q}}\right) \right], \quad \text{with} \quad \phi(x) = x \log(x).$$

Since ϕ is convex on $\mathbb{R}^+$, Jensen inequality ensures that

$$KL(\mathbb{P}, \mathbb{Q}) \geq \phi\left(\mathbb{E}_{\mathbb{Q}} \left[\frac{d\mathbb{P}}{d\mathbb{Q}} \right] \right) = \phi(1) = 0.$$

2. **Tensorisation.** For $\mathbb{P}_1 \ll \mathbb{Q}_1$ and $\mathbb{P}_2 \ll \mathbb{Q}_2$, we have

$$
\begin{aligned}
KL(\mathbb{P}_1 \otimes \mathbb{P}_2, \mathbb{Q}_1 \otimes \mathbb{Q}_2) \\
&= \int_{x_1,x_2} \left(\log\left(\frac{d\mathbb{P}_1}{d\mathbb{Q}_1}(x_1)\right) + \log\left(\frac{d\mathbb{P}_2}{d\mathbb{Q}_2}(x_2)\right)\right) d\mathbb{P}_1(x_1) d\mathbb{P}_2(x_2) \\
&= KL(\mathbb{P}_1, \mathbb{Q}_1) + KL(\mathbb{P}_2, \mathbb{Q}_2).
\end{aligned}
$$

This property is very handy in order to compute the KL divergence between distributions of data, when the observations are independent.

3.2.1 Fano's Lemma

The next result is a central tool in deriving lower bounds in statistics.

Theorem 3.1 Fano's lemma.
Let $(\mathbb{P}_j)_{j=1,\ldots,N}$ be a set of probability distributions on $\mathcal{Y}$. For any probability distribution $\mathbb{Q}$ such that $\mathbb{P}_j \ll \mathbb{Q}$, for $j = 1, \ldots, N$, we have

$$
\min_{\hat{J}:\mathcal{Y}\to\{1,\ldots,N\}} \frac{1}{N} \sum_{j=1}^{N} \mathbb{P}_j\left[\hat{J}(Y) \neq j\right] \geq 1 - \frac{1 + \frac{1}{N}\sum_{j=1}^{N} KL(\mathbb{P}_j, \mathbb{Q})}{\log(N)}, \tag{3.2}
$$

where $KL(\mathbb{P}, \mathbb{Q})$ is the Kullback–Leibler divergence between $\mathbb{P}$ and $\mathbb{Q}$.

A classical choice for $\mathbb{Q}$ is

$$
\mathbb{Q} = \frac{1}{N} \sum_{j=1}^{N} \mathbb{P}_j,
$$

but some other choices are sometimes more handy, depending on the problem.

Proof of Fano's lemma.
First, we observe that

$$
\min_{\hat{J}:\mathcal{Y}\to\{1,\ldots,N\}} \frac{1}{N} \sum_{j=1}^{N} \mathbb{P}_j\left[\hat{J}(Y) \neq j\right] = 1 - \max_{\hat{J}:\mathcal{Y}\to\{1,\ldots,N\}} \frac{1}{N} \sum_{j=1}^{N} \mathbb{P}_j\left[\hat{J}(Y) = j\right].
$$

Next lemma provides an explicit formula for the best average error.

Lemma 3.2 Best average risk.
We have

$$
\max_{\hat{J}:\mathcal{Y}\to\{1,\ldots,N\}} \frac{1}{N} \sum_{j=1}^{N} \mathbb{P}_j\left[\hat{J}(Y) = j\right] = \frac{1}{N} \mathbb{E}_{\mathbb{Q}}\left[\max_{j=1,\ldots,N} \frac{d\mathbb{P}_j}{d\mathbb{Q}}(Y)\right].
$$

Proof of Lemma 3.2. We have

$$\begin{aligned}\sum_{j=1}^{N} \mathbb{P}_j\left[\hat{J}(Y)=j\right] &= \int_{\mathscr{Y}} \sum_{j=1}^{N} \mathbf{1}_{\hat{J}(y)=j} \frac{d\mathbb{P}_j}{d\mathbb{Q}}(y)\, d\mathbb{Q}(y) \\ &\leq \int_{\mathscr{Y}} \underbrace{\sum_{j=1}^{N} \mathbf{1}_{\hat{J}(y)=j}}_{=1} \max_{j'=1,\ldots,N} \frac{d\mathbb{P}_{j'}}{d\mathbb{Q}}(y)\, d\mathbb{Q}(y) \\ &= \mathbb{E}_{\mathbb{Q}}\left[\max_{j'=1,\ldots,N} \frac{d\mathbb{P}_{j'}}{d\mathbb{Q}}(Y)\right].\end{aligned}$$

In addition, the inequality above is an equality for

$$\hat{J}(y) \in \underset{j=1,\ldots,N}{\operatorname{argmax}} \frac{d\mathbb{P}_j}{d\mathbb{Q}}(y).$$

The proof of Lemma 3.2 is complete. □

In the proof above, it is worth noticing that the best $\hat{J}$ corresponds to the maximum likelihood estimator for j.

So far, we have obtained that

$$\min_{\hat{J}:\mathscr{Y}\to\{1,\ldots,N\}} \frac{1}{N}\sum_{j=1}^{N} \mathbb{P}_j\left[\hat{J}(Y)\neq j\right] = 1-\frac{1}{N}\mathbb{E}_{\mathbb{Q}}\left[\max_{j=1,\ldots,N} \frac{d\mathbb{P}_j}{d\mathbb{Q}}(Y)\right]. \tag{3.3}$$

It remains to bound $\mathbb{E}_{\mathbb{Q}}\left[\max_{j=1,\ldots,N} \frac{d\mathbb{P}_j}{d\mathbb{Q}}(Y)\right]$ from above in terms of the $KL(\mathbb{P}_j,\mathbb{Q})$.

(Too) naive bound. When we have positive random variables $Z_1,\ldots,Z_N$, a simple way to bound the deviations or the expectation of $\max_{j=1,\ldots,N} Z_j$ is to use a union bound, which amounts to replace the maximum $\max_j$ by the sum $\sum_j$ in the expectation

$$\mathbb{E}\left[\max_j Z_j\right] \leq \sum_j \mathbb{E}\left[Z_j\right].$$

If we apply this simple bound in our case, we obtain

$$\frac{1}{N}\mathbb{E}_{\mathbb{Q}}\left[\max_{j=1,\ldots,N} \frac{d\mathbb{P}_j}{d\mathbb{Q}}(Y)\right] \leq \frac{1}{N}\sum_{j=1,\ldots,N} \underbrace{\mathbb{E}_{\mathbb{Q}}\left[\frac{d\mathbb{P}_j}{d\mathbb{Q}}(Y)\right]}_{=\int_{\mathscr{Y}} d\mathbb{P}_j(y)=1} = 1.$$

So, at the end, we have upper-bounded $\frac{1}{N}\sum_{j=1}^{N} \mathbb{P}_j\left[\hat{J}(Y)=j\right]$ by one, which could have been done more directly!
Hence, we need to bound more carefully the expectation of the maximum.

Powerful variant. A simple but powerful variant of the max/sum bound is to combine it with Jensen inequality. We state the result as a lemma in order to highlight the generality of the bound.

Lemma 3.3 Bounding a maximum.

Let $Z_1, \dots, Z_N$ be N random variables with values in an interval $I \subset \mathbb{R}$. Then, for any convex function $\varphi : I \to \mathbb{R}^+$ we have

$$\varphi\left(\mathbb{E}\left[\max_{j=1,\dots,N} Z_j\right]\right) \le \sum_{j=1}^{N} \mathbb{E}\left[\varphi(Z_j)\right]. \tag{3.4}$$

Proof of Lemma 3.3. We have

$$\begin{aligned}
\varphi\left(\mathbb{E}\left[\max_{j=1,\dots,N} Z_j\right]\right) &\le \mathbb{E}\left[\varphi\left(\max_{j=1,\dots,N} Z_j\right)\right] && \text{(Jensen inequality)}\\
&\le \mathbb{E}\left[\max_{j=1,\dots,N} \varphi(Z_j)\right]\\
&\le \sum_{j=1,\dots,N} \mathbb{E}\left[\varphi(Z_j)\right] && (\varphi \text{ non-negative}),
\end{aligned}$$

which gives (3.4). □

In order to bound $\mathbb{E}_{\mathbb{Q}}\left[\max_{j=1,\dots,N} \frac{d\mathbb{P}_j}{d\mathbb{Q}}(Y)\right]$ from above with Lemma 3.3, it remains to choose the function φ. Setting $\varphi(u) = u\log(u) - u + 1$, we observe that

$$\begin{aligned}
\mathbb{E}_{\mathbb{Q}}\left[\varphi\left(\frac{d\mathbb{P}_j}{d\mathbb{Q}}(Y)\right)\right] &= \int_{\mathscr{Y}} \log\left(\frac{d\mathbb{P}_j}{d\mathbb{Q}}(y)\right) \frac{d\mathbb{P}_j}{d\mathbb{Q}}(y)\, d\mathbb{Q}(y) - \int_{\mathscr{Y}} \frac{d\mathbb{P}_j}{d\mathbb{Q}}(y)\, d\mathbb{Q}(y) + 1\\
&= \int_{\mathscr{Y}} \log\left(\frac{d\mathbb{P}_j}{d\mathbb{Q}}(y)\right) d\mathbb{P}_j(y) - \underbrace{\int_{\mathscr{Y}} d\mathbb{P}_j(y)}_{=1} + 1\\
&= KL(\mathbb{P}_j, \mathbb{Q}).
\end{aligned}$$

The function φ is non-negative and convex, as can been visualized in Figure 3.1.

Applying Lemma 3.3, we get

$$\varphi\left(\mathbb{E}_{\mathbb{Q}}\left[\max_{j=1,\dots,N} \frac{d\mathbb{P}_j}{d\mathbb{Q}}(Y)\right]\right) \le \sum_{j=1}^{N} \mathbb{E}_{\mathbb{Q}}\left[\varphi\left(\frac{d\mathbb{P}_j}{d\mathbb{Q}}(Y)\right)\right] = \sum_{j=1}^{N} KL(\mathbb{P}_j, \mathbb{Q}).$$

Let us set

$$u = \frac{1}{N}\, \mathbb{E}_{\mathbb{Q}}\left[\max_{j=1,\dots,N} \frac{d\mathbb{P}_j}{d\mathbb{Q}}(Y)\right]. \tag{3.5}$$

We have

$$\begin{aligned}
\varphi(Nu) &= Nu(\log(N) + \log(u)) - Nu + 1\\
&= Nu\log(N) + \underbrace{N(u\log(u) - u + 1)}_{=N\varphi(u)\ge 0} - (N-1)\\
&\ge Nu\log(N) - N,
\end{aligned}$$

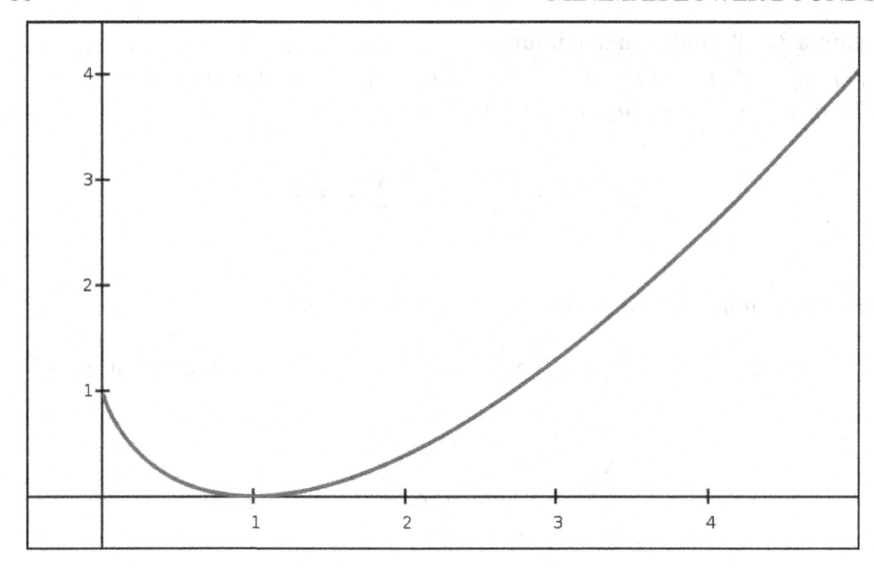

Figure 3.1 *Plot of the function* $\varphi(u) = u\log(u) - u + 1$.

so replacing u by its value (3.5), we get

$$\mathbb{E}_{\mathbb{Q}}\left[\max_{j=1,\dots,N}\frac{d\mathbb{P}_j}{d\mathbb{Q}}(Y)\right] \times \log(N) \leq N + \varphi\left(\mathbb{E}_{\mathbb{Q}}\left[\max_{j=1,\dots,N}\frac{d\mathbb{P}_j}{d\mathbb{Q}}(Y)\right]\right)$$

$$\leq N + \sum_{j=1}^{N} KL(\mathbb{P}_j, \mathbb{Q}). \tag{3.6}$$

Combining (3.3) and (3.6), we get Fano's inequality (3.2). □

3.2.2 From Fano's Lemma to a Lower Bound over a Finite Set

Let $\{f_1, \dots, f_N\} \subset \mathscr{F}$ be any discretization of $\mathscr{F}$. Let us now explain how we can get a lower bound on

$$\overline{\mathscr{R}}(\mathscr{F}, f_1, \dots, f_N) := \min_{\hat{f}:\mathscr{Y} \xrightarrow{\text{measurable}} \mathscr{F}} \frac{1}{N}\sum_{j=1}^{N} \mathbb{E}_{f_j}\left[d(\hat{f}(Y), f_j)^q\right]$$

from Fano's inequality (3.2).
Fano's inequality provides a lower bound on

$$\min_{\tilde{f}:\mathscr{Y} \xrightarrow{\text{measurable}} \{f_1,\dots,f_N\}} \frac{1}{N}\sum_{j=1}^{N} \mathbb{P}_{f_j}\left[\tilde{f}(Y) \neq f_j\right].$$

So, to get a lower bound on $\overline{\mathscr{R}}(\mathscr{F}, f_1, \dots, f_N)$ from Fano's inequality, we must reduce two problems. First, we must reduce the minimum over $\hat{f} : \mathscr{Y} \xrightarrow{\text{measurable}} \mathscr{F}$ to the minimum over $\tilde{f} : \mathscr{Y} \xrightarrow{\text{measurable}} \{f_1, \dots, f_N\}$, and second we must lower bound $\mathbb{E}_{f_j}\left[d(\hat{f}(Y), f_j)^q\right]$ in terms of $\mathbb{P}_{f_j}\left[\tilde{f}(Y) \neq f_j\right]$. As explained in the next paragraph, these two reductions can be easily obtained by "projecting" $\hat{f}(y)$ over the finite set $\{f_1(y), \dots, f_N(y)\}$ for all $y \in \mathscr{Y}$.

For any measurable $\hat{f} : \mathscr{Y} \to \mathscr{F}$, let us define

$$\tilde{J}(y) \in \operatorname*{argmin}_{j=1,\dots,N} d(\hat{f}(y), f_j).$$

By triangular inequality and the definition of $\tilde{J}(y)$, we have for any $j = 1, \dots, N$

$$\begin{aligned}
\min_{i \neq k} d(f_i, f_k) \mathbf{1}_{\tilde{J}(y) \neq j} &\leq d(f_j, f_{\tilde{J}(y)}) \\
&\leq d(f_j, \hat{f}(y)) + d(\hat{f}(y), f_{\tilde{J}(y)}) \\
&\leq 2d(f_j, \hat{f}(y)).
\end{aligned}$$

So, for any measurable $\hat{f} : \mathscr{Y} \to \mathscr{F}$ we have

$$\begin{aligned}
\frac{2^q}{N} \sum_{j=1}^{N} \mathbb{E}_{f_j}\left[d(\hat{f}(Y), f_j)^q\right] &\geq \min_{i \neq k} d(f_i, f_k)^q \times \frac{1}{N} \sum_{j=1}^{N} \mathbb{P}_{f_j}\left[\tilde{J}(Y) \neq j\right] \\
&\geq \min_{i \neq k} d(f_i, f_k)^q \times \min_{\hat{J} : \mathscr{Y} \to \{1,\dots,N\}} \frac{1}{N} \sum_{j=1}^{N} \mathbb{P}_{f_j}\left[\hat{J}(Y) \neq j\right].
\end{aligned}$$

Combining this last bound with Fano's inequality (3.2), we have proved the following result.

Corollary 3.4 Lower bound for discrete problem.
For any $\{f_1, \dots, f_N\} \subset \mathscr{F}$ and for any probability distribution $\mathbb{Q}$ such that $\mathbb{P}_{f_j} \ll \mathbb{Q}$, for $j = 1, \dots, N$, we have

$$\begin{aligned}
\min_{\hat{f} : \mathscr{Y} \xrightarrow{\text{meas.}} \mathscr{F}} \frac{1}{N} \sum_{j=1}^{N} \mathbb{E}_{f_j}\left[d(\hat{f}(Y), f_j)^q\right] & \\
\geq 2^{-q} \left(1 - \frac{1 + \frac{1}{N} \sum_{j=1}^{N} KL(\mathbb{P}_{f_j}, \mathbb{Q})}{\log(N)}\right) & \min_{i \neq k} d(f_i, f_k)^q, \qquad (3.7)
\end{aligned}$$

where $KL(\mathbb{P}, \mathbb{Q})$ denotes the Kullback–Leibler divergence between $\mathbb{P}$ and $\mathbb{Q}$.

3.2.3 Back to the Original Problem: Finding a Good Discretization

Since for any $\{f_1,\dots,f_N\} \subset \mathscr{F}$, we have

$$\begin{aligned}\mathscr{R}^*(\mathscr{F}) &= \min_{\hat{f}:\mathscr{Y}\xrightarrow{\text{meas.}}\mathscr{F}} \max_{f\in\mathscr{F}} \mathbb{E}_f\left[d(\hat{f}(Y),f)^q\right] \\ &\geq \min_{\hat{f}:\mathscr{Y}\xrightarrow{\text{meas.}}\mathscr{F}} \max_{j=1,\dots,N} \mathbb{E}_{f_j}\left[d(\hat{f}(Y),f_j)^q\right] \\ &\geq \min_{\hat{f}:\mathscr{Y}\xrightarrow{\text{meas.}}\mathscr{F}} \frac{1}{N}\sum_{j=1}^{N} \mathbb{E}_{f_j}\left[d(\hat{f}(Y),f_j)^q\right],\end{aligned}$$

the lower bound (3.7) provides a lower bound on $\mathscr{R}^*(\mathscr{F})$. As we have lower bounded a maximum by an average in the last step, this inequality is tight only when all the expectations $\mathbb{E}_{f_j}\left[d(\hat{f}(Y),f_j)^q\right]$ are of similar size.

Corollary 3.4 is general, and in practice all the art is to find a good discretization of $\mathscr{F}$ so that the lower bound (3.7) is as large as possible. We observe that we must find a discrete set $\{f_1,\dots,f_N\} \subset \mathscr{F}$ with

1. $\min_{i\neq k} d(f_i,f_k)$ as large as possible
2. and $\log(N)^{-1}\left(1+\frac{1}{N}\sum_{j=1}^{N} KL(\mathbb{P}_{f_j},\mathbb{Q})\right)$ bounded away from above from 1.

Both conditions are antagonistic, as the first condition requires the f_j to be as spread as possible, while the second condition asks for the f_j to be close enough to each other (with respect to the KL divergence). Hence, the size of the minimax risk $\mathscr{R}^*(\mathscr{F})$, depends on how much we can separate points in $\mathscr{F}$ (with respect to d), within a region of fixed diameter (with respect to the KL divergence).

It is best to illustrate the choice of a good discretization on a (simple) example.

3.3 Illustration

As an illustration, let us consider the simple case where $\mathscr{F} = \mathbb{R}^d$ and $\mathbb{P}_f$ is the Gaussian $\mathscr{N}(f,\sigma^2 I_d)$ distribution on $\mathbb{R}^d$. In the Gaussian case, the KL divergence has the simple expression

$$KL(\mathbb{P}_f,\mathbb{P}_g) = \frac{\|f-g\|^2}{2\sigma^2},$$

see Exercise 3.6.1. Below, we consider the distance $d(f,g) = \|f-g\|$.

For $\varepsilon, R > 0$, a set N_ε is called an ε-packing of $B(0,R)$ if $N_\varepsilon \subset B(0,R)$ and if $\|f-f'\|^2 > \varepsilon$ for any $f,f' \in N_\varepsilon$.
Let us consider a maximal ε-packing N_ε of $B(0,R)$. We observe that

$$B(0,R) \subset \bigcup_{f\in N_\varepsilon} B(f,\varepsilon), \tag{3.8}$$

as otherwise

- there would exist $f' \in B(0,R)$ fulfilling $\|f-f'\| > \varepsilon$ for all $f \in N_\varepsilon$,

- $N'_\varepsilon = N_\varepsilon \cup \{f'\}$ would be an ε-packing of $B(0,R)$, which contradicts the fact that N_ε is a maximal ε-packing of $B(0,R)$.

Comparing the volume on both sides of (3.8), we get

$$R^d \leq |N_\varepsilon| \varepsilon^d. \tag{3.9}$$

Let us set $R^2 = d\sigma^2$ and $\varepsilon = e^{-3}R$. We have

$$1 + \frac{R^2}{2\sigma^2} \leq \frac{3d}{2} = \frac{d}{2}\log(R/\varepsilon) \leq \frac{1}{2}\log(|N_\varepsilon|).$$

Hence, Corollary 3.4 with $\mathbb{Q} = \mathbb{P}_0$ gives

$$\begin{aligned}
\mathscr{R}^*(\mathscr{F}) &\geq \min_{\hat{f}:\mathscr{Y} \xrightarrow{\text{measurable}} \mathscr{F}} \max_{f \in N_\varepsilon} \mathbb{E}_f\left[\|\hat{f}(Y) - f\|^2\right] \\
&\geq \min_{\hat{f}:\mathscr{Y} \xrightarrow{\text{measurable}} \mathscr{F}} \frac{1}{|N_\varepsilon|} \sum_{f \in N_\varepsilon} \mathbb{E}_f\left[\|\hat{f}(Y) - f\|^2\right] \\
&\geq \frac{1}{4}\left(1 - \frac{1 + \frac{1}{|N_\varepsilon|}\sum_{f \in N_\varepsilon} \frac{\|f-0\|^2}{2\sigma^2}}{\log(|N_\varepsilon|)}\right) \min_{f \neq f' \in N_\varepsilon} \|f - f'\|^2 \\
&\geq \frac{1}{4}\left(1 - \frac{1 + \frac{R^2}{2\sigma^2}}{\log(|N_\varepsilon|)}\right) \varepsilon^2 \\
&\geq \frac{1}{8}\varepsilon^2 = \frac{1}{8e^6} d\sigma^2.
\end{aligned}$$

Hence, we have proved that the minimax risk for estimating the mean of a Gaussian distribution can be lower-bounded by a constant times $d\sigma^2$. As the risk of the estimator $\hat{f}(y) = y$ is equal to $d\sigma^2$ (check it!), we obtain that the minimax risk $\mathscr{R}^*(\mathscr{F})$ is proportional to $d\sigma^2$.

3.4 Minimax Risk for Coordinate-Sparse Regression

We investigate in this section the optimality of the model selection estimator (2.9), defined on page 35. For simplicity, we focus on the example of coordinate sparsity. Similar results can be proved in the other sparse settings. For $D \in \{1, \ldots, p\}$, we define

$$V_D(\mathbf{X}) = \{\mathbf{X}\beta : \beta \in \mathbb{R}^p, |\beta|_0 = D\} \quad \text{where } |\beta|_0 = \text{Card}\{j : \beta_j \neq 0\}.$$

We address in this section the two following issues:

1. What is the best risk that an estimator $\widehat{f}$ can achieve uniformly on $f^* \in V_D(X)$?
2. Is the risk of $\widehat{f} = \widehat{f}_{\widehat{m}}$ given by (2.9) of the same size as this optimal risk?

3.4.1 Lower Bound on the Minimax Risk

As in the previous illustration, we write $\mathbb{P}_f$ for the Gaussian distribution $\mathcal{N}(f, \sigma^2 I_n)$ and $\mathbb{E}_f$ for the expectation when the vector of observations Y is distributed according to $\mathbb{P}_f$. We consider the minimax risk on $V_D(\mathbf{X})$

$$\mathbf{R}^*[\mathbf{X}, D] = \inf_{\widehat{f}} \sup_{f^* \in V_D(\mathbf{X})} \mathbb{E}_{f^*}\left[\|\widehat{f} - f^*\|^2\right],$$

where the infimum is taken over all the estimators. It corresponds to the best risk that an estimator $\widehat{f}$ can achieve uniformly on $f^* \in V_D(X)$. In the theorem below, we implement the analysis presented in Section 3.2 in order to derive a lower bound on $\mathbf{R}^*[\mathbf{X}, D]$.

For any integer $D_{\max}$ not larger than $p/2$, we introduce the restricted isometry constants

$$\underline{c}_{\mathbf{X}} := \inf_{\beta : |\beta|_0 \leq 2D_{\max}} \frac{\|\mathbf{X}\beta\|}{\|\beta\|} \leq \sup_{\beta : |\beta|_0 \leq 2D_{\max}} \frac{\|\mathbf{X}\beta\|}{\|\beta\|} =: \overline{c}_{\mathbf{X}}. \tag{3.10}$$

Theorem 3.5 Minimax risk for coordinate-sparse regression

Let us fix some $D_{\max} \leq p/5$. For any $D \leq D_{\max}$, we have the lower bound

$$\mathbf{R}^*[\mathbf{X}, D] \geq \frac{e}{4(2e+1)^2} \left(\frac{\underline{c}_{\mathbf{X}}}{\overline{c}_{\mathbf{X}}}\right)^2 D \log\left(\frac{p}{5D}\right) \sigma^2, \tag{3.11}$$

with $\underline{c}_{\mathbf{X}}$ and $\overline{c}_{\mathbf{X}}$ defined by (3.10).

Proof of Theorem 3.5.
To prove Theorem 3.5, we rely on a slight variant of Corollary 3.4.

Corollary 3.6 (variant of Corollary 3.4)

For any finite set $\mathcal{V} \subset \mathbb{R}^n$, when

$$\max_{f \neq f' \in \mathcal{V}} \|f - f'\|^2 \leq \frac{4e}{2e+1} \sigma^2 \log(|\mathcal{V}|), \tag{3.12}$$

we have the lower bound

$$\inf_{\widehat{f}} \max_{f \in \mathcal{V}} \mathbb{E}_f\left[\|\widehat{f} - f\|^2\right] \geq \frac{1}{4(2e+1)} \times \min_{f \neq f' \in \mathcal{V}} \|f - f'\|^2. \tag{3.13}$$

We use Corollary 3.6 instead of Corollary 3.4 in order to have cleaner constants, but there is no substantial differences between the two corollaries.

Proof of Corollary 3.6.
The proof of Corollary 3.6 is the same as the proof of Corollary 3.4, except that Fano's lemma is replaced by Birgé's lemma (Theorem B.13, Appendix B, page 309), which is a variant of Fano's lemma.

With the notation of Fano's lemma on page 57, for any measurable $\hat{J} : \mathscr{Y} \to \{1,\ldots,N\}$, the events $A_j = \{\hat{J}(Y) = j\}$, for $j = 1,\ldots,N$ are disjoint. So, Birgé's lemma (Theorem B.13, Appendix B, page 309) ensures that

$$\min_{j=1,\ldots,N} \mathbb{P}_j\left[\hat{J}(Y) = j\right] \leq \frac{2e}{2e+1} \bigvee \max_{j \neq j'} \frac{KL(\mathbb{P}_j, \mathbb{P}_{j'})}{\log(N)}.$$

Hence, if

$$KL(\mathbb{P}_j, \mathbb{P}_{j'}) \leq \frac{2e}{2e+1} \log(N), \tag{3.14}$$

we have

$$\min_{\hat{J}:\mathscr{Y} \to \{1,\ldots,N\}} \max_{j=1,\ldots,N} \mathbb{P}_j\left[\hat{J}(Y) \neq j\right] \geq \frac{1}{2e+1}. \tag{3.15}$$

This variant of Fano's lemma being established, the proof of Corollary 3.6 follows the same lines as the proof of Corollary 3.4, with (3.2) replaced by (3.15).

To conclude, since $KL(\mathbb{P}_f, \mathbb{P}_g) = \|f - g\|^2/(2\sigma^2)$ (see Exercise 3.6.1), Condition (3.14) is equivalent to (3.12). □

Let us now prove Theorem 3.5 from Corollary 3.6. In light of Corollary 3.6, we will build a subset $\mathscr{V} \subset V_D(\mathbf{X})$ of well-separated points fulfilling (3.12). We rely on the following combinatorial lemma.

Lemma 3.7 Spreading points in a sparse hypercube

For any positive integer D less than $p/5$, there exists a set $\mathscr{C}$ in $\{0,1\}^p_D := \{x \in \{0,1\}^p : |x|_0 = D\}$, fulfilling

$$|\beta - \beta'|_0 > D, \quad \textit{for all } \beta \neq \beta' \in \mathscr{C}$$
$$\textit{and} \quad \log|\mathscr{C}| \geq \frac{D}{2} \log\left(\frac{p}{5D}\right).$$

We refer to Exercise 3.6.2, page 69, for a proof of this lemma.

The set $\mathscr{C}$ of Lemma 3.7 gathers coordinate-sparse vectors, which are well separated. The set $\mathscr{V} = \mathbf{X}\mathscr{C}$ is included in $V_D(\mathbf{X})$, yet it may not fulfill (3.12). In order to fulfill this condition, we rescale this set by an appropriate factor $r > 0$.

We set

$$r^2 = \frac{e}{2e+1} \times \frac{\sigma^2}{\overline{c}_{\mathbf{X}}^2} \times \log\left(\frac{p}{5D}\right),$$

with $\bar{c}_{\mathbf{X}}$ defined by (3.10) and $\mathscr{V} = \{r\mathbf{X}\beta : \beta \in \mathscr{C}\}$. For any $\beta, \beta' \in \mathscr{C}$, we have $\|\beta - \beta'\|^2 = |\beta - \beta'|_0$ and $|\beta - \beta'|_0 \leq 2D$, so

$$\begin{aligned}
\max_{f \neq f' \in \mathscr{V}} \frac{\|f - f'\|^2}{2\sigma^2} &= r^2 \max_{\beta \neq \beta' \in \mathscr{C}} \frac{\|\mathbf{X}(\beta - \beta')\|^2}{2\sigma^2} \\
&\leq \bar{c}_{\mathbf{X}}^2 r^2 \max_{\beta \neq \beta' \in \mathscr{C}} \frac{\|\beta - \beta'\|^2}{2\sigma^2} \\
&\leq \bar{c}_{\mathbf{X}}^2 r^2 \frac{2D}{2\sigma^2} \leq \frac{2e}{2e+1} \times \frac{D}{2} \log\left(\frac{p}{5D}\right) \leq \frac{2e}{2e+1} \log(|\mathscr{C}|).
\end{aligned}$$

Hence, $\mathscr{V} = \{r\mathbf{X}\beta : \beta \in \mathscr{C}\}$ fulfills (3.12).

When $\underline{c}_{\mathbf{X}} = 0$, Theorem 3.5 is obvious, so we focus in the following on the case where $\underline{c}_{\mathbf{X}} > 0$. Then $|\mathscr{V}| = |\mathscr{C}|$ and combining Corollary 3.6 with the above inequality gives

$$\begin{aligned}
\inf_{\widehat{f}} \max_{f \in \mathscr{V}} \mathbb{E}_f\left[\|\widehat{f} - f\|^2\right] &\geq \frac{1}{4(2e+1)} \min_{f \neq f' \in \mathscr{V}} \|f - f'\|^2 \\
&\geq \frac{\underline{c}_{\mathbf{X}}^2 r^2}{4(2e+1)} \min_{\beta \neq \beta' \in \mathscr{C}} \|\beta - \beta'\|^2.
\end{aligned}$$

Since $\mathscr{C} \subset \{0,1\}^D$ and $|\beta - \beta'|_0 > D$ for all $\beta \neq \beta' \in \mathscr{C}$, we have $\|\beta - \beta'\|^2 \geq D$ for all $\beta \neq \beta' \in \mathscr{C}$, and then

$$\inf_{\widehat{f}} \max_{f \in \mathscr{V}} \mathbb{E}_f\left[\|\widehat{f} - f\|^2\right] \geq \frac{\underline{c}_{\mathbf{X}}^2 r^2 D}{4(2e+1)} \geq \frac{e}{4(2e+1)^2} \left(\frac{\underline{c}_{\mathbf{X}}}{\bar{c}_{\mathbf{X}}}\right)^2 D \log\left(\frac{p}{5D}\right) \sigma^2.$$

The proof of Theorem 3.5 is complete. □

3.4.2 Minimax Optimality of the Model Selection Estimator

So far, we have derived a lower bound (3.11) on the minimax risk $\mathbf{R}^*[\mathbf{X}, D]$ over $V_D(\mathbf{X})$, but we have not answered our initial question "does the model selection estimator (2.9) from Chapter 2, page 35, fulfill some optimality properties?". As we have not exhibited an estimator $\widehat{f}$ whose risk is upper-bounded on $V_D(\mathbf{X})$ by $CD\log(p/D)\sigma^2$ for some constant $C > 0$, we have even not proved that the minimax risk $\mathbf{R}^*[\mathbf{X}, D]$ is of the order of the lower bound (3.11).

Below, we show that for some suitable choices of the models S_m, and with probability π_m, the model selection estimator (2.9) has a risk upper-bounded by $CD\log(p/D)\sigma^2$ for some constant $C > 0$. As a consequence, the minimax risk $\mathbf{R}^*[\mathbf{X}, D]$ is of the order of $D\log(p/D)\sigma^2$ up to a multiplicative constant, possibly depending on the design $\mathbf{X}$. In addition, the maximum risk of the estimator (2.9) matches the minimax risk $\mathbf{R}^*[\mathbf{X}, D]$, up to a multiplicative constant, possibly depending on the design $\mathbf{X}$.

Let us upper-bound the risk of the estimator (2.9) when $f^* \in V_D(\mathbf{X})$. We choose the collection of models as in Section 2.2.1 for the coordinate-sparse setting and the probability

$$\pi_m = \left(C_p^{|m|}\right)^{-1} e^{-|m|} \frac{e-1}{e-e^{-p}}.$$

Then, the risk Bound (2.12), page 37, ensures that there exists a constant $C'_K > 1$, such that $\mathbb{E}_{f^*}\left[\|\widehat{f}-f^*\|^2\right] \leq C'_K D \log(p/D)\,\sigma^2$ uniformly on all $f^* \in V_D(\mathbf{X})$, all matrices $\mathbf{X}$, all $n, p \in \mathbb{N}$, and all $D \leq p/2$. In particular, for all $n, p \in \mathbb{N}$ and $D \leq p/2$, we have

$$\sup_{\mathbf{X}} \sup_{f^* \in V_D(\mathbf{X})} \mathbb{E}_{f^*}\left[\|\widehat{f}-f^*\|^2\right] \leq C'_K D \log\left(\frac{p}{D}\right)\sigma^2. \tag{3.16}$$

Let us compare (3.16) and (3.11). We observe that the lower bound (3.11) and the upper bound (3.16) are similar, except that the lower bound involves the ratio $(\underline{c}_{\mathbf{X}}/\overline{c}_{\mathbf{X}})^2$. We emphasize that this ratio can be equal to 0 for $D_{\max}$ large, for example, when $2D_{\max} \geq 1 + \mathrm{rank}(\mathbf{X})$. Yet, there exists some designs $\mathbf{X}$ for which this ratio is non-vanishing for $D_{\max}$ of the order of $n/\log(p)$. For example, if the matrix $\mathbf{X} = [X_{i,j}]_{i,j}$ is obtained by sampling each entry $X_{i,j}$ independently according to a standard Gaussian distribution, then Lemma 9.4, page 191, ensures that with large probability, for $D_{\max} \leq n/(32\log(p))$, we have $\underline{c}_{\mathbf{X}}/\overline{c}_{\mathbf{X}} \geq 1/4$. As a consequence, there exists a numerical constant $C > 0$, such that for all $\sigma^2 > 0$, for all $n, p \in \mathbb{N}$, and for all D smaller than $p/5$ and $n/(32\log(p))$, we have

$$\sup_{\mathbf{X}} \mathbf{R}^*[\mathbf{X}, D] \geq C D \log\left(\frac{p}{5D}\right)\sigma^2. \tag{3.17}$$

Combining (3.17) with (3.16), we obtain for any D smaller than $p/5$ and $n/(32\log(p))$

$$C D \log\left(\frac{p}{5D}\right)\sigma^2 \leq \sup_{\mathbf{X}} \mathbf{R}^*[\mathbf{X}, D] \leq \sup_{\mathbf{X}} \sup_{f^* \in V_D(\mathbf{X})} \mathbb{E}_{f^*}\left[\|\widehat{f}-f^*\|^2\right] \leq C'_K D \log\left(\frac{p}{D}\right)\sigma^2.$$

Up to the size of the constants C and C'_K, the lower and upper bounds have the same form. In this sense, the estimation procedure (2.9) is optimal.

3.4.3 Frontier of Estimation in High Dimensions

The minimax lower Bound (3.11) provides insightful information on the frontier between the statistical problems that can be successfully solved and those that are hopeless. Again, we only discuss the sparse coordinate case.

The prediction risk $\mathbb{E}\left[\|\widehat{f}-f^*\|^2\right]$ considered in this chapter involves the expectation of the square of the Euclidean norm of an n-dimensional vector. Since the square of the Euclidean norm of an n-dimensional vector with entries of order $\varepsilon > 0$ grows like $n\varepsilon^2$ with n, it is meaningful to discuss the accuracy of an estimator in terms of the size of the scaled risk $\mathbb{E}\left[n^{-1}\|\widehat{f}-f^*\|^2\right]$.

From Bounds (3.16) and (3.11), we know that the minimax scaled-risk

$$\inf_{\widehat{f}} \sup_{f^* \in V_D(\mathbf{X})} \mathbb{E}_{f^*}\left[n^{-1}\|\widehat{f}-f^*\|^2\right] \quad \text{is of order} \quad \frac{D}{n}\log\left(\frac{p}{D}\right)\sigma^2,$$

as long as $\underline{c}_{\mathbf{X}}/\overline{c}_{\mathbf{X}} \approx 1$ and $D \leq p/5$. In practice, it means that when $f^* \in V_D(\mathbf{X})$ with $D\sigma^2 \log(p/D)$ small compared to n, a procedure like (2.9) will produce an accurate estimation of f^*. On the contrary, when $D\sigma^2 \log(p/D)$ is large compared to n, no estimator can provide a reliable estimation uniformly on $f^* \in V_D(\mathbf{X})$. In particular, if the dimension p is larger than e^n, accurate estimation over all $V_D(\mathbf{X})$ is hopeless.

3.5 Discussion and References

3.5.1 Take-Home Message

Minimax risk is a popular notion in mathematical statistics in order to assess the optimality of an estimator. The typical objective of a theoretical statistician is to design an estimation procedure with a small computational complexity and which is minimax optimal (up to constants). Hence, for any new statistical problem, deriving a (sharp) lower bound on the minimax risk is an important part of the analysis of the problem. The path for deriving the lower bound in Section 3.4.1 or Exercise 3.6.2 is the most standard one.

3.5.2 References

The main principle for deriving lower bounds dates back to Le Cam [103]. Fano's inequality was introduced in [71], and the proof presented in this chapter is taken from Baraud [17], see also Guntuboyina [87]. The proof of the generalized Fano inequality presented in Exercise 3.6.5 is adapted from Gerchinovitz *et al.* [77]. Birgé's inequality (Theorem B.13, Appendix B, page 309) was introduced in [30].
Has'minskii [90] proposed to use Fano's inequality for deriving lower bounds in nonparametric estimation problems and the analysis of Exercise 3.6.4 is due to Yang and Barron [168]. We refer the reader to the last chapter of Wainwright's book [163] and to the second chapter of Tsybakov's book [153] for a more comprehensive treatment of the topic.

Finally, we refer to Verzelen [160] and Jin [96] for some discussion on the frontier of successful estimation in high-dimensional settings.

3.6 Exercises

3.6.1 Kullback–Leibler Divergence between Gaussian Distribution

In this exercise, we write $\mathbb{P}_{f,\Sigma}$ for the Gaussian distribution $\mathcal{N}(f,\Sigma)$ and we compute the Kullback–Leibler divergence $KL(\mathbb{P}_{f_0,\Sigma_0},\mathbb{P}_{f_1,\Sigma_1})$ for $f_0, f_1 \in \mathbb{R}^n$ and Σ_0, Σ_1 two $n \times n$ symmetric positive definite matrices.

1. As a warm-up, let us start with the spherical case where $\Sigma_0 = \Sigma_1 = \sigma^2 I_n$. Prove

the sequence of equalities below

$$KL\left(\mathbb{P}_{f_0,\sigma^2 I_n},\mathbb{P}_{f_1,\sigma^2 I_n}\right)=\mathbb{E}_{f_0,\sigma^2 I_n}\left[\log\left(\frac{\exp(-\|Y-f_0\|^2/(2\sigma^2))}{\exp(-\|Y-f_1\|^2/(2\sigma^2))}\right)\right]$$
$$=\frac{1}{2\sigma^2}\mathbb{E}_{f_0,\sigma^2 I_n}\left[\|Y-f_1\|^2-\|Y-f_0\|^2\right]=\frac{\|f_0-f_1\|^2}{2\sigma^2}.$$

2. Following the same lines, prove for any $f_0, f_1 \in \mathbb{R}^n$ and Σ_0, Σ_1 symmetric positive definite matrices

$$KL\left(\mathbb{P}_{f_0,\Sigma_0},\mathbb{P}_{f_1,\Sigma_1}\right)=\frac{1}{2}\left[\log|\Sigma_1\Sigma_0^{-1}|+(f_0-f_1)^T\Sigma_1^{-1}(f_0-f_1)+\mathrm{Tr}(\Sigma_1^{-1}\Sigma_0-I_n)\right].$$

3.6.2 Spreading Points in a Sparse Hypercube

We prove in this part Lemma 3.7, page 65. Let D be any positive integer less than $p/5$ and consider a maximal set $\mathscr{C}$ in $\{0,1\}^p_D := \{x \in \{0,1\}^p : |x|_0 = D\}$, fulfilling

$$|\beta-\beta'|_0 > D, \quad \text{for all } \beta \neq \beta' \in \mathscr{C}.$$

In the next questions, we will prove that

$$\log|\mathscr{C}| \geq \frac{D}{2}\log\left(\frac{p}{5D}\right). \tag{3.18}$$

1. Check that

$$\{0,1\}^p_D = \bigcup_{\beta\in\mathscr{C}}\left\{x\in\{0,1\}^p_D : |x-\beta|_0 \leq D\right\}.$$

Deduce from this covering the upper bound $C_p^D \leq |\mathscr{C}|\, \max_{\beta\in\mathscr{C}}|B(\beta,D)|$, where $B(\beta,D)=\left\{x\in\{0,1\}^p_D : |x-\beta|_0\leq D\right\}$.

2. Writing d for the integer part of $D/2$, check the sequence of inequalities

$$\frac{|B(\beta,D)|}{C_p^D}=\sum_{k=0}^{d}\frac{C_D^k C_{p-D}^k}{C_p^D}\leq 2^D\frac{C_p^d}{C_p^D}\leq 2^D\left(\frac{D}{p-D+1}\right)^{D-d}\leq\left(\frac{5D}{p}\right)^{D/2}.$$

3. Conclude the proof of (3.18).

3.6.3 Some Other Minimax Lower Bounds

Let d be a distance on $\mathbb{R}^p$ and fix $q \geq 1$. As in Section 3.4.1, we write $\mathbb{P}_f$ for the Gaussian distribution $\mathscr{N}(f,\sigma^2 I_n)$ and the expectation $\mathbb{E}_f$ will refer to the expectation when the vector of observations Y is distributed according to $\mathbb{P}_f$.

1. Let $\mathscr{C}$ be any finite subset of $\mathbb{R}^p$. Replacing Fano's lemma by Birgé's lemma (stated and proved in Theorem B.13, Appendix B, page 309) in the proof of Corollary 3.4, prove that when

$$\max_{\beta\neq\beta'\in\mathscr{C}}\|\mathbf{X}(\beta-\beta')\|^2\leq\frac{4e}{2e+1}\sigma^2\log(|\mathscr{C}|),$$

we have

$$\inf_{\widehat{\beta}} \max_{\beta \in \mathscr{C}} \mathbb{E}_{\mathbf{X}\beta}\left[d(\widehat{\beta},\beta)^q\right] \geq \frac{1}{2^q}\frac{1}{2e+1} \times \min_{\beta \neq \beta' \in \mathscr{C}} d(\beta,\beta')^q.$$

We fix $D_{\max} \leq p/5$. For $D \leq D_{\max}$, we set

$$r^2 = \frac{e}{2e+1} \times \frac{\sigma^2}{\overline{c}_{\mathbf{X}}^2} \times \log\left(\frac{p}{5D}\right),$$

with $\overline{c}_{\mathbf{X}}$ defined by (3.10). We consider below the set $\mathscr{C}_r = \{r\beta : \beta \in \mathscr{C}\}$, with $\mathscr{C}$ from the previous exercise and the distance d induced by the norm $|x|_q = \left(\sum_j |x_j|^q\right)^{1/q}$.

2. Prove that

$$\inf_{\widehat{\beta}} \max_{\beta \in \mathscr{C}_r} \mathbb{E}_{\mathbf{X}\beta}\left[|\widehat{\beta}-\beta|_q^q\right] \geq \frac{r^q D}{2^q(2e+1)}.$$

3. Conclude that for any $q \geq 1$, we have the lower bound

$$\inf_{\widehat{\beta}} \sup_{\beta:|\beta|_0=D} \mathbb{E}_{\mathbf{X}\beta}\left[|\widehat{\beta}-\beta|_q^q\right] \geq \frac{e^{q/2}}{2^q(2e+1)^{1+q/2}}\left(\frac{\sigma}{\overline{c}_{\mathbf{X}}}\right)^q D\left(\log\left(\frac{p}{5D}\right)\right)^{q/2}.$$

3.6.4 Non-Parametric Lower Bounds

Let (S,d) be a metric space and $(\mathbb{P}_f)_{f\in S}$ be a collection of distributions on a measurable space $(\mathscr{Y},\mathscr{A})$. We assume in the following that for any $f,g \in S$ the Kullback–Leibler divergence between $\mathbb{P}_f$ and $\mathbb{P}_g$ fulfills

$$KL(\mathbb{P}_f,\mathbb{P}_g) \leq n\,d(f,g)^2,$$

with n an integer larger than 2. A set $\{f_1,\ldots,f_N\} \subset S$ is said to be δ-separated with respect to d, if $d(f_i,f_j) \geq \delta$ for any $i \neq j$. We denote by $N_d(\delta)$ the maximal number of δ-separated points in (S,d). We also assume that there exist $\alpha > 0$ and $0 < C_- < C_+$ such that, for any $\delta > 0$,

$$C_-\delta^{-\alpha} \leq \log(N_d(\delta)) \leq C_+\delta^{-\alpha}.$$

1. The covering number $N_{cov}(\delta)$ of S corresponds to the minimal number of balls centered in S and with radius δ (relative to the distance d) needed to cover S. Check that $N_{cov}(\delta)$ is smaller than $N_d(\delta)$.
2. Let $r > 0$ and $\{g_1,\ldots,g_{N_{cov}}(r)\} \subset S$ be such that the union of the balls (with respect to the distance d) $\left\{B_d(g_j,r) : j = 1,\ldots,N_{cov}(r)\right\}$ centered in g_j and with radius r covers S. Let us define

$$\mathbb{Q} = \frac{1}{N_{cov}(r)}\sum_{j=1}^{N_{cov}(r)} \mathbb{P}_{g_j}.$$

Prove that for any $f \in S$

$$KL(P_f,\mathbb{Q}) \leq nr^2 + \log N_d(r).$$

3. Prove that there exists a distribution $\mathbb{Q}$ on $(\mathscr{Y},\mathscr{A})$ such that for any $N \geq 1$ and any $f_1,\dots,f_N \in S$

$$\frac{1}{N}\sum_{j=1}^{N} KL(P_{f_j},\mathbb{Q}) \leq \inf_{r>0}\left\{nr^2 + \log N_d(r)\right\}. \tag{3.19}$$

4. Prove that there exists a constant $\rho(\alpha,C_-,C_+) > 0$ depending only on α,C_-,C_+ such that

$$\inf_{\hat{f}:\mathscr{Y}\to S}\ \max_{f\in S}\mathbb{E}_f\left[d(\hat{f},f)^2\right] \geq \rho(\alpha,C_-,C_+)\, n^{-\frac{2}{2+\alpha}},$$

where the infimum is over all measurable maps from $\mathscr{Y}$ to S.

3.6.5 Data Processing Inequality and Generalized Fano Inequality

Our goal in this problem is to show that Fano inequality can be derived from the simple data processing inequality. The latter inequality gives a meaning to the intuitive statement "for any random variable X, the image measures $\mathbb{P}^X$ and $\mathbb{Q}^X$ are closer to each other than $\mathbb{P}$ and $\mathbb{Q}$". We start by proving the data processing inequality in a general form, before specializing it to the Kullback–Leibler divergence and deriving a generalized version of Fano inequality (3.2), page 57.

A) Data Processing Inequality

Let $f:[0,+\infty)\to\mathbb{R}$ be any convex function fulfilling $f(1)=0$. For two probability distributions $\mathbb{P}$ and $\mathbb{Q}$ on a common measurable space $(\Omega,\mathscr{F})$, with $\mathbb{P}\ll\mathbb{Q}$, we define the f-divergence

$$D_f(\mathbb{P},\mathbb{Q}) := \mathbb{E}_{\mathbb{Q}}\left[f\left(\frac{d\mathbb{P}}{d\mathbb{Q}}\right)\right] = \int_\Omega f\left(\frac{d\mathbb{P}}{d\mathbb{Q}}\right)d\mathbb{Q}, \tag{3.20}$$

where $\mathbb{E}_{\mathbb{Q}}$ denotes the expectation with respect to $\mathbb{Q}$.

1. As a warm-up, prove with Jensen inequality that, for any $\mathbb{P},\mathbb{P}_1,\mathbb{P}_2 \ll \mathbb{Q}$ and $\lambda \in [0,1]$, we have

$$D_f(\mathbb{P},\mathbb{Q})\geq 0,\ \text{ and } D_f(\lambda\mathbb{P}_1+(1-\lambda)\mathbb{P}_2,\mathbb{Q}) \leq \lambda D_f(\mathbb{P}_1,\mathbb{Q})+(1-\lambda)D_f(\mathbb{P}_2,\mathbb{Q}).$$

The data processing inequality states that, for any probability distributions $\mathbb{P}\ll\mathbb{Q}$ on a common measurable space $(\Omega,\mathscr{F})$, and any random variable $X:(\Omega,\mathscr{F})\to(E,\mathscr{A})$, we have

$$D_f(\mathbb{P}^X,\mathbb{Q}^X)\leq D_f(\mathbb{P},\mathbb{Q}), \tag{3.21}$$

where $\mathbb{P}^X$ (respectively $\mathbb{Q}^X$) is the distribution of X under $\mathbb{P}$ (resp. $\mathbb{Q}$) defined by $\mathbb{P}^X(A)=\mathbb{P}(X\in A)$ for any $A\in\mathscr{A}$. The distribution $\mathbb{P}^X$ is also called the push-forward of $\mathbb{P}$ (resp. $\mathbb{Q}$) by X. The intuition behind (3.21), is that the distributions $\mathbb{P}$ and $\mathbb{Q}$ can only become closer after being processed by X.

A first step in order to prove (3.21) is to relate the Radon–Nikodym derivative $\frac{d\mathbb{P}^X}{d\mathbb{Q}^X}$ to the conditional expectation $\mathbb{E}_{\mathbb{Q}}\left[\frac{d\mathbb{P}}{d\mathbb{Q}}|X\right]$.

2. Let g be any measurable function such that $\mathbb{Q}$-almost surely $g(X) = \mathbb{E}_{\mathbb{Q}}\left[\frac{d\mathbb{P}}{d\mathbb{Q}}|X\right]$. Prove that for any $A \in \mathscr{A}$

$$\mathbb{P}^X(A) = \mathbb{E}_{\mathbb{Q}}\left[\mathbf{1}_A(X)\frac{d\mathbb{P}}{d\mathbb{Q}}\right] = \mathbb{E}_{\mathbb{Q}}\left[\mathbf{1}_A(X)g(X)\right] = \int_E \mathbf{1}_A\, g\, d\mathbb{Q}^X.$$

3. Conclude that $\mathbb{P}^X \ll \mathbb{Q}^X$ and that $\mathbb{Q}^X$-almost surely $\frac{d\mathbb{P}^X}{d\mathbb{Q}^X} = g$.
4. Conclude the proof of (3.21) by proving the sequence of inequalities

$$\begin{aligned} D_f(\mathbb{P}^X, \mathbb{Q}^X) &= \int_\Omega f\left(\mathbb{E}_{\mathbb{Q}}\left[\frac{d\mathbb{P}}{d\mathbb{Q}}|X\right]\right) d\mathbb{Q} \\ &\leq \int_\Omega \mathbb{E}_{\mathbb{Q}}\left[f\left(\frac{d\mathbb{P}}{d\mathbb{Q}}\right)|X\right] d\mathbb{Q} = D_f(\mathbb{P}, \mathbb{Q}). \end{aligned}$$

B) Two Corollaries of the Data Processing Inequality

i) Joint Convexity

We have seen in the warm-up that the divergence D_f is convex in the first variable. We will now prove that it is jointly convex in the two variables. More precisely, we will prove that for any distributions $\mathbb{P}_1 \ll \mathbb{Q}_1$ and $\mathbb{P}_2 \ll \mathbb{Q}_2$ on $(\Omega, \mathscr{F})$, and any $\lambda \in [0,1]$ we have

$$D_f(\lambda\mathbb{P}_1 + (1-\lambda)\mathbb{P}_2, \lambda\mathbb{Q}_1 + (1-\lambda)\mathbb{Q}_2) \leq \lambda D_f(\mathbb{P}_1, \mathbb{Q}_1) + (1-\lambda)D_f(\mathbb{P}_2, \mathbb{Q}_2). \tag{3.22}$$

We will derive this inequality directly from the data processing inequality. Define the probabilities $\mathbb{P}$ and $\mathbb{Q}$ on $\{1,2\} \times \Omega$ by

$$\mathbb{P}(\{j\} \times B) = \begin{cases} \lambda\mathbb{P}_1(B) & \text{if } j = 1, \\ (1-\lambda)\mathbb{P}_2(B) & \text{if } j = 2, \end{cases}$$

and the same for $\mathbb{Q}$ with $\mathbb{P}_j(B)$ replaced by $\mathbb{Q}_j(B)$. Let us define the random variable $X : \{1,2\} \times \Omega \to \Omega$, by $X(j, \omega) = \omega$.

1. Check that $\mathbb{P}^X = \lambda\mathbb{P}_1 + (1-\lambda)\mathbb{P}_2$ and $\mathbb{Q}^X = \lambda\mathbb{Q}_1 + (1-\lambda)\mathbb{Q}_2$.
2. Notice that

$$\frac{d\mathbb{P}}{d\mathbb{Q}}(j, \omega) = \mathbf{1}_{j=1}\frac{d\mathbb{P}_1}{d\mathbb{Q}_1}(\omega) + \mathbf{1}_{j=2}\frac{d\mathbb{P}_2}{d\mathbb{Q}_2}(\omega),$$

and prove (3.22) with the sequence of inequalities

$$\begin{aligned} D_f(\lambda\mathbb{P}_1 + (1-\lambda)\mathbb{P}_2, \lambda\mathbb{Q}_1 + (1-\lambda)\mathbb{Q}_2) &= D_f(\mathbb{P}^X, \mathbb{Q}^X) \\ &\leq D_f(\mathbb{P}, \mathbb{Q}) = \lambda D_f(\mathbb{P}_1, \mathbb{Q}_1) + (1-\lambda)D_f(\mathbb{P}_2, \mathbb{Q}_2). \end{aligned}$$

ii) Variant of the Data Processing Inequality

We next explain how the data processing inequality (3.21) can be used to upper bound the difference between two expectations $\mathbb{E}_{\mathbb{P}}[Z]$ and $\mathbb{E}_{\mathbb{Q}}[Z]$ in terms of $D_f(\mathbb{P}, \mathbb{Q})$.

More precisely, we will prove the following corollary of the data processing inequality (3.21). Let $\mathscr{B}(q)$ denote the Bernoulli distribution with parameter q. For any random variable $Z : \Omega \to [0,1]$ and any probability distributions $\mathbb{P} \ll \mathbb{Q}$ on Ω, we have

$$D_f(\mathscr{B}(\mathbb{E}_{\mathbb{P}}[Z]), \mathscr{B}(\mathbb{E}_{\mathbb{Q}}[Z])) \leq D_f(\mathbb{P}, \mathbb{Q}). \tag{3.23}$$

1. Let ℓ be the Lebesgue measure on $[0,1]$ and let A be any event on $\Omega \times [0,1]$. Applying the data processing inequality (3.21) with $X = 1_A$ prove that

$$\begin{aligned} D_f(\mathbb{P}, \mathbb{Q}) &= D_f(\mathbb{P} \otimes \ell, \mathbb{Q} \otimes \ell) \\ &\geq D_f(\mathscr{B}(\mathbb{P} \otimes \ell(A)), \mathscr{B}(\mathbb{Q} \otimes \ell(A))). \end{aligned}$$

2. Notice that for the event $A = \{(\omega, x) \in \Omega \times [0,1] : x \leq Z(\omega)\}$, we have $\mathbb{P} \otimes \ell(A) = \mathbb{E}_{\mathbb{P}}[Z]$, and conclude the proof of (3.23).

C) Generalized Fano Inequalities

We specialize the above results for $f(x) = x\log(x)$. For this choice of f, we have $D_f(\mathbb{P}, \mathbb{Q}) = KL(\mathbb{P}, \mathbb{Q})$. For $p, q \in [0,1]$, we define

$$kl(p,q) := KL(\mathscr{B}(p), \mathscr{B}(q)) = p\log\left(\frac{p}{q}\right) + (1-p)\log\left(\frac{1-p}{1-q}\right).$$

1. Prove that $p\log(p) + (1-p)\log(1-p) \geq -\log(2)$ and $kl(p,q) \geq p\log\left(\frac{1}{q}\right) - \log(2)$.
2. Let $Z_1, \ldots, Z_N$ be N random variables taking values in $[0,1]$ and let $\mathbb{P}_j \ll \mathbb{Q}_j$, for $j = 1, \ldots, N$ be distributions on Ω. By combining the joint convexity (3.22) of kl and the variant (3.23) of the data processing inequality, prove that

$$kl\left(\frac{1}{N}\sum_{i=1}^{N} \mathbb{E}_{\mathbb{P}_i}[Z_i], \frac{1}{N}\sum_{i=1}^{N} \mathbb{E}_{\mathbb{Q}_i}[Z_i]\right) \leq \frac{1}{N}\sum_{i=1}^{N} KL(\mathbb{P}_i, \mathbb{Q}_i).$$

3. Combining the two previous questions, prove the generalized Fano inequality

$$\frac{1}{N}\sum_{i=1}^{N} \mathbb{E}_{\mathbb{P}_i}[Z_i] \leq \frac{\frac{1}{N}\sum_{i=1}^{N} KL(\mathbb{P}_i, \mathbb{Q}_i) + \log(2)}{-\log\left(\frac{1}{N}\sum_{i=1}^{N} \mathbb{E}_{\mathbb{Q}_i}[Z_i]\right)}.$$

4. In the special case where $\mathbb{Q}_i = \mathbb{Q}$ for $i = 1, \ldots, N$ and $Z_1 + \ldots + Z_N \leq 1$, $\mathbb{Q}$ almost-surely, prove that

$$\frac{1}{N}\sum_{i=1}^{N} \mathbb{E}_{\mathbb{P}_i}[Z_i] \leq \frac{\frac{1}{N}\sum_{i=1}^{N} KL(\mathbb{P}_i, \mathbb{Q}) + \log(2)}{\log N}. \tag{3.24}$$

The Fano inequality can be recovered from (3.24) as follows. Let $A_1, \ldots, A_N$ be N disjoint events in Ω and let $\mathbb{P}_1, \ldots, \mathbb{P}_N \ll \mathbb{Q}$ be $N+1$ probability distribution on Ω. Then, taking $Z_i = \mathbf{1}_{A_i}$ the Inequality (3.24) gives

$$\frac{1}{N}\sum_{i=1}^{N} \mathbb{P}_i(A_i) \leq \frac{\frac{1}{N}\sum_{i=1}^{N} KL(\mathbb{P}_i, \mathbb{Q}) + \log(2)}{\log N}.$$

You may notice that compared to (3.2), page 57, the constant 1 has been improved into $\log(2)$.

Chapter 4

Aggregation of Estimators

Estimator aggregation is an alternative to model selection for exploiting possible unknown structures in the data. The main idea is to use a convex combination of a collection of estimators instead of selecting one among them. Estimator aggregation shares all the good properties of model selection. Unfortunately, it suffers from the same high-computational cost. Some approximate computations can be implemented; some of them are described at the end of the chapter.

In the following, we present the aggregation of estimators in the same statistical setting as in Chapter 2.

4.1 Introduction

The model $\widehat{m}$ selected in the model selection procedure (2.9) can provide some interesting information on the data. Yet, when the objective is only to predict at best f^* (as in Examples 2, 4, and 5 in Section 2.1), we only want to have a ℓ^2-risk (2.4) as small as possible, and we do not care to select (or not) a model. In this case, instead of selecting an estimator $\widehat{f}_{\widehat{m}}$ in $\{\widehat{f}_m,\ m \in \mathcal{M}\}$, we may prefer to estimate f^* by a convex (or linear) combination $\widehat{f}$ of the $\{\widehat{f}_m,\ m \in \mathcal{M}\}$

$$\widehat{f} = \sum_{m \in \mathcal{M}} w_m \widehat{f}_m, \quad \text{with} \quad w_m \geq 0 \quad \text{and} \quad \sum_{m \in \mathcal{M}} w_m = 1. \tag{4.1}$$

Of course, selecting a model is a special case of convex combination of the $\{\widehat{f}_m,\ m \in \mathcal{M}\}$, with the weights $w_{\widehat{m}} = 1$ and $w_m = 0$ for $m \neq \widehat{m}$. Removing the requirement that all the weights w_m are equal to 0 except one of them allows for more flexibility and can possibly provide more stable estimators (in the sense that a small perturbation of the data Y only induces a small change in the estimator $\widehat{f}$).

4.2 Gibbs Mixing of Estimators

As in Chapter 2, page 31, we consider a collection of models $\{S_m,\ m \in \mathcal{M}\}$. We remind the reader that an unbiased estimator of the risk $r_m = \mathbb{E}\left[\|f^* - \widehat{f}_m\|^2\right]$ of the estimator $\widehat{f}_m = \text{Proj}_{S_m} Y$ is

$$\widehat{r}_m = \|Y - \widehat{f}_m\|^2 + 2d_m\sigma^2 - n\sigma^2, \quad \text{with } d_m = \dim(S_m);$$

see (2.7) in Chapter 2, page 33. We associate to $\beta > 0$ and a probability distribution π on $\mathcal{M}$ the Gibbs mixing $\widehat{f}$ of the collection of estimators $\{\widehat{f}_m,\ m \in \mathcal{M}\}$

$$\widehat{f} = \sum_{m\in\mathcal{M}} w_m \widehat{f}_m, \quad \text{with } w_m = \frac{\pi_m e^{-\beta\widehat{r}_m/\sigma^2}}{\mathcal{Z}}, \quad \text{where } \mathcal{Z} = \sum_{m\in\mathcal{M}} \pi_m e^{-\beta\widehat{r}_m/\sigma^2}. \tag{4.2}$$

The Gibbs distribution w corresponds to the distribution minimizing over the set of probability q on $\mathcal{M}$ the functional

$$\mathcal{G}(q) = \sum_m q_m \widehat{r}_m + \frac{\sigma^2}{\beta} KL(q,\pi), \tag{4.3}$$

where $KL(q,\pi) = \sum_{m\in\mathcal{M}} q_m \log(q_m/\pi_m) \geq 0$ is the Kullback–Leibler divergence between the probabilities q and π; see Exercise 4.7.1. We will use this property in the next section for analyzing the risk of $\widehat{f}$.

Remark. The estimator $\widehat{f}_{\widetilde{m}}$ with the largest weight $w_{\widetilde{m}}$ corresponds to the model $\widetilde{m}$ minimizing the criterion

$$\widehat{r}_m + \frac{\sigma^2}{\beta}\log\left(\frac{1}{\pi_m}\right) = \|Y - \widehat{f}_m\|^2 + 2d_m\sigma^2 + \frac{\sigma^2}{\beta}\log\left(\frac{1}{\pi_m}\right) - n\sigma^2, \tag{4.4}$$

which is quite similar to the model selection Criterion (2.9), since $n\sigma^2$ plays no role in the minimization. Since the weights w_m are inversely proportional to the exponential of (4.4), when one of the models has a Criterion (4.4) much smaller than the others, the estimator $\widehat{f}$ is very close to the estimator $\widehat{f}_{\widetilde{m}}$, with $\widetilde{m}$ minimizing (4.4). Estimator aggregation then significantly differs from estimator selection only when there are several models approximately minimizing Criterion (4.4). We refer to Exercise 4.7.4 for a brief analysis of the virtues of estimator aggregation in this context.

4.3 Oracle Risk Bound

The analysis of the risk of the estimator $\widehat{f}$ relies on the famous Stein's formula.

Proposition 4.1 Stein's formula

Let Y be an n-dimensional Gaussian vector with mean μ and covariance matrix $\sigma^2 I_n$. For any function $F : \mathbb{R}^n \to \mathbb{R}^n$, $F(x) = (F_1(x),\dots,F_n(x))$ continuous, with piecewise continuous partial derivatives, fulfilling for all $i \in \{1,\dots,n\}$

(a) $\lim_{|y_i|\to\infty} F_i(y_1,\dots,y_n)e^{-(y_i-\mu_i)^2/2\sigma^2} = 0$ *for all* $(y_1,\dots,y_{i-1},y_{i+1},\dots,y_n) \in \mathbb{R}^{n-1}$

(b) $\mathbb{E}\left[|\partial_i F_i(Y)|\right] < \infty$,

we have

$$\mathbb{E}\left[\|F(Y)-\mu\|^2\right] = \mathbb{E}\left[\|F(Y)-Y\|^2 - n\sigma^2 + 2\sigma^2 \mathrm{div}(F)(Y)\right], \tag{4.5}$$

where $\mathrm{div}(F) = \sum_i \partial_i F_i$.

Proof.
An integration by parts gives

$$\sigma^2 \int_{y_i \in \mathbb{R}} \partial_i F_i(y_1, \ldots, y_n) e^{-(y_i-\mu_i)^2/2\sigma^2} \, dy_i$$
$$= \int_{y_i \in \mathbb{R}} (y_i - \mu_i) F_i(y_1, \ldots, y_n) e^{-(y_i-\mu_i)^2/2\sigma^2} \, dy_i,$$

for all $(y_1, \ldots, y_{i-1}, y_{i+1}, \ldots, y_n) \in \mathbb{R}^{n-1}$. Taking the expectation with respect to $Y_1, \ldots, Y_{i-1}, Y_{i+1}, \ldots, Y_n$, it follows that $\mathbb{E}[(Y_i - \mu_i)F_i(Y)] = \sigma^2 \mathbb{E}[\partial_i F_i(Y)]$, and finally $\mathbb{E}[\langle Y - \mu, F(Y) \rangle] = \sigma^2 \mathbb{E}[\operatorname{div}(F)(Y)]$.

We conclude by expanding

$$\begin{aligned}
\mathbb{E}\left[\|F(Y) - \mu\|^2\right] &= \mathbb{E}\left[\|F(Y) - Y\|^2\right] + \mathbb{E}\left[\|Y - \mu\|^2\right] + 2\mathbb{E}[\langle F(Y) - Y, Y - \mu \rangle] \\
&= \mathbb{E}\left[\|F(Y) - Y\|^2\right] + n\sigma^2 + 2\sigma^2 \mathbb{E}[\operatorname{div}(F)(Y)] - 2n\sigma^2,
\end{aligned}$$

where the last equality comes from $\mathbb{E}[\langle Y, Y - \mu \rangle] = \mathbb{E}\left[\|Y - \mu\|^2\right] = n\sigma^2$, since Y is a Gaussian vector with mean μ and covariance matrix $\sigma^2 I_n$. The proof of Stein's formula is complete. □

We consider a collection of models $\{S_m, \, m \in \mathcal{M}\}$, a probability distribution $\{\pi_m, \, m \in \mathcal{M}\}$ on $\mathcal{M}$, a constant $\beta > 0$, and the estimator $\widehat{f}$ given by (4.2). We have the following risk bound on $R(\widehat{f})$.

Theorem 4.2 Oracle risk bound

For $\beta \leq 1/4$ we have

$$\mathbb{E}\left[\|\widehat{f} - f^*\|^2\right] \leq \min_{m \in \mathcal{M}} \left\{ \mathbb{E}\left[\|\widehat{f}_m - f^*\|^2\right] + \frac{\sigma^2}{\beta} \log\left(\frac{1}{\pi_m}\right) \right\}. \tag{4.6}$$

Before proving this result, let us comment on it. The risk Bound (4.6) is very similar to the risk Bound (2.12), page 37, for model selection. The main difference lies in the constant in front of $\mathbb{E}\left[\|\widehat{f}_m - f\|^2\right]$, which is exactly one in (4.6). For this reason, Bound (4.6) is better than Bound (2.12). In particular, the aggregated estimator $\widehat{f}$ given by (4.2) fulfills all the good properties of the estimator (2.9) obtained by model selection.

Proof.
The Stein formula applied to $F(Y) = \widehat{f}$ and the linearity of the divergence operator ensure that

$$\widehat{r} = \|\widehat{f} - Y\|^2 - n\sigma^2 + 2\sigma^2 \sum_{m \in \mathcal{M}} \operatorname{div}(w_m \widehat{f}_m) \quad \text{fulfills} \quad \mathbb{E}[\widehat{r}] = \mathbb{E}\left[\|\widehat{f} - f^*\|^2\right].$$

After a simplification of the expression of $\widehat{r}$, the result will follow from the fact that w minimizes (4.3).

Upper bound on $\widehat{r}$. We first expand

$$\begin{aligned}
\|Y-\widehat{f}\|^2 &= \left\langle \sum_{m\in\mathcal{M}} w_m(Y-\widehat{f}_m)\,,\,Y-\widehat{f}\right\rangle \\
&= \sum_{m\in\mathcal{M}} w_m\langle Y-\widehat{f}_m, Y-\widehat{f}_m\rangle + \sum_{m\in\mathcal{M}} w_m\langle Y-\widehat{f}_m, \widehat{f}_m-\widehat{f}\rangle \\
&= \sum_{m\in\mathcal{M}} w_m\|Y-\widehat{f}_m\|^2 + \underbrace{\sum_{m\in\mathcal{M}} w_m\langle Y-\widehat{f}, \widehat{f}_m-\widehat{f}\rangle}_{=0} - \sum_{m\in\mathcal{M}} w_m\|\widehat{f}-\widehat{f}_m\|^2,
\end{aligned}$$

and notice that the divergence of $w_m\widehat{f}_m$ is given by

$$\operatorname{div}(w_m\widehat{f}_m) = w_m\operatorname{div}(\widehat{f}_m) + \langle \widehat{f}_m, \nabla w_m\rangle.$$

Plugging these two formulas in the definition of $\widehat{r}$, we obtain

$$\begin{aligned}
\widehat{r} = \sum_{m\in\mathcal{M}} w_m\left(\|Y-\widehat{f}_m\|^2 - n\sigma^2 + 2\sigma^2\operatorname{div}(\widehat{f}_m)\right) \\
+ \sum_{m\in\mathcal{M}} w_m\left(2\sigma^2\left\langle \widehat{f}_m, \frac{\nabla w_m}{w_m}\right\rangle - \|\widehat{f}-\widehat{f}_m\|^2\right).
\end{aligned}$$

For a linear function $G(Y) = AY$, we have $\partial_i G_i(Y) = A_{ii}$, so $\operatorname{div}(G)(Y) = \operatorname{Tr}(A)$. Since $\widehat{f}_m = \operatorname{Proj}_{S_m}(Y)$, the divergence of $\widehat{f}_m$ is the trace of $\operatorname{Proj}_{S_m}$, which equals the dimension d_m of the space S_m. Then, we have

$$\|Y-\widehat{f}_m\|^2 - n\sigma^2 + 2\sigma^2\operatorname{div}(\widehat{f}_m) \;=\; \|Y-\widehat{f}_m\|^2 + (2d_m-n)\sigma^2 \;=\; \widehat{r}_m.$$

In addition, we have

$$\begin{aligned}
\frac{\nabla w_m}{w_m} &= -2\frac{\beta}{\sigma^2}(Y-\widehat{f}_m) - \frac{\nabla\mathcal{Z}}{\mathcal{Z}} = -2\frac{\beta}{\sigma^2}\Big((Y-\widehat{f}_m) - \sum_{m'\in\mathcal{M}} w_{m'}(Y-\widehat{f}_{m'})\Big) \\
&= -2\frac{\beta}{\sigma^2}\left(\widehat{f}-\widehat{f}_m\right),
\end{aligned}$$

so

$$\widehat{r} = \sum_{m\in\mathcal{M}} w_m\widehat{r}_m - 4\beta\sum_{m\in\mathcal{M}} w_m\langle \widehat{f}_m, \widehat{f}-\widehat{f}_m\rangle - \sum_{m\in\mathcal{M}} w_m\|\widehat{f}-\widehat{f}_m\|^2.$$

We notice that $\sum_{m\in\mathcal{M}} w_m\langle \widehat{f}, \widehat{f}-\widehat{f}_m\rangle = 0$ so

$$\begin{aligned}
\widehat{r} &= \sum_{m\in\mathcal{M}} w_m\widehat{r}_m + (4\beta-1)\sum_{m\in\mathcal{M}} w_m\|\widehat{f}-\widehat{f}_m\|^2 \\
&\leq \sum_{m\in\mathcal{M}} w_m\widehat{r}_m, \qquad (4.7)
\end{aligned}$$

where we used for the last line the condition $\beta \leq 1/4$.

Optimality condition. Since $KL(w,\pi) \geq 0$ and w minimizes $\mathscr{G}$ over the set of probability on $\mathscr{M}$ (see Exercise 4.7.1, page 83), we have

$$\widehat{r} \leq \mathscr{G}(w) \leq \mathscr{G}(q) = \sum_{m\in\mathscr{M}} q_m \widehat{r}_m + \frac{\sigma^2}{\beta} KL(q,\pi), \quad \text{for any probability } q \text{ on } \mathscr{M}.$$

Taking the expectation, we obtain for any probability q on $\mathscr{M}$

$$\mathbb{E}\left[\|\widehat{f} - f^*\|^2\right] = \mathbb{E}\left[\widehat{r}\right] \leq \mathbb{E}\left[\mathscr{G}(q)\right] = \sum_{m\in\mathscr{M}} q_m r_m + \frac{\sigma^2}{\beta} KL(q,\pi), \tag{4.8}$$

with $r_m = \mathbb{E}\left[\widehat{r}_m\right] = \mathbb{E}\left[\|\widehat{f}_m - f^*\|^2\right]$. The right-hand side of the above bound is minimum for $q_m = \pi_m e^{-\beta r_m/\sigma^2}/\mathscr{Z}'$, with $\mathscr{Z}' = \sum_m \pi_m e^{-\beta r_m/\sigma^2}$. Since $\mathscr{Z}' \geq \pi_m e^{-\beta r_m/\sigma^2}$ for all $m \in \mathscr{M}$, we observe that

$$\sum_{m\in\mathscr{M}} q_m r_m + \frac{\sigma^2}{\beta} KL(q,\pi) = -\frac{\sigma^2}{\beta} \log(\mathscr{Z}') \leq \min_{m\in\mathscr{M}} \left\{ r_m - \frac{\sigma^2}{\beta} \log(\pi_m) \right\}. \tag{4.9}$$

Combining (4.8) and (4.9), we obtain (4.6). □

4.4 Numerical Approximation by Metropolis–Hastings

Similarly to model selection, when the cardinality of $\mathscr{M}$ is large, the Gibbs mixing of estimators (4.2) is computationally intractable, since it requires at least card$(\mathscr{M})$ computations. To overcome this issue, one possibility is to compute an approximation of (4.2) by using Markov Chain Monte Carlo (MCMC) techniques.

Let $F : \mathscr{M} \to \mathbb{R}$ be a real function on $\mathscr{M}$ and w be an arbitrary probability distribution on $\mathscr{M}$ fulfilling $w_m > 0$ for all $m \in \mathscr{M}$. The Metropolis–Hastings algorithm is a classical MCMC technique to approximate

$$\mathbb{E}_w[F] := \sum_{m\in\mathscr{M}} w_m F(m). \tag{4.10}$$

The underlying idea of the Metropolis–Hastings algorithm is to generate an ergodic[1] Markov chain $(M_t)_{t\in\mathbb{N}}$ in $\mathscr{M}$ with stationary distribution w and to approximate $\mathbb{E}_w[F]$ by the average $T^{-1}\sum_{t=1}^T F(M_t)$, with T large. This approximation relies on the fact that for any ergodic Markov chain $(M_t)_{t\in\mathbb{N}}$ in $\mathscr{M}$ with stationary distribution w, we have almost surely the convergence

$$\frac{1}{T}\sum_{t=1}^{T} F(M_t) \stackrel{T\to\infty}{\to} \mathbb{E}_w[F]. \tag{4.11}$$

[1] We refer to Norris book [127] for lecture notes on Markov chains.

We remind the reader that a Markov chain on $\mathcal{M}$ with transition probability $Q(m,m')$ is said to be *reversible according to the probability distribution* w if

$$w_m Q(m,m') = w_{m'} Q(m',m) \quad \text{for all } m,m' \in \mathcal{M}. \tag{4.12}$$

An aperiodic and irreducible Markov chain that is reversible according to the probability w automatically fulfills (4.11). The Metropolis–Hastings algorithm proposes a way to build a Markov chain $(M_t)_{t\in\mathbb{N}}$ fulfilling the requirements that

- $(M_t)_{t\in\mathbb{N}}$ is aperiodic, irreducible, and reversible according to the probability w,
- the numerical generation of the chain $(M_t)_{t=1,\ldots,T}$ is computationally efficient.

To generate such a Markov chain, the principle of the Metropolis–Hastings algorithm is the following. Let $\Gamma(m,m')$ be the transition probability of an aperiodic irreducible random walk on $\mathcal{M}$ and define

$$\begin{aligned} Q(m,m') &= \Gamma(m,m') \wedge \frac{w_{m'}\Gamma(m',m)}{w_m} \quad \text{for all } m \neq m' \\ \text{and } Q(m,m) &= 1 - \sum_{m':\, m'\neq m} Q(m,m') \quad \text{for all } m \in \mathcal{M}. \end{aligned} \tag{4.13}$$

With such a definition, w and Q obviously fulfill (4.12). Therefore, the Markov chain $(M_t)_{t\in\mathbb{N}}$ with transition Q fulfills for any initial condition

$$\frac{1}{T}\sum_{t=1}^{T} F(M_t) \overset{T\to\infty}{\to} \sum_{m\in\mathcal{M}} w_m F(m) \quad \text{a.s.}$$

From a practical point of view, the Markov chain $(M_t)_t$ can be generated as follows. We start from the transition $\Gamma(m,m')$, which is the transition probability of an aperiodic irreductible random walk on $\mathcal{M}$, and then implement the following algorithm.

Metropolis–Hastings algorithm

Initialization: Pick an arbitrary $M_1 \in \mathcal{M}$ and choose a burn-in time T_0.

Iterate: For $t = 1,\ldots,T$

1. From the current state M_t, generate M'_{t+1} according to the distribution $\Gamma(M_t,.)$
2. Set
$$p_{t+1} = 1 \wedge \frac{w_{M'_{t+1}}\Gamma(M'_{t+1},M_t)}{w_{M_t}\Gamma(M_t,M'_{t+1})}$$
3. $$\begin{cases} \text{With probability } p_{t+1} & : \ \text{set } M_{t+1} = M'_{t+1}, \\ \text{otherwise} & : \ \text{set } M_{t+1} = M_t. \end{cases}$$

Output:

$$\frac{1}{T-T_0}\sum_{t=T_0+1}^{T} F(M_t)$$

Let us come back to the problem of approximating the Gibbs aggregation (4.2) of estimators. In this case, the function F is $F(m) = \widehat{f}_m$, and the probability distribution is

$$w_m = \frac{\pi_m e^{-\beta \widehat{r}_m/\sigma^2}}{\mathscr{Z}}, \quad \text{where } \mathscr{Z} = \sum_{m \in \mathscr{M}} \pi_m e^{-\beta \widehat{r}_m/\sigma^2}.$$

Computing a single w_m is very time consuming when $\mathscr{M}$ is very large, since computing $\mathscr{Z}$ requires to sum card$(\mathscr{M})$ terms. Fortunately, the Metropolis–Hastings algorithm does not require computing the weights w_m but only some ratios,

$$\frac{w_{m'}}{w_m} = \frac{\pi_{m'} e^{-\beta \widehat{r}_{m'}/\sigma^2}}{\pi_m e^{-\beta \widehat{r}_m/\sigma^2}},$$

that can be easily evaluated, since the term $\mathscr{Z}$ cancels.

To implement the Metropolis–Hastings algorithm for evaluating (4.2), we must choose a transition probability Γ. The choice of this transition probability Γ is delicate in practice, since it can dramatically change the speed of the convergence in (4.11). Ideally, the transition Γ must be simple enough so that we can efficiently generate M'_{t+1} from M_t, and it must lead to a "good" exploration of $\mathscr{M}$ in order to avoid getting trapped in a small neighborhood. We refer to Robert and Casella [134, 135] for discussions on the choice of Γ and much more on the Metropolis–Hastings algorithm. Below, we give an example of implementation of the Metropolis–Hastings algorithm for evaluating (4.2) in the coordinate-sparse setting.

Example: Approximating (4.2) in the sparse regression setting

We consider the collection of estimators $\{\widehat{f}_m : m \in \mathscr{M}\}$ and the probability π introduced in Sections 2.2.1 and 2.2.2 for the coordinate-sparse setting. We write $\mathscr{V}(m)$ for the set of all the subsets m' that can be obtained from m by adding or removing one integer $j \in \{1, \ldots, p\}$ to m. If we take the uniform distribution on $\mathscr{V}(m)$ as proposal distribution $\Gamma(m,.)$, then $\Gamma(m,m') = \Gamma(m',m)$ for any $m' \in \mathscr{V}(m)$. As a consequence, we only need to compute $w_{M'_{t+1}}/w_{M'_t}$ at the second step of the Metropolis–Hastings algorithm. It is crucial to compute this ratio efficiently, since it will be iterated many times. We have seen that

$$\frac{w_{m'}}{w_m} = \frac{\pi_{m'} e^{-\beta \widehat{r}_{m'}/\sigma^2}}{\pi_m e^{-\beta \widehat{r}_m/\sigma^2}},$$

so we need to compute $\pi_{m'}/\pi_m$ efficiently when $m' \in \mathscr{V}(m)$. Let us consider the choice $\pi_m \propto e^{-|m|}/C_p^{|m|}$ proposed for the coordinate-sparse setting. We have the simple formulas

$$\frac{\pi_{m'}}{\pi_m} = \frac{|m|+1}{e(p-|m|)} \text{ for } m' = m \cup \{j\} \quad \text{and} \quad \frac{\pi_{m'}}{\pi_m} = \frac{e(p-|m|+1)}{|m|} \text{ for } m' = m \setminus \{j\}.$$

Convergence rate of the approximation

The main weakness of this Metropolis–Hastings approximation of $\widehat{f}$ is that we do not know the convergence rate in (4.11). In particular, we have no guarantee that we can achieve a reasonably good approximation of (4.2) by the Metropolis–Hastings algorithm in a reasonable time. We emphasize that there is no hope to evaluate precisely all the weights $\{w_m : m \in \mathcal{M}\}$ in less than card($\mathcal{M}$) iterations since the Markov chain $(M_t)_{t\in\mathbb{N}}$ needs to visit each $m \in \mathcal{M}$ many times in order to evaluate precisely each w_m. Yet, if only a few weights w_m are significantly positive, we only need to evaluate properly these few ones, which may happen in a short amount of time. However, we emphasize that no result guarantees that these few larger weights w_m are estimated accurately after a reasonable amount of time.

To sum up, the Metropolis–Hastings algorithm presented above can be viewed as a stochastic version of the forward–backward algorithm presented in Section 2.4, where we average the estimators $(\widehat{f}_{M_t})_{t=1,\dots,T}$ instead of selecting the one with smallest Criterion (2.9).

4.5 Numerical Illustration

Let us illustrate the estimator aggregation on the simulated example of Section 2.5. We use again the models and probability distribution suited for the coordinate-sparse setting. The Gibbs mixing $\widehat{f}$ defined by (4.2) with $\beta = 1/4$ is plotted in Figure 4.1. Comparing Figure 4.1 with Figure 2.2, we observe that the estimators (4.2) and (2.9) are very similar in this example. We refer to Exercise 4.7.2 for a comparison of the Gibbs mixing and model selection in the orthonormal setting.

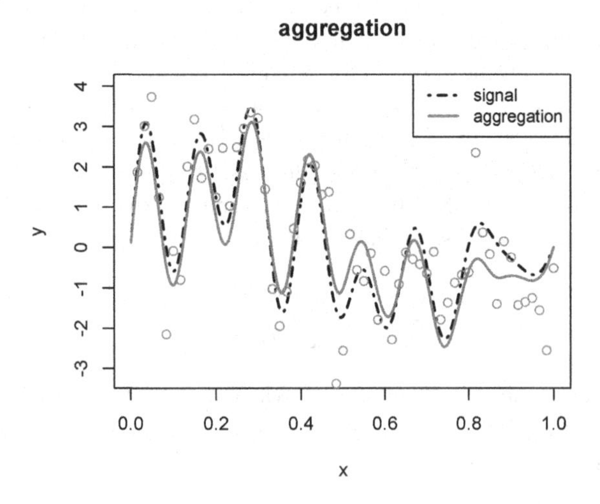

Figure 4.1 *Dotted line: Unknown signal. Gray dots: Noisy observations. Gray line: Estimator (4.2).*

4.6 Discussion and References

4.6.1 Take-Home Message

The Gibbs aggregation of estimators shares all the nice statistical properties of model selection estimators, with a slightly better risk Bound (4.6). Unfortunately, it suffers from the same computational complexity, and in many cases it is not possible to directly implement it in practice. A possible direction to overcome this issue is to approximate (4.2) by a Metropolis–Hastings algorithm, which is some kind of stochastic version of the forward–backward algorithm of Section 2.4 with averaging. No precise convergence rate is known for this algorithm, so we have no guarantee on the quality of the finite-time approximation.

4.6.2 References

Ideas on aggregation of estimators in statistics date back at least to Barron [21] and Catoni [50, 51]. Most of the material presented here comes from the seminal paper of Leung and Barron [109]. The computational aspects presented in Section 4.4 have been proposed by Rigollet and Tsybakov [132]. We refer to Rigollet and Tsybakov [133] for a recent review on the topic and more examples of the use of the Metropolis–Hastings algorithm. We also point out the paper by Sanchez-Perez [139], which provides a convergence bound on the Metropolis–Hastings approximation in a slightly different setting.

The case of unknown variance is more tricky and has attracted less attention. We refer to Giraud [79] and Gerchinovitz [76] for two points of view on this case. Finally, the Stein formula for non-Gaussian noise has been investigated in Dalalyan and Tsybakov [59].

4.7 Exercises

4.7.1 Gibbs Distribution

In this exercise, we first check that the Kullback–Leibler divergence $KL(q,\pi)$ is non-negative on the set $P(\mathcal{M})$ of probability distributions on $\mathcal{M}$. Then, we prove that the Gibbs distribution w defined by (4.2), page 76, minimizes the function $\mathcal{G}$ defined by (4.3).

For $\lambda \in \mathbb{R}$ and $q \in (\mathbb{R}^+)^{\mathcal{M}}$ we set,

$$\mathscr{L}^{(\lambda)}(q) = \mathcal{G}(q) + \lambda \sum_{m \in \mathcal{M}} q_m .$$

1. From the convexity of $x \to -\log(x)$ on $\mathbb{R}^+$, prove that the Kullback–Leibler divergence $KL(q,\pi) = \sum_m q_m \log(q_m/\pi_m)$ is non-negative for $q, \pi \in P(\mathcal{M})$.
2. Prove that $x \to x\log(x)$ is convex on $\mathbb{R}^+$ and $q \to \mathscr{L}^{(\lambda)}(q)$ is convex on $(\mathbb{R}^+)^{\mathcal{M}}$.
3. Prove that the minimum of $\mathscr{L}^{(\lambda)}$ on $(\mathbb{R}^+)^{\mathcal{M}}$ is achieved for

$$q_m^{(\lambda)} = \pi_m \exp(-\beta \widehat{r}_m/\sigma^2) \exp(-1 - \beta\lambda/\sigma^2).$$

4. Conclude that

$$\left(\frac{\pi_m e^{-\beta \widehat{r}_m/\sigma^2}}{\sum_{m' \in \mathcal{M}} \pi_{m'} e^{-\beta \widehat{r}_{m'}/\sigma^2}} : m \in \mathcal{M} \right) \in \underset{q \in P(\mathcal{M})}{\operatorname{argmin}} \mathcal{G}(q).$$

4.7.2 Orthonormal Setting with Power Law Prior

We consider the linear regression setting (2.2), in the case where the columns of $\mathbf{X}$ are orthonormal. We choose the collection of models described in Section 2.2.1 for the sparse-coordinate setting and take the distribution π on $\mathcal{M}$ given by

$$\pi_m = \left(1 + \frac{1}{p}\right)^{-p} p^{-|m|}.$$

1. Let us write $Z_j = \langle Y, \mathbf{X}_j \rangle$, for $j = 1, \ldots, p$. Prove that

$$\widehat{f} = \sum_{j=1}^{p} \frac{\exp\left(\beta Z_j^2/\sigma^2\right)}{\exp(2\beta + \log(p)) + \exp\left(\beta Z_j^2/\sigma^2\right)} Z_j \mathbf{X}_j .$$

2. What is the consequence in terms of computational complexity?
3. Compare qualitatively this mixing procedure to the model selection procedure in this setting.

4.7.3 Group-Sparse Setting

In the spirit of the algorithm for the coordinate-sparse setting, propose a Metropolis–Hastings algorithm for computing $\widehat{f}$ in the group-sparse setting (with the collection of models and the probability π given in Chapter 2).

4.7.4 Gain of Combining

For $\beta \le 1/4$ and $\delta > 0$, we set

$$\mathcal{M}_\delta = \left\{ m \in \mathcal{M} : r_m - \frac{\sigma^2}{\beta} \log(\pi_m) \le \inf_{m \in \mathcal{M}} \left\{ r_m - \frac{\sigma^2}{\beta} \log(\pi_m) \right\} + \delta \right\}.$$

By adapting the end the proof of Theorem 4.2, prove that

$$\mathbb{E}\left[\|\widehat{f} - f^*\|^2 \right] \le \min_{m \in \mathcal{M}} \left\{ \mathbb{E}\left[\|\widehat{f}_m - f^*\|^2 \right] + \frac{\sigma^2}{\beta} \log\left(\frac{1}{\pi_m}\right) \right\} + \inf_{\delta > 0} \left\{ \delta - \frac{\sigma^2}{\beta} \log|\mathcal{M}_\delta| \right\}.$$

In particular, when M estimators are similarly good, the risk bound is reduced by a term of order $\sigma^2 \log(M)$.

4.7.5 Online Aggregation

We consider here a slightly different statistical setting. We assume that we observe some temporal data $(X_t, Y_t)_{t=1,\dots,T}$, with $Y_t \in \mathbb{R}$ and $X_t \in \mathbb{R}^p$. For example, Y_t can be an air pollution indicator (as the concentration of ozone), and X_t can gather some atmospheric measures, as well as some past values of Y. Our goal is to predict at each time t the value Y_t based on the observations X_t and $(X_i, Y_i)_{i=1,\dots,t-1}$.

We assume that we have a collection $\left\{\widehat{f}_m = \left[\widehat{f}_m(t)\right]_{t=1,\dots,T},\ m \in \mathcal{M}\right\}$ of estimators with $\widehat{f}_m(t) = F_{m,t}\left(X_t, (X_i, Y_i)_{i=1,\dots,t-1}\right)$ for some measurable function $F_{m,t}$. We denote by

$$r_m(T) = \sum_{t=1}^{T} \left(Y_t - \widehat{f}_m(t)\right)^2$$

the cumulated prediction risk of the estimator $\widehat{f}_m$. We want to combine the estimators $\{\widehat{f}_m,\ m \in \mathcal{M}\}$ in order to have a cumulated prediction risk $r(T)$ almost as good as the best $r_m(T)$. We consider a probability distribution π on $\mathcal{M}$, and we denote by

$$\widehat{r}_m(t) = \sum_{i=1}^{t-1} \left(Y_i - \widehat{f}_m(i)\right)^2$$

the estimated prediction risk at time t, with $\widehat{r}_m(1) = 0$. At each time t, we build the Gibbs mixture as in (4.2)

$$\widehat{f}(t) = \sum_{m \in \mathcal{M}} w_m(t) \widehat{f}_m(t), \quad \text{with } w_m(t) = \frac{\pi_m e^{-\beta \widehat{r}_m(t)}}{\mathcal{Z}_t} \quad \text{and } \mathcal{Z}_t = \sum_{m \in \mathcal{M}} \pi_m e^{-\beta \widehat{r}_m(t)}.$$

We emphasize that the weights $w_m(t)$ only depend on the observations available at time t.

We assume in the following that the observations Y_t belong to a bounded interval $[-B, B]$, which is known. We also assume that the predictors $\widehat{f}_m(t)$ belong to this interval. We will prove that in this setting, for a suitable choice of β, we have an oracle inequality similar (4.6).

1. Prove that

$$\begin{aligned}
&\sum_{t=1}^{T} \log\left(\sum_{m \in \mathcal{M}} w_m(t) \exp\left(-\beta \left(Y_t - \widehat{f}_m(t)\right)^2\right)\right) \\
&= \log\left(\sum_{m \in \mathcal{M}} \pi_m \exp\left(-\beta \sum_{t=1}^{T} \left(Y_t - \widehat{f}_m(t)\right)^2\right)\right) \\
&\geq \max_{m \in \mathcal{M}} \left\{\log(\pi_m) - \beta r_m(T)\right\}.
\end{aligned}$$

2. Check that $x \to \exp(-x^2)$ is concave on the interval $[-2^{-1/2}, 2^{-1/2}]$. Prove that for $\beta \leq 1/(8B^2)$ we have

$$\sum_{m \in \mathcal{M}} w_m(t) \exp\left(-\beta \left(Y_t - \widehat{f}_m(t)\right)^2\right) \leq \exp\left(-\beta \left(Y_t - \widehat{f}(t)\right)^2\right).$$

3. Combining the above results, conclude that for $\beta \leq 1/(8B^2)$

$$\sum_{t=1}^{T}\left(Y_t-\widehat{f}(t)\right)^2 \leq \min_{m\in\mathcal{M}}\left\{\sum_{t=1}^{T}\left(Y_t-\widehat{f}_m(t)\right)^2+\frac{1}{\beta}\log\left(\frac{1}{\pi_m}\right)\right\}.$$

We emphasize that up to this point, the only assumption on the distribution of the data is that the variables Y_t are bounded. In order to state a result similar to (4.6), we consider in the following the setting

$$Y_t = f(X_t)+\varepsilon_t, \quad t=1,\ldots,T,$$

where ε_t has mean zero, finite variance σ_t^2 and is independent of X_t and $(X_i,Y_i)_{i=1,\ldots,t-1}$. We still assume that $|Y_t| \leq B$.

4. Prove that for $\beta = 1/(8B^2)$

$$\mathbb{E}\left[\frac{1}{T}\sum_{t=1}^{T}\left(\widehat{f}(t)-f(X_t)\right)^2\right] \leq \min_{m\in\mathcal{M}}\left\{\mathbb{E}\left[\frac{1}{T}\sum_{t=1}^{T}\left(\widehat{f}_m(t)-f(X_t)\right)^2\right]+\frac{8B^2}{T}\log\left(\frac{1}{\pi_m}\right)\right\}.$$

4.7.6 Aggregation with Laplace Prior

This exercise is adapted from Dalalyan *et al.* [60]. We recommend to have a look at Chapter 5, Section 5.2 before starting this exercise.

We observe $Y \in \mathbb{R}^n$ and $\mathbf{X} \in \mathbb{R}^{n\times p}$. We assume that $Y = \mathbf{X}\beta^* + \varepsilon$, with ε a random variable following a subgaussian$(\sigma^2 I)$ distribution, which means that for any $u \in \mathbb{R}^n$ we have $\mathbb{P}[|\langle u,\varepsilon\rangle| \geq \sigma t\|u\|] \leq e^{-t^2/2}$. We assume that the p columns $X_1,\ldots,X_p$ of $\mathbf{X}$ have unit norm.

In this chapter, we have considered the aggregation of a finite number of estimators. Here, we will aggregate a continuous set of estimators. We consider all vectors $u \in \mathbb{R}^p$ as an estimator of β^*, and we will aggregate them with a weight decreasing exponentially with

$$V(u) = \|Y-\mathbf{X}u\|^2+\lambda|u|_1.$$

More precisely, for $\alpha > 0$, we define the probability distribution on $\mathbb{R}^p$ by

$$d\pi(u) = \pi(u)\,du \quad \text{and} \quad \pi(u) = \frac{e^{-\alpha V(u)}}{\int_{\mathbb{R}^p} e^{-\alpha V(w)}\,dw},$$

and our goal is to investigate the risk $\|\mathbf{X}(\widehat{\beta}-\beta^*)\|^2$ of the estimator

$$\widehat{\beta} = \int_{\mathbb{R}^p} u\,d\pi(u)$$

of β^*. We observe that when $\alpha \to \infty$, the probability distribution $d\pi$ is concentrated on $\operatorname{argmin}_u V(u)$ and we recover the Lasso estimator (5.4), page 91.

Let $\delta \in (0,1)$. In this exercise, you will prove that, for $\lambda = 3\sqrt{2\sigma^2 \log(p/\delta)}$, with probability at least $1-\delta$, we have

$$\|\mathbf{X}(\widehat{\beta} - \beta^*)\|^2 \leq \inf_{\beta \in \mathbb{R}^p} \left\{ \|\mathbf{X}(\beta - \beta^*)\|^2 + \frac{\lambda^2 |\beta|_0}{\kappa(\beta)^2} \right\} + p/\alpha, \qquad (4.14)$$

where $\kappa(\beta)$ is the compatibility constant defined by (5.8), page 95.

Again, we notice that when α goes to infinity, we recover the result of Theorem 5.1, page 95. We also observe that to get the last term p/α small enough, we need to take $\alpha \geq p/\sigma^2$.

The key of the proof of (4.14) is the following lemma, proved in the parts (B) and (C) of the exercise.

Lemma 4.3

For all $\beta \in \mathbb{R}^p$ we have

$$V(\widehat{\beta}) - V(\beta) \leq p/\alpha - \|\mathbf{X}\widehat{\beta} - \mathbf{X}\beta\|^2. \qquad (4.15)$$

A) Conclusion Based on Lemma 4.3

1. Assume that $\lambda = 3\sqrt{2\sigma^2 \log(p/\delta)}$. Prove that

$$\mathbb{P}\left[\max_{j=1,\ldots,p} |\langle X_j, \varepsilon\rangle| \geq \lambda/3\right] \leq \delta.$$

2. Admitting Lemma 4.3, check that

$$\|\mathbf{X}(\widehat{\beta} - \beta^*)\|^2 - \|\mathbf{X}(\beta - \beta^*)\|^2 \leq \\ p/\alpha - \|\mathbf{X}(\widehat{\beta} - \beta)\|^2 + 2\langle \varepsilon, \mathbf{X}(\widehat{\beta} - \beta)\rangle + \lambda \left[|\beta|_1 - |\widehat{\beta}|_1\right].$$

3. Following the same arguments as in the proof of Theorem 5.1 (page 95), conclude that (4.14) holds with probability at least $1-\delta$.

The two next sections are dedicated to the proof of Lemma 4.3. A first step is to prove the intermediate result

$$V(\widehat{\beta}) - \int_{\mathbb{R}^p} V(u)\, d\pi(u) - \int_{\mathbb{R}^p} \|\mathbf{X}(u-\beta)\|^2 d\pi(u) \leq -\|\mathbf{X}\widehat{\beta} - \mathbf{X}\beta\|^2. \qquad (4.16)$$

B) Proof of the Intermediate Inequality (4.16)

1. Check that

$$\int_{\mathbb{R}^p} \left(\|\mathbf{X}(\beta^* - u)\|^2 + \|\mathbf{X}(\beta - u)\|^2\right) d\pi(u) \\ = \|\mathbf{X}(\beta^* - \widehat{\beta})\|^2 + \|\mathbf{X}(\widehat{\beta} - \beta)\|^2 + 2\int_{\mathbb{R}^p} \|\mathbf{X}(\widehat{\beta} - u)\|^2 d\pi(u).$$

2. Prove that

$$V(\widehat{\beta})-\int_{\mathbb{R}^p}V(u)\,d\pi(u)-\int_{\mathbb{R}^p}\|\mathbf{X}(u-\beta)\|^2d\pi(u)$$
$$=2\|\mathbf{X}\widehat{\beta}\|^2+\lambda|\widehat{\beta}|_1-\int_{\mathbb{R}^p}\left(2\|\mathbf{X}u\|^2+\lambda|u|_1\right)d\pi(u)-\|\mathbf{X}\widehat{\beta}-\mathbf{X}\beta\|^2.$$

3. Conclude the proof of (4.16).

C) Proof of Lemma 4.3

1. Prove that for all $\beta\in\mathbb{R}^p$ and all u such that V is differentiable in u, we have

$$V(\beta)-\|\mathbf{X}(u-\beta)\|^2\geq V(u)+\langle\nabla V(u),\beta-u\rangle.$$

2. With (4.16) and the above inequality, prove the inequality (4.15).

Chapter 5

Convex Criteria

We have seen in Chapter 2 that the model selection procedure (2.9), page 35, has some nice statistical properties but suffers from a prohibitive computational cost in many cases. For example, in the coordinate-sparse setting, the algorithmic complexity for computing (2.9) is exponential in p, so it cannot be implemented in moderate or high-dimensional settings. To circumvent this issue, a standard trick is to derive from the NP-hard problem (2.9) a convex criterion that can be minimized efficiently. The main point is then to check that the estimator derived from this convex criterion is almost as good as the estimator (2.9), at least for some classes of matrix $\mathbf{X}$. This chapter is devoted to this issue and to some related computational aspects.

5.1 Reminder on Convex Multivariate Functions

In this chapter, we will investigate some estimators that are obtained by minimizing some convex criteria. In order to analyze these estimators, we will need some basic results from convex analysis. This section is a brief reminder on the subdifferentials of convex functions. We refer to Appendix D for the details.

5.1.1 Subdifferentials

A function $F:\mathbb{R}^n \to \mathbb{R}$ is convex if $F(\lambda x + (1-\lambda)y) \leq \lambda F(x) + (1-\lambda)F(y)$ for all $x, y \in \mathbb{R}^n$ and $\lambda \in [0,1]$. When a function F is convex and differentiable, we have

$$F(y) \geq F(x) + \langle \nabla F(x), y - x \rangle, \quad \text{for all } x, y \in \mathbb{R}^n;$$

see Lemma D.1 in Appendix D, page 321. For any convex function F (possibly non-differentiable), we introduce the subdifferential ∂F of F, defined by

$$\partial F(x) = \{ w \in \mathbb{R}^n : F(y) \geq F(x) + \langle w, y - x \rangle \text{ for all } y \in \mathbb{R}^n \}. \tag{5.1}$$

A vector $w \in \partial F(x)$ is called a subgradient of F in x. It is straightforward to check (see Lemma D.2 in Appendix D, page 321) that F is convex if and only if the set $\partial F(x)$ is non-empty for all $x \in \mathbb{R}^n$. Furthermore, when F is convex and differentiable, $\partial F(x) = \{\nabla F(x)\}$; see again Lemma D.2 in Appendix D.

Examples of subdifferentials

We refer to Lemma D.5, page 323, for the derivation of the following subdifferentials.

1. The subdifferential of the ℓ^1 norm $|x|_1 = \sum_j |x_j|$ is given by

$$\partial|x|_1 = \left\{w \in \mathbb{R}^n : \ w_j = \text{sign}(x_j) \text{ for } x_j \neq 0, \ w_j \in [-1,1] \text{ for } x_j = 0\right\},$$

where $\text{sign}(x) = \mathbf{1}_{x>0} - \mathbf{1}_{x\leq 0}$.
Equivalently, $\partial|x|_1 = \{\phi : \langle \phi, x\rangle = |x|_1 \text{ and } |\phi|_\infty \leq 1\}$.

2. The subdifferential of the ℓ^∞-norm $|x|_\infty = \max_j |x_j|$ is given by $\partial|x|_\infty = \{w \in \mathbb{R}^n : \ |w|_1 \leq 1 \text{ and } \langle w,x\rangle = |x|_\infty\}$. For $x \neq 0$, writing $J_* = \left\{j : \ |x_j| = |x|_\infty\right\}$, a vector w is a subgradient of $|x|_\infty$ if and only if it fulfills

$$w_j = 0 \text{ for } j \notin J_* \text{ and } w_j = \lambda_j \text{sign}(x_j) \text{ for } j \in J_* \text{ where } \lambda_j \geq 0 \text{ and } \sum_{j\in J_*} \lambda_j = 1.$$

5.1.2 Two Useful Properties

We recall two useful properties of convex functions.

1. The subdifferential of a convex function $F : \mathbb{R}^n \to \mathbb{R}$ is monotone:

$$\langle w_x - w_y, x - y\rangle \geq 0, \quad \text{for all } w_x \in \partial F(x) \text{ and } w_y \in \partial F(y).$$

Actually, by definition we have $F(y) \geq F(x) + \langle w_x, y-x\rangle$ and $F(x) \geq F(y) + \langle w_y, x-y\rangle$. Summing these two inequalities gives $\langle w_x - w_y, x-y\rangle \geq 0$.

2. The minimizers of a convex function $F : \mathbb{R}^n \to \mathbb{R}$ are characterized by

$$x_* \in \underset{x\in\mathbb{R}^n}{\text{argmin}}\, F(x) \iff 0 \in \partial F(x_*). \tag{5.2}$$

This immediately follows from the fact that $F(y) \geq F(x_*) + \langle 0, y - x_*\rangle$ for all $y \in \mathbb{R}^n$ in both cases.

5.2 Lasso Estimator

In this section, we explain the main ideas of the chapter in the coordinate-sparse setting described in Chapter 2, Section 2.1. The other settings are investigated in Section 5.3. In particular, we focus on the linear model $Y = \mathbf{X}\beta^* + \varepsilon$, and we will assume for simplicity that the columns of $\mathbf{X}$ have ℓ^2-norm 1.

Let us consider the family of estimators $\left\{\widehat{f}_m : m \in \mathcal{M}\right\}$ introduced in Chapter 2, Section 2.2, page 31, for the coordinate-sparse setting. We remind the reader that the model selection estimator obtained by minimizing (2.9) in the coordinate-sparse setting has some very good statistical properties, but the minimization of (2.9) has a prohibitive computational cost for moderate to large p. Our goal below is to derive from (2.9) a convex criterion that can be minimized efficiently even for large p. For deriving this convex criterion, we will start from a slight variation of the selection Criterion (2.9) in the coordinate-sparse setting with $\pi_m = (1+1/p)^{-p}p^{-|m|}$, namely

$$\widehat{m} \in \underset{m\in\mathcal{M}}{\operatorname{argmin}} \left\{ \|Y - \mathbf{X}\widehat{\beta}_m\|^2 + \lambda|m| \right\}, \quad \text{with } \lambda = \left(1+\sqrt{2\log(p)}\right)^2 \sigma^2,$$

where $\widehat{\beta}_m$ is defined by $\widehat{f}_m = \mathbf{X}\widehat{\beta}_m$. We write $\operatorname{supp}(\beta) = \{j : \beta_j \neq 0\}$ for the support of $\beta \in \mathbb{R}^p$, and we observe that for all $m \subset \{1,\ldots,p\}$, the models S_m for the coordinate-sparse setting (defined in Section 2.2.1, page 32) can be written as $S_m = \{\mathbf{X}\beta : \operatorname{supp}(\beta) = m\}$. The estimator $\widehat{f}_m = \operatorname{Proj}_{S_m} Y$ is then equal to $\mathbf{X}\widehat{\beta}_m$, with $\widehat{\beta}_m \in \operatorname{argmin}_{\beta:\, \operatorname{supp}(\beta)=m} \|Y - \mathbf{X}\beta\|^2$, so we have

$$\widehat{m} \;\in\; \underset{m\in\mathcal{M}}{\operatorname{argmin}} \; \min_{\beta:\, \operatorname{supp}(\beta)=m} \left\{ \|Y - \mathbf{X}\beta\|^2 + \lambda|\beta|_0 \right\},$$

with $|\beta|_0 = \operatorname{card}(\operatorname{supp}(\beta))$. Slicing the minimization of $\beta \to \|Y - \mathbf{X}\beta\|^2 + \lambda|\beta|_0$ according to the β with support in $m \subset \{1,\ldots,p\}$, we obtain the identity

$$\widehat{\beta}_{\widehat{m}} \in \underset{\beta\in\mathbb{R}^p}{\operatorname{argmin}} \left\{ \|Y - \mathbf{X}\beta\|^2 + \lambda|\beta|_0 \right\}. \tag{5.3}$$

The function $\beta \to \|Y - \mathbf{X}\beta\|^2$ is smooth and convex, so it can be handled easily in the minimization of (5.3). The troubles come from $|\beta|_0$, which is non-smooth and non-convex. The main idea for deriving from (5.3) a criterion easily amenable to minimization is

$$\text{to replace } |\beta|_0 = \sum_{j=1}^p \mathbf{1}_{\beta_j\neq 0} \text{ by } |\beta|_1 = \sum_{j=1}^p |\beta_j|, \text{ which is convex.}$$

For $\lambda > 0$, we can then relax the minimization problem (5.3) by considering the convex surrogate

$$\widehat{\beta}_\lambda \in \underset{\beta\in\mathbb{R}^p}{\operatorname{argmin}} \mathcal{L}(\beta), \quad \text{where } \mathcal{L}(\beta) = \|Y - \mathbf{X}\beta\|^2 + \lambda|\beta|_1. \tag{5.4}$$

The estimator $\widehat{\beta}_\lambda$ is called the Lasso estimator. The solution to the minimization problem (5.4) may not be unique, but the resulting estimator $\widehat{f}_\lambda = \mathbf{X}\widehat{\beta}_\lambda$ is always unique. We refer to Exercise 5.5.1 for a proof of this result and for a criterion ensuring the uniqueness of the solution to (5.4). Criterion (5.4) is convex, and we will describe in Section 5.2.4 some efficient procedures for minimizing it. So the Lasso estimator has the nice feature that it can be computed even for large p. Yet, does the Lasso estimator have some good statistical properties? We will see below that the support of this estimator is a subset of $\{1,\ldots,p\}$ for λ large enough, and Theorem 5.1 will provide a risk bound for the resulting estimator $\widehat{f}_\lambda = \mathbf{X}\widehat{\beta}_\lambda$.

5.2.1 Geometric Insights

Let us denote by $B_{\ell^1}(R)$ the ℓ^1-ball of radius R defined by $B_{\ell^1}(R) = \{\beta \in \mathbb{R}^p : |\beta|_1 \leq R\}$. We set $\widehat{R}_\lambda = |\widehat{\beta}_\lambda|_1$, and we notice that $\widehat{R}_\lambda$ decreases when λ increases. By Lagrangian duality, the Lasso estimator $\widehat{\beta}_\lambda$ is solution of

$$\widehat{\beta}_\lambda \in \underset{\beta \in B_{\ell^1}(\widehat{R}_\lambda)}{\operatorname{argmin}} \|Y - \mathbf{X}\beta\|^2. \tag{5.5}$$

In Figure 5.1, the level sets of the function $\beta \to \|Y - \mathbf{X}\beta\|^2$ are plotted (dashed lines), together with the constraint $|\beta|_1 \leq R$ for decreasing values of R in an example with $p = 2$. We remark that for R small enough (which corresponds to λ large enough), some coordinate $[\widehat{\beta}_\lambda]_j$ are set to 0. This illustrates the fact that the Lasso estimator selects variables for λ large enough.

Remark. We point out that the selection of variables for R small enough comes from the non-smoothness of the ℓ^1-ball $B_{\ell^1}(R)$. Actually, replacing the ℓ^1-ball by a ℓ^p-ball $B_{\ell^p}(R)$ with $1 < p < +\infty$ would not lead to variable selection. We refer to Exercise 5.5.7 for the analysis of the estimator (5.4) when $|\beta|_1$ is replaced by $\|\beta\|^2$.

5.2.2 Analytic Insights

Let us better understand the variable selection observed above by analyzing the shape of the solution of (5.4). The subdifferential of the function $\mathscr{L}$ is

$$\partial \mathscr{L}(\beta) = \left\{-2\mathbf{X}^T(Y - \mathbf{X}\beta) + \lambda z : z \in \partial |\beta|_1\right\},$$

so the first-order optimality condition (5.2) ensures the existence of $\widehat{z} \in \partial |\widehat{\beta}_\lambda|_1$ fulfilling $-2\mathbf{X}^T(Y - \mathbf{X}\widehat{\beta}_\lambda) + \lambda \widehat{z} = 0$. According to the description of the subdifferential of the ℓ^1-norm given in Section 5.1, we obtain

$$\mathbf{X}^T\mathbf{X}\widehat{\beta}_\lambda = \mathbf{X}^T Y - \frac{\lambda}{2}\widehat{z}. \tag{5.6}$$

for some $\widehat{z} \in \mathbb{R}^p$, fulfilling $\widehat{z}_j = \operatorname{sign}([\widehat{\beta}_\lambda]_j)$ when $[\widehat{\beta}_\lambda]_j \neq 0$ and $\widehat{z}_j \in [-1, 1]$ when $[\widehat{\beta}_\lambda]_j = 0$. Let us investigate the selection of variables induced by this formula.

Orthonormal setting

We first consider the simple case where the columns of $\mathbf{X}$ are orthonormal[1] and thus $\mathbf{X}^T\mathbf{X} = I_p$. In this case, Equation (5.6) gives

$$[\widehat{\beta}_\lambda]_j + \frac{\lambda}{2}\operatorname{sign}([\widehat{\beta}_\lambda]_j) = \mathbf{X}_j^T Y \quad \text{for } [\widehat{\beta}_\lambda]_j \neq 0,$$

[1] It means that the columns of $\mathbf{X}$ are orthogonal with norm 1. Notice that this enforces $p \leq n$.

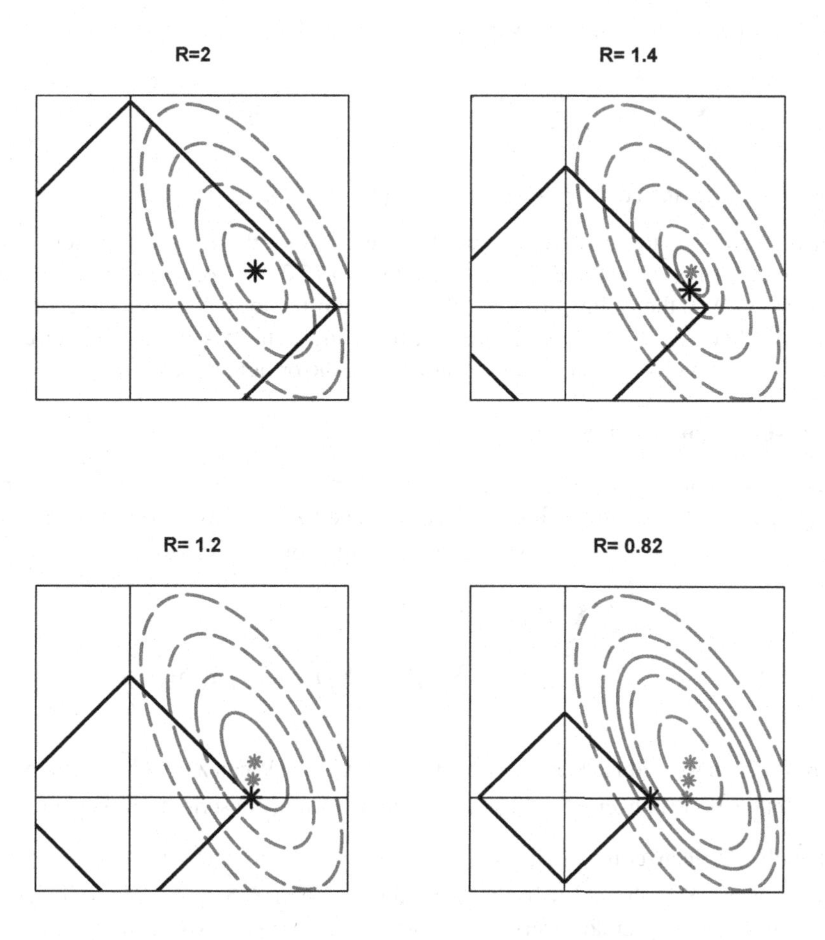

Figure 5.1 *Dashed gray lines represent the level sets of the function* $\beta \to \|Y - \mathbf{X}\beta\|^2$. *Black plain lines represent the* ℓ^1*-balls* $B_{\ell^1}(R)$ *for* $R = 2$, $R = 1.4$, $R = 1.2$ *and* $R = 0.82$. *The dark stars represent* $\widehat{\beta}_R = \text{argmin}_{|\beta|_1 \leq R} \|Y - \mathbf{X}\beta\|^2$ *for the current value of R. When* $R = 2$, *the* ℓ^1*-norm of* $\widehat{\beta}_R$ *is smaller than R, so* $\widehat{\beta}_R$ *coincides with* $\widehat{\beta}^{ols} = \text{argmin}_\beta \|Y - \mathbf{X}\beta\|^2$. *When R is smaller than 1.2, the second coordinate of* $\widehat{\beta}_R$ *is equal to 0.*

which enforces both $\text{sign}([\widehat{\beta}_\lambda]_j) = \text{sign}(\mathbf{X}_j^T Y)$ and $[\widehat{\beta}_\lambda]_j = \mathbf{X}_j^T Y - \lambda\,\text{sign}(\mathbf{X}_j^T Y)/2$ when $[\widehat{\beta}_\lambda]_j \neq 0$. In particular, we notice that we cannot have $[\widehat{\beta}_\lambda]_j \neq 0$ when $|\mathbf{X}_j^T Y| \leq \lambda/2$. Therefore, we have $[\widehat{\beta}_\lambda]_j = 0$ when $|\mathbf{X}_j^T Y| \leq \lambda/2$ and $[\widehat{\beta}_\lambda]_j = \mathbf{X}_j^T Y - \lambda\,\text{sign}(\mathbf{X}_j^T Y)/2$ otherwise. To sum up the above analysis, in the orthonormal setting, the Lasso estimator (5.4) is given by

$$[\widehat{\beta}_\lambda]_j = \mathbf{X}_j^T Y \left(1 - \frac{\lambda}{2|\mathbf{X}_j^T Y|}\right)_+, \quad j = 1, \ldots, p, \quad \text{with } (x)_+ = \max(x, 0). \tag{5.7}$$

It then selects the coordinates j such that $|\langle \mathbf{X}_j, Y \rangle| > \lambda/2$.

It is interesting to compare the variables selected by the Lasso estimator to those selected by the initial model-selection estimator (5.3). According to Exercise 2.8.1, Question A.2, the estimator (5.3) selects in the orthonormal setting the coordinates j such that $|\langle \mathbf{X}_j, Y \rangle| > \sqrt{\lambda}$. Therefore, replacing λ in (5.4) by $2\sqrt{\lambda}$ both estimators (5.3) and (5.4) select the same variables in the orthonormal setting.

Non-orthogonal setting

When the columns of $\mathbf{X}$ are not orthogonal, there is no analytic formula for $\widehat{\beta}_\lambda$, and the Lasso estimator will not select the same variables as (5.3) in general. Write $\widehat{m}_\lambda = \{j : [\widehat{\beta}_\lambda]_j \neq 0\}$ for the support of $\widehat{\beta}_\lambda$. Equation (5.6) gives

$$\begin{aligned} 0 \leq \widehat{\beta}_\lambda^T \mathbf{X}^T \mathbf{X} \widehat{\beta}_\lambda &= \langle \widehat{\beta}_\lambda, \mathbf{X}^T Y - \lambda \widehat{z}/2 \rangle \\ &= \sum_{j \in \widehat{m}_\lambda} [\widehat{\beta}_\lambda]_j \left(\mathbf{X}_j^T Y - \frac{\lambda}{2} \operatorname{sign}([\widehat{\beta}_\lambda]_j) \right), \end{aligned}$$

from which we deduce that $\widehat{\beta}_\lambda = 0$ for $\lambda \geq 2|\mathbf{X}^T Y|_\infty$. When $\lambda < 2|\mathbf{X}^T Y|_\infty$, the Lasso estimator $\widehat{\beta}_\lambda$ is nonzero, but there is no simple formula describing its support.

Finally, we can compare the estimator $\widehat{f}_\lambda = \mathbf{X}\widehat{\beta}_\lambda$ to the estimator $\widehat{f}_{\widehat{m}_\lambda}$, where $\widehat{f}_m = \operatorname{Proj}_{S_m} Y$ with S_m, as in the beginning of Section 5.2. We denote by A^+ the Moore–Penrose pseudo-inverse of a matrix A (see Appendix C.2 for a reminder on this pseudo-inverse). The matrix $(A^T)^+ A^T$ equals the projection on the range of A, so $A = (A^T)^+ A^T A$. Accordingly, for $\lambda < 2|\mathbf{X}^T Y|_\infty$, we derive from (5.6)

$$\begin{aligned} \widehat{f}_\lambda = \mathbf{X}_{\widehat{m}_\lambda} [\widehat{\beta}_\lambda]_{\widehat{m}_\lambda} &= \left(\mathbf{X}_{\widehat{m}_\lambda}^T\right)^+ \left(\mathbf{X}_{\widehat{m}_\lambda}^T Y - \frac{\lambda}{2} \operatorname{sign}([\widehat{\beta}_\lambda]_{\widehat{m}_\lambda})\right) \\ &= \operatorname{Proj}_{\widehat{S}_\lambda} Y - \frac{\lambda}{2} \left(\mathbf{X}_{\widehat{m}_\lambda}^T\right)^+ \operatorname{sign}([\widehat{\beta}_\lambda]_{\widehat{m}_\lambda}), \end{aligned}$$

where $\widehat{S}_\lambda = \operatorname{range}(\mathbf{X}_{\widehat{m}_\lambda}) = \operatorname{span}\{\mathbf{X}_j : j \in \widehat{m}_\lambda\}$ and where $\operatorname{sign}(v)$ represents the vector with coordinates $\operatorname{sign}(v_j)$. We observe in particular that $\widehat{f}_\lambda$ differs from $\widehat{f}_{\widehat{m}_\lambda} = \operatorname{Proj}_{\widehat{S}_\lambda} Y$ by an additional term proportional to λ. As we will discuss in Section 5.2.5, this additional term induces a shrinkage of the estimator $\widehat{f}_{\widehat{m}_\lambda}$ toward 0. The intensity of this shrinkage is proportional to λ.

In the next two sections, we state a risk bound for the Lasso estimator, and we describe two numerical schemes for computing it.

5.2.3 Oracle Risk Bound

We have proved in Chapter 2 that the risk of the model selection estimator (2.9) can be nicely bounded; see Theorem 2.2 and Exercise 2.8.2, part A. We derive in this section a risk bound for the Lasso estimator $\widehat{f}_\lambda = \mathbf{X}\widehat{\beta}_\lambda$, which is similar, at least for some classes of design matrix $\mathbf{X}$.

The best risk bounds available in the literature involve the so-called compatibility constant

$$\kappa(\beta) = \min_{v \in \mathscr{C}(\beta)} \left\{ \frac{\sqrt{|m|}\, \|\mathbf{X}v\|}{|v_m|_1} \right\},$$
$$\text{where } m = \text{supp}(\beta) \text{ and } \mathscr{C}(\beta) = \{ v \in \mathbb{R}^p : 5|v_m|_1 > |v_{m^c}|_1 \}. \quad (5.8)$$

This compatibility constant is a measure of the lack of orthogonality of the columns of $\mathbf{X}_m$; see Exercise 5.5.3. We emphasize that it can be very small for some matrices $\mathbf{X}$. We refer again to Exercise 5.5.3 for a simple lower bound on $\kappa(\beta)$.

A deterministic bound

We first state a deterministic bound and then derive a risk bound from it.

Theorem 5.1 A deterministic bound

For $\lambda \geq 3|\mathbf{X}^T\varepsilon|_\infty$ we have

$$\|\mathbf{X}(\widehat{\beta}_\lambda - \beta^*)\|^2 \leq \inf_{\beta \in \mathbb{R}^p \setminus \{0\}} \left\{ \|\mathbf{X}(\beta - \beta^*)\|^2 + \frac{\lambda^2}{\kappa(\beta)^2} |\beta|_0 \right\}, \quad (5.9)$$

with $\kappa(\beta)$ defined by (5.8).

Proof. The proof mainly relies on the optimality condition (5.2) for (5.4) and some simple (but clever) algebra.

Optimality condition: We have $0 \in \partial\mathscr{L}(\widehat{\beta}_\lambda)$. Since any $\widehat{w} \in \partial\mathscr{L}(\widehat{\beta}_\lambda)$ can be written as $\widehat{w} = -2\mathbf{X}^T(Y - \mathbf{X}\widehat{\beta}_\lambda) + \lambda\widehat{z}$ with $\widehat{z} \in \partial|\widehat{\beta}_\lambda|_1$, using $Y = \mathbf{X}\beta^* + \varepsilon$ we obtain that there exists $\widehat{z} \in \partial|\widehat{\beta}_\lambda|_1$ such that $2\mathbf{X}^T(\mathbf{X}\widehat{\beta}_\lambda - \mathbf{X}\beta^*) - 2\mathbf{X}^T\varepsilon + \lambda\widehat{z} = 0$. In particular, for all $\beta \in \mathbb{R}^p$

$$2\langle \mathbf{X}(\widehat{\beta}_\lambda - \beta^*), \mathbf{X}(\widehat{\beta}_\lambda - \beta) \rangle - 2\langle \mathbf{X}^T\varepsilon, \widehat{\beta}_\lambda - \beta \rangle + \lambda\langle \widehat{z}, \widehat{\beta}_\lambda - \beta \rangle = 0. \quad (5.10)$$

Convexity: Since $|.|_1$ is convex, the subgradient monotonicity ensures that $\langle \widehat{z}, \widehat{\beta}_\lambda - \beta \rangle \geq \langle z, \widehat{\beta}_\lambda - \beta \rangle$ for all $z \in \partial|\beta|_1$. Therefore, Equation (5.10) gives

for all $\beta \in \mathbb{R}^p$ and for all $z \in \partial|\beta|_1$ we have,

$$2\langle \mathbf{X}(\widehat{\beta}_\lambda - \beta^*), \mathbf{X}(\widehat{\beta}_\lambda - \beta) \rangle \leq 2\langle \mathbf{X}^T\varepsilon, \widehat{\beta}_\lambda - \beta \rangle - \lambda\langle z, \widehat{\beta}_\lambda - \beta \rangle. \quad (5.11)$$

The next lemma provides an upper bound on the right-hand side of (5.11).

Lemma 5.2

We set $m = supp(\beta)$. There exists $z \in \partial|\beta|_1$, such that for $\lambda \geq 3|\mathbf{X}^T\varepsilon|_\infty$ we have

1. *the inequality* $2\langle \mathbf{X}^T\varepsilon, \widehat{\beta}_\lambda - \beta\rangle - \lambda\langle z, \widehat{\beta}_\lambda - \beta\rangle \leq 2\lambda|(\widehat{\beta}_\lambda - \beta)_m|_1$,
2. *and* $5|(\widehat{\beta}_\lambda - \beta)_m|_1 > |(\widehat{\beta}_\lambda - \beta)_{m^c}|_1$ *when* $\langle \mathbf{X}(\widehat{\beta}_\lambda - \beta^*), \mathbf{X}(\widehat{\beta}_\lambda - \beta)\rangle > 0$.

Proof of the lemma

1. Since $\partial|z|_1 = \{z \in \mathbb{R}^p : z_j = \text{sign}(\beta_j) \text{ for } j \in m \text{ and } z_j \in [-1,1] \text{ for } j \in m^c\}$, we can choose $z \in \partial|\beta|_1$, such that $z_j = \text{sign}([\widehat{\beta}_\lambda - \beta]_j) = \text{sign}([\widehat{\beta}_\lambda]_j)$ for all $j \in m^c$. Using the duality bound $\langle x, y\rangle \leq |x|_\infty |y|_1$, we have for this choice of z

$$
\begin{aligned}
&2\langle \mathbf{X}^T\varepsilon, \widehat{\beta}_\lambda - \beta\rangle - \lambda\langle z, \widehat{\beta}_\lambda - \beta\rangle \\
&= 2\langle \mathbf{X}^T\varepsilon, \widehat{\beta}_\lambda - \beta\rangle - \lambda\langle z_m, (\widehat{\beta}_\lambda - \beta)_m\rangle - \lambda\langle z_{m^c}, (\widehat{\beta}_\lambda - \beta)_{m^c}\rangle \\
&\leq 2|\mathbf{X}^T\varepsilon|_\infty|\widehat{\beta}_\lambda - \beta|_1 + \lambda|(\widehat{\beta}_\lambda - \beta)_m|_1 - \lambda|(\widehat{\beta}_\lambda - \beta)_{m^c}|_1 \\
&\leq \frac{5\lambda}{3}|(\widehat{\beta}_\lambda - \beta)_m|_1 - \frac{\lambda}{3}|(\widehat{\beta}_\lambda - \beta)_{m^c}|_1 \qquad (5.12) \\
&\leq 2\lambda|(\widehat{\beta}_\lambda - \beta)_m|_1,
\end{aligned}
$$

where we used $3|\mathbf{X}^T\varepsilon|_\infty \leq \lambda$ and $|\widehat{\beta}_\lambda - \beta|_1 = |(\widehat{\beta}_\lambda - \beta)_{m^c}|_1 + |(\widehat{\beta}_\lambda - \beta)_m|_1$ for the Bound (5.12).

2. When $\langle \mathbf{X}(\widehat{\beta}_\lambda - \beta^*), \mathbf{X}(\widehat{\beta}_\lambda - \beta)\rangle > 0$, combining (5.11) with (5.12) gives the inequality $5|(\widehat{\beta}_\lambda - \beta)_m|_1 > |(\widehat{\beta}_\lambda - \beta)_{m^c}|_1$. □

We now conclude the proof of Theorem 5.1. Al-Kashi formula gives

$$
2\langle \mathbf{X}(\widehat{\beta}_\lambda - \beta^*), \mathbf{X}(\widehat{\beta}_\lambda - \beta)\rangle = \|\mathbf{X}(\widehat{\beta}_\lambda - \beta^*)\|^2 + \|\mathbf{X}(\widehat{\beta}_\lambda - \beta)\|^2 - \|\mathbf{X}(\beta - \beta^*)\|^2.
$$

When this quantity is non-positive, we have directly (5.9). When this quantity is positive, we can combine it with (5.11) and apply successively the first part of the above lemma, the second part of the lemma with (5.8), and finally $2ab \leq a^2 + b^2$ to get that for all $\beta \in \mathbb{R}^p$

$$
\begin{aligned}
\|\mathbf{X}(\widehat{\beta}_\lambda - \beta^*)\|^2 + \|\mathbf{X}(\widehat{\beta}_\lambda - \beta)\|^2 &\leq \|\mathbf{X}(\beta - \beta^*)\|^2 + 2\lambda|(\widehat{\beta}_\lambda - \beta)_m|_1 \\
&\leq \|\mathbf{X}(\beta - \beta^*)\|^2 + \frac{2\lambda\sqrt{|\beta|_0}}{\kappa(\beta)}\|\mathbf{X}(\widehat{\beta}_\lambda - \beta)\| \\
&\leq \|\mathbf{X}(\beta - \beta^*)\|^2 + \frac{\lambda^2|\beta|_0}{\kappa(\beta)^2} + \|\mathbf{X}(\widehat{\beta}_\lambda - \beta)\|^2.
\end{aligned}
$$

The proof of Theorem 5.1 is complete. □

If the tuning parameter λ of the Lasso estimator is such that $\lambda \geq 3|\mathbf{X}^T\varepsilon|_\infty$ with high probability, then (5.9) holds true with high probability for this choice of λ. We state in the next corollary such a risk bound in the Gaussian setting (2.3).

Corollary 5.3 Risk bound for the Lasso

Assume that all the columns of $\mathbf{X}$ *have norm 1 and that the noise* $(\varepsilon_i)_{i=1,\dots,n}$ *is i.i.d. with* $\mathcal{N}(0,\sigma^2)$ *distribution.*

Then, for any $L>0$*, the Lasso estimator with tuning parameter*

$$\lambda = 3\sigma\sqrt{2\log(p)+2L}$$

fulfills with probability at least $1-e^{-L}$ *the risk bound*

$$\|\mathbf{X}(\widehat{\beta}_\lambda-\beta^*)\|^2 \leq \inf_{\beta\neq 0}\left\{\|\mathbf{X}(\beta-\beta^*)\|^2+\frac{18\sigma^2(L+\log(p))}{\kappa(\beta)^2}|\beta|_0\right\}, \tag{5.13}$$

with $\kappa(\beta)$ *defined by (5.8).*

Proof. All we need is to prove that $|\mathbf{X}^T\varepsilon|_\infty=\max_{j=1,\dots,p}|\mathbf{X}_j^T\varepsilon|$ is smaller than $\lambda/3=\sigma\sqrt{2\log(p)+2L}$ with probability at least $1-e^{-L}$. Combining the union bound with the fact that each $\mathbf{X}_j^T\varepsilon$ is distributed according to a $\mathcal{N}(0,\sigma^2)$ Gaussian distribution, we obtain

$$\begin{aligned}\mathbb{P}\left(|\mathbf{X}^T\varepsilon|_\infty>\sigma\sqrt{2\log(p)+2L}\right) &\leq \sum_{j=1}^{p}\mathbb{P}\left(|\mathbf{X}_j^T\varepsilon|>\sigma\sqrt{2\log(p)+2L}\right)\\ &\leq p\,\mathbb{P}(\sigma|Z|>\sigma\sqrt{2\log(p)+2L}),\end{aligned}$$

with Z distributed according to a $\mathcal{N}(0,1)$ Gaussian distribution. From Lemma B.4, page 298, in Appendix B, we have $\mathbb{P}(|Z|\geq x)\leq e^{-x^2/2}$ for all $x\geq 0$, so the probability $\mathbb{P}\left(|\mathbf{X}^T\varepsilon|_\infty>\sigma\sqrt{2\log(p)+2L}\right)$ is upper-bounded by e^{-L}, which concludes the proof of (5.13). □

Discussion of Corollary 5.3

We can compare directly the risk Bound (5.13) for the Lasso estimator to the risk Bound (2.12), page 37, for model selection in the coordinate-sparse setting. Actually, from Inequality (2.24) in Exercise 2.8.2, page 49, we know that there exists a constant $C_K>1$ depending only on $K>1$, such that the model selection estimator $\widehat{\beta}$ defined by (2.9) fulfills the inequality

$$\mathbb{E}\left[\|\mathbf{X}(\widehat{\beta}-\beta^*)\|^2\right]\leq C_K\inf_{\beta\neq 0}\left\{\|\mathbf{X}(\beta-\beta^*)\|^2+|\beta|_0\left[1+\log\left(\frac{p}{|\beta|_0}\right)\right]\sigma^2\right\}.$$

Compared to this bound, the risk Bound (5.13) has the nice feature to have a constant one in front of the term $\|\mathbf{X}(\beta-\beta^*)\|^2$, but the complexity term $|\beta|_0\log(p)\sigma^2$ is inflated by a factor $\kappa(\beta)^{-2}$, which can be huge, even infinite, when the columns of $\mathbf{X}$ are far from being orthogonal. The Lasso estimator can actually behave very poorly when $\kappa(\beta^*)$ is small; see, e.g., the second example described in Section 6.3

of Baraud *et al.* [19]. Recent results by Zhang *et al.* [170] suggest that this constant $\kappa(\beta^*)$ is unavoidable in the sense that for some matrices $\mathbf{X}$ the constant $\kappa(\beta^*)^{-2}$ necessarily appears in an upper bound of $\|\mathbf{X}\widehat{\beta} - \mathbf{X}\beta^*\|^2$ for any estimator $\widehat{\beta}$ with polynomial algorithmic complexity (see the original paper for a precise statement).

To sum up the above discussion, compared to the model selection estimator (2.9), the Lasso estimator (5.4) is not universally optimal, but it is good in many cases, and, crucially, it can be computed efficiently even for p large. Compared to the forward–backward algorithm described in Section 2.4 and the Metropolis–Hastings algorithm described in Section 4.4 (stopped after T iterations), we can provide a risk bound for the Lasso estimator in a non-asymptotic setting for any design matrix $\mathbf{X}$. We can also give some conditions that ensure that the support of the Lasso estimator $\widehat{\beta}_\lambda$ is equal to the support of β^*; see Exercise 5.5.2.

5.2.4 Computing the Lasso Estimator

To compute the Lasso estimator (5.4), we need to minimize the function $\beta \to \mathscr{L}(\beta)$, which is convex but non-differentiable. We briefly describe below three numerical schemes for computing $\widehat{\beta}_\lambda$.

Coordinate descent

Repeatedly minimizing $\mathscr{L}(\beta_1, \ldots, \beta_p)$ with respect to each coordinate β_j is a simple and efficient scheme for minimizing (5.4). This algorithm converges to the Lasso estimator thanks to the convexity of $\mathscr{L}$.

We remind the reader that the columns $\mathbf{X}_j$ of $\mathbf{X}$ are assumed to have norm 1. Setting $R_j = \mathbf{X}_j^T(Y - \sum_{k \neq j} \beta_k \mathbf{X}_k)$, the partial derivative of the function $\mathscr{L}$, with respect to the variable β_j, is

$$\partial_j \mathscr{L}(\beta) = -2\mathbf{X}_j^T(Y - \mathbf{X}\beta) + \lambda \frac{\beta_j}{|\beta_j|} = 2\beta_j - 2R_j + \lambda \frac{\beta_j}{|\beta_j|}, \quad \text{for all } \beta_j \neq 0.$$

Since $\mathscr{L}$ is convex, the minimizer of $\beta_j \to \mathscr{L}(\beta_1, \ldots, \beta_{j-1}, \beta_j, \beta_{j+1}, \ldots, \beta_p)$ is the solution in β_j of $\partial_j \mathscr{L}(\beta_1, \ldots, \beta_{j-1}, \beta_j, \beta_{j+1}, \ldots, \beta_p) = 0$ when such a solution exists, and it is $\beta_j = 0$ otherwise. Therefore, the function $\beta_j \to \mathscr{L}(\beta_1, \ldots, \beta_{j-1}, \beta_j, \beta_{j+1}, \ldots, \beta_p)$ is minimum in

$$\beta_j = R_j \left(1 - \frac{\lambda}{2|R_j|}\right)_+ \quad \text{with } R_j = \mathbf{X}_j^T \left(Y - \sum_{k \neq j} \beta_k \mathbf{X}_k\right). \tag{5.14}$$

Repeatedly computing $\beta_1, \ldots, \beta_p, \beta_1, \ldots, \beta_p, \ldots$ according to (5.14) gives the coordinate descent algorithm summarized below.

Coordinate descent algorithm

Initialization: $\beta = \beta_{\text{init}}$ with $\beta_{\text{init}} \in \mathbb{R}^p$ arbitrary.

Repeat, until convergence of β, the loop:
for $j = 1, \ldots, p$

$$\beta_j = R_j \left(1 - \frac{\lambda}{2|R_j|}\right)_+, \quad \text{with } R_j = \mathbf{X}_j^T \left(Y - \sum_{k \neq j} \beta_k \mathbf{X}_k\right).$$

Output: β

When we want to compute $\{\widehat{\beta}_\lambda : \lambda \in \Lambda\}$ for a grid $\Lambda = \{\lambda_1, \ldots, \lambda_T\}$ of values ranked in decreasing order, it is advised to compute first $\widehat{\beta}_{\lambda_1}$ starting from $\beta_{\text{init}} = 0$, then $\widehat{\beta}_{\lambda_2}$ starting from $\beta_{\text{init}} = \widehat{\beta}_{\lambda_1}$, then $\widehat{\beta}_{\lambda_3}$ starting from $\beta_{\text{init}} = \widehat{\beta}_{\lambda_2}$, etc.

The coordinate descent algorithm is implemented in the R package glmnet available at http://cran.r-project.org/web/packages/glmnet/. For illustration, we give below the R code for analyzing the data set diabetes, which records the age, sex, body mass index, average blood pressure, some serum measurements, and a quantitative measure of disease progression for $n = 442$ diabetes patients. The goal is to predict from the other variables the measure of disease progression.

```
data(diabetes, package="lars")
library(glmnet)
attach(diabetes)
fit = glmnet(x,y)
plot(fit)
coef(fit,s=1) # extract coefficients at a single value of lambda
predict(fit,newx=x[1:10,],s=1) # make predictions
```

The instruction plot(fit) produces a plot of the values of the coordinates of the Lasso estimator $\widehat{\beta}_\lambda$ when λ decreases: The abscissa in the plot corresponds to $|\widehat{\beta}_\lambda|_1$ and the line number j corresponds to the set of points $\{(|\widehat{\beta}_\lambda|_1, [\widehat{\beta}_\lambda]_j) : \lambda \geq 0\}$. It is displayed in Figure 5.2. The left-hand side corresponds to $\lambda = +\infty$, the right-hand side corresponds to $\lambda = 0$. We observe that only a few coefficients are nonzero for λ large (left-hand side), and that enlarging λ tends to shrink all the coefficients toward 0. We refer to Chapter 7 for the issue of choosing at best λ.

FISTA algorithm

For $\alpha \in \mathbb{R}^p$, we have noticed in Section 5.2.2, page 94, that the minimization problem

$$\min_{\beta \in \mathbb{R}^p} \left\{ \frac{1}{2} \|\beta - \alpha\|^2 + \lambda |\beta|_1 \right\}$$

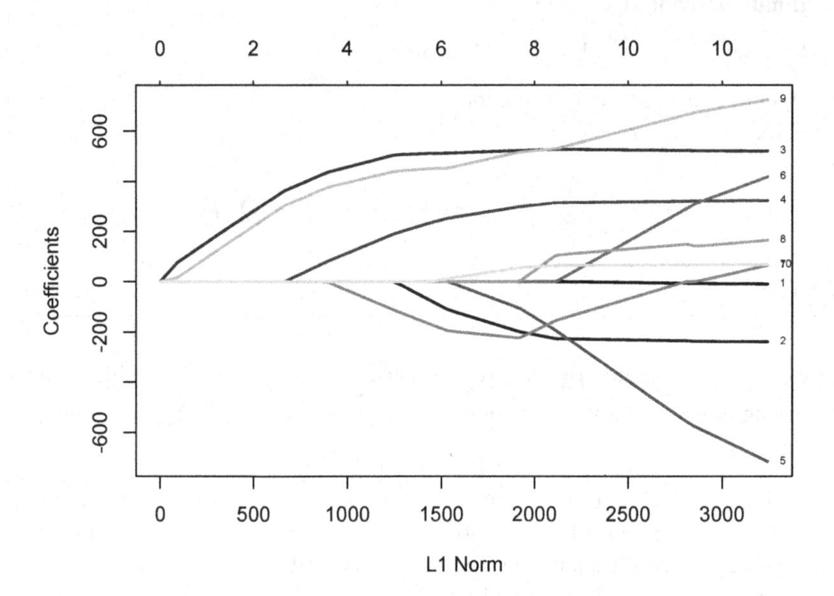

Figure 5.2 *The line j represents the value of the jth coordinate $[\widehat{\beta}_\lambda]_j$ of the Lasso estimator $\widehat{\beta}_\lambda$, when λ decreases from $+\infty$ to 0 on the* `diabetes` *data set.*

has an explicit solution given by

$$S_\lambda(\alpha) = \begin{bmatrix} \alpha_1(1-\lambda/|\alpha_1|)_+ \\ \vdots \\ \alpha_p(1-\lambda/|\alpha_p|)_+ \end{bmatrix}.$$

The Fast Iterative Shrinkage Thresholding Algorithm (FISTA) builds on this formula for computing recursively an approximation of the solution to the minimization problem (5.4). Setting $F(\beta) = \|Y - \mathbf{X}\beta\|^2$, we have for any $b, \beta \in \mathbb{R}^p$

$$\mathcal{L}(b) = F(b) + \lambda|b|_1 = F(\beta) + \langle b - \beta, \nabla F(\beta)\rangle + O(\|b-\beta\|^2) + \lambda|b|_1.$$

For a small $\eta > 0$, starting from $\beta_1 = 0$, we can then iterate until convergence

$$\begin{aligned}\beta_{t+1} &= \underset{b\in\mathbb{R}^p}{\operatorname{argmin}} \left\{ F(\beta_t) + \langle b - \beta_t, \nabla F(\beta_t)\rangle + \frac{1}{2\eta}\|b - \beta_t\|^2 + \lambda|b|_1 \right\} \\ &= S_{\lambda\eta}\left(\beta_t - \eta\nabla F(\beta_t)\right).\end{aligned}$$

When $\lambda = 0$, since $S_0(\alpha) = \alpha$, the above algorithm simply amounts to a minimization of F by gradient descent with step size η. This algorithm can be accelerated by using Nesterov's acceleration trick [126] leading to FISTA algorithm described below.

FISTA algorithm

Initialization: $\beta_1 = \alpha_1 = 0 \in \mathbb{R}^p$, $\mu_1 = 1$, $t = 1$, and $\eta = |2\mathbf{X}^T\mathbf{X}|_{\text{op}}^{-1}$.

Iterate until convergence in $\mathbb{R}^p$ of the sequence $\beta_1, \beta_2, \ldots$

$$\mu_{t+1} = 2^{-1}\left(1+\sqrt{1+4\mu_t^2}\right) \quad \text{and} \quad \delta_t = (1-\mu_t)\mu_{t+1}^{-1}$$
$$\beta_{t+1} = S_{\lambda\eta}\left((I-2\eta\mathbf{X}^T\mathbf{X})\alpha_t + 2\eta\mathbf{X}^T Y\right)$$
$$\alpha_{t+1} = (1-\delta_t)\beta_{t+1} + \delta_t\beta_t$$

increase t of one unit

Output: β_t

FISTA algorithm has been proved to fulfill some very good convergence properties. We refer to Bubeck [39] for a recent review on convex optimization for machine learning, including an analysis of FISTA algorithm.

LARS algorithm

LARS algorithm is an alternative to coordinate descent and FISTA algorithms. We observe in Figure 5.2 that the map $\lambda \to \widehat{\beta}_\lambda$ is piecewise linear. This observation can be easily explained from the first-order optimality condition (5.6), which gives

$$\mathbf{X}_{\widehat{m}_\lambda}^T \mathbf{X}_{\widehat{m}_\lambda} [\widehat{\beta}_\lambda]_{\widehat{m}_\lambda} = \mathbf{X}_{\widehat{m}_\lambda}^T Y - \frac{\lambda}{2}\,\text{sign}([\widehat{\beta}_\lambda]_{\widehat{m}_\lambda}), \quad \text{where } \widehat{m}_\lambda = \text{supp}(\widehat{\beta}_\lambda). \tag{5.15}$$

For the values of λ where $\widehat{m}_\lambda$ remains constant, the above equation enforces that $\widehat{\beta}_\lambda$ depends linearly on λ. The LARS algorithm computes the sequence $\{\widehat{\beta}_{\lambda_1}, \widehat{\beta}_{\lambda_2}, \ldots\}$ of Lasso estimators, with $\lambda_1 > \lambda_2 > \ldots$ corresponding to the breakpoints of the path $\lambda \to \widehat{\beta}_\lambda$. At each breakpoint λ_k two situations may occur. Either one of the coordinate of $\widehat{\beta}_\lambda$ tends to 0 when λ tends to λ_k from above, in which case the support of $\widehat{\beta}_{\lambda_k}$ is obtained from the support of $\widehat{\beta}_\lambda$ with $\lambda > \lambda_k$ by removing this coordinate. Or Equation (5.15) requires to add one coordinate in $\widehat{m}_\lambda$ when λ becomes smaller than λ_k.

The LARS algorithm is implemented in the `R` package `lars` available on the CRAN `http://cran.r-project.org/web/packages/lars/`.
It has been reported to be computationally less efficient than the coordinate descent and FISTA algorithms when p is very large. We give below the `R` code for analyzing the `diabetes` data set with LARS.

```
library(lars)
data(diabetes)
attach(diabetes)
fit = lars(x,y)
plot(fit)
```

5.2.5 Removing the Bias of the Lasso Estimator

Let us come back to our simulated example of Section 2.5, page 43. In Figure 5.3, we plot the Lasso estimator $\widehat{f}_\lambda = \mathbf{X}\widehat{\beta}_\lambda$, with $\lambda = 3\sigma\sqrt{2\log(p)}$ as suggested by Corollary 5.3.

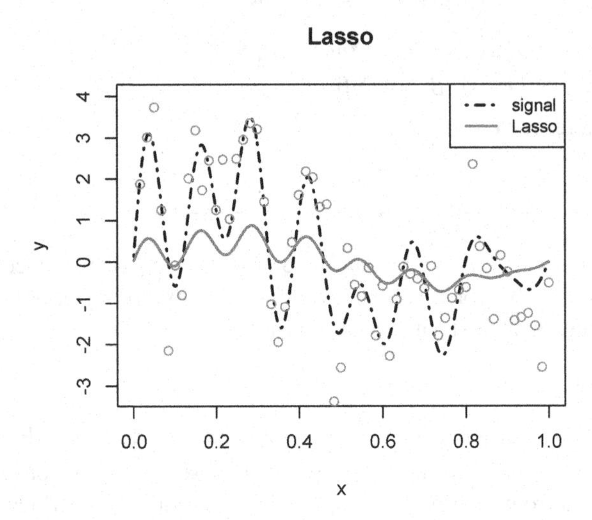

Figure 5.3 *Dotted line: Unknown signal. Gray dots: Noisy observations. Gray line: Lasso estimator with* $\lambda = 3\sigma\sqrt{2\log(p)}$.

We observe that the Lasso estimator reproduces the oscillations of the signal f, but these oscillations are shrunk toward zero. This shrinkage is easily understood when considering the minimization problem (5.4). Actually, the ℓ^1 penalty has the nice feature to favor sparse solution, but it does also favor the β with small ℓ^1 norm, and thus it induces a shrinkage of the signal. This shrinkage can be seen explicitly in the orthonormal setting described in Section 5.2.2: The ℓ^1 penalty selects the j such that $|\mathbf{X}_j^T Y| > \lambda/2$, but it does also shrink the coordinates $\mathbf{X}_j^T Y$ by a factor $\big(1 - \lambda/(2|\mathbf{X}_j^T Y|)\big)_+$; see Equation (5.7).

A common trick to remove this shrinkage is to use as final estimator the so-called Gauss-Lasso estimator

$$\widehat{f}_\lambda^{\text{Gauss}} = \text{Proj}_{\widehat{S}_\lambda} Y, \quad \text{where} \quad \widehat{S}_\lambda = \text{span}\left\{\mathbf{X}_j : j \in \widehat{m}_\lambda\right\}.$$

In other words, with the notations of Chapter 2, the Lasso estimator (5.4) is computed in order to select the model $\widehat{m}_\lambda = \text{supp}(\widehat{\beta}_\lambda)$, and the signal is estimated by $\widehat{f}_{\widehat{m}_\lambda} = \text{Proj}_{S_{\widehat{m}_\lambda}} Y$. The result for our example is displayed in Figure 5.4. We notice that the shrinkage effect is completely removed.

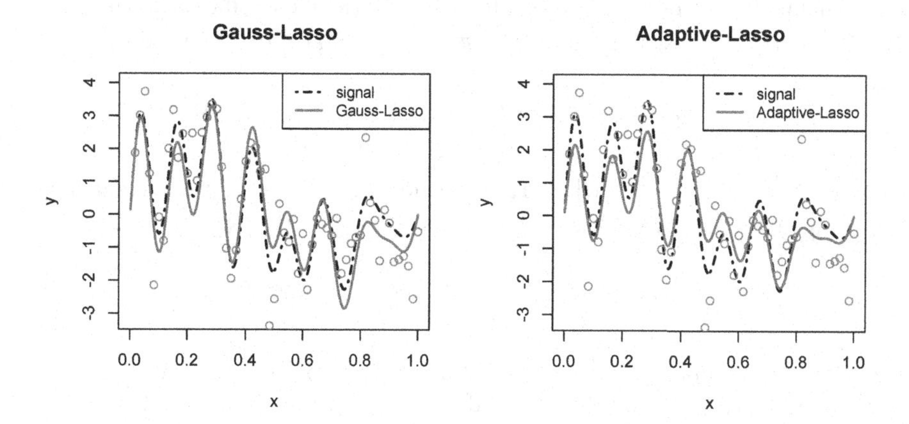

Figure 5.4 *Dotted line: Unknown signal. Gray dots: Noisy observations. Left: Gauss-Lasso estimator (gray line). Right: Adaptive-Lasso estimator (gray line).*

Another trick for reducing the shrinkage is to compute first the Gauss-Lasso estimator $\widehat{f}_\lambda^{\text{Gauss}} = \mathbf{X}\widehat{\beta}_\lambda^{\text{Gauss}}$ and then estimate β^* with the so-called Adaptive-Lasso estimator

$$\widehat{\beta}_{\lambda,\mu}^{\text{adapt}} \in \underset{\beta \in \mathbb{R}^p}{\operatorname{argmin}} \left\{ \|Y - \mathbf{X}\beta\|^2 + \mu \sum_{j=1}^{p} \frac{|\beta_j|}{|(\widehat{\beta}_\lambda^{\text{Gauss}})_j|} \right\}.$$

The above minimization problem remains convex, and it can be solved easily by a coordinate descent algorithm. Let us give a heuristic for considering this estimator. When $\beta \approx \widehat{\beta}_\lambda^{\text{Gauss}}$, we have $\sum_j |\beta_j|/|(\widehat{\beta}_\lambda^{\text{Gauss}})_j| \approx |\beta|_0$, so the above minimization problem can be viewed as an approximation of the initial minimization problem (5.3). This analogy suggests to choose $\mu = (1 + \sqrt{2\log(p)})^2\sigma^2$ as in the initial problem (5.3). The Adaptive-Lasso estimator for this value of μ is displayed in Figure 5.4. In this example, the Gauss-Lasso and the Adaptive-Lasso are very similar. In practice, the Adaptive-Lasso is very popular since it tends to select more accurately the variables than the Gauss-Lasso estimator.

5.3 Convex Criteria for Various Sparsity Patterns

The Lasso estimator provides a computationally efficient estimator for the coordinate-sparse setting. We describe below some variations on the Lasso estimator, suited for the other sparsity patterns described in Section 2.1, Chapter 2.

5.3.1 Group-Lasso for Group Sparsity

We focus in this section on the group-sparse setting described in Section 2.1, Chapter 2. We consider the collection of models described on page 32 for this setting, and we define $\beta_{G_k} = (\beta_j)_{j\in G_k}$. We start from a slight variation of the selection Criterion (2.9) in the group-sparse setting with $\pi_m = (1+1/M)^{-M}M^{-|m|}$, namely

$$\widehat{m} \in \operatorname*{argmin}_m \left\{ \|Y - \widehat{f}_m\|^2 + \lambda|m| \right\}, \quad \text{with } \lambda = \left(1+\sqrt{2\log(M)}\right)^2 \sigma^2.$$

We write $\mathscr{K}(\beta) = \{k : \beta_{G_k} \neq 0\}$, and as in Section 5.2, we observe that the estimator $\widehat{f}_m = \text{Proj}_{S_m} Y$ is equal to $\mathbf{X}\widehat{\beta}_m$, with $\widehat{\beta}_m \in \operatorname{argmin}_{\beta:\ \mathscr{K}(\beta)=m} \|Y - \mathbf{X}\beta\|^2$. Therefore,

$$\widehat{m} \ \in \ \operatorname*{argmin}_m \ \min_{\beta:\ \mathscr{K}(\beta)=m} \left\{\|Y-\mathbf{X}\beta\|^2 + \lambda|\mathscr{K}(\beta)|\right\},$$

and slicing the minimization of $\beta \to \|Y - \mathbf{X}\beta\|^2 + \lambda|\mathscr{K}(\beta)|$ according to the β with $\mathscr{K}(\beta) = m \subset \{1,\ldots,M\}$, we obtain the identity

$$\widehat{\beta}_{\widehat{m}} \in \operatorname*{argmin}_{\beta\in\mathbb{R}^p} \left\{\|Y-\mathbf{X}\beta\|^2 + \lambda|\mathscr{K}(\beta)|\right\}. \tag{5.16}$$

As in Section 5.2, we want to replace the non-convex term $|\mathscr{K}(\beta)| = \sum_k \mathbf{1}_{\beta_{G_k}\neq 0}$ by a convex surrogate. In the coordinate-sparse setting, we have replaced the indicator function $\mathbf{1}_{\beta_j\neq 0}$ by $|\beta_j|$. Following the same idea, we can replace $\mathbf{1}_{\beta_{G_k}\neq 0}$ by $\|\beta_{G_k}\|/\sqrt{|G_k|}$. We divide the norm $\|\beta_{G_k}\|$ by the square-root of the size of the group in order to penalize similarly large and small groups.

For $\lambda = (\lambda_1,\ldots,\lambda_M) \in (\mathbb{R}^+)^M$, the group-Lasso estimator $\widehat{\beta}_\lambda$ is defined as the minimizer of the convex criterion

$$\widehat{\beta}_\lambda \in \operatorname*{argmin}_{\beta\in\mathbb{R}^p} \mathscr{L}(\beta), \quad \text{where } \mathscr{L}(\beta) = \|Y-\mathbf{X}\beta\|^2 + \sum_{k=1}^M \lambda_k \|\beta_{G_k}\|. \tag{5.17}$$

Let us understand why Criterion (5.17) promotes solutions where some groups of coordinates β_{G_k} are zero, leading to group selection.

We first observe that the geometric picture is the same as in Figure 5.1 for the Lasso, except that here the axes will represent the linear subspaces $\mathbb{R}^{G_1}$ and $\mathbb{R}^{G_2}$. When the minimum is achieved at one of the vertices, the group corresponding to the vertex is selected. From an analytical point of view, the subdifferential of $\sum_k \lambda_k \|\beta_{G_k}\|$ is given by

$$\partial \sum_k \lambda_k \|\beta_{G_k}\| = \left\{z \in \mathbb{R}^p : \ z_{G_k} = \lambda_k \beta_{G_k}/\|\beta_{G_k}\| \text{ if } \|\beta_{G_k}\| > 0,\ \|z_{G_k}\| \leq \lambda_k \text{ else}\right\}.$$

Similarly to the Lasso, the rigidity of the subdifferential $\partial\|\beta_{G_k}\|$ for $\|\beta_{G_k}\| > 0$ will enforce some $[\widehat{\beta}_\lambda]_{G_k}$ to be zero for λ_k large enough.

Risk bound

We can state a risk bound for the group-Lasso estimator similar to Corollary 5.3 for the Lasso estimator. For simplicity, we assume that each group G_k has the same cardinality $T = p/M$, and we focus on the case where $\lambda_1 = \ldots = \lambda_M = \lambda$.

Similarly to the coordinate-sparse case, we introduce the group-compatibility constant

$$\kappa_G(\beta) = \min_{v \in \mathscr{C}_G(\mathscr{K}(\beta))} \left\{ \frac{\sqrt{\mathrm{card}(\mathscr{K}(\beta))}\,\|\mathbf{X}v\|}{\sum_{k \in \mathscr{K}(\beta)} \|v_{G_k}\|} \right\}, \quad \text{where } \mathscr{K}(\beta) = \{k : \beta_{G_k} \neq 0\},$$

$$\text{and } \mathscr{C}_G(\mathscr{K}) = \left\{ v : \sum_{k \in \mathscr{K}^c} \|v_{G_k}\| < 5 \sum_{k \in \mathscr{K}} \|v_{G_k}\| \right\}. \quad (5.18)$$

We have the following risk bound for the group-Lasso estimator, which is similar to Corollary 5.3 for the Lasso estimator. We write henceforth $\mathbf{X}_{G_k}$ for the submatrix obtained by keeping only the columns of $\mathbf{X}$ with index in G_k.

Theorem 5.4 Risk bound for the group-Lasso

Assume that all the columns of $\mathbf{X}$ *have norm 1 and that the noise* $(\varepsilon_i)_{i=1,\ldots,n}$ *is i.i.d. with* $\mathcal{N}(0, \sigma^2)$ *distribution. We set*

$$\phi_G = \max_{k=1,\ldots,M} \frac{|\mathbf{X}_{G_k}|_{\mathrm{op}}}{\sqrt{|G_k|}},$$

and we assume that all groups have the same cardinality $T = p/M$.

Then, for any $L > 0$ *the group-Lasso estimator with tuning parameter*

$$\lambda = 3\sigma\sqrt{T}\left(1 + \phi_G\sqrt{2L + 2\log M}\right) \quad (5.19)$$

fulfills with probability at least $1 - e^{-L}$ *the risk bound*

$$\|\mathbf{X}(\widehat{\beta}_\lambda - \beta^*)\|^2 \leq \inf_{\beta \neq 0} \left\{ \|\mathbf{X}(\beta - \beta^*)\|^2 + \frac{18\sigma^2}{\kappa_G(\beta)^2} \mathrm{card}(\mathscr{K}(\beta)) T \left(1 + 2\phi_G^2 (L + \log M)\right) \right\}, \quad (5.20)$$

with $\kappa_G(\beta)$ *and* $\mathscr{K}(\beta)$ *defined in (5.18).*

We refer to Exercise 5.5.4 for a proof of this result.

Let us compare this risk bound to the risk Bound (5.13) for the Lasso estimator. We first observe that when β has a pure group-sparse structure, we have $T\mathrm{card}(\mathscr{K}(\beta)) = |\beta|_0$. Hence the term $|\beta|_0 \log(p)$ has been replaced by $|\beta|_0 \log(M)$, which can be much smaller if the number M of groups is much smaller than p. Inequality (5.20) then gives a tighter bound than (5.13) when the vector β has a group-sparse structure.

Computing the group-Lasso

Similarly to the Lasso, for computing the group-Lasso estimator, we can apply a block coordinate descent algorithm. The principle is simply to alternate minimization over each block of variables β_{G_k}. For $\beta_{G_k} \neq 0$, the gradient with respect to the block G_k of the function $\mathscr{L}$ defined in (5.17) is

$$\nabla_{\beta_{G_k}} \mathscr{L}(\beta) = -2\mathbf{X}_{G_k}^T R_k + 2\mathbf{X}_{G_k}^T \mathbf{X}_{G_k} \beta_{G_k} + \lambda_k \frac{\beta_{G_k}}{\|\beta_{G_k}\|}, \quad \text{with } R_k = Y - \sum_{j \notin G_k} \beta_j \mathbf{X}_j.$$

Since $\beta_{G_k} \to \mathscr{L}(\beta_{G_1}, \ldots, \beta_{G_M})$ is convex and tends to $+\infty$ when $\|\beta_{G_k}\|$ tends to $+\infty$, the minimum of $\beta_{G_k} \to \mathscr{L}(\beta_{G_1}, \ldots, \beta_{G_M})$ is either the solution of

$$\beta_{G_k} = \left(\mathbf{X}_{G_k}^T \mathbf{X}_{G_k} + \frac{\lambda_k}{2\|\beta_{G_k}\|} I\right)^{-1} \mathbf{X}_{G_k}^T R_k$$

if it exists or it is 0. For $\alpha > 0$, let us define $x_\alpha = \left(\mathbf{X}_{G_k}^T \mathbf{X}_{G_k} + \alpha I\right)^{-1} \mathbf{X}_{G_k}^T R_k$. The minimum of $\beta_{G_k} \to \mathscr{L}(\beta_{G_1}, \ldots, \beta_{G_M})$ is nonzero if and only if there exists $\alpha > 0$, such that $\alpha \|x_\alpha\| = \lambda_k/2$, and then $\beta_{G_k} = x_\alpha$. According to Exercise 5.5.4, there exists $\alpha > 0$ fulfilling $\alpha\|x_\alpha\| = \lambda_k/2$ if and only if $\|\mathbf{X}_{G_k}^T R_k\| > \lambda_k/2$. Let us summarize the resulting minimization algorithm.

Block descent algorithm

Initialization: $\beta = \beta_{\text{init}}$ with β_{init} arbitrary.

Iterate until convergence
for $k = 1, \ldots, M$

- $R_k = Y - \sum_{j \notin G_k} \beta_j \mathbf{X}_j$
- if $\|\mathbf{X}_{G_k}^T R_k\| \leq \lambda_k/2$ then $\beta_{G_k} = 0$
- if $\|\mathbf{X}_{G_k}^T R_k\| > \lambda_k/2$, solve $\beta_{G_k} = \left(\mathbf{X}_{G_k}^T \mathbf{X}_{G_k} + \frac{\lambda_k}{2\|\beta_{G_k}\|} I\right)^{-1} \mathbf{X}_{G_k}^T R_k$

Output: β

An implementation of the group-Lasso is available in the `R` package `gglasso` at `http://cran.r-project.org/web/packages/gglasso/`.

5.3.2 Sparse-Group Lasso for Sparse-Group Sparsity

In the case of sparse-group sparsity, as described in Section 2.1 of Chapter 2, the nonzero groups β_{G_k} are coordinate sparse. To obtain such a sparsity pattern, we can add a ℓ^1 penalty to the group-Lasso criterion, leading to the Sparse–Group Lasso

$$\widehat{\beta}_{\lambda,\mu} \in \underset{\beta \in \mathbb{R}^p}{\operatorname{argmin}} \mathscr{L}(\beta), \quad \text{where } \mathscr{L}(\beta) = \|Y - \mathbf{X}\beta\|^2 + \sum_{k=1}^{M} \lambda_k \|\beta_{G_k}\| + \mu|\beta|_1. \tag{5.21}$$

A similar analysis as for the Lasso and group-Lasso shows that this estimator has a sparse-group sparsity pattern. Risk bounds similar to (5.13) and (5.20) can be proved. An implementation of the Sparse–Group Lasso is available in the `R` package `SGL` at `http://cran.r-project.org/web/packages/SGL/`.

5.3.3 Fused-Lasso for Variation Sparsity

In the case of variation sparsity, only a few increments $\beta^*_{j+1} - \beta^*_j$ are nonzero. Therefore, we can penalize the residual sum of squares by the ℓ^1 norm of the increments of β, leading to the so-called fused-Lasso estimator

$$\widehat{\beta}_\lambda \in \operatorname*{argmin}_{\beta\in\mathbb{R}^p} \mathscr{L}(\beta), \quad \text{where } \mathscr{L}(\beta) = \|Y - \mathbf{X}\beta\|^2 + \lambda \sum_{j=1}^{p-1} |\beta_{j+1} - \beta_j|. \tag{5.22}$$

Setting $\Delta_j = \beta_{j+1} - \beta_j$ for $j = 1, \ldots, p-1$ and $\Delta_0 = \beta_1$, we observe that

$$\mathscr{L}(\beta) = \left\| Y - \sum_{j=0}^{p-1} \Big(\sum_{k=j+1}^{p} \mathbf{X}_k \Big) \Delta_j \right\|^2 + \lambda \sum_{j=1}^{p-1} |\Delta_j|.$$

So computing the fused-Lasso estimator essentially amounts to solving a Lasso problem after a change of variables.

5.4 Discussion and References

5.4.1 Take-Home Message

A successful strategy to bypass the prohibitive computational complexity of model selection is to convexify the model selection criteria. The resulting estimators (Lasso, group-Lasso, fused-Lasso, etc.) are not universally optimal, but they are good in many cases (both in theory and in practice). They can be computed in high-dimensional settings, and they are widely used in science. Furthermore, some risk bounds have been derived for these estimators, providing guarantees on their performances.

5.4.2 References

The Lasso estimator has been introduced conjointly by Tibshirani [150] and Chen *et al.* [56]. The variants presented above and in the next exercises have been proposed by Zou and Hastie[171], Tibshirani *et al.* [151], Yuan and Lin [167], Friedman *et al.* [74], Candès and Tao [47], and Zou *et al.* [172]. We refer to Bach *et al.* [14] for more complex sparsity patterns.

The theoretical analysis of the performance of the Lasso estimator has given rise to many papers, including Bickel *et al.* [29], Bunea *et al.* [45], Meinshausen and Bühlmann [120], Van de Geer [154], and Wainwright [162]. The analysis presented in this chapter is an adaptation of the results of Koltchinskii, Lounici, and Tsybakov [99] on trace regression. The analysis of the support recovery presented in Exercise 5.5.2 is adapted from Wainwright [162].

We observe that a sensible choice of the tuning parameter λ of the Lasso estimator (or its variants) depends on the variance σ^2 of the noise, which is usually unknown. We will describe in the next chapter some procedures for selecting among different estimators. In particular, these procedures will allow to select (almost) optimally the tuning parameter λ. We also point out some variants of the Lasso estimator, which do not require the knowledge of the variance for selecting their tuning parameter. The most popular variant is probably the square-root/scaled Lasso estimator [4, 25, 149] described in Section 7.4, Chapter 7; see Giraud *et al.* [82] for a review.

Finally, the numerical aspects presented in Section 5.2.4 are from Efron *et al.* [70], Friedman *et al.* [72], and Beck and Teboulle [24]. We refer to Bubeck [39] for a recent review on convex optimization.

5.5 Exercises

5.5.1 When Is the Lasso Solution Unique?

The solution $\widehat{\beta}_\lambda$ of the minimization problem (5.4) is not always unique. In this exercise, we prove that the fitted value $\widehat{f}_\lambda = \mathbf{X}\widehat{\beta}_\lambda$ is always unique, and we give a criterion that enables us to check whether a solution is unique or not.

1. Let $\widehat{\beta}_\lambda^{(1)}$ and $\widehat{\beta}_\lambda^{(2)}$ be two solutions of (5.4) and set $\widehat{\beta} = \left(\widehat{\beta}_\lambda^{(1)} + \widehat{\beta}_\lambda^{(2)}\right)/2$. From the strong convexity of $x \to \|x\|^2$ prove that if $\mathbf{X}\widehat{\beta}_\lambda^{(1)} \neq \mathbf{X}\widehat{\beta}_\lambda^{(2)}$, then we have

$$\|Y - \mathbf{X}\widehat{\beta}\|^2 + \lambda|\widehat{\beta}|_1 < \frac{1}{2}\left(\|Y - \mathbf{X}\widehat{\beta}^{(1)}\|^2 + \lambda|\widehat{\beta}^{(1)}|_1 + \|Y - \mathbf{X}\widehat{\beta}^{(2)}\|^2 + \lambda|\widehat{\beta}^{(2)}|_1\right).$$

   Conclude that $\mathbf{X}\widehat{\beta}_\lambda^{(1)} = \mathbf{X}\widehat{\beta}_\lambda^{(2)}$, so the fitted value $\widehat{f}_\lambda$ is unique.

2. Let again $\widehat{\beta}_\lambda^{(1)}$ and $\widehat{\beta}_\lambda^{(2)}$ be two solutions of (5.4) with $\lambda > 0$. From the optimality Condition (5.2), there exists $\widehat{z}^{(1)}$ and $\widehat{z}^{(2)}$, such that

$$-2\mathbf{X}^T(Y - \mathbf{X}\widehat{\beta}_\lambda^{(1)}) + \lambda\widehat{z}^{(1)} = 0 \quad \text{and} \quad -2\mathbf{X}^T(Y - \mathbf{X}\widehat{\beta}_\lambda^{(2)}) + \lambda\widehat{z}^{(2)} = 0.$$

   Check that $\widehat{z}^{(1)} = \widehat{z}^{(2)}$. We write henceforth $\widehat{z}$ for this common value.

3. Set $J = \{j : |\widehat{z}_j| = 1\}$. Prove that any solution $\widehat{\beta}_\lambda$ to (5.4) fulfills

$$[\widehat{\beta}_\lambda]_{J^c} = 0 \quad \text{and} \quad \mathbf{X}_J^T\mathbf{X}_J[\widehat{\beta}_\lambda]_J = \mathbf{X}_J^T Y - \frac{\lambda}{2}\widehat{z}_J.$$

4. Conclude that

   "when $\mathbf{X}_J^T\mathbf{X}_J$ is non-singular, the solution to (5.4) is unique."

In practice we can check the uniqueness of the Lasso solution by first computing a solution $\widehat{\beta}_\lambda$, then computing $J = \left\{ j : |\mathbf{X}_j^T(Y - \mathbf{X}\widehat{\beta}_\lambda)| = \lambda/2 \right\}$, and finally checking that $\mathbf{X}_J^T\mathbf{X}_J$ is non-singular.

The uniqueness of the fitted value $\widehat{f}_\lambda = \mathbf{X}\widehat{\beta}_\lambda$ can be easily understood geometrically. Let us consider the constrained version of the Lasso estimator

$$\widehat{\beta}^{(R)} \in \operatorname*{argmin}_{|\beta|_1 \leq R} \|Y - \mathbf{X}\beta\|^2,$$

and let us denote by $\mathbf{X}B_{\ell^1}(R) = \{\mathbf{X}\beta : |\beta|_1 \leq R\}$, the image by $\mathbf{X}$ of the ℓ^1-ball $B_{\ell^1}(R)$ of radius R in $\mathbb{R}^p$. The fitted value $\widehat{f}^{(R)} := \mathbf{X}\widehat{\beta}^{(R)}$ associated to $\widehat{\beta}^{(R)}$ is solution to

$$\widehat{f}^{(R)} \in \operatorname*{argmin}_{f \in \mathbf{X}B_{\ell^1}(R)} \|Y - f\|^2.$$

Hence, the fitted value $\widehat{f}^{(R)}$ is simply the projection of Y onto the closed convex set $\mathbf{X}B_{\ell^1}(R)$. Since this projection is unique, the fitted value $\widehat{f}^{(R)}$ is unique.

This geometrical interpretation is illustrated in Figure 5.5. To understand the illustration of Figure 5.5, it is important to notice that $\mathbf{X}B_{\ell^1}(R)$ is equal to the convex hull of the set $\{RX_1, \ldots, RX_p, -RX_1, \ldots, -RX_p\}$, where $X_1, \ldots, X_p$ are the columns of $\mathbf{X}$. We also emphasize that the plot in Figure 5.5 represents the fitted value $\widehat{f}^{(R)} := \mathbf{X}\widehat{\beta}^{(R)}$ in $\mathbb{R}^n$, with $n = 2$, while Figure 5.1 represents the (constrained) Lasso solution $\widehat{\beta}^{(R)}$ in $\mathbb{R}^p$, with $p = 2$. Hence, the two figures propose two different views of the Lasso estimator in $\mathbb{R}^n$ and $\mathbb{R}^p$.

5.5.2 Support Recovery via the Witness Approach

The goal here is to give some simple conditions for ensuring that the support $\widehat{m}_\lambda$ of the Lasso estimator $\widehat{\beta}_\lambda$ coincides with the support m^* of β^*. It is adapted from Wainwright [162]. The main idea is to compare the solution of Criterion (5.4) with the solution $\widetilde{\beta}_\lambda$ of the same minimization problem restricted to the $\beta \in \mathbb{R}^p$, with support in m^*

$$\widetilde{\beta}_\lambda \in \operatorname*{argmin}_{\beta : \text{supp}(\beta) \subset m^*} \left\{ \|Y - \mathbf{X}\beta\|^2 + \lambda|\beta|_1 \right\}. \tag{5.23}$$

For the sake of simplicity, we assume henceforth that $\mathscr{Q}(\beta) = \|Y - \mathbf{X}\beta\|^2$ is strictly convex ($\text{rank}(\mathbf{X}) = p$), even if the weaker condition $\text{rank}(\mathbf{X}_{m^*}) = |m^*|$ is actually sufficient.

The optimality Condition (5.2) for Problem (5.23) ensures the existence of $\widetilde{z} \in \partial|\widetilde{\beta}_\lambda|_1$, fulfilling $(\nabla\mathscr{Q}(\widetilde{\beta}_\lambda)) + \lambda\widetilde{z})_j = 0$ for all $j \in m^*$. We define $\widehat{z}$ from $\widetilde{z}$ by

$$\widehat{z}_j = \widetilde{z}_j \text{ for } j \in m^* \quad \text{and} \quad \widehat{z}_j = -\frac{1}{\lambda}(\nabla\mathscr{Q}(\widetilde{\beta}_\lambda))_j \text{ for } j \notin m^*.$$

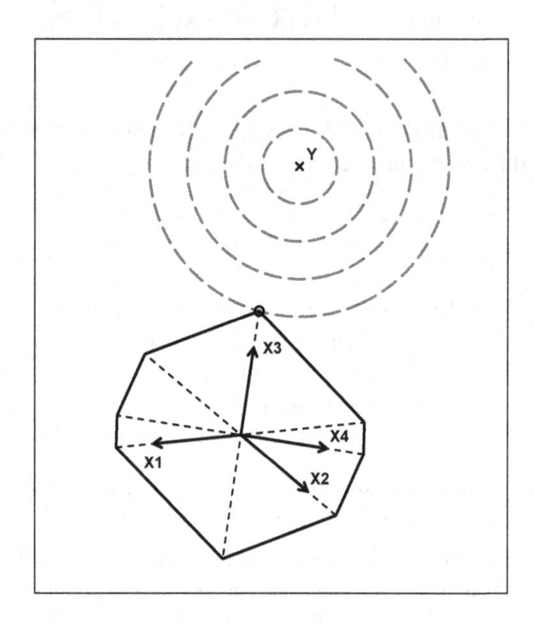

Figure 5.5 *Illustration of the fitted value of the constrained Lasso estimator, for* $n = 2$ *and* $p = 4$. *The dashed gray lines represent the level sets of the function* $f \to \|Y - f\|^2$. *The dark convex polytope represents* $\mathbf{X}B_{\ell^1}(R) = $ *convex hull*$\{RX_1, \ldots, RX_4, -RX_1, \ldots, -RX_4\}$. *The black circle, at the intersection of the polytope with the level sets, represents the fitted value* $\widehat{f}^{(R)}$.

A) The witness approach

1. Check that $\nabla \mathscr{Q}(\widetilde{\beta}_\lambda) + \lambda \widehat{z} = 0$.
2. Check that if $|\widehat{z}_j| \leq 1$ for all $j \notin m^*$, then $\widehat{z} \in \partial |\widetilde{\beta}_\lambda|_1$ and $\widetilde{\beta}_\lambda$ is solution of (5.4). Prove that in this case the support $\widehat{m}_\lambda$ of $\widehat{\beta}_\lambda$ is included in m^*.

B) Checking the dual feasibility condition

We assume henceforth that $\mathbf{X}$ fulfills the incoherence condition

$$\left|\mathbf{X}_{(m^*)^c}^T \mathbf{X}_{m^*} (\mathbf{X}_{m^*}^T \mathbf{X}_{m^*})^{-1}\right|_{\ell^\infty \to \ell^\infty} \leq 1 - \gamma, \quad \text{with } 0 < \gamma < 1, \tag{5.24}$$

where $|A|_{\ell^\infty \to \ell^\infty} = \sup_{|x|_\infty = 1} |Ax|_\infty$ is the operator norm of the matrix A with respect to the ℓ^∞ norm. We also assume that the columns of $\mathbf{X}$ are normalized: $\|\mathbf{X}_j\| = 1$. We set $\lambda = 2\gamma^{-1}\sigma\sqrt{2(1+A)\log(p)}$, with $A > 0$. We prove in this part that when the incoherence condition (5.24) is met, $|\widehat{z}_j| \leq 1$ for all $j \notin m^*$ with large probability.

For a subset $m \subset \{1,\ldots,p\}$, we write P_m for the orthogonal projection on the linear span $\{\mathbf{X}\beta : \beta \in \mathbb{R}^p \text{ and } \mathrm{supp}(\beta) \subset m\}$, with respect to the canonical scalar product in $\mathbb{R}^p$.

1. Check that $(\widetilde{\beta}_\lambda)_{m^*} = (\mathbf{X}_{m^*}^T\mathbf{X}_{m^*})^{-1}(\mathbf{X}_{m^*}^T Y - \lambda \widetilde{z}_{m^*}/2)$.
2. Check that $\widehat{z}_j = \frac{2}{\lambda}\mathbf{X}_j^T(I - P_{m^*})\varepsilon + \mathbf{X}_j^T\mathbf{X}_{m^*}(\mathbf{X}_{m^*}^T\mathbf{X}_{m^*})^{-1}\widetilde{z}_{m^*}$ for all $j \notin m^*$.
3. Prove that $|\mathbf{X}_j^T\mathbf{X}_{m^*}(\mathbf{X}_{m^*}^T\mathbf{X}_{m^*})^{-1}\widetilde{z}_{m^*}| \leq 1-\gamma$, for all $j \notin m^*$.
4. Prove that with probability at least $1 - p^{-A}$, we have $\max_{j \notin m^*} |\mathbf{X}_j^T(I - P_{m^*})\varepsilon| \leq \lambda\gamma/2$, and conclude that $|\widehat{z}_j| \leq 1$ for all $j \notin m^*$ with the same probability.

C) Support recovery

1. Check that $(\widetilde{\beta}_\lambda - \beta^*)_{m^*} = (\mathbf{X}_{m^*}^T\mathbf{X}_{m^*})^{-1}(\mathbf{X}_{m^*}^T\varepsilon - \lambda\widetilde{z}_{m^*}/2)$.
2. Prove that with probability at least $1 - p^{-A}$

$$\max_{j=1,\ldots,p} \left|(\widetilde{\beta}_\lambda - \beta^*)_j\right| \leq \frac{3\lambda}{4}\left|(\mathbf{X}_{m^*}^T\mathbf{X}_{m^*})^{-1}\right|_{\ell^\infty \to \ell^\infty}.$$

3. Assume that $\min_{j \in m^*} |\beta_j^*| > \frac{3\lambda}{4}\left|(\mathbf{X}_{m^*}^T\mathbf{X}_{m^*})^{-1}\right|_{\ell^\infty \to \ell^\infty}$. Under the hypotheses of Part B, prove by combining the above results that the support $\widehat{m}_\lambda$ of the Lasso estimator $\widehat{\beta}_\lambda$ defined by (5.4) coincides with the support m^* of β^*, with probability at least $1 - 2p^{-A}$.

5.5.3 Lower Bound on the Compatibility Constant

We will give a simple lower bound on the compatibility constant $\kappa(\beta)$ defined by (5.8). In the following, m refers to the support of β, and $\mathbf{X}_m$ is the matrix obtained by keeping only the column of $\mathbf{X}$ with index in m.

We assume that the norms of the columns $\mathbf{X}_j$ are normalized to one, and we write

$$\theta = \max_{i \neq j} \left|\langle \mathbf{X}_i, \mathbf{X}_j \rangle\right|$$

for the maximum correlation between the columns in $\mathbf{X}$. We prove below that when $|m|$ fulfills $|m| < (11\theta)^{-1}$, the compatibility constant $\kappa(\beta)$ is positive.

1. Considering apart the coordinates in m and the coordinates in m^c, prove that for any $v \in \mathbb{R}^p$ we have

$$\|\mathbf{X}v\|^2 \geq \|\mathbf{X}_m v_m\|^2 - 2\left|v_m^T\mathbf{X}_m^T\mathbf{X}_{m^c}v_{m^c}\right|.$$

2. Check that $\|\mathbf{X}_m v_m\|^2 \geq \|v_m\|^2 - \theta|v_m|_1^2$.
3. Prove that for $v \in \mathscr{C}(\beta)$, where $\mathscr{C}(\beta)$ is defined in (5.8), we have $\left|v_m^T\mathbf{X}_m^T\mathbf{X}_{m^c}v_{m^c}\right| \leq 5\theta|v_m|_1^2$.
4. Prove that $\kappa(\beta)^2 \geq 1 - 11\theta|m|$ and conclude.

5.5.4 On the Group-Lasso

In parts A and B, we prove Theorem 5.4, page 105. The proof follows the same lines as the proof of Theorem 5.1 for the Lasso estimator. In part C, we check the conditions for solving $\alpha\|(A^TA+\alpha I)^{-1}A^TY\| = \lambda/2$ as needed for the block-gradient algorithm described in Section 5.3.1.

A) Deterministic bound

We first prove that for $\lambda \geq 3\max_{k=1,\ldots,M}\|\mathbf{X}_{G_k}^T\varepsilon\|$, we have

$$\|\mathbf{X}(\widehat{\beta}_\lambda-\beta^*)\|^2 \leq \inf_{\beta\in\mathbb{R}^p\setminus\{0\}}\left\{\|\mathbf{X}(\beta-\beta^*)\|^2+\frac{\lambda^2\mathrm{card}\,(\mathcal{K}(\beta))}{\kappa_G(\beta)^2}\right\}. \tag{5.25}$$

For any $\mathcal{K}\subset\{1,\ldots,M\}$ and $\beta\in\mathbb{R}^p$, we introduce the notation

$$\|\beta\|_{(\mathcal{K})}=\sum_{k\in\mathcal{K}}\|\beta_{G_k}\|.$$

1. Following the same lines as in the proof of Theorem 5.1, prove that

$$\text{for all } \beta\in\mathbb{R}^p \text{ and for all } z\in\partial\sum_{k=1}^{M}\|\beta_{G_k}\|, \text{ we have}$$
$$2\langle\mathbf{X}(\widehat{\beta}_\lambda-\beta^*),\mathbf{X}(\widehat{\beta}_\lambda-\beta)\rangle \;\leq\; 2\langle\mathbf{X}^T\varepsilon,\widehat{\beta}_\lambda-\beta\rangle-\lambda\langle z,\widehat{\beta}_\lambda-\beta\rangle.$$

2. Let us fix some $\beta\in\mathbb{R}^p$ and write $\mathcal{K}=\mathcal{K}(\beta)=\{k:\beta_{G_k}\neq 0\}$. Prove that for a clever choice of $z\in\partial\sum_k\|\beta_{G_k}\|$, we have

$$-\lambda\langle z,\widehat{\beta}_\lambda-\beta\rangle\leq\lambda\|\widehat{\beta}_\lambda-\beta\|_{(\mathcal{K})}-\lambda\|\widehat{\beta}_\lambda-\beta\|_{(\mathcal{K}^c)}.$$

3. Prove that for $\lambda\geq 3\max_{k=1,\ldots,M}\|\mathbf{X}_{G_k}^T\varepsilon\|$, we have

$$2\langle\mathbf{X}(\widehat{\beta}_\lambda-\beta^*),\mathbf{X}(\widehat{\beta}_\lambda-\beta)\rangle\;\leq\;\frac{\lambda}{3}\left(5\|\widehat{\beta}_\lambda-\beta\|_{(\mathcal{K})}-\|\widehat{\beta}_\lambda-\beta\|_{(\mathcal{K}^c)}\right).$$

4. With Al-Kashi's formula and Definition (5.18) of $\kappa_G(\beta)$, prove that for $\beta\neq 0$

$$\|\mathbf{X}(\widehat{\beta}_\lambda-\beta^*)\|^2+\|\mathbf{X}(\widehat{\beta}_\lambda-\beta)\|^2\leq\|\mathbf{X}(\beta^*-\beta)\|^2+\frac{2\lambda\sqrt{\mathrm{card}(\mathcal{K})}\,\|\mathbf{X}(\widehat{\beta}_\lambda-\beta)\|}{\kappa_G(\beta)}$$

and conclude the proof of (5.25).

B) Stochastic control

It remains to prove that with probability at least $1-e^{-L}$, we have

$$\max_{k=1,\ldots,M}\|\mathbf{X}_{G_k}^T\varepsilon\|\leq\sigma\sqrt{T}\left(1+\phi_G\sqrt{2L+2\log M}\right), \tag{5.26}$$

where $\phi_G=\max_{k=1,\ldots,M}|\mathbf{X}_{G_k}|_{\mathrm{op}}/\sqrt{|G_k|}$.

1. Prove that for each $k = 1, \dots, M$, there exists some exponential random variable ξ_k such that
$$\|\mathbf{X}_{G_k}^T \varepsilon\| \leq \|\mathbf{X}_{G_k}\|_F \sigma + |\mathbf{X}_{G_k}|_{\text{op}} \sigma \sqrt{2\xi_k}.$$
2. Check that $\|\mathbf{X}_{G_k}\|_F^2 = T$ when the columns of $\mathbf{X}$ have norm 1, and check that we have (5.26) with probability at least $1 - e^{-L}$.
3. Conclude the proof of (5.20).

C) On the block descent algorithm

Let λ be a positive real number. The block descent algorithm described in Section 5.3.1, page 103, requires to solve in $\alpha > 0$

$$\alpha \|x_\alpha\| = \lambda/2, \quad \text{where} \quad x_\alpha = (A^T A + \alpha I)^{-1} A^T Y, \tag{5.27}$$

when such a solution exists. In this subsection, we prove that there exists a solution $\alpha > 0$ to the problem (5.27) if and only if $\|A^T Y\| > \lambda/2$.

1. Let $A = \sum_{k=1}^r \sigma_k u_k v_k^T$ be a singular value decomposition of A (see Appendix C for a reminder on the singular value decomposition). Prove that $A^T Y \in \text{span}\{v_1, \dots, v_r\}$.
2. Check that
$$\alpha x_\alpha = \sum_{k=1}^r \frac{\alpha}{\sigma_k^2 + \alpha} \langle A^T Y, v_k \rangle v_k.$$
3. Check that the map
$$\alpha \to \alpha^2 \|x_\alpha\|^2 = \sum_{k=1}^r \frac{\alpha^2}{(\sigma_k^2 + \alpha)^2} \langle A^T Y, v_k \rangle^2$$
is non-decreasing from 0 to $\|A^T y\|^2 = \sum_{k=1}^r \langle A^T Y, v_k \rangle^2$ when α goes from 0 to $+\infty$.
4. Conclude that there exists a solution $\alpha > 0$ to the problem (5.27) if and only if $\|A^T Y\| > \lambda/2$.

5.5.5 Dantzig Selector

The Dantzig selector is a variant of the Lasso that has been proposed by Candès and Tao [47]. It is obtained by solving the minimization problem

$$\widehat{\beta}_\lambda^D \in \operatorname*{argmin}_{\beta : |\mathbf{X}^T (Y - \mathbf{X}\beta)|_\infty \leq \lambda/2} |\beta|_1. \tag{5.28}$$

The Dantzig selector is implemented in the `R` package `flare` available at `http://cran.r-project.org/web/packages/flare/`.

1. Check that the minimization Problem (5.28) is a convex problem that can be recast in a linear program

$$\widehat{\beta}^D_\lambda \in \underset{\beta:-\lambda\mathbf{1}\leq 2\mathbf{X}^T(Y-\mathbf{X}\beta)\leq\lambda\mathbf{1}}{\operatorname{argmin}} \quad \min_{u\in\mathbb{R}^p_+:-u\leq\beta\leq u} \mathbf{1}^T u$$

where $\mathbf{1}$ is a p-dimensional vector with all components equal to 1.

In the following, we will prove that $\widehat{\beta}^D_\lambda$ coincides with the Lasso estimator $\widehat{\beta}_\lambda$ defined by (5.4) when the matrix $(\mathbf{X}^T\mathbf{X})^{-1}$ is diagonal dominant

$$\sum_{k:k\neq j} |(\mathbf{X}^T\mathbf{X})^{-1}_{kj}| < (\mathbf{X}^T\mathbf{X})^{-1}_{jj}, \quad \text{for all } j=1,\ldots,p.$$

It is adapted from Meinshausen *et al.* [121].

2. Check that the estimator $\widehat{\beta}^D_\lambda$ is a solution of (5.4) if it fulfills
   (a) $2|\mathbf{X}^T(Y-\mathbf{X}\widehat{\beta}^D_\lambda)|_\infty \leq \lambda$
   (b) $2\mathbf{X}^T_j(Y-\mathbf{X}\widehat{\beta}^D_\lambda) = \lambda\,\text{sign}([\widehat{\beta}^D_\lambda]_j)$ for all $j \in \text{supp}(\widehat{\beta}^D_\lambda)$.
3. Assume that the Condition (b) is not satisfied for some $j \in \text{supp}(\widehat{\beta}^D_\lambda)$. Write e_j for the jth vector of the canonical basis in $\mathbb{R}^p$. Prove that the vector $\widetilde{\beta}_\lambda = \widehat{\beta}^D_\lambda - \eta\,\text{sign}([\widehat{\beta}^D_\lambda]_j)(\mathbf{X}^T\mathbf{X})^{-1}e_j$ fulfills Condition (a) for $\eta > 0$ small enough.
4. Prove the inequalities for $\eta > 0$ small enough

$$\begin{aligned} |\widetilde{\beta}_\lambda|_1 &= |[\widetilde{\beta}_\lambda]_{-j}|_1 + |[\widehat{\beta}^D_\lambda]_j| - \eta(\mathbf{X}^T\mathbf{X})^{-1}_{jj} \\ &\leq |\widehat{\beta}^D_\lambda|_1 + \eta\sum_{k:k\neq j}|(\mathbf{X}^T\mathbf{X})^{-1}_{kj}| - \eta(\mathbf{X}^T\mathbf{X})^{-1}_{jj}. \end{aligned}$$

5. Conclude that $\widehat{\beta}^D_\lambda$ must fulfill Condition (b) and finally that $\widehat{\beta}^D_\lambda = \widehat{\beta}_\lambda$.

We emphasize that the Dantzig selector and the Lasso do not coincide in general.

5.5.6 Projection on the ℓ^1-Ball

We present in this exercise a simple algorithm for computing the projection of a vector $\beta \in \mathbb{R}^p$ on the ℓ^1-ball of radius $R > 0$. When $|\beta|_1 \leq R$, the projection is simply β. We assume in the following that $|\beta|_1 > R$. We write $\beta_{(j)}$ for the jth-largest coordinate of β in absolute value, and for $\lambda \geq 0$ we set

$$S_\lambda(\beta) = \left[\beta_j\left(1-\frac{\lambda}{|\beta_j|}\right)_+\right]_{j=1,\ldots,p}.$$

1. Check that $S_\lambda(\beta) \in \text{argmin}_{\alpha\in\mathbb{R}^p}\left\{\|\beta-\alpha\|^2 + 2\lambda|\alpha|_1\right\}$.
2. Prove that the projection of β on the ℓ^1-ball of radius $R > 0$ is given by $S_{\widehat{\lambda}}(\beta)$, where $\widehat{\lambda} > 0$ is such that $|S_{\widehat{\lambda}}(\beta)|_1 = R$.

3. Let $\widehat{J} \in \{1,\ldots,p\}$ be such that $|\beta_{(\widehat{J}+1)}| \leq \widehat{\lambda} < |\beta_{(\widehat{J})}|$, with the convention $\beta_{(p+1)} = 0$. Check that

$$|S_{\widehat{\lambda}}(\beta)|_1 = \sum_{j \leq \widehat{J}} |\beta_{(j)}| - \widehat{J}\widehat{\lambda}.$$

4. Prove that $\widehat{J}$ then fulfills the two conditions

$$\sum_{j \leq \widehat{J}} |\beta_{(j)}| - \widehat{J}|\beta_{(\widehat{J})}| < R \quad \text{and} \quad \sum_{j \leq \widehat{J}+1} |\beta_{(j)}| - (\widehat{J}+1)|\beta_{(\widehat{J}+1)}| \geq R.$$

5. Conclude that the projection of β on the ℓ^1-ball of radius $R > 0$ is given by $S_{\widehat{\lambda}}(\beta)$, where $\widehat{\lambda}$ is given by

$$\widehat{\lambda} = \widehat{J}^{-1} \left(\sum_{j \leq \widehat{J}} |\beta_{(j)}| - R \right) \quad \text{with} \quad \widehat{J} = \max \left\{ J : \sum_{j \leq J} |\beta_{(j)}| - J|\beta_{(J)}| < R \right\}.$$

5.5.7 Ridge and Elastic-Net

We consider the linear model $Y = \mathbf{X}\beta + \varepsilon$, with $Y, \varepsilon \in \mathbb{R}^n$ et $\beta \in \mathbb{R}^p$. We assume that $\mathbf{E}[\varepsilon] = 0$ and $\mathbf{Cov}(\varepsilon) = \sigma^2 I_n$.

A) Ridge Regression

For $\lambda > 0$, the Ridge estimator $\widehat{\beta}_\lambda$ is defined by

$$\widehat{\beta}_\lambda \in \underset{\beta \in \mathbb{R}^p}{\operatorname{argmin}} \mathscr{L}_1(\beta) \quad \text{with} \quad \mathscr{L}_1(\beta) = \|Y - \mathbf{X}\beta\|^2 + \lambda \|\beta\|^2. \tag{5.29}$$

1. Check that $\mathscr{L}_1$ is strictly convex and has a unique minimum.
2. Prove that $\widehat{\beta}_\lambda = A_\lambda Y$ with $A_\lambda = (\mathbf{X}^T\mathbf{X} + \lambda I_p)^{-1}\mathbf{X}^T$.
3. Let $\sum_{k=1}^r \sigma_k u_k v_k^T$ be a singular value decomposition of $\mathbf{X}$ (see Theorem (C.1), page 311, in Appendix C). Prove that

$$A_\lambda = \sum_{k=1}^r \frac{\sigma_k}{\sigma_k^2 + \lambda} v_k u_k^T \overset{\lambda \to 0+}{\to} \mathbf{X}^+$$

   where A^+ is the Moore–Penrose pseudo-inverse of A (see Appendix C for a reminder on the Moore–Penrose pseudo-inverse).
4. Check that we have

$$\mathbf{X}\widehat{\beta}_\lambda = \sum_{k=1}^r \frac{\sigma_k^2}{\sigma_k^2 + \lambda} \langle u_k, Y \rangle \, u_k. \tag{5.30}$$

5. Let us denote by $P = \sum_{j=1}^r v_j v_j^T$ the projection on the range of $\mathbf{X}^T$. Check that we have

$$\mathbb{E}\left[\widehat{\beta}_\lambda\right] = \sum_{k=1}^r \frac{\sigma_k^2}{\sigma_k^2 + \lambda} \langle v_k, \beta \rangle v_k,$$

and

$$\left\|\beta - \mathbb{E}\left[\widehat{\beta}_\lambda\right]\right\|^2 = \|\beta - P\beta\|^2 + \sum_{k=1}^{r} \left(\frac{\lambda}{\lambda+\sigma_k^2}\right)^2 \langle v_k, \beta\rangle^2.$$

6. Check that the variance of the Ridge estimator is given by

$$\operatorname{var}\left(\widehat{\beta}_\lambda\right) = \sigma^2 \operatorname{Tr}(A_\lambda^T A_\lambda) = \sigma^2 \sum_{k=1}^{r} \left(\frac{\sigma_k}{\sigma_k^2+\lambda}\right)^2.$$

7. How does the size of the bias and the variance of $\widehat{\beta}_\lambda$ vary when λ increases?

Remark. We notice from (5.30) that the Ridge estimator shrinks Y in the directions u_k where $\sigma_k \ll \lambda$, whereas it leaves Y almost unchanged in the directions u_k where $\sigma_k \gg \lambda$.

B) Elastic-Net

From Question A.2 we observe that the Ridge regression does not select variables. Actually, the difference between the Lasso estimator and the Ridge estimator is that the ℓ^1 penalty is replaced by a ℓ^2 penalty. We have seen in Section 5.2.1 that the selection property of the Lasso estimator is induced by the non-smoothness of the ℓ^1 ball. Since the ℓ^2 ball is smooth, it is not surprising that the Ridge estimator does not select variables.

The Elastic-Net estimator involves both a ℓ^2 and a ℓ^1 penalty. It is meant to improve the Lasso estimator when the columns of $\mathbf{X}$ are strongly correlated. It is defined for $\lambda, \mu \geq 0$ by

$$\widetilde{\beta}_{\lambda,\mu} \in \underset{\beta\in\mathbb{R}^p}{\operatorname{argmin}}\, \mathscr{L}_2(\beta) \quad \text{with} \quad \mathscr{L}_2(\beta) = \|Y - \mathbf{X}\beta\|^2 + \lambda\|\beta\|^2 + \mu|\beta|_{\ell^1}.$$

In the following, we assume that the columns of $\mathbf{X}$ have norm 1.

1. Check that the partial derivative of $\mathscr{L}_2$ with respect to $\beta_j \neq 0$ is given by

$$\partial_j \mathscr{L}_2(\beta) = 2\left((1+\lambda)\beta_j - R_j + \frac{\mu}{2}\operatorname{sign}(\beta_j)\right) \quad \text{with} \quad R_j = \mathbf{X}_j^T\Big(Y - \sum_{k\neq j} \beta_k \mathbf{X}_k\Big).$$

2. Prove that the minimum of $\beta_j \to \mathscr{L}_2(\beta_1, \ldots, \beta_j, \ldots, \beta_p)$ is reached at

$$\beta_j = \frac{R_j}{1+\lambda}\left(1 - \frac{\mu}{2|R_j|}\right)_+.$$

3. What is the difference between the coordinate descent algorithm for the Elastic-Net and the one for the Lasso estimator?

The Elastic-Net procedure is implemented in the `R` package `glmnet` available at `http://cran.r-project.org/web/packages/glmnet/`.

5.5.8 Approximately Sparse Linear Regression

This exercise is adapted from Cevid *et al.* [52]. We observe $Y \in \mathbb{R}^n$ and $\mathbf{X} \in \mathbb{R}^{n\times p}$. We assume that

$$Y = \mathbf{X}(\beta^* + b^*) + \varepsilon, \tag{5.31}$$

with $\beta^* \in \mathbb{R}^p$ coordinate sparse and $b^* \in \mathbb{R}^p$ with small ℓ^2 norm. The noise term ε is a random variable following a subgaussian($\sigma^2 I$) distribution, which means that for any $u \in \mathbb{R}^n$ we have $\mathbb{P}[\langle u, \varepsilon\rangle \geq \sigma t\|u\|] \leq e^{-t^2/2}$. We assume that the p columns $X_1, \ldots, X_p$ of $\mathbf{X}$ are deterministic and have norm $\sqrt{n}$.

A) A Convex Estimator

For $\lambda_1, \lambda_2 > 0$, we consider the estimator of (β^*, b^*)

$$(\widehat{\beta}, \widehat{b}) \in \underset{\beta, b \in \mathbb{R}^p}{\operatorname{argmin}} \left\{ \frac{1}{n}\|Y - \mathbf{X}(\beta + b)\|^2 + \lambda_1|\beta|_1 + \lambda_2\|b\|^2 \right\}. \tag{5.32}$$

We emphasize that this estimator does not correspond to the Elastic-Net, as the ℓ^1 penalization is on β and the ℓ^2 penalization is on b.

1. Why does this estimator make sense in this context?
2. Prove that there exists a matrix G, to be made explicit, such that

$$\widehat{\beta} \in \underset{\beta \in \mathbb{R}^p}{\operatorname{argmin}} \left\{ \frac{1}{n}\|(I - \mathbf{X}G)^{1/2}(Y - \mathbf{X}\beta)\|^2 + \lambda_1|\beta|_1 \right\} \quad \text{and} \quad \widehat{b} = G(Y - \mathbf{X}\widehat{\beta}).$$

3. Assume that $n \leq p$ and $\operatorname{rank}(\mathbf{X}) = n$. Let $\mathbf{X} = \sum_{k=1}^n \sigma_k u_k v_k^T$ be a singular value decomposition of $\mathbf{X}$ (Theorem C.1, page 311). Compute the eigenvalues of $F := (I - \mathbf{X}G)^{1/2}$ in terms of the singular values of $\mathbf{X}$.

B) Linearly Transformed Lasso

Let $F \in \mathbb{R}^n$ be *any* symmetric matrix and set $\widetilde{\mathbf{X}} = F\mathbf{X}$, $\widetilde{Y} = FY$ and $\widetilde{\varepsilon} = F\varepsilon$. We analyze the estimator

$$\widehat{\beta} \in \underset{\beta \in \mathbb{R}^p}{\operatorname{argmin}} \left\{ \frac{1}{n}\|\widetilde{Y} - \widetilde{\mathbf{X}}\beta\|^2 + \lambda|\beta|_1 \right\},$$

in the model (5.31), for the choice $\lambda = A\sigma\sqrt{\frac{\log(p)}{n}}\,\lambda_{\max}(F^2)$, with $A > \sqrt{32}$ and $\lambda_{\max}(F^2)$ the largest eigenvalue of F^2. Our goal is to prove that we have

$$\mathbb{P}\left[|\widehat{\beta} - \beta^*|_1 \leq \frac{9\lambda|\beta|_0}{\phi(\widetilde{\mathbf{X}}, S)^2} + \frac{6}{n\lambda}\|\widetilde{\mathbf{X}}b^*\|^2\right] \geq 1 - 2p^{1-A^2/32}, \tag{5.33}$$

where $S = \left\{j : \beta_j^* \neq 0\right\}$,

$$\phi(\widetilde{\mathbf{X}}, S) = \min_{x \in R(S)} \frac{\sqrt{|S|}\,\|\widetilde{\mathbf{X}}x\|}{\sqrt{n}|x_S|_1}, \quad \text{with } R(S) = \{x \in \mathbb{R}^p : |x_{S^c}|_1 \leq 5|x_S|_1\}.$$

The proof technique is somewhat different from the one of Theorem 5.1 (page 95).

1. Prove that

$$\frac{1}{n}\|\widetilde{\mathbf{X}}(\widehat{\beta}-\beta^*-b^*)\|^2+\lambda|\widehat{\beta}|_1 \leq \frac{1}{n}\|\widetilde{\mathbf{X}}b^*\|^2+\frac{2}{n}\langle\widetilde{\varepsilon},\widetilde{\mathbf{X}}(\widehat{\beta}-\beta^*)\rangle+\lambda|\beta^*|_1.$$

2. In Questions 2 to 4, we assume that the event $\Omega_\lambda = \left\{\frac{1}{n}|\widetilde{\mathbf{X}}^T\widetilde{\varepsilon}|_\infty \leq \lambda/4\right\}$ holds. Prove that on Ω_λ

$$\frac{1}{n}\|\widetilde{\mathbf{X}}(\widehat{\beta}-\beta^*-b^*)\|^2+\frac{\lambda}{2}|\widehat{\beta}_{S^c}|_1 \leq \frac{3\lambda}{2}|\widehat{\beta}_S-\beta^*_S|_1+\frac{1}{n}\|\widetilde{\mathbf{X}}b^*\|^2.$$

3. In the case where $\|\widetilde{\mathbf{X}}b^*\|^2 \geq n\lambda|\widehat{\beta}_S-\beta^*_S|_1$, prove that on Ω_λ

$$|\widehat{\beta}-\beta^*|_1 \leq \frac{6}{n\lambda}\|\widetilde{\mathbf{X}}b^*\|^2.$$

4. In the case where $\|\widetilde{\mathbf{X}}b^*\|^2 \leq n\lambda|\widehat{\beta}_S-\beta^*_S|_1$, prove that on Ω_λ

$$\frac{1}{n}\|\widetilde{\mathbf{X}}(\widehat{\beta}-\beta^*-b^*)\|^2+\frac{\lambda}{2}|\widehat{\beta}-\beta^*|_1 \leq 3\frac{\lambda\sqrt{|\beta|_0}\|\widetilde{\mathbf{X}}(\widehat{\beta}-\beta^*)\|}{\sqrt{n}\phi(\widetilde{\mathbf{X}},S)}.$$

5. Conclude the proof of (5.33).

In light of (5.33), we observe that a good transformation F is a transformation such that $|\mathbf{X}^TF^T\varepsilon|_\infty$ and $\|F\mathbf{X}b\|^2$ are not too large and $\phi(F\mathbf{X},S)$ is not too small. Shrinking the largest singular values of $\mathbf{X}$ can help for getting such results. For example, the estimator (5.32) produces the following adjustment

$$\widetilde{X}=F\mathbf{X}=\sum_{k=1}^n\sqrt{\frac{n\lambda_2}{n\lambda_2+\sigma_k^2}}\,\sigma_k u_k v_k^T,$$

since $F=\sum_{k=1}^n\sqrt{\frac{n\lambda_2}{n\lambda_2+\sigma_k^2}}\,u_k u_k^T$. Another possible choice is to cap the largest singular values of $\mathbf{X}$ at some level $\tau>0$, with

$$F=\sum_{k=1}^n\frac{\sigma_k\wedge\tau}{\sigma_k}u_k u_k^T,$$

so that $F\mathbf{X}=\sum_{k=1}^n(\sigma_k\wedge\tau)u_k v_k^T$. We refer to [52] for a detailed discussion of the choice of F.

5.5.9 Slope Estimator

We recommend reading Chapter 10 before starting this exercise.

We consider again the linear regression model $Y=\mathbf{X}\beta^*+\varepsilon$ where ε has a Gaussian $\mathcal{N}(0,I_n\sigma^2)$ distribution and $\beta^*\in\mathbb{R}^p$ is coordinate-sparse. The Lasso estimator does not provide guarantees on the FDR of the selected variables. In order to get such a control, Bogdan *et al.* [36] propose to replace the ℓ^1 norm by a sorted ℓ^1-norm with adaptive weights. In this exercise, we present some basic properties of the resulting estimator.

For any $\beta \in \mathbb{R}^p$, we write $|\beta|_{[1]} \geq |\beta|_{[2]} \geq \ldots \geq |\beta|_{[p]} \geq 0$ for the values $|\beta_1|, \ldots, |\beta_p|$ ranked in non-increasing order. For $\lambda_1 \geq \ldots \geq \lambda_p > 0$, the Slope estimator (introduced by [36]) is defined by

$$\widehat{\beta} \in \underset{\beta \in \mathbb{R}^p}{\operatorname{argmin}} \mathscr{L}(\beta) \quad \text{where } \mathscr{L}(\beta) = \frac{1}{2}\|Y - \mathbf{X}\beta\|^2 + \sum_{j=1}^{p} \lambda_j |\beta|_{[j]}. \tag{5.34}$$

A) Benjamini-Hochberg Procedure

We assume here that $\mathbf{X}^T\mathbf{X} = I_p$ and $\sigma^2 = 1$. We set $z = \mathbf{X}^T Y$ and $\tau(x) = \mathbb{P}(|N| > x)$ where N follows a Gaussian $\mathcal{N}(0,1)$ distribution.

1. What is the distribution of z?
2. We consider the family of hypotheses testing $\mathscr{H}_{0,j} : \beta_j = 0$ against $\mathscr{H}_{1,j} : \beta_j \neq 0$ and the associated tests $\widehat{\mathscr{T}_j} = \mathbf{1}_{\widehat{S}_j \geq s_j}$ with $\widehat{S}_j = |z_j|$ and where $s_j \geq 0$ is a given threshold. Compute in terms of z_j the p-value $\widehat{p}_j$ associated to the test $\widehat{\mathscr{T}_j}$.
3. Prove that the Benjamini-Hochberg procedure (Corollary 10.5, page 213) associated to a level $\alpha > 0$, amounts to reject $\mathscr{H}_{0,j}$ when $|z_j| \geq |z|_{[\widehat{j}_{HB}]}$ with $\widehat{j}_{HB}$ a function of $|z|_{[1]}, \ldots, |z|_{[p]}$ which will be made explicit.

B) Link Between the two Procedures

We assume again that $\mathbf{X}^T\mathbf{X} = I_p$ and $\sigma^2 = 1$. In this part, we assume also that

$$z_1 > \ldots > z_p \geq 0.$$

In the following, we set $\lambda_j = \tau^{-1}(\alpha j/p)$ with τ defined in part (A).

1. Check that $\mathscr{L}(\beta) = \frac{1}{2}\|Y - \mathbf{X}z\|^2 + \frac{1}{2}\|z - \beta\|^2 + \sum_{j=1}^{p} \lambda_j |\beta|_{[j]}$.
2. Let $1 \leq i < j \leq p$ and $\beta \in \mathbb{R}^p$ be such that $\beta_i < \beta_j$. We define $\tilde{\beta} \in \mathbb{R}^p$ by $\tilde{\beta}_k = \beta_k$ if $k \neq i, j$, $\tilde{\beta}_i = \beta_j$ and $\tilde{\beta}_j = \beta_i$. Prove that $\mathscr{L}(\tilde{\beta}) < \mathscr{L}(\beta)$.
3. Prove that $\widehat{\beta}_1 \geq \ldots \geq \widehat{\beta}_p \geq 0$.
4. Define $\widehat{\beta}'$ by $\widehat{\beta}'_j = \widehat{\beta}_j$ if $j \leq \widehat{j}_{HB}$ and $\widehat{\beta}'_j = 0$ else. By comparing $\mathscr{L}(\widehat{\beta}')$ to $\mathscr{L}(\widehat{\beta})$ prove that $\widehat{\beta}_j = 0$ for $j > \widehat{j}_{HB}$.

C) Subgradients

We set

$$C = \bigcap_{j=1}^{p} \left\{ w \in \mathbb{R}^p : |w|_{[1]} + \ldots + |w|_{[j]} \leq \lambda_1 + \ldots + \lambda_j \right\}.$$

1. Prove that for all $\alpha, \beta \in \mathbb{R}^p$, with $\alpha_1 \geq \ldots \geq \alpha_p \geq 0$ we have

$$\sum_{j=1}^{p} \alpha_j \beta_j = \alpha_p(\beta_1 + \ldots + \beta_p) + (\alpha_{p-1} - \alpha_p)(\beta_1 + \ldots + \beta_{p-1}) + \ldots + (\alpha_1 - \alpha_2)\beta_1$$
$$\leq \sum_{j=1}^{p} \alpha_j |\beta|_{[j]}.$$

2. For all $w \in C$ and $\beta \in \mathbb{R}^p$ check that

$$\langle w, \beta \rangle \leq \sum_{j=1}^{p} |w|_{[j]} |\beta|_{[j]} \leq \sum_{j=1}^{p} \lambda_j |\beta|_{[j]}.$$

3. Conclude that

$$\sum_{j=1}^{p} \lambda_j |\beta|_{[j]} = \sup_{w \in C} \langle w, \beta \rangle.$$

4. Prove that $\mathscr{L}$ is convex and compute its subdifferential $\partial \mathscr{L}(\beta)$.

5.5.10 Compress Sensing

In this exercise, our goal is to compress efficiently sparse signals in $\mathbb{R}^p$. The strategy proposed by Candès, Romberg and Tao [48] and Donoho [67] is to multiply the signal by a matrix $A \in \mathbb{R}^{m \times p}$ with m smaller than p. We have in mind that the signals are coordinate sparse, with at most k non-zero entries. We will answer the two questions:
- How can we decompress efficiently the signals?
- For which value of m are we able to compress / decompress any k-sparse signal with no loss?

Let us formalize this problem. Let $x^* \in \mathbb{R}^p$ be a vector with at most k nonzero entries. Write $\text{supp}(x^*) = \{j : x_j^* \neq 0\}$ for the (unknown) support of x^*. Let $y = Ax^*$ be the compressed signal and define the decompression algorithm

$$\hat{x} \in \underset{x \in \mathbb{R}^p :\, Ax = y}{\text{argmin}} |x|_1, \tag{5.35}$$

with $|x|_1$ the ℓ^1-norm of x.

A) Null Space Property

In this problem, for any set $S \subset \{1, \ldots, p\}$ and any vector $v \in \mathbb{R}^p$, we denote by v_S the vector in $\mathbb{R}^p$ with entries

$$[v_S]_j = v_j \mathbf{1}_{j \in S}, \quad \text{for } j = 1, \ldots, p.$$

The matrix $A \in \mathbb{R}^{m \times p}$ is said to fulfill the null space property, NSP(k), if

$$\text{for any } v \in \ker(A) \setminus \{0\} \text{ and any } S \subset \{1, \ldots, p\} \text{ with cardinality } |S| = k,$$
$$\text{we have } |v_S|_1 < |v_{S^c}|_1. \tag{5.36}$$

1. Prove that if A fulfills the nullspace property, then there is a unique solution to (5.35), which is $\hat{x} = x^*$.

B) Restricted Isometry Property Enforces Null Space Property

Let $\delta \in (0,1)$. The matrix A is said to fulfill the (d,δ)-Restricted Isometry Property, abbreviated (d,δ)-RIP, if

$$1-\delta \leq \|Au\|^2 \leq 1+\delta, \quad \text{for all } u \in \mathbb{R}^p \text{ such that } |u|_0 \leq d \text{ and } \|u\| = 1, \tag{5.37}$$

where $\|u\|$ is the Euclidean norm of u and $|u|_0 = |\text{supp}(u)|$. In this part, we will prove that the $(2k,\delta)$-RIP property with $\delta < 1/3$ enforces the Null Space Property (5.36).

Let A be a matrix fulfilling the $(2k,\delta)$-RIP property with $\delta < 1/3$. Let $v \in \ker(A) \setminus \{0\}$ and let

$$|v_{i_1}| \geq |v_{i_2}| \geq \ldots \geq |v_{i_p}|,$$

be the absolute values of the entries of v ranked in decreasing order. Let J denote the integer part of p/k and define $S_j = \{i_{kj+1},\ldots,i_{k(j+1)}\}$ for $j = 0,\ldots,J-1$ and $S_J = \{i_{kJ+1},\ldots,i_p\}$.

1. For $j = 0,\ldots,J$, set $\bar{v}_{S_j} = v_{S_j}/\|v_{S_j}\|$ and prove that for $j = 1,\ldots,J$

$$-\langle A\bar{v}_{S_0}, A\bar{v}_{S_j}\rangle = \frac{1}{4}\left(\|A(\bar{v}_{S_0}-\bar{v}_{S_j})\|^2 - \|A(\bar{v}_{S_0}+\bar{v}_{S_j})\|^2\right) \leq \delta.$$

2. Prove that

$$\|Av_{S_0}\|^2 = -\sum_{j=1}^{J}\langle Av_{S_0}, Av_{S_j}\rangle \leq \delta \sum_{j=1}^{J}\|v_{S_0}\|\|v_{S_j}\|.$$

3. Check that $\|v_{S_j}\| \leq k^{-1/2}|v_{S_{j-1}}|_1$ for $j = 1,\ldots,J$ and $|v_{S_0}|_1 \leq \sqrt{k}\|v_{S_0}\|$.
4. Putting pieces together, derive the bound

$$|v_{S_0}|_1 \leq \frac{\delta}{1-\delta}\sum_{j=1}^{J}|v_{S_{j-1}}|_1.$$

5. Conclude that the null space property NSP(k) holds for any matrix A fulfilling the $(2k,\delta)$-RIP property with $\delta < 1/3$.

C) RIP by Random Sampling

It remains to build a matrix A fulfilling the $(2k,\delta)$-RIP property with $\delta < 1/3$. The goal of this part is to show that such a matrix can be obtained by random sampling. More precisely, let $B \in \mathbb{R}^{m\times p}$ be a matrix with entries B_{ij} i.i.d. with $\mathcal{N}(0,1)$ Gaussian distribution. We will prove in this part, that for any $\delta \in (0,1/3)$ and any

$$m \geq \frac{9}{\delta^2}\left(\sqrt{d}+\sqrt{2L+2d\log(ep/d)}\right)^2, \quad \text{with } L \geq 0, \tag{5.38}$$

the matrix $A = m^{-1/2}B$ fulfills the (d,δ)-RIP property with probability at least $1-2e^{-L}$.

The first question below relies on the following bound (Lemma 8.3, page 163; see also Exercise 12.9.6, page 288): for $E \in \mathbb{R}^{d_1\times d_2}$ with i.i.d. entries with $\mathcal{N}(0,1)$ Gaussian distribution, we have $\mathbb{E}\left[|E|_{\text{op}}\right] \leq \sqrt{d_1}+\sqrt{d_2}$.

1. Let $S \subset \{1,\ldots,p\}$ with cardinality $|S| = d$. Prove that for any $t > 0$

$$\mathbb{P}\left[\max\{\|Bu\| : \|u\| = 1,\ \mathrm{supp}(u) \subset S\} \geq \sqrt{m} + \sqrt{d} + \sqrt{2t}\right] \leq e^{-t}.$$

2. Prove that for m fulfilling (5.38), we have

$$\mathbb{P}\left[\|Au\|^2 \leq 1 + \delta,\ \text{ for all } u \in \mathbb{R}^p \text{ such that } |u|_0 \leq d \text{ and } \|u\| = 1\right] \geq 1 - e^{-L}.$$

The lower bound $\|Au\|^2 \geq 1 - \delta$ can be proved in the same way.

Chapter 6

Iterative Algorithms

As discussed in Chapters 2 and 5, the model selection paradigm provides statistically optimal estimators, but with a prohibitive computational cost. A classical recipe is then to convexify the minimization problem issued from model selection, in order to get estimators that can be easily computed with standard tools from convex optimization. This approach has been successfully implemented in many settings, including the coordinate-sparse setting (Lasso estimator) and the group-sparse setting (Group-Lasso estimator). As illustrated on page 102, the bias introduced by the convexification is a recurrent issue with this approach, and for some complex problems, even if the minimization problem is convex, the minimization cost can be high, especially in high dimension. In order to handle these two issues, there has been a renewed interest in iterative methods in the recent years. In the coordinate-sparse setting, these methods include the Forward-Backward algorithm described on page 43, and the Iterative Hard Thresholding algorithm analyzed in this chapter.

6.1 Iterative Hard Thresholding

Let us focus again on the coordinate-sparse linear regression setting

$$Y = \mathbf{X}\beta^* + \varepsilon,$$

where $|\beta^*|_0$ is small. As in Chapter 5, we assume in this section that the columns of $\mathbf{X}$ have a unit ℓ^2-norm.
The take home message of Chapter 2 is that the estimator solution to the optimization problem

$$\min_{\beta \in \mathbb{R}^p} \|Y - \mathbf{X}\beta\|^2 + \lambda^2 |\beta|_0, \tag{6.1}$$

with $\lambda^2 = c\sigma^2 \log(p)$ is (almost) statistically optimal in this setting. As the optimization problem (6.1) cannot be solved efficiently in general, the recipe described in Chapter 5 is to replace $|\beta|_0$ by the ℓ^1-norm, leading to the Lasso algorithm solution to

$$\min_{\beta \in \mathbb{R}^p} \underbrace{\|Y - \mathbf{X}\beta\|^2 + \lambda |\beta|_1}_{=F(\beta)}. \tag{6.2}$$

This minimization problem being convex, it can be solved efficiently.

6.1.1 Reminder on the Proximal Method

As reminded in the next paragraph, the minimization of (6.2) can be performed with the proximal method, which amounts to iteratively
(i) apply a gradient step relative to the function $F(\beta) = \|Y - \mathbf{X}\beta\|^2$, and
(ii) apply the soft-thresholding operator

$$S_\lambda(\alpha) = \begin{bmatrix} \alpha_1(1-\lambda/|\alpha_1|)_+ \\ \vdots \\ \alpha_p(1-\lambda/|\alpha_p|)_+ \end{bmatrix}. \tag{6.3}$$

Indeed, if, in the minimization of the Lasso problem (6.2), we replace at each iteration the function $F(\beta)$ by the quadratic approximation

$$F(\beta^t) + \langle \beta - \beta^t, \nabla F(\beta^t)\rangle + \frac{1}{2\eta}\|\beta - \beta^t\|^2 \tag{6.4}$$

around the current estimate β^t, we obtain at the next estimate

$$\beta^{t+1} \in \underset{\beta\in\mathbb{R}^p}{\operatorname{argmin}} \left\{ F(\beta^t) + \langle \beta - \beta^t, \nabla F(\beta^t)\rangle + \frac{1}{2\eta}\|\beta - \beta^t\|^2 + \lambda|\beta|_1 \right\}.$$

The update at step $t+1$ is then given by the soft-thresholding of a gradient step

$$\beta^{t+1} = S_{\lambda\eta}(\beta^t - \eta\nabla F(\beta^t)), \tag{6.5}$$

with $\nabla F(\beta^t) = 2\mathbf{X}^T(\mathbf{X}\beta^t - Y)$ and S_λ the soft-thresholding operator (6.3).
Since the Lasso problem (6.2) is convex, the proximal algorithm can be shown to converge to the solution of (6.2), as long as the gradient step η is small enough.

6.1.2 Iterative Hard-Thresholding Algorithm

As proved in Section 5.2.2, page 92, the soft-thresholding operator is the solution to the minimization problem

$$S_\lambda(\alpha) \in \underset{\beta\in\mathbb{R}^p}{\operatorname{argmin}} \left\{ \frac{1}{2}\|\alpha - \beta\|^2 + \lambda|\beta|_1 \right\}.$$

In order to avoid the bias of the Lasso (Section 5.2.5, page 102), we may consider to replace in step (ii) the soft-thresholding operator by the hard-thresholding operator

$$H_\lambda(\alpha) = \begin{bmatrix} \alpha_1 \mathbf{1}_{|\alpha_1|>\lambda} \\ \vdots \\ \alpha_p \mathbf{1}_{|\alpha_p|>\lambda} \end{bmatrix}, \tag{6.6}$$

which is the solution to the minimization problem (see Exercise 2.8.1, page 46)

$$H_\lambda(\alpha) \in \underset{\beta\in\mathbb{R}^p}{\operatorname{argmin}} \left\{ \|\alpha - \beta\|^2 + \lambda^2|\beta|_0 \right\}.$$

Let us check that this amounts to consider the proximal algorithm for the initial problem (6.1). Indeed, replacing again $F(\beta)$ by the quadratic approximation (6.4) in (6.1), the proximal update is

$$\begin{aligned}\beta^{t+1} &= \underset{\beta\in\mathbb{R}^p}{\operatorname{argmin}}\left\{F(\beta^t)+\langle\beta-\beta^t,\nabla F(\beta^t)\rangle+\frac{1}{2\eta}\|\beta-\beta^t\|^2+\lambda^2|\beta|_0\right\}\\ &= \underset{\beta\in\mathbb{R}^p}{\operatorname{argmin}}\left\{\frac{1}{2\eta}\|\beta-\beta^t+\eta\nabla F(\beta^t)\|^2+\lambda^2|\beta|_0\right\}\\ &= H_{\lambda\sqrt{2\eta}}(\beta^t-\eta\nabla F(\beta^t)),\end{aligned} \tag{6.7}$$

with H_λ the hard-thresholding operator (6.6).

Does the algorithm (6.7) provide a good estimator for a sparse β^*? While the proximal algorithm (6.5) for the Lasso can be shown to converge to the minimizer of (6.2) for η small enough, we do not have such a result for (6.7). The original criterion (6.1) being highly non-convex, the iterates (6.7) may not converge, and, in particular, they do not converge to the solution of problem (6.1) in general.

We can prove yet some strong results for a variant of (6.7), where $\widehat{\beta}^0=0$, $\eta=1/2$ and where the thresholding level λ_t is updated at each time step

$$\widehat{\beta}^{t+1}=H_{\lambda_{t+1}}\left(\widehat{\beta}^t-\frac{1}{2}\nabla F(\widehat{\beta}^t)\right)=H_{\lambda_{t+1}}\left((I-\mathbf{X}^T\mathbf{X})\widehat{\beta}^t+\mathbf{X}^TY\right). \tag{6.8}$$

This algorithm is called the Iterative Hard Thresholding (IHT) algorithm. The recipe is to choose a thresholding level λ_t decreasing from a high-value to the target value $\lambda=c\sigma\sqrt{\log(p)}$ corresponding to the optimal level in (6.1). Similarly as for the forward-backward algorithm, this recipe allows to keep under control the sparsity of the iterates $\widehat{\beta}^t$ and hence the variance of the estimator.

In the following, we consider, for some $A,B>0$ and $a>1$, the sequence of threshold levels

$$\lambda_t=a^{-t}A+B. \tag{6.9}$$

In particular, the sequence (6.9) is solution to the recurrence equation $\lambda_{t+1}=a^{-1}\lambda_t+\frac{a-1}{a}B$, with initialization $\lambda_0=A+B$. As discussed after Theorem 6.1, this specific form with well-chosen A and B, allows to keep under control the sparsity of $\widehat{\beta}^t$ at each iterate, which is the key for controlling the reconstruction error.

6.1.3 Reconstruction Error

We provide in this section some theoretical results on the reconstruction error $\|\widehat{\beta}^t-\beta^*\|$. Before stating the results, let us discuss their nature.

The classical approach in statistics or machine learning is
(i) to define an estimator, very often as the solution of a minimization problem, and
(ii) to define an optimization algorithm, in order to approximate the solution of the minimization problem.

For example, the Lasso estimator $\widehat{\beta}_{Lasso}$ is defined as the solution of the minimization problem (6.2), and then proximal iterations (6.5) are used to numerically approximate $\widehat{\beta}_{Lasso}$. The estimator actually used is then β^t obtained after t iterations of (6.5). The reconstruction error of β^t can then be decomposed in terms of an optimization error $\|\beta^t - \widehat{\beta}_{Lasso}\|$ and a statistical error $\|\widehat{\beta}_{Lasso} - \beta^*\|$

$$\|\beta^t - \beta^*\| \leq \underbrace{\|\beta^t - \widehat{\beta}_{Lasso}\|}_{\text{optim. error}} + \underbrace{\|\widehat{\beta}_{Lasso} - \beta^*\|}_{stat.error}.$$

The two errors can be analyzed apart, using tools from convex optimization for the optimization error and tools from statistics for the second term.

We emphasize that this classical approach does not make sense for the analysis of the IHT estimator $\widehat{\beta}^t$ obtained after t iterations of (6.8). The estimator $\widehat{\beta}^t$ is not the numerical approximation of an "ideal" estimator, and it may even not converge when t goes to infinity. So, we have to directly analyze the reconstruction error $\|\widehat{\beta}^t - \beta^*\|$ without using an optimization / statistical error decomposition. Hence, the optimization and statistics must be tackled altogether. This corresponds to a recent trend in machine learning and this approach is recurrent in the analysis of iterative algorithms. A second example can be found in this book, in Chapter 12, for the analysis of the Lloyd algorithm.

Let us start with a deterministic bound on the reconstruction error $\|\widehat{\beta}^t - \beta^*\|$.

Theorem 6.1 Deterministic error bound for IHT.
We set $\Lambda = I - \mathbf{X}^T\mathbf{X}$ *and* $m^* = supp(\beta^*)$. *Assume that for some* $0 < \delta < 1$ *and some* $c \geq 1$ *such that* $c^2|\beta^*|_0 \in \mathbb{N}$,

$$\max_{S \subset \{1,\ldots,p\}:\ |S| \leq \bar{k}} |\Lambda_{SS}|_{op} \leq \delta, \quad \textit{with} \quad \bar{k} = (1+2c^2)|\beta^*|_0. \tag{6.10}$$

Assume also that,

$$1 < a \leq \frac{c}{\delta(1+2c)}, \quad A \geq \frac{\|\beta^*\|}{(1+2c)\sqrt{|\beta^*|_0}} \quad \textit{and} \quad B > \frac{a}{a-1}|\mathbf{X}^T\varepsilon|_\infty. \tag{6.11}$$

Then, for all $t \geq 0$, *the estimator* $\widehat{\beta}^t$ *defined by (6.8) with threshold levels given by (6.9) fulfills*

$$|\widehat{\beta}^t_{\bar{m}^*}|_0 \leq c^2|\beta^*|_0, \quad \textit{where} \quad \bar{m}^* = \{1,\ldots,p\} \setminus m^*, \tag{6.12}$$

and

$$\|\widehat{\beta}^t - \beta^*\| \leq (1+2c)\lambda_t\sqrt{|\beta^*|_0}. \tag{6.13}$$

Before proving Theorem 6.1, let us explain informally how the IHT algorithm works.

At step $t+1$, the estimator $\widehat{\beta}^{t+1}$ is given by a hard thresholding at level λ_{t+1} of the gradient step

$$\Lambda\widehat{\beta}^t + \mathbf{X}^T Y = \beta^* + \Lambda(\widehat{\beta}^t - \beta^*) + \mathbf{X}^T \varepsilon.$$

We observe that the value after the gradient step is given by the target β^* perturbed by two terms $\Lambda(\widehat{\beta}^t - \beta^*)$ and $\mathbf{X}^T\varepsilon$.

The first term $\Lambda(\widehat{\beta}^t - \beta^*)$ reflects the propagation of the error $\widehat{\beta}^t - \beta^*$ in the iterative scheme. We observe that it has the nice following contraction property. As long as (6.12) holds, we have $|\widehat{\beta}^t - \beta^*|_0 \leq \bar{k}$ and hence, according to (6.10), we have a contraction of the error

$$\|\Lambda(\widehat{\beta}^t - \beta^*)\| \leq \delta \|\widehat{\beta}^t - \beta^*\|. \tag{6.14}$$

In addition to this iterative error, we have the noise term $\mathbf{X}^T\varepsilon$, which maintains some noise in the iterations, at level $|\mathbf{X}^T\varepsilon|_\infty$. This noise level, which is the same (up to constant) as the one appearing in Theorem 5.1 (page 95) for the Lasso, corresponds, up-to-constant, to the threshold level λ_t for t large.

It is worth emphasizing the important role of the hard thresholding at each step. The choice (6.11) of the threshold level λ_t ensures that the property (6.12) holds at each step, and hence, that we have the contraction (6.14) of the residual error at each step. We refer to Exercise 6.4.1 for a comparison with the case where the hard thresholding is not applied at each iteration, but only at the final stage.

Let us now formalize the arguments exposed above.

Proof of Theorem 6.1.

We set $|\beta^*|_0 = k^*$, $Z = \mathbf{X}^T\varepsilon$, and $b^{t+1} = \beta^* + \Lambda(\widehat{\beta}^t - \beta^*) + Z$, so that $\widehat{\beta}^{t+1} = H_{\lambda_{t+1}}(b^{t+1})$. We prove Theorem 6.1 by induction. The properties (6.12) and (6.13) hold at $t = 0$. Let us assume that they hold at step t and let us prove that, then, they also hold at step $t+1$.

Next lemma will be used repeatedly in the proof.

Lemma 6.2 Contraction property

If the property (6.12) holds at step t, then, we have

$$\max_{S\subset\{1,\ldots,p\}:\ |S|\leq c^2k^*} \left(\|(b^{t+1} - \beta^*)_S\| - \sqrt{|S|}\,|Z|_\infty \right) \leq \delta \|\beta^* - \widehat{\beta}^t\|. \tag{6.15}$$

Proof of Lemma 6.2. Let us set $\widetilde{S}^t = m^* \cup \mathrm{supp}(\widehat{\beta}^t_{\overline{m}^*})$, and $S' = \widetilde{S}^t \cup S$. We have $|\widetilde{S}^t| \leq (1+c^2)k^*$ by (6.12) at step t, and $|S'| \leq (1+2c^2)k^* = \bar{k}$. Hence

$$\begin{aligned}
\|(b^{t+1} - \beta^*)_S\| &\leq \|(\Lambda(\widehat{\beta}^t - \beta^*))_S\| + \|Z_S\| \\
&\leq \|(\Lambda(\widehat{\beta}^t - \beta^*))_{S'}\| + \sqrt{|S|}\,|Z|_\infty \\
&\leq \|\Lambda_{S'S'}(\widehat{\beta}^t - \beta^*)_{S'}\| + \sqrt{|S|}\,|Z|_\infty \\
&\leq \delta\|(\widehat{\beta}^t - \beta^*)_{S'}\| + \sqrt{|S|}\,|Z|_\infty \;=\; \delta\|\widehat{\beta}^t - \beta^*\| + \sqrt{|S|}\,|Z|_\infty.
\end{aligned}$$

The proof of Lemma 6.2 is complete. □

We have

$$\|\widehat{\beta}^{t+1}-\beta^*\| \leq \|\widehat{\beta}^{t+1}_{m^*}-\beta^*_{m^*}\| + \|\widehat{\beta}^{t+1}_{\bar{m}^*}\|. \tag{6.16}$$

The first term in the hand-right side can be upper-bounded by Lemma 6.2, the Bound (6.13) at step t and $(1+2c)\delta \leq ca^{-1}$,

$$\begin{aligned}\|\widehat{\beta}^{t+1}_{m^*}-\beta^*_{m^*}\| &\leq \|b^{t+1}_{m^*}-H_{\lambda_{t+1}}(b^{t+1}_{m^*})\| + \|b^{t+1}_{m^*}-\beta^*_{m^*}\| \\ &\leq \sqrt{k^*}\,\lambda_{t+1}+\delta\|\widehat{\beta}^t-\beta^*\|+\sqrt{k^*}\,|Z|_\infty. \\ &\leq \sqrt{k^*}\left(\lambda_{t+1}+ca^{-1}\lambda_t+|Z|_\infty\right).\end{aligned}$$

The Condition (6.11) ensures that $\lambda_{t+1} > a^{-1}\lambda_t + |Z|_\infty$, and we then get

$$\|\widehat{\beta}^{t+1}_{m^*}-\beta^*_{m^*}\| \leq (1+c)\sqrt{k^*}\lambda_{t+1}. \tag{6.17}$$

For the second term, we first prove that $|\widehat{\beta}^{t+1}_{\bar{m}^*}|_0 \leq c^2k^*$. For any $S \subset \mathrm{supp}(\widehat{\beta}^{t+1}_{\bar{m}^*})$, with $|S| \leq c^2k^*$, we have from Lemma 6.2

$$\lambda_{t+1}\sqrt{|S|} \leq \|\widehat{\beta}^{t+1}_S\| = \|b^{t+1}_S\| = \|b^{t+1}_S-\beta^*_S\| \leq \sqrt{|S|}\,|Z|_\infty+\delta\|\beta^*-\widehat{\beta}^t\|. \tag{6.18}$$

Hence, since $\lambda_{t+1} > a^{-1}\lambda_t + |Z|_\infty$,

$$\sqrt{|S|} \leq \frac{\delta\|\beta^*-\widehat{\beta}^t\|}{\lambda_{t+1}-|Z|_\infty} \leq \frac{a^{-1}c\sqrt{k^*}\lambda_t}{\lambda_{t+1}-|Z|_\infty} < \frac{a^{-1}c\sqrt{k^*}\lambda_t}{a^{-1}\lambda_t} = c\sqrt{k^*}.$$

It follows that $|\widehat{\beta}^{t+1}_{\bar{m}^*}|_0 < c^2k^*$, so (6.12) is proved at step $t+1$.
In addition, the Inequality (6.18) with $S = \mathrm{supp}(\widehat{\beta}^{t+1}_{\bar{m}^*})$ gives the upper bound

$$\|\widehat{\beta}^{t+1}_{\bar{m}^*}\| \leq c\sqrt{k^*}\,|Z|_\infty+\delta\|\beta^*-\widehat{\beta}^t\| \leq c\sqrt{k^*}\,|Z|_\infty+a^{-1}c\lambda_t\sqrt{k^*} \leq c\lambda_{t+1}\sqrt{k^*},$$

where the last inequality again follows from $\lambda_{t+1} \geq a^{-1}\lambda_t + |Z|_\infty$. Combining this bound with the Bounds (6.16) and (6.17), we get

$$\|\widehat{\beta}^{t+1}-\beta^*\| \leq (1+2c)\lambda_{t+1}\sqrt{k^*}.$$

Hence, (6.13) holds at step $t+1$ and the proof of Theorem 6.1 is complete. □

Let us build on Theorem 6.1, in order to provide a risk bound in the case where the noise ε follows a $\mathcal{N}(0,\sigma^2 I_n)$ Gaussian distribution. In order to have A, B fulfilling the condition (6.11) with large probability, we set for $L > 0$

$$A = \frac{\|\mathbf{X}^TY\|+\sigma|\mathbf{X}|_{\mathrm{op}}\sqrt{2}(1+\sqrt{L})}{3(1-\delta)} \quad \text{and} \quad B = \frac{a\sigma}{a-1}\sqrt{2\log(p)+2L}. \tag{6.19}$$

Then, we choose a number $\hat{t}$ of iterations such that the first term $a^{-\hat{t}}A$ is smaller than the second term B in (6.9). Accordingly, we define $\hat{t}$ as the smallest integer larger than[1] $\log_a(A/B)$,

$$\hat{t} := \min\{k \in \mathbb{N} : k \geq \log_a(A/B)\}. \tag{6.20}$$

Corollary 6.3 Error bound in the Gaussian setting for IHT.
Assume that the columns of $\mathbf{X}$ *have unit* ℓ^2*-norm, and that the noise* ε *follows a* $\mathcal{N}(0, \sigma^2 I_n)$ *Gaussian distribution. Assume also that Assumption (6.10) holds and that* $1 < a \leq \frac{c}{\delta(1+2c)}$.
Then, for any iteration t *larger than* $\hat{t}$ *defined by (6.20), with probability larger than* $1 - 2e^{-L}$*, the estimator* $\widehat{\beta}^t$*, with* A, B *given by (6.19), fulfills* $|\widehat{\beta}^t_{\bar{m}^*}|_0 \leq c^2 |\beta^*|_0$ *and*

$$\|\widehat{\beta}^t - \beta^*\|^2 \leq C_{a,c} |\beta^*|_0 \sigma^2 (\log(p) + L), \tag{6.21}$$

with $C_{a,c} := 2\left(\frac{2a(1+2c)}{a-1}\right)^2$.

Before proceeding to the proof of this corollary, let us comment on the result. First, we observe that the reconstruction error $\|\widehat{\beta}^t - \beta^*\|^2$ is $O(\sigma^2 \log(p))$, which is (almost) optimal according to Exercise 3.6.3, page 69. In addition, we observe that the error bound holds for a number $\hat{t}$ of iterations that can be explicitly computed from A and B.

The stronger assumption in Theorem 6.1 and Corollary 6.3 is Condition (6.10), which ensures that any subset $\{\mathbf{X}_j : j \in S\}$ of $\bar{k}$ columns of $\mathbf{X}$ are close to be orthogonal. This condition is somewhat strong, and it is seldom satisfied in practice. This is one of the main weaknesses of this result.

As a side remark, we notice that under (6.10), we have for any $u \in \mathbb{R}^p$ with $|u|_0 \leq \bar{k}$

$$(1-\delta)\|u\|^2 \leq \langle \mathbf{X}^T \mathbf{X} u, u \rangle = \|\mathbf{X} u\|^2 \leq (1+\delta)\|u\|^2.$$

This property is usually called the Restricted Isometry Property. Hence, under the conditions of Corollary 6.3, we also have with probability larger than $1 - 2e^{-L}$,

$$\|\mathbf{X}(\widehat{\beta}^t - \beta^*)\|^2 \leq (1+\delta)\|\widehat{\beta}^t - \beta^*\|^2 \leq (1+\delta) C_{a,c} |\beta^*|_0 \sigma^2 (\log(p) + L).$$

This bound is comparable to the risk bound (5.13) obtained for the Lasso estimator, on page 97.

Proof of Corollary 6.3.
Let Ω_L be the event

$$\Omega_L = \left\{ \frac{\|\mathbf{X}^T Y\| + \sigma |\mathbf{X}|_{op} \sqrt{2}(1+\sqrt{L})}{3(1-\delta)} \geq \frac{\|\beta^*\|}{1+2c} \text{ and } |Z|_\infty < \sigma\sqrt{2\log(p) + 2L} \right\}.$$

[1] $\log_a(x) = \log(x/a)$

Let us observe that $\lambda_t = a^{-t}A + B \leq 2B$ for $t \geq \hat{t}$. On the event Ω_L, Condition (6.11) is fulfilled, so according to Theorem 6.1, we have $|\widehat{\beta}^t_{\bar{m}^*}|_0 \leq c^2 k^*$ and

$$\begin{aligned}\|\widehat{\beta}^t - \beta^*\|^2 &\leq (1+2c)^2 k^* (a^{-t}A + B)^2 \\ &\leq 4(1+2c)^2 k^* B^2 \\ &\leq C_{a,c}\, k^* \sigma^2 (\log(p) + L).\end{aligned}$$

So, to conclude the proof of Corollary 6.3, all we need is to prove that $\mathbb{P}[\Omega_L] \geq 1 - 2e^{-L}$.
From the proof of Corollary 5.3, page 97, we already have that

$$\mathbb{P}\left[|Z|_\infty \geq \sigma\sqrt{2\log(p) + 2L}\right] \leq e^{-L}.$$

Let us now bound $\|\mathbf{X}^T Y\|$ from below. Since the application $\varepsilon \to \|\mathbf{X}^T(\mathbf{X}\beta^* + \varepsilon)\|$ is $|\mathbf{X}|_{\text{op}}$-Lipschitz, Gaussian concentration inequality (Theorem B.7, page 301) ensures that with probability at least $1 - e^{-L}$ we have $\|\mathbf{X}^T Y\| \geq \mathbb{E}\left[\|\mathbf{X}^T Y\|\right] - |\mathbf{X}|_{\text{op}}\sigma\sqrt{2L}$. While there is no simple formula for $\mathbb{E}\left[\|\mathbf{X}^T Y\|\right]$, we have the simple lower bound

$$\mathbb{E}\left[\|\mathbf{X}^T Y\|^2\right] = \|\mathbf{X}^T\mathbf{X}\beta^*\|^2 + \mathbb{E}\left[\|\mathbf{X}^T\varepsilon\|^2\right] \geq \|\mathbf{X}^T\mathbf{X}\beta^*\|^2.$$

Using again the Gaussian concentration inequality, we can lower-bound $\mathbb{E}\left[\|\mathbf{X}^T Y\|\right]$ in terms of $\mathbb{E}\left[\|\mathbf{X}^T Y\|^2\right]$. Indeed, there exists a standard exponential variable ξ, such that $\|\mathbf{X}^T Y\| \leq \mathbb{E}\left[\|\mathbf{X}^T Y\|\right] + \sigma|\mathbf{X}|_{\text{op}}\sqrt{2\xi}$. Since the function $\xi \to \left(\mathbb{E}\left[\|\mathbf{X}^T Y\|\right] + \sigma|\mathbf{X}|_{\text{op}}\sqrt{2\xi}\right)^2$ is concave, Jensen inequality gives

$$\begin{aligned}\mathbb{E}\left[\|\mathbf{X}^T Y\|^2\right] &\leq \mathbb{E}\left[\left(\mathbb{E}\left[\|\mathbf{X}^T Y\|\right] + \sigma|\mathbf{X}|_{\text{op}}\sqrt{2\xi}\right)^2\right] \\ &\leq \left(\mathbb{E}\left[\|\mathbf{X}^T Y\|\right] + \sigma|\mathbf{X}|_{\text{op}}\sqrt{2\mathbb{E}[\xi]}\right)^2.\end{aligned}$$

Since $\mathbb{E}[\xi] = 1$, we get with probability at least $1 - e^{-L}$

$$\begin{aligned}\|\mathbf{X}^T Y\| &\geq \mathbb{E}\left[\|\mathbf{X}^T Y\|\right] - |\mathbf{X}|_{\text{op}}\sigma\sqrt{2L} \\ &\geq \sqrt{\mathbb{E}\left[\|\mathbf{X}^T Y\|^2\right]} - |\mathbf{X}|_{\text{op}}\sigma\sqrt{2}\left(1 + \sqrt{L}\right) \\ &\geq \|\mathbf{X}^T\mathbf{X}\beta^*\| - |\mathbf{X}|_{\text{op}}\sigma\sqrt{2}\left(1 + \sqrt{L}\right).\end{aligned}$$

Let us denote by m^* the support of β^*. To conclude, it remains to notice that, according to (6.10), we have

$$\begin{aligned}\|\mathbf{X}^T\mathbf{X}\beta^*\| &\geq \|(\mathbf{X}^T\mathbf{X}\beta^*)_{m^*}\| \\ &\geq \|\beta^*_{m^*}\| - \|\Lambda_{m^*m^*}\beta^*_{m^*}\| \\ &\geq (1-\delta)\|\beta^*_{m^*}\| = (1-\delta)\|\beta^*\|.\end{aligned}$$

So, with probability at least $1-e^{-L}$, we have

$$\frac{\|\beta^*\|}{1+2c} \leq \frac{\|\mathbf{X}^T Y\| + \sigma|\mathbf{X}|_{\text{op}}\sqrt{2}(1+\sqrt{L})}{3(1-\delta)}.$$

The proof of Corollary 6.3 is complete. □

6.2 Iterative Group Thresholding

In this section, we extend the methodology developed above to the group-sparse setting described in Section 2.1, Chapter 2. In this setting, the indices $\{1,\ldots,p\}$ are partitioned into M groups $\{1,\ldots,p\} = \cup_{j=1}^M G_j$, and our goal is to recover groups G_j of variables such that $\beta^*_{G_j} \neq 0$. For this task, we can consider the generalization of the minimization problem (5.16), on page 104,

$$\min_{\beta\in\mathbb{R}^p}\left\{\|Y-\mathbf{X}\beta\|^2 + \sum_{j=1}^M (\lambda^{(j)})^2 \mathbf{1}_{\beta_{G_j}\neq 0}\right\}. \tag{6.22}$$

Let us write a proximal algorithm related to this problem. Replacing again in (6.22) the function $F(\beta) = \|Y-\mathbf{X}\beta\|^2$ by the quadratic approximation

$$F(\beta^t) + \langle \beta-\beta^t, \nabla F(\beta^t)\rangle + \frac{1}{2\eta}\|\beta-\beta^t\|^2$$

around the current estimate β^t, we get the update

$$\begin{aligned}
\beta^{t+1} &\in \operatorname*{argmin}_{\beta\in\mathbb{R}^p}\left\{F(\beta^t) + \langle \beta-\beta^t, \nabla F(\beta^t)\rangle + \frac{1}{2\eta}\|\beta-\beta^t\|^2 + \sum_{j=1}^M (\lambda^{(j)})^2 \mathbf{1}_{\beta_{G_j}\neq 0}\right\} \\
&= \operatorname*{argmin}_{\beta\in\mathbb{R}^p}\left\{\|\beta-\beta^t+\eta\nabla F(\beta^t)\|^2 + 2\eta\sum_{j=1}^M (\lambda^{(j)})^2 \mathbf{1}_{\beta_{G_j}\neq 0}\right\}.
\end{aligned}$$

This minimization problem can be solved explicitly, see Exercise 6.4.2, page 135. The solution is given by $\beta^{t+1} = H^G_{\lambda\sqrt{2\eta}}(\beta^t - \eta\nabla F(\beta^t))$, where the operator H^G_λ is the group thresholding operator defined by

$$[H^G_\lambda(\alpha)]_{G_j} = \alpha_{G_j}\mathbf{1}_{\|\alpha_{G_j}\|>\lambda^{(j)}}, \quad \text{for } j=1,\ldots,M.$$

So, we again have iterations involving a gradient step and a thresholding operator.

As before, the sequence $(\beta^t)_{t=1,2,\ldots}$ does not converge to the solution of (6.22) in general. We consider instead a variant of the Iterative Hard Thresholding algorithm (6.8) defined by $\widehat{\beta}^0 = 0$ and

$$\widehat{\beta}^{t+1} = H^G_{\lambda_{t+1}}\left(\widehat{\beta}^t - \frac{1}{2}\nabla F(\widehat{\beta}^t)\right) = H^G_{\lambda_{t+1}}\left((I-\mathbf{X}^T\mathbf{X})\widehat{\beta}^t + \mathbf{X}^T Y\right) \tag{6.23}$$

for a sequence of threshold levels $(\lambda_t^{(j)})_{t=1,2,\ldots}$, $j=1,\ldots,M$, decreasing with t. This algorithm is called hereafter Iterative Group Thresholding (IGT). The update for $\widehat{\beta}_{G_j}^{t+1}$ is then given by

$$\widehat{\beta}_{G_j}^{t+1} = \left(\widehat{\beta}_{G_j}^{t} + \mathbf{X}_{G_j}^T(Y-\mathbf{X}\widehat{\beta}^t)\right)\mathbf{1}_{\|\widehat{\beta}_{G_j}^{t}+\mathbf{X}_{G_j}^T(Y-\mathbf{X}\widehat{\beta}^t)\|>\lambda_{t+1}^{(j)}}.$$

In the following, we consider threshold levels of the form

$$\lambda_t^{(j)} = \sqrt{|G_j|}\,\gamma_t, \quad \text{where} \quad \gamma_t = a^{-t}A+B, \tag{6.24}$$

with $a>1$ and $A,B>0$. When each group is a singleton, we recover the IHT algorithm described in the previous section.

Before stating the counterpart of Theorem 6.1 for the Iterative Group Thresholding algorithm (6.23), we need to introduce a couple of notations. We define the group-support of β as

$$\mathrm{supp}_G(\beta) = \bigcup_{j:\beta_{G_j}\neq 0} G_j,$$

and its cardinality by $|\beta|_0^G = \mathrm{card}(\mathrm{supp}_G(\beta))$. We also set

$$|Z|_\infty^G := \max_{j=1,\ldots,M}\frac{\|Z_{G_j}\|}{\sqrt{|G_j|}}. \tag{6.25}$$

Theorem 6.4 Deterministic error bound for IGT
Let us set $\Lambda = I-\mathbf{X}^T\mathbf{X}$ and $m_G^* = \mathrm{supp}_G(\beta^*)$. Let us assume that all the groups have the same cardinality q. Let us also assume that for some $0<\delta<1$ and some $c\geq 1$ such that $c^2|\beta^*|_0^G\in q\mathbb{N}$,

$$\max_{S=\cup_{j\in J}G_j:\ |J|\leq \bar{J}_G} |\Lambda_{SS}|_{op}\leq\delta, \quad \textit{with} \quad \bar{J}_G=(1+2c^2)|\beta^*|_0^G/q. \tag{6.26}$$

Assume also that,

$$1<a\leq\frac{c}{\delta(1+2c)}, \quad A\geq\frac{\|\beta^*\|}{(1+2c)\sqrt{|\beta^*|_0^G}} \quad \textit{and} \quad B>\frac{a}{a-1}|\mathbf{X}^T\varepsilon|_\infty^G, \tag{6.27}$$

with $|\cdot|_\infty^G$ defined in (6.25).

Then, for all $t\geq 0$, the estimator $\widehat{\beta}^t$ defined by (6.23) with threshold levels given by (6.24) fulfills

$$|\widehat{\beta}_{\bar{m}_G^*}^t|_0^G\leq c^2|\beta^*|_0^G, \quad \textit{where} \quad \bar{m}_G^*=\{1,\ldots,p\}\setminus m_G^*, \tag{6.28}$$

and

$$\|\widehat{\beta}^t-\beta^*\|\leq(1+2c)\gamma_t\sqrt{|\beta^*|_0^G}. \tag{6.29}$$

This result is proved in Exercise 6.4.3, page 136.
Let us consider the case where the noise ε follows a $\mathcal{N}(0,\sigma^2 I_n)$ Gaussian distribution. In order to have A,B fulfilling the condition (6.27) with large probability, we set for $L>0$

$$A=\frac{\|\mathbf{X}^T Y\|+\sigma|\mathbf{X}|_{\mathrm{op}}\sqrt{2}(1+\sqrt{L})}{3(1-\delta)}\quad\text{and}\quad B=\frac{a\sigma}{a-1}\left(1+\phi_G\sqrt{2\log(M)+2L}\right), \tag{6.30}$$

with

$$\phi_G=\max_{j=1,\dots,M}\frac{|\mathbf{X}_{G_j}|_{\mathrm{op}}}{\sqrt{|G_j|}}.$$

As in the coordinate-sparse setting, we define $\hat{t}$ as the smallest integer larger than $\log_a(A/B)$. We then have the next corollary of Theorem 6.4.

Corollary 6.5 Error bound in the Gaussian setting for IGT.
Assume that the columns of $\mathbf{X}$ *have unit* ℓ^2*-norm, and that the noise* ε *follows a* $\mathcal{N}(0,\sigma^2 I_n)$ *Gaussian distribution. Assume also that Assumption (6.26) holds and that* $1<a\leq\frac{c}{\delta(1+2c)}$.
Then, for any iteration t *larger than* $\hat{t}$, *with probability larger than* $1-2e^{-L}$, *the estimator* $\widehat{\beta}^t$, *with* A,B *given by (6.30), fulfills* $|\widehat{\beta}^t_{\widehat{m}^*}|_0^G\leq c^2|\beta^*|_0^G$ *and*

$$\|\widehat{\beta}^t-\beta^*\|^2\leq C_{a,c}|\beta^*|_0^G\sigma^2\left(1+2\phi_G^2(\log(M)+L)\right), \tag{6.31}$$

with $C_{a,c}:=2\left(\frac{2a(1+2c)}{a-1}\right)^2$.

Corollary 6.5 is proved in Exercise 6.4.3, page 136. We can observe in the bound (6.31) the benefit of using IGT instead of IHT in the group-sparse setting. Comparing (6.31) and (6.21), we observe that the $\log(p)$ term has been replaced by a $\log(M)$ term, which can be much smaller if the number M of groups is much smaller than p. We can also notice that the bound (6.31) is very similar to the bound (5.20), page 105, for the group-Lasso.

6.3 Discussion and References

6.3.1 Take-Home Message

A robust strategy in statistics and machine learning is
(i) to start from a statistically optimal, but computationally intractable estimator, often defined through a minimization problem; and
(ii) to convexify the minimization problem as in Chapter 5, in order to define an estimator which is amenable to numerical computations in high dimensions.

A explained in Chapter 5, theoretical guarantees can be derived for such estimators, and this approach has been successful in many different settings. This approach suffers yet from two drawbacks. First, convex estimators suffer from some undesirable shrinkage bias (see Section 5.2.5, page 102), and this bias can sometimes strongly deteriorate the statistical properties, see for example [148]. Second, in some complex settings, as in Chapter 12, Section 12.4, minimizing the convexified problem can be computationally intensive.

As an alternative to this approach, there is a renewed interest in statistics and machine learning for iterative algorithms, which are often greedy algorithms related to the initial statistically optimal estimator. Such algorithms are typically computationally efficient, and they do not suffer from a shrinkage bias. While they do not converge to the solution to the original problem in general, many such algorithms have been shown to enjoy good statistical properties, as good as those of convex estimators. It is worth emphasizing that, while, for convex estimators, the optimization error and the statistical error can be evaluated separately, for iterative algorithms like IHT or IGT, the optimization and statistical errors cannot be separated and need to be tackled together. This corresponds to a recent trend in statistics and machine learning.

6.3.2 References

The Iterative Hard Thresholding algorithm has been introduced by Blumensath and Davies [35], and it has been analyzed by Jain, Tewari and Kar [95] and Liu and Foygel Barber [111], among others. The version analyzed in this chapter has been proposed by Ndaoud [125], and most of the material of this chapter is adapted from Ndaoud [125].

6.4 Exercices

6.4.1 Linear versus Non-Linear Iterations

In the proximal iterations (6.7), the non-linear operator

$$\beta \to H_{\lambda\sqrt{2\eta}}((I-\eta\mathbf{X}^T\mathbf{X})\beta + 2\eta\mathbf{X}^T Y))$$

is applied iteratively. This non-linear operator corresponds to the succession of a gradient step and a hard thresholding. In order to understand the benefit of the non-linearity induced by the hard thresholding, it is insightful to compare the proximal algorithm with the linearized version, where the iterations

$$\widetilde{\beta}^{t+1} = (I-2\eta\mathbf{X}^T\mathbf{X})\widetilde{\beta}^t + 2\eta\mathbf{X}^T Y$$

are applied successively, started from $\widetilde{\beta}^0 = 0$, and the hard thresholding is applied only at the final step $\widehat{\beta}_L = H_\lambda(\widetilde{\beta}^\infty)$, where $\widetilde{\beta}^\infty = \lim_{t\to\infty}\widetilde{\beta}^t$.

1. Prove that $\widetilde{\beta}^t$ is given by

$$\widetilde{\beta}^t = \sum_{k=0}^{t-1}(I-2\eta\mathbf{X}^T\mathbf{X})^k 2\eta\mathbf{X}^T Y.$$

2. Let $\mathbf{X} = \sum_{j=1}^{r} \sigma_j u_j v_j^T$ be a singular value decomposition (Theorem C.1, page 311) of $\mathbf{X}$. Check that

$$\widetilde{\beta}^t = L_t Y, \quad \text{with} \quad L_t = \sum_{j=1}^{r} \frac{1-(1-2\eta\sigma_j^2)^t}{\sigma_j} v_j u_j^T.$$

3. Prove that for $2\eta < |\mathbf{X}|_{op}^{-2}$, we have $L_t \to \mathbf{X}^+$, where $\mathbf{X}^+$ is the Moore–Penrose pseudo-inverse of $\mathbf{X}$, defined by (C.2), page 312.
4. Conclude that $\widehat{\beta}_L = H_\lambda(\mathbf{X}^+ Y)$.

Comment. We have $\mathbf{X}^+\mathbf{X} = P^\perp$, where $P^\perp$ is the projection onto the orthogonal complement of $\ker(\mathbf{X})$, see (C.3), page 313. We observe that if $Y = \mathbf{X}\beta^* + \varepsilon$, then $\widehat{\beta}_L = H_\lambda(P^\perp\beta^* + \mathbf{X}^+\varepsilon)$. In the favorable case where $P^\perp\beta^* = \beta^*$, we then get that $\widehat{\beta}_L$ is the hard thresholding at level λ of $\mathbf{X}^+Y = \beta^* + \mathbf{X}^+\varepsilon$.

As highlighted in Theorem 6.1 (page 126) for the IHT algorithm (6.8), the thresholding level is tightly linked to the infinite norm $|\mathbf{X}^T\varepsilon|_\infty = \max_j |\mathbf{X}_j^T\varepsilon|$. When the noise ε has a $\mathcal{N}(0,\sigma^2 I_n)$ distribution, and the matrix $\mathbf{X}$ has columns with unit ℓ^2-norm, then $|\mathbf{X}^T\varepsilon|_\infty = O(\sigma\sqrt{\log(p)})$ with high probability. Similarly, the choice of the threshold λ for $\widehat{\beta}_L$ should be tightly linked to the infinite norm $|\mathbf{X}^+\varepsilon|_\infty = \max_j |\mathbf{X}_j^+\varepsilon|$. The switch from the matrix $\mathbf{X}^T$ to the matrix $\mathbf{X}^+$ has a strong impact on the size of the infinite norm. Actually, since $\|\mathbf{X}_j\| = 1$, each variable $\mathbf{X}_j^T\varepsilon$ follows a Gaussian $\mathcal{N}(0,\sigma^2)$ distribution. In comparison, the variable $\mathbf{X}_j^+\varepsilon$ follows a Gaussian $\mathcal{N}(0,\|\mathbf{X}_j^+\|^2\sigma^2)$ distribution and the norms $\|\mathbf{X}_j^+\|$ can be huge, especially in high dimension, when the matrix $\mathbf{X}$ is badly conditioned.

We then observe one of the benefits of applying the non-linear thresholding at each step: instead of having a noise level related to $\mathbf{X}^+\varepsilon$ as for $\widehat{\beta}_L$, we have a noise level related to $\mathbf{X}^T\varepsilon$ for the IHT algorithm (6.8).

6.4.2 Group Thresholding

Let the indices $\{1,\ldots,p\}$ be partitioned into M groups $\{1,\ldots,p\} = \cup_{j=1}^M G_j$ and let $\lambda_1,\ldots,\lambda_M$ be M positive real numbers. We give in this exercise an explicit solution to the minimization problem

$$\widehat{\beta} \in \operatorname*{argmin}_{\beta\in\mathbb{R}^p} \left\{ \|\alpha-\beta\|^2 + \sum_{j=1}^{M} (\lambda^{(j)})^2 \mathbf{1}_{\beta_{G_j}\neq 0} \right\}. \tag{6.32}$$

1. Check that a solution to the minimization problem $\min_{\beta\in\mathbb{R}^p}\left\{\|\alpha-\beta\|^2 + \lambda^2\mathbf{1}_{\beta\neq 0}\right\}$ is given by $\alpha\mathbf{1}_{\|\alpha\|>\lambda}$.
2. Using the decomposition,

$$\|\alpha-\beta\|^2 = \sum_{j=1}^{M} \|\alpha_{G_j} - \beta_{G_j}\|^2,$$

conclude that $\widehat{\beta} = H_\lambda^G(\alpha)$ where $[H_\lambda^G(\alpha)]_{G_j} = \alpha_{G_j}\mathbf{1}_{\|\alpha_{G_j}\|>\lambda^{(j)}}$.

6.4.3 Risk Bound for Iterative Group Thresholding

In this exercise, we prove Theorem 6.4 and Corollary 6.5. The proofs of these two results are closely related to the proofs of Theorem 6.1 and Corollary 6.3.

As in the proof of Theorem 6.1, we set $k_G^* = |\beta^*|_0^G$, $Z = \mathbf{X}^T\varepsilon$ and $b^{t+1} = \beta^* + \Lambda(\widehat{\beta}^t - \beta^*) + Z$, so that $\widehat{\beta}^{t+1} = H^G_{\lambda_{t+1}}(b^{t+1})$. The next lemma is central in the proofs of Theorem 6.4.

Lemma 6.6 Contraction property.

If the properties (6.28) and (6.29) hold at step t, then we have

$$\max_{S=\cup_{j\in J}G_j:\ |J|\leq c^2k_G^*/q}\left(\|(b^{t+1}-\beta^*)_S\| - \sqrt{|S|}\,|Z|_\infty^G\right) \leq \delta\|\beta^* - \widehat{\beta}^t\|. \tag{6.33}$$

A) Proof of Lemma 6.6.

The proof of Lemma 6.6 is similar to the proof of Lemma 6.2. For $S = \cup_{j\in J}G_j$ with $|J| \leq c^2k_G^*/q$, we set $S' = S \cup m_G^* \cup \text{supp}_G(\widehat{\beta}^t_{\bar{m}_G^*})$.

1. Check that $|S'| \leq (1+2c^2)k_G^*$.
2. Prove the inequalities

$$\begin{aligned}\|(b^{t+1}-\beta^*)_S\| &\leq \|\Lambda(\widehat{\beta}^t - \beta^*)_S\| + \sqrt{|S|}\,|Z|_\infty^G \\ &\leq \delta\|\widehat{\beta}^t - \beta^*\| + \sqrt{|S|}|Z|_\infty^G.\end{aligned}$$

3. Conclude the proof of Lemma 6.6.

B) Proof of Theorem 6.4.

As for Theorem 6.1, we proceed by induction.

1. Check that (6.28) and (6.29) hold for $t = 0$.

In the following, we assume that (6.28) and (6.29) hold at step t. We will prove that they then hold at step $t+1$.

2. Prove that

$$\begin{aligned}\|(\widehat{\beta}^{t+1}-\beta^*)_{m_G^*}\| &\leq \|[H^G_{\lambda_{t+1}}(b^{t+1})]_{m_G^*} - b^{t+1}_{m_G^*}\| + \|b^{t+1}_{m_G^*} - \beta^*_{m_G^*}\| \\ &\leq \gamma_{t+1}\sqrt{k_G^*} + \delta\|\widehat{\beta}^t - \beta^*\| + \sqrt{k_G^*}|Z|_\infty^G.\end{aligned}$$

3. With the induction property, prove that

$$\|(\widehat{\beta}^{t+1}-\beta^*)_{m_G^*}\| \leq \sqrt{k_G^*}\left(\gamma_{t+1} + ca^{-1}\gamma_t + |Z|_\infty^G\right).$$

4. For $S = \cup_{j\in J}G_j \subset \text{supp}_G(\widehat{\beta}^{t+1}_{\bar{m}_G^*})$, with $|S| \leq c^2k_G^*$, check that

$$\gamma_{t+1}\sqrt{|S|} \leq \|\widehat{\beta}_S^{t+1}\| \leq \delta\|\widehat{\beta}^t - \beta^*\| + \sqrt{|S|}\,|Z|_\infty^G.$$

5. Prove that we cannot have $|S| = c^2 k_G^*$, and hence (6.28) holds at step $t+1$.
6. As a consequence, prove the inequality

$$\|\widehat{\beta}^{t+1}_{\bar{m}^*_G}\| \leq \sqrt{k_G^*}\left(ca^{-1}\gamma_t + c|Z|^G_\infty\right).$$

7. Combining Questions 3 and 6, prove that (6.29) holds at step $t+1$.

C) Proof of Corollary 6.5.

All we need is to prove that (6.27) holds with probability larger than $1-2e^{-L}$. We remind the reader that all the columns of $\mathbf{X}$ have unit ℓ^2-norm.

1. Check that $\|\mathbf{X}_{G_j}\|_F^2 = |G_j|$.
2. Check that the application $x \to \|\mathbf{X}_{G_j}^T x\|$ is $|\mathbf{X}_{G_j}|_{op}$-Lipschitz.
3. With the Gaussian concentration inequality (Theorem B.7, page 301), prove that for $u \geq 0$

$$\mathbb{P}\left[\|Z_{G_j}\| \geq \sigma\left(\sqrt{|G_j|} + |\mathbf{X}_{G_j}|_{op}\sqrt{2u}\right)\right] \leq e^{-u}.$$

4. Prove that for $L \geq 0$

$$\mathbb{P}\left[\max_{j=1,\dots,M} \frac{\|Z_{G_j}\|}{\sqrt{|G_j|}} \geq \sigma\left(1+\phi_G\sqrt{2\log(M)+2L}\right)\right] \leq e^{-L},$$

and conclude the proof of Corollary 6.5.

Chapter 7

Estimator Selection

Which modeling and which estimator shall I use? These are eternal questions of the statistician when facing data.

The first step is to write down a statistical model suited to analyze the data. This step requires some deep discussions with the specialists of the field where the data come from, some basic data mining to detect some possible key features of the data, and some... experience. This process is crucial for the subsequent work, yet it seems hardly amenable to a theorization.

The second step is to choose at best an estimator. Assume, for example, that after some pretreatments of the data, you end up with a linear model (2.2). In some cases, you know from expert knowledge the class of structures (coordinate sparsity, group sparsity, variation-sparsity, etc.) hidden in the data. But in many cases, this class is unknown, and we need to select this class from the data. Even if you know the class of structures, say the coordinate sparsity, many estimators have been proposed for this setting, including the Lasso estimator (5.4), the mixing estimator (4.2), the forward–backward estimator of Section 2.4, the Dantzig selector (5.28), the Ridge estimator (5.29), and many others. Which of these estimators shall you use? By the way, all these estimators depend on some "tuning parameter" (λ for the Lasso, β for the mixing, etc.), and we have seen that a suitable choice of these estimators requires the knowledge of the variance σ^2 of the noise. Unfortunately, this variance is usually unknown. So even if you are a Lasso enthusiast, you miss some key information to apply properly the Lasso procedure.

Of course, you should start by removing the estimators whose computational cost exceeds your computational resources. But then you will still have to decide among different estimation schemes and different tuning parameters.

We formalize this issue in Section 7.1, and we then describe different approaches for selecting the estimators and their tuning parameters. The problem of choosing the parameter λ for the Lasso estimator (5.4) will be use as a prototypal example of parameter tuning. We describe three approaches for estimator selection. The first two approaches are quite universal. The first one is based on some data resampling, and the second one is inspired by model selection techniques. The last approach is specifically designed for the estimators introduced in Chapter 5. The theory for analyzing these selection procedures is somewhat more involved than the theory presented in the previous chapters, so we mainly focus in this chapter on the practical description of the procedures, and we refer to the original papers for details on the theoretical aspects.

7.1 Estimator Selection

Let us formalize the issues discussed above in the regression setting

$$Y_i = f(x^{(i)}) + \varepsilon_i, \quad i = 1, \ldots, n,$$

with $f : \mathbb{R}^p \to \mathbb{R}$, for the problem of estimating $f^* = [f(x^{(i)})]_{i=1,\ldots,n}$. We want to choose at best one estimator among many different estimators corresponding to different classes of structures (coordinate sparsity, group sparsity, etc.), to different estimation schemes (Lasso, mixing, etc.) and to different values of their tuning parameters.

For an estimation scheme s and a tuning parameter h (possibly multidimensional), we denote by $\widehat{f}_{(s,h)}$ the corresponding estimator of f^*. Let $\mathscr{S}$ be a (finite) family of estimation schemes. At each scheme $s \in \mathscr{S}$, we associate a (finite) set $\mathscr{H}_s$ of tuning parameters. Pooling together all these estimators $\{\widehat{f}_{(s,h)} : s \in \mathscr{S},\ h \in \mathscr{H}_s\}$, our ideal goal is to select among this family an estimator $\widehat{f}_{(s,h)}$ whose risk

$$R(\widehat{f}_{(s,h)}) = \mathbb{E}\left[\|\widehat{f}_{(s,h)} - f^*\|^2\right]$$

is almost as small as the oracle risk $\min_{\{s\in\mathscr{S}, h\in\mathscr{H}_s\}} R(\widehat{f}_{(s,h)})$.

In order to lighten the notations, let us denote by λ the couple (s,h). Setting

$$\Lambda = \bigcup_{s\in\mathscr{S}} \bigcup_{h\in\mathscr{H}_s} \{(s,h)\},$$

this ideal objective can be rephrased as the problem of selecting among $\{\widehat{f}_\lambda,\ \lambda \in \Lambda\}$ an estimator $\widehat{f}_{\widehat{\lambda}}$ whose risk $R(\widehat{f}_{\widehat{\lambda}})$ is almost as small as the oracle risk $\min_{\lambda\in\Lambda} R(\widehat{f}_\lambda)$.

Tuning the Lasso Estimator

The Lasso estimator is a prototypal example of estimator requiring a data-driven selection of its tuning parameter λ. Actually, the choice of λ proposed in Corollary 5.3 is proportional to the standard deviation σ of Y, which is usually unknown. This issue is due to the fact that the Lasso estimator is not invariant by rescaling, as explained below.

Let $\widehat{\beta}(Y,\mathbf{X})$ be any estimator of β^* in the linear model $Y = \mathbf{X}\beta^* + \varepsilon$. Assume that we change the unit of measurement of Y, which amounts to rescale Y by a scaling factor $s > 0$. We then observe $sY = \mathbf{X}(s\beta) + s\varepsilon$ instead of Y. A proper estimator of β should be scale-invariant, which means that it fulfills

$$\widehat{\beta}(sY,\mathbf{X}) = s\widehat{\beta}(Y,\mathbf{X}) \quad \text{for any } s > 0.$$

It turns out that the Lasso estimator (5.4) is not scale-invariant. Actually, let us denote by $\widehat{\beta}_{\lambda}^{\text{lasso}}(Y,\mathbf{X})$ the solution of the minimization problem (5.4) (assuming for simplicity that there is a unique solution). Since

$$\|sY - \mathbf{X}\beta\|^2 + \lambda|\beta|_1 = s^2\left(\|Y - \mathbf{X}(\beta/s)\|^2 + \frac{\lambda}{s}|\beta/s|_1\right),$$

we have

$$\widehat{\beta}_{\lambda}^{\text{lasso}}(sY,\mathbf{X}) = s\widehat{\beta}_{\lambda/s}^{\text{lasso}}(Y,\mathbf{X}),$$

so the Lasso estimator is not scale-invariant, since we need to rescale the parameter λ by $1/s$. This explains why a sensible choice of the tuning parameter λ of the Lasso estimator (5.4) should be proportional to the standard deviation σ of the response Y.

In practice, to choose the parameter λ, we choose a finite grid Λ of $\mathbb{R}^+$ and select a Lasso estimator $\widehat{f}_{\widehat{\lambda}} = \mathbf{X}\widehat{\beta}_{\widehat{\lambda}}^{\text{lasso}}(Y,\mathbf{X})$ among the collection $\{\widehat{f}_{\lambda} = \mathbf{X}\widehat{\beta}_{\lambda}^{\text{lasso}}(Y,\mathbf{X}) : \lambda \in \Lambda\}$ with one of the procedures described below.

7.2 Cross-Validation Techniques

The cross-validation (CV) schemes are nearly universal in the sense that they can be implemented in most statistical frameworks and for most estimation procedures. The principle of the cross-validation schemes is to keep aside a fraction of the data in order to evaluate the prediction accuracy of the estimators. More precisely, the data is split into a *training* set and a *validation* set: the estimators are computed with the *training* set only, and the *validation* set is used for estimating their prediction risk. This training / validation splitting is eventually repeated several times. The cross-validated estimator then selects the estimator with the smallest estimated risk (see below for details). The most popular cross-validation schemes are:

- *Hold-out CV*, which is based on a single split of the data for *training* and *validation.*
- *V-fold CV*, where the data is split into V subsamples. Each subsample is successively removed for *validation*, the remaining data being used for *training*; see Figure 7.1.
- *Leave-one-out*, which corresponds to n-fold CV.
- *Leave-q-out* (or *delete-q-CV*) where every possible subset of cardinality q of the data is removed for *validation*, the remaining data being used for *training*.

V-fold cross-validation is arguably the most popular cross-validation scheme: It is more stable than the hold-out thanks to the repeated subsampling, and for small V it is computationally much less intensive than the Leave-one-out or Leave-q-out. We describe below the V-fold CV selection procedure in the regression setting of Section 7.1.

train	train	train	train	test
train	train	train	test	train
train	train	test	train	train
train	test	train	train	train
test	train	train	train	train

Figure 7.1 *Recursive data splitting for 5-fold CV.*

V-Fold Cross-Validation

Cross-validation schemes are naturally suited to estimators $\widehat{f}_\lambda$ of f^* of the following form. Let us denote by $\mathscr{D} = (y_i, x^{(i)})_{i=1,\ldots,n} \in \mathbb{R}^{(p+1)n}$ our data and assume that there is a map $F_\lambda : \cup_{n\in\mathbb{N}}\mathbb{R}^{(p+1)n} \times \mathbb{R}^p \to \mathbb{R}$ such that $\widehat{f}_\lambda = [F_\lambda(\mathscr{D}, x^{(i)})]_{i=1,\ldots,n}$. In the case of Lasso estimator (5.4), the map F_λ is given by $F_\lambda((Y,\mathbf{X}),x) = \langle \widehat{\beta}_\lambda^{\text{lasso}}(Y,\mathbf{X}), x\rangle$ (with the notations of Section 7.1).

Let $\{1,\ldots,n\} = I_1 \cup \ldots \cup I_V$ be a partition of $\{1,\ldots,n\}$. For a given $k \in \{1,\ldots,V\}$ we denote by $\mathscr{D}_{-I_k}$ the partial data set $\mathscr{D}_{-I_k} = \left\{(y_i, x^{(i)}) : i \in \{1,\ldots,n\} \setminus I_k\right\}$, where we have removed the data points with index in I_k. For each k, the partial data set $\mathscr{D}_{-I_k}$ is used to compute the map $F_\lambda(\mathscr{D}_{-I_k}, \cdot)$, and the remaining data $\mathscr{D}_{I_k} = (y_i, x^{(i)})_{i\in I_k}$ is used to evaluate the accuracy of $F_\lambda(\mathscr{D}_{-I_k}, \cdot)$ for predicting the response y. More precisely, in the setting of Section 7.1, this evaluation will be performed by computing for each $i \in I_k$ the square-prediction error $(y_i - F_\lambda(\mathscr{D}_{-I_k}, x^{(i)}))^2$. Averaging these prediction errors, we obtain the V-fold CV ℓ^2-risk of F_λ

$$\widehat{R}_{CV}[F_\lambda] = \frac{1}{V}\sum_{k=1}^{V} \frac{1}{|I_k|} \sum_{i\in I_k} \left(y_i - F_\lambda(\mathscr{D}_{-I_k}, x^{(i)})\right)^2. \tag{7.1}$$

The V-fold CV procedure then selects $\widehat{\lambda}_{CV}$ by minimizing this estimated risk

$$\widehat{\lambda}_{CV} \in \operatorname*{argmin}_{\lambda\in\Lambda} \widehat{R}_{CV}[F_\lambda].$$

The final estimator of f^* is $\widehat{f}_{\widehat{\lambda}_{CV}}$.

V-fold CV is widely used for estimator selection. It is in particular very popular for tuning procedure such as the Lasso or the group-Lasso. It usually gives good results in practice, yet it suffers from the two following caveats.

Caveat 1. The V-fold CV selection procedure requires to compute V times the maps $\{F(\mathscr{D}_{-I_k}, \cdot) : \lambda \in \Lambda\}$ and also the final estimator $\widehat{f}_{\widehat{\lambda}_{CV}}$. The computational cost of the V-fold CV procedure is therefore roughly $V \times \text{card}(\Lambda)$ times the computational cost of computing a single estimator $\widehat{f}_\lambda$. For large V, this computational cost can overly exceed the available computational resources. A remedy for this issue is to choose a small V. Yet, for V small, we lose the stabilizing effect of the repeated subsampling, and the V-fold CV procedure can be unstable in this case. A suitable choice of V must then achieve a good balance between the computational complexity and the stability. A common choice in practice is $V = 10$, but some choose smaller values (like $V = 5$) when the computations resources are limited. We point out that when several Central Processing Units (CPUs) are available, the maps $\{F(\mathscr{D}_{-I_k}, \cdot) : \lambda \in \Lambda,\ k = 1, \ldots, V\}$ can be computed in parallel.

Caveat 2. Despite its wide popularity and its simplicity, there is no general theoretical result that guarantees the validity of the V-fold CV procedure in high-dimensional settings. Actually, while it is not hard to identify the expected values of $\widehat{R}_{CV}[F_\lambda]$ in a random design setting (see Exercise 7.6.1), the analysis of the fluctuations of $\widehat{R}_{CV}[F_\lambda]$ around its expectation is much more involved in a high-dimensional setting. In practice, data splitting can be an issue when n is small, since the estimators $F_\lambda(\mathscr{D}_{-I_k}, \cdot)$ are only based on the partial data set $\mathscr{D}_{-I_k}$ and they can be unstable. This instability shrinks when n increases.

7.3 Complexity Selection Techniques

In order to avoid the two above caveats, an alternative point of view is to adapt to estimator selection the ideas of model selection introduced in Chapter 2. Similarly to the model selection estimator (2.9), we introduce a collection $\{S_m,\ m \in \mathscr{M}\}$ of models designed to approximate the estimators $(\widehat{f}_\lambda)_{\lambda \in \Lambda}$ and a probability distribution π on $\mathscr{M}$. We choose exactly the same models and probability distributions as in Chapter 2, depending on the setting under study (coordinate sparsity, group sparsity, variation sparsity, etc.).

Compared to model selection as described in Chapter 2, we have to address three issues. First, for computational efficiency, we cannot explore a huge collection of models. For this reason, we restrict to a subcollection of models $\{S_m,\ m \in \widehat{\mathscr{M}}\}$, possibly data-dependent. The choice of this subcollection depends on the statistical problem; see Section 7.3.1 for the coordinate-sparse setting and 7.3.2 for the group-sparse regression. Second, we have to take into account the fact that a good model S_m for an estimator $\widehat{f}_\lambda$ is a model that achieves a good balance between the approximation error $\|\widehat{f}_\lambda - \text{Proj}_{S_m} \widehat{f}_\lambda\|^2$ and the complexity measured by $\log(1/\pi_m)$. Therefore, the selection criterion involves this approximation error term in addition to the penalty. Finally, the criterion shall not depend on the unknown variance σ^2. For this reason, we replace σ^2 in front of the penalty term by the estimator

$$\widehat{\sigma}_m^2 = \frac{\|Y - \text{Proj}_{S_m} Y\|^2}{n - \dim(S_m)}. \tag{7.2}$$

Combining these three features, we can select the estimator $\widehat{f}_{\widehat{\lambda}}$ by minimizing the criterion

$$\text{Crit}(\widehat{f}_\lambda) = \inf_{m \in \widehat{\mathscr{M}}} \left[\|Y - \text{Proj}_{S_m} \widehat{f}_\lambda\|^2 + \frac{1}{2} \|\widehat{f}_\lambda - \text{Proj}_{S_m} \widehat{f}_\lambda\|^2 + \text{pen}_\pi(m)\, \widehat{\sigma}_m^2 \right], \tag{7.3}$$

where $\widehat{\sigma}_m^2$ is given by (7.2) and $\text{pen}_\pi(m)$ is defined as the unique solution of

$$\mathbb{E}\left[\left(U - \frac{\text{pen}_\pi(m)}{1.1(n-\dim(S_m))}V\right)_+\right] = \pi_m, \tag{7.4}$$

where U and V are two independent chi-square random variables with $\dim(S_m)+1$ and $n-\dim(S_m)-1$ degrees of freedom, respectively. In the cases we will consider here, the penalty $\text{pen}_\pi(m)$ is roughly of the order of $\log(1/\pi_m)$ (see Exercise 7.6.3), and therefore it penalizes S_m according to its complexity $\log(1/\pi_m)$. An implementation of Criterion (7.3) is available in the `R` package `LINselect` `http://cran.r-project.org/web/packages/LINselect/`.

Compared to V-fold CV, this selection criterion enjoys two nice properties:

(i) It does not involve any data splitting. The estimators are then built on the whole data set (which is wise when n is small) and they are computed once.

(ii) The risk of the procedure is controlled by a bound similar to Theorem 2.2 in Chapter 2 for model selection. Deriving such a risk bound involves the same ideas as for the model selection procedure of Chapter 2, but it is somewhat more technical. Therefore, we will not detail these theoretical aspects and we will only focus below on the practical aspects. The references for the theoretical details are given in Section 7.5; see also Exercise 7.6.4.

Caveat. The major caveat of the selection procedure (7.3) is that it is designed for the Gaussian setting and there is no guarantee when the noise does not have a Gaussian distribution.

In the remainder of this section, we describe the instantiation of Criterion (7.3) in the coordinate-sparse setting, in the group-sparse setting, and in a mixed setting.

7.3.1 Coordinate-Sparse Regression

Let us consider the problem of selecting among sparse linear regressors $\{\widehat{f}_\lambda = \mathbf{X}\widehat{\beta}_\lambda : \lambda \in \Lambda\}$ in the linear regression model $Y = \mathbf{X}\beta^* + \varepsilon$ in the coordinate-sparse setting described in Chapter 2. We consider the collection of models $\{S_m,\ m \in \mathcal{M}\}$ and the probability π designed in Chapter 2 for the coordinate-sparse regression setting. For $\lambda \in \Lambda$, we define $\widehat{m}_\lambda = \text{supp}(\widehat{\beta}_\lambda)$ and

$$\widehat{\mathcal{M}} = \{\widehat{m}_\lambda :\ \lambda \in \Lambda \ \text{ and } \ 1 \leq |\widehat{m}_\lambda| \leq n/(3\log p)\}. \tag{7.5}$$

The estimator $\widehat{f}_{\widehat{\lambda}} = \mathbf{X}\widehat{\beta}_{\widehat{\lambda}}$ is then selected by minimizing (7.3).

As mentioned earlier, the risk of the estimator $\widehat{f}_{\widehat{\lambda}}$ can be compared to the risk of the best estimator in the collection $\{\widehat{f}_\lambda : \lambda \in \Lambda\}$. Let us describe such a result in the case where $\widehat{\beta}_\lambda$ is a Lasso-estimator and $\Lambda = \mathbb{R}^+$.

For $m \subset \{1,\ldots,p\}$, we define ϕ_m as the largest eigenvalue of $\mathbf{X}_m^T\mathbf{X}_m$, where $\mathbf{X}_m$ is the matrix derived from $\mathbf{X}$ by only keeping the columns with index in m. The following proposition involves the restricted eigenvalue $\phi^* = \max\{\phi_m : \text{card}(m) \leq n/(3\log p)\}$ and the compatibility constant $\kappa(\beta)$ defined by (5.8) in Chapter 5, page 95.

Proposition 7.1 *There exist positive numerical constants C, C_1, C_2, and C_3, such that for any β^* fulfilling*

$$|\beta^*|_0 \leq C \frac{\kappa^2(\beta^*)}{\phi^*} \times \frac{n}{\log(p)},$$

the Lasso estimator $\widehat{f}_{\widehat{\lambda}} = \mathbf{X}\widehat{\beta}_{\widehat{\lambda}}$ selected according to (7.3) fulfills

$$\|\mathbf{X}(\widehat{\beta}_{\widehat{\lambda}} - \beta^*)\|^2 \leq C_3 \inf_{\beta \neq 0} \left\{ \|\mathbf{X}(\beta^* - \beta)\|^2 + \frac{\phi^* |\beta|_0 \log(p)}{\kappa^2(\beta)} \sigma^2 \right\}, \tag{7.6}$$

with probability at least $1 - C_1 p^{-C_2}$.

The risk Bound (7.6) is similar to (5.13) except that

- it does not require the knowledge of the variance σ^2;
- it requires an upper bound on the cardinality of the support of β^*; and
- it involves two constants C_3, ϕ^* larger than 1.

The proof of this result relies on arguments close to those of Theorem 2.2, yet it is slightly lengthy and we omit it (see Section 7.5 for a reference).

7.3.2 Group-Sparse Regression

Let us now describe the procedure (7.3) in the group-sparse setting of Chapter 2. For simplicity, we restrict to the specific case where all the groups G_k have the same cardinality $T = p/M$ and the columns of $\mathbf{X}$ are normalized to 1. For any $m \subset \{1,\ldots,M\}$, we define the submatrix $\mathbf{X}_{(m)}$ of $\mathbf{X}$ by only keeping the columns of $\mathbf{X}$ with index in $\bigcup_{k\in m} G_k$. We also write $\mathbf{X}_{G_k}$ for the submatrix of $\mathbf{X}$ built from the columns with index in G_k. The collection $\{S_m,\ m \in \mathcal{M}\}$ and the probability π are those defined in Chapter 2 for group sparsity. For a given $\lambda > 0$, similarly to the coordinate-sparse case, we define $\widehat{m}(\lambda) = \{k :\ [\widehat{\beta}_\lambda]_{G_k} \neq 0\}$ and

$$\widehat{\mathcal{M}} = \left\{ \widehat{m}(\lambda) :\ \lambda > 0 \text{ and } 1 \leq |\widehat{m}(\lambda)| \leq \frac{n-2}{2T \vee 3\log(M)} \right\}. \tag{7.7}$$

The estimator $\widehat{f}_{\widehat{\lambda}} = \mathbf{X}\widehat{\beta}_{\widehat{\lambda}}$ is then selected by minimizing (7.3).

As above, let us give a risk bound for the case where the estimator $\widehat{\beta}_\lambda$ is the group-Lasso estimator (5.17) with $\lambda_1 = \ldots = \lambda_M = \lambda \in \mathbb{R}^+$. For $m \subset \{1,\ldots,M\}$, we define $\phi_{(m)}$ as the largest eigenvalue of $\mathbf{X}_{(m)}^T \mathbf{X}_{(m)}$. The following risk bound involves the cardinality $|\mathcal{K}(\beta)|$ of the set $\mathcal{K}(\beta) = \{k : \beta_{G_k} \neq 0\}$, the group-restricted eigenvalue

$$\phi_G^* = \max\left\{ \phi_{(m)} :\quad 1 \leq |m| \leq \frac{n-2}{2T \vee 3\log(M)} \right\}, \tag{7.8}$$

and the group-compatibility constant $\kappa_G(\beta)$ defined by (5.18) in Chapter 5, page 105.

Proposition 7.2 *There exist positive numerical constants C, C_1, C_2, and C_3, such that when*

$$T \leq (n-2)/4 \quad \text{and} \quad 1 \leq |\mathscr{K}(\beta^*)| \leq C \frac{\kappa_G^2(\beta^*)}{\phi_G^*} \times \frac{n-2}{\log(M) \vee T},$$

the group-Lasso estimator $\widehat{f}_{\widehat{\lambda}} = \mathbf{X}\widehat{\beta}_{\widehat{\lambda}}$ selected according to (7.3) fulfills

$$\|\mathbf{X}(\widehat{\beta}_{\widehat{\lambda}} - \beta^*)\|^2 \leq C_3 \inf_{\beta \neq 0} \left\{ \|\mathbf{X}(\beta - \beta^*)\|^2 + \frac{\phi_G^*}{\kappa_G^2(\beta)} |\mathscr{K}(\beta)| \, (T + \log(M))\, \sigma^2 \right\}, \tag{7.9}$$

with probability at least $1 - C_1 M^{-C_2}$.

Again, for conciseness, we omit the proof of this result (see Section 7.5 for a reference).

7.3.3 Multiple Structures

When we ignore which class of structures (coordinate-sparse, group-sparse, variation-sparse, etc.) is hidden in the data, we wish to select the best estimator among a collection of estimators corresponding to various classes of structures. To illustrate this point, let us assume that our family $\{\widehat{f}_\lambda = \mathbf{X}\widehat{\beta}_\lambda : \lambda \in \Lambda\}$ gathers some estimators $\{\widehat{f}_\lambda = \mathbf{X}\widehat{\beta}_\lambda : \lambda \in \Lambda^{\text{coord}}\}$ with $\widehat{\beta}_\lambda$ coordinate-sparse, and some estimators $\{\widehat{f}_\lambda = \mathbf{X}\widehat{\beta}_\lambda : \lambda \in \Lambda^{\text{group}}\}$ with $\widehat{\beta}_\lambda$ group-sparse. Let us denote by $\{S_m^{\text{coord}} : m \in \mathscr{M}^{\text{coord}}\}$ and π^{coord} (respectively, $\{S_m^{\text{group}} : m \in \mathscr{M}^{\text{group}}\}$ and π^{group}) the models and the probability distribution defined in Chapter 2 for the coordinate-sparse setting (respectively, for the group-sparse setting). Writing $\widehat{\mathscr{M}}^{\text{coord}}$ (respectively, $\widehat{\mathscr{M}}^{\text{group}}$) for the family defined by (7.5) (respectively, by (7.7)), the selection procedure (7.3) can then be applied with the collection of models $\{S_m^{\text{coord}} : m \in \widehat{\mathscr{M}}^{\text{coord}}\} \cup \{S_m^{\text{group}} : m \in \widehat{\mathscr{M}}^{\text{group}}\}$, and with the probability distribution $\pi_m = \pi_m^{\text{coord}}/2$ if $m \in \mathscr{M}^{\text{coord}}$ and $\pi_m = \pi_m^{\text{group}}/2$ if $m \in \mathscr{M}^{\text{group}}$. The theoretical guarantees on the selection procedure (7.3) then ensures that the selected estimator $\widehat{f}_{\widehat{\lambda}}$ is almost as good as the best of the estimators in $\{\widehat{f}_\lambda : \lambda \in \Lambda\}$. Again, we refer to the original papers (see Section 7.5) for details.

7.4 Scale-Invariant Criteria

In this section, we describe an alternative to the cross-validation techniques and the complexity selection techniques for the specific problem of choosing the tuning parameter of procedures like the Lasso. In the linear regression setting $Y = \mathbf{X}\beta^* + \varepsilon$, as explained in Section 7.1, a sensible estimator $\widehat{\beta}(Y,\mathbf{X})$ of β^* should be scale-invariant, which means that $\widehat{\beta}(sY,\mathbf{X}) = s\widehat{\beta}(Y,\mathbf{X})$ for all $s > 0$. Let us consider an estimator $\widehat{\beta}(Y,\mathbf{X})$, obtained by minimizing some function $\beta \to \mathscr{L}(Y,\mathbf{X},\beta)$. The estimator $\widehat{\beta}(Y,\mathbf{X})$ is scale-invariant if

$$\underset{\beta}{\operatorname{argmin}}\, \mathscr{L}(sY,\mathbf{X},s\beta) = \underset{\beta}{\operatorname{argmin}}\, \mathscr{L}(Y,\mathbf{X},\beta).$$

Such a condition is met, for example, when there exists some functions $a(Y,\mathbf{X},s)$ and $b(Y,\mathbf{X},s)$, such that

$$\mathscr{L}(sY,\mathbf{X},s\beta) = a(Y,\mathbf{X},s)\mathscr{L}(Y,\mathbf{X},\beta) + b(Y,\mathbf{X},s).$$

The estimators introduced in Chapter 5 are obtained by minimizing a function

$$\mathscr{L}(Y,\mathbf{X},\beta) = \|Y - \mathbf{X}\beta\|^2 + \lambda\Omega(\beta), \tag{7.10}$$

which is the sum of a quadratic term $\|Y - \mathbf{X}\beta\|^2$ and a convex function Ω homogeneous with degree 1. A prototypal example is the Lasso estimator where $\Omega(\beta) = |\beta|_1$. Such estimators are not scale-invariant. We observe that the standard deviation σ of the noise ε is equal to the standard deviation of $Y = \mathbf{X}\beta^* + \varepsilon$. Setting $\lambda = \mu\sigma$, we derive from (7.10) a scale-invariant estimator by minimizing

$$\sigma^{-1}\|Y - \mathbf{X}\beta\|^2 + \mu\Omega(\beta), \quad \text{with } \sigma = \operatorname{sdev}(Y) = \operatorname{sdev}(\varepsilon)$$

instead of (7.10). Yet, the standard deviation σ of the noise is usually unknown, so we cannot compute the above criterion. An idea is to estimate σ with the norm of the residuals, namely $\|Y - \mathbf{X}\beta\|/\sqrt{n}$. Following this idea, we obtain for $\mu > 0$ the scale-invariant criterion

$$\overline{\mathscr{L}}(Y,\mathbf{X},\beta) = \sqrt{n}\|Y - \mathbf{X}\beta\| + \mu\Omega(\beta). \tag{7.11}$$

Since any norm is convex, the function $\overline{\mathscr{L}}(Y,\mathbf{X},\beta)$ is convex and $\overline{\mathscr{L}}(sY,\mathbf{X},s\beta) = s\overline{\mathscr{L}}(Y,\mathbf{X},\beta)$, so the minimizer $\widehat{\beta}(Y,\mathbf{X})$ is scale-invariant.

Square-Root Lasso

Let us investigate the properties of the estimator $\widehat{\beta}$ obtained by minimizing $\overline{\mathscr{L}}(Y,\mathbf{X},\beta)$ when $\Omega(\beta) = |\beta|_1$ (we can follow the same lines with the other criteria). For $\mu > 0$, the resulting estimator is the so-called *square-root Lasso* estimator

$$\widehat{\beta} \in \underset{\beta\in\mathbb{R}^p}{\operatorname{argmin}} \left\{ \sqrt{n}\|Y - \mathbf{X}\beta\| + \mu|\beta|_1 \right\}. \tag{7.12}$$

The minimization problem (7.12) is scale-invariant and it can be computed in high-dimensional settings since it is convex. The next theorem shows that the resulting estimator enjoys some nice theoretical properties.

Theorem 7.3 Risk bound for the square-root Lasso

For $L > 0$, we set $\mu = 4\sqrt{\log(p)+L}$. Assume that for some $\delta \in (0,1)$ we have

$$|\beta^*|_0 \leq \delta^2 \frac{n\kappa^2(\beta^*)}{64\mu^2}, \tag{7.13}$$

with the compatibility constant $\kappa(\beta)$ defined in (5.8). Then, with probability at least $1 - e^{-L} - (1+e^2)e^{-n/24}$, the estimator (7.12) fulfills

$$\|\mathbf{X}(\widehat{\beta} - \beta^*)\|^2 \leq \frac{482(\log(p)+L)}{\delta(1-\delta)\kappa(\beta^*)^2} |\beta^*|_0 \sigma^2. \tag{7.14}$$

Proof. We first prove a deterministic bound on $\|\mathbf{X}(\widehat{\beta} - \beta^*)\|^2$.

Lemma 7.4 Deterministic bound

Let us consider some $\delta \in (0,1)$ and $\mu > 0$. Assume that (7.13) holds and that ε fulfills

$$2|\mathbf{X}^T\varepsilon|_\infty \leq \mu\|\varepsilon\|/\sqrt{n}. \tag{7.15}$$

Then, for $\varepsilon \neq 0$ we have the upper bound

$$\|\mathbf{X}(\widehat{\beta} - \beta^*)\|^2 \leq \frac{18\mu^2\|\varepsilon\|^2}{\delta(1-\delta)\kappa(\beta^*)^2 n} |\beta^*|_0. \tag{7.16}$$

Proof of Lemma 7.4

We set $m_* = \text{supp}(\beta^*)$. We first prove that

$$|(\widehat{\beta} - \beta^*)_{m_*^c}|_1 \leq 3|(\widehat{\beta} - \beta^*)_{m_*}|_1. \tag{7.17}$$

Since $\widehat{\beta}$ minimizes (7.12), we have

$$\begin{aligned}\|Y - \mathbf{X}\widehat{\beta}\| - \|Y - \mathbf{X}\beta^*\| &\leq \frac{\mu}{\sqrt{n}}\left(|\beta^*|_1 - |\widehat{\beta}|_1\right) = \frac{\mu}{\sqrt{n}}\left(|\beta^*_{m^*}|_1 - |\widehat{\beta}_{m_*}|_1 - |\widehat{\beta}_{m_*^c}|_1\right) \\ &\leq \frac{\mu}{\sqrt{n}}\left(|(\widehat{\beta} - \beta^*)_{m^*}|_1 - |(\widehat{\beta} - \beta^*)_{m_*^c}|_1\right).\end{aligned} \tag{7.18}$$

For $\varepsilon \neq 0$, the differential of the map $\beta \to \|Y - \mathbf{X}\beta\|$ at β^* is

$$\nabla_\beta \|Y - \mathbf{X}\beta^*\| = \frac{-\mathbf{X}^T(Y - \mathbf{X}\beta^*)}{\|Y - \mathbf{X}\beta^*\|} = -\frac{\mathbf{X}^T\varepsilon}{\|\varepsilon\|}.$$

Since $\beta \to \|Y - \mathbf{X}\beta\|$ is convex, we have (Lemma D.1, page 321, in Appendix D)

$$\begin{aligned}\|Y - \mathbf{X}\widehat{\beta}\| - \|Y - \mathbf{X}\beta^*\| &\geq -\left\langle \frac{\mathbf{X}^T\varepsilon}{\|\varepsilon\|}, \widehat{\beta} - \beta^* \right\rangle \\ &\geq -\frac{|\mathbf{X}^T\varepsilon|_\infty}{\|\varepsilon\|} |\widehat{\beta} - \beta^*|_1 \geq -\frac{\mu}{2\sqrt{n}} |\widehat{\beta} - \beta^*|_1,\end{aligned} \tag{7.19}$$

where the last inequality follows from (7.15). Combining (7.19) with (7.18), we get (7.17).

We now prove (7.16). Since Criterion (7.12) is convex, when $Y \neq \mathbf{X}\widehat{\beta}$, the first-order optimality condition (Lemma D.4, page 323, in Appendix D) ensures the existence of $\widehat{z} \in \partial|\widehat{\beta}|_1$, such that

$$\frac{-\sqrt{n}\mathbf{X}^T(Y-\mathbf{X}\widehat{\beta})}{\|Y-\mathbf{X}\widehat{\beta}\|} + \mu\widehat{z} = 0.$$

In particular,

$$-\langle Y-\mathbf{X}\widehat{\beta}, \mathbf{X}(\widehat{\beta}-\beta^*)\rangle + \frac{\mu}{\sqrt{n}}\|Y-\mathbf{X}\widehat{\beta}\|\langle \widehat{z}, \widehat{\beta}-\beta^*\rangle = 0.$$

This inequality still holds when $Y = \mathbf{X}\widehat{\beta}$. Since $|\widehat{z}|_\infty \leq 1$ (Lemma D.5 in Appendix D), we have according to (7.15) and (7.17)

$$\begin{aligned}
\|\mathbf{X}(\widehat{\beta}-\beta^*)\|^2 &\leq \langle \varepsilon, \mathbf{X}(\widehat{\beta}-\beta^*)\rangle + \mu\frac{\|Y-\mathbf{X}\widehat{\beta}\|}{\sqrt{n}}|\widehat{\beta}-\beta^*|_1 \\
&\leq \left(|\mathbf{X}^T\varepsilon|_\infty + \mu\frac{\|Y-\mathbf{X}\widehat{\beta}\|}{\sqrt{n}}\right)|\widehat{\beta}-\beta^*|_1 \\
&\leq 4\mu\left(\frac{\|\varepsilon\|/2 + \|Y-\mathbf{X}\widehat{\beta}\|}{\sqrt{n}}\right)|(\widehat{\beta}-\beta^*)_{m_*}|_1.
\end{aligned}$$

We first observe that $\|Y-\mathbf{X}\widehat{\beta}\| \leq \|\varepsilon\| + \|\mathbf{X}(\widehat{\beta}-\beta^*)\|$. Furthermore, according to (7.17) and the definition of the compatibility constant (5.8), we have the upper bound $|(\widehat{\beta}-\beta^*)_{m_*}|_1 \leq \sqrt{|\beta^*|_0}\,\|\mathbf{X}(\widehat{\beta}-\beta^*)\|/\kappa(\beta^*)$, so

$$\|\mathbf{X}(\widehat{\beta}-\beta^*)\|^2 \leq 6\mu\frac{\|\varepsilon\|}{\sqrt{n}}\frac{\sqrt{|\beta^*|_0}\,\|\mathbf{X}(\widehat{\beta}-\beta^*)\|}{\kappa(\beta^*)} + 4\mu\frac{\sqrt{|\beta^*|_0}\,\|\mathbf{X}(\widehat{\beta}-\beta^*)\|^2}{\kappa(\beta^*)\sqrt{n}}.$$

For the second right-hand-side term, Condition (7.13) ensures that the factor $4\mu|\beta^*|_0^{1/2}/(\sqrt{n}\kappa(\beta^*))$ is smaller than $\delta/2$. Applying the inequality $ab \leq (2\delta)^{-1}a^2 + \delta b^2/2$ to the first right-hand-side term then gives

$$\|\mathbf{X}(\widehat{\beta}-\beta^*)\|^2 \leq \frac{18\mu^2\|\varepsilon\|^2}{\delta\kappa(\beta^*)^2 n}|\beta^*|_0 + \frac{\delta}{2}\|\mathbf{X}(\widehat{\beta}-\beta^*)\|^2 + \frac{\delta}{2}\|\mathbf{X}(\widehat{\beta}-\beta^*)\|^2,$$

from which (7.16) follows. □

Let us now conclude the proof of Theorem 7.3. On the event

$$\mathscr{A} = \left\{|\mathbf{X}^T\varepsilon|_\infty^2 \leq 2\sigma^2(\log(p)+L) \text{ and } |\sigma - \|\varepsilon\|/\sqrt{n}| \leq (1-2^{-1/2})\sigma\right\},$$

Condition (7.15) holds with $\mu^2 = 16(\log(p)+L)$, and the Bound (7.16) enforces

Bound (7.14). So all we need is to check that the probability of the event $\mathscr{A}$ is at least $1-e^{-L}-(1+e^2)e^{-n/24}$. According to the Gaussian concentration Inequality (B.2) and Bound (B.4) in Appendix B, we have

$$\begin{aligned}
&\mathbb{P}\left(|\sigma-\|\varepsilon\|/\sqrt{n}|\geq(1-2^{-1/2})\sigma\right)\\
&\leq\mathbb{P}\left(\|\varepsilon\|\geq\mathbb{E}[\|\varepsilon\|]+(\sqrt{n}-\sqrt{n/2})\sigma\right)+\mathbb{P}\left(\|\varepsilon\|\leq\mathbb{E}[\|\varepsilon\|]-(\sqrt{n-4}-\sqrt{n/2})\sigma\right)\\
&\leq e^{-\left(\sqrt{n}-\sqrt{n/2}\right)^2/2}+e^{-\left(\sqrt{n-4}-\sqrt{n/2}\right)^2/2}\leq(1+e^2)e^{-n/24}.
\end{aligned}$$

Finally, the proof of Corollary 5.3 ensures that

$$\mathbb{P}\left(|\mathbf{X}^T\varepsilon|_\infty^2\geq 2\sigma^2(\log(p)+L)\right)\leq e^{-L},$$

so the probability of the event $\mathscr{A}$ is at least $1-e^{-L}-(1+e^2)e^{-n/24}$. The proof of Theorem 7.3 is complete. □

Since Criterion (7.12) has been derived from the Lasso Criterion (5.4), it is worth it to investigate the possible connections between the square-root Lasso $\widehat{\beta}$ and the Lasso $\widehat{\beta}_\lambda^{\text{lasso}}(Y,\mathbf{X})$. We first observe that

$$\sqrt{n}\,\|Y-\mathbf{X}\beta\|=\min_{\sigma>0}\left\{\frac{n\sigma}{2}+\frac{\|Y-\mathbf{X}\beta\|^2}{2\sigma}\right\},$$

and the minimum is achieved for $\sigma=\|Y-\mathbf{X}\beta\|/\sqrt{n}$. As a consequence, the estimator $\widehat{\beta}$ defined by (7.12) and the standard-deviation estimator $\widehat{\sigma}=\|Y-\mathbf{X}\widehat{\beta}\|/\sqrt{n}$ are solution of the convex minimization problem

$$(\widehat{\beta},\widehat{\sigma})\in\operatorname*{argmin}_{(\beta,\sigma)\in\mathbb{R}^p\times\mathbb{R}^+}\left\{\frac{n\sigma}{2}+\frac{\|Y-\mathbf{X}\beta\|^2}{2\sigma}+\mu|\beta|_1\right\}.\tag{7.20}$$

In particular, defining $\widehat{\sigma}$ by (7.20), the square-root Lasso estimator $\widehat{\beta}$ is a solution of $\widehat{\beta}\in\operatorname{argmin}_\beta\left\{\|Y-\mathbf{X}\beta\|^2+2\mu\widehat{\sigma}|\beta|_1\right\}$. In other words, in the notations of Section 7.1, we have

$$\widehat{\beta}=\widehat{\beta}_{2\mu\widehat{\sigma}}^{\text{lasso}}(Y,\mathbf{X}).\tag{7.21}$$

This link between the square-root Lasso estimator and the Lasso estimator has some nice practical and theoretical consequences. From a practical point of view, we observe that Criterion (7.20) is convex in (β,σ), so the estimator $\widehat{\beta}$ can be efficiently computed by alternating minimization in β and σ. For a given σ, the minimization in β amounts to solve a Lasso problem with tuning parameter $2\mu\sigma$, while for a fixed β the minimization in σ has a closed-form solution $\sigma=\|Y-\mathbf{X}\beta\|/\sqrt{n}$. The resulting algorithm is the so-called scaled-Lasso algorithm described below. An implementation of this algorithm is available in the `R` package `scalreg`
`http://cran.r-project.org/web/packages/scalreg/`.

Scaled-Lasso algorithm

Initialization: $\widehat{\beta} = 0$

Repeat until convergence

- $\widehat{\sigma} = \|Y - \mathbf{X}\widehat{\beta}\| / \sqrt{n}$
- $\widehat{\beta} = \widehat{\beta}^{\text{lasso}}_{2\mu\widehat{\sigma}}(Y, \mathbf{X})$

Output: $\widehat{\beta}$

From a theoretical point of view, the link (7.21) allows to improve Bound (7.14) by transposing the results for the Lasso estimator. The only issue is to get a control on the size of $\widehat{\sigma}$. Such a control can be derived from Theorem 7.3. The next corollary provides a risk bound for the square-root Lasso similar to Corollary 5.3, page 97, for the Lasso.

Corollary 7.5 A tighter risk bound for the square-root Lasso

For $L > 0$, let us set $\mu = 3\sqrt{2\log(p) + 2L}$ and assume that β^* fulfills (7.13) with $\delta = 1/5$. Then, with probability at least $1 - e^{-L} - (1 + e^2)e^{-n/24}$, we have

$$\|\mathbf{X}(\widehat{\beta}_\lambda - \beta^*)\|^2 \le \inf_{\beta \neq 0} \left\{ \|\mathbf{X}(\beta - \beta^*)\|^2 + \frac{202\sigma^2(L + \log(p))}{\kappa(\beta)^2} |\beta|_0 \right\}. \quad (7.22)$$

We refer to Exercise 7.6.2 for a proof of this corollary.

7.5 Discussion and References

7.5.1 Take-Home Message

When analyzing a data set, after writing down a statistical model (say, a linear regression model), the statistician faces three issues:

- What is the underlying class of structures hidden in the data? (for example, coordinate sparsity, group sparsity, etc.)
- For a given class of structures, which estimator is the best? For example, for coordinate-sparse structures, shall we use the Lasso estimator, the Dantzig estimator, or the Elastic-Net? No estimator is universally better than the others, so the choice must be adapted to the data.
- For a given estimator, which tuning parameter should be chosen? Many estimators (like the Lasso) are not scale-invariant, so any sensible choice of the tuning parameter depends on the variance of the response Y. Since the variance of Y is usually unknown, the choice of the tuning parameter must be adapted to the data. Furthermore, even for scale-invariant estimators $\mathbf{X}\widehat{\beta}_\mu$ like the square-root Lasso

estimator (7.12), the oracle tuning parameter

$$\mu^*_{\text{oracle}} \in \operatorname*{argmin}_{\mu} \mathbb{E}\left[\|\mathbf{X}(\widehat{\beta}_\mu - \beta^*)\|^2\right]$$

depends on $\mathbf{X}\beta^*$, so it is wise to also tune μ according to the data.

These three issues can be handled all together by gathering a collection of estimators $\{\widehat{f}_\lambda : \lambda \in \Lambda\}$ corresponding to different estimation schemes (adapted to various classes of structures) with different tuning parameters. The ideal objective is then to choose $\widehat{f}_{\widehat{\lambda}}$ in this collection with a risk almost as small as the risk of the oracle estimator $\widehat{f}_{\lambda^*_{\text{oracle}}}$, where

$$\lambda^*_{\text{oracle}} \in \operatorname*{argmin}_{\lambda\in\Lambda} \mathbb{E}\left[\|\widehat{f}_\lambda - f^*\|^2\right].$$

Different approaches have been developed in order to handle this issue.

A first class of selection procedures is based on subsampling: part of the data is used to compute the estimators, while the remaining data is used to evaluate their prediction accuracy. This process is possibly repeated several times. This approach, including V-fold CV, is very popular and widely used. It usually provides satisfying results in practice, yet it suffers from two caveats. First, the repeated subsampling can lead to intensive computations exceeding the available computing resources. Second, there are no theoretical guarantees in general on the outcome of the selection process in high-dimensional settings. In particular, the subsampling device can suffer from instability for very small sample sizes.

A second estimator selection approach is inspired by the model selection techniques of Chapter 2. This approach has the nice features to avoid data splitting and to enjoy some non-asymptotic theoretical guarantees on the outcome of the selection process. Yet, it is limited to the setting where the noise $\varepsilon_1, \ldots, \varepsilon_n$ is i.i.d. with Gaussian distribution.

Finally, for tuning estimators based on the minimization of a non-homogeneous criterion, a third strategy is to modify this criterion in order to obtain scale-invariant estimators like the square-root Lasso. This approach is computationally efficient and enjoys some nice theoretical properties. Yet, it is limited to the tuning of some specific estimators like those introduced in Chapter 5, and it does not allow to compare different estimation schemes.

There are a few other estimator selection procedures that have been proposed recently in the literature. We point some of them out in the references below. The mathematical analysis of general estimator selection procedures is somewhat more involved than the theory described in the previous chapters. We also observe that the results stated in Proposition 7.1, Proposition 7.2, and Theorem 7.3 involve some extra conditions on the sparsity of the regressor β^* compared to Theorem 5.1 and Theorem 5.4 in Chapter 5 for the Lasso or group-Lasso estimators. For example, in the coordinate-sparse setting, these conditions roughly require that $|\beta^*|_0$ remains small compared to

$n/\log(p)$. This unpleasant condition is unfortunately unavoidable when the variance σ^2 is unknown, as has been shown by Verzelen [160].

The theory of estimator selection has to be strengthened, yet some tools are already available, each providing various interesting features.

7.5.2 References

Cross-validation techniques date back at least to the '60s; classical references include the papers by Mosteller and Tukey [123], Devroye and Wagner [63], Geisser [75], Stone [147], and Shao [142]. The asymptotic analysis of V-fold-CV was carried out in the '80s by Li [110], among others. The non asymptotic analysis of V-fold CV is much more involved. Some non asymptotic results have been derived by Arlot, Célisse, and Lerasle [5, 8, 108] for some specific settings. We refer to Arlot and Celisse [7] for a recent review on the topic.

The complexity selection Criterion (7.3) has been introduced and analyzed by Baraud, Giraud, Huet, and Verzelen [19, 20, 82]. An associated `R` package `LINselect` is available on the CRAN archive network
`http://cran.r-project.org/web/packages/LINselect/`.

The square-root Lasso estimator (7.12) has been introduced and analyzed by Belloni, Chernozhukov and Wang [25], Antoniadis [4], and Sun and Zhang [149] (see Städler, Bühlmann, and van de Geer [145] for a variant based on a penalized maximum likelihood criterion). An associated `R` package `scalreg` is also available on the CRAN archive network `http://cran.r-project.org/web/packages/scalreg/`. An analog of the square-root Lasso for the group-sparse setting is analyzed in Bunea *et al.* [42].

Some other non-asymptotic approaches have been developed for estimator selection in the regression setting with unknown variance σ^2. The *slope-heuristic* developed by Birgé and Massart [33], Arlot and Massart [9], and Arlot and Bach [6] builds on the following idea. Consider $\widehat{f}_{\widehat{\lambda}}$ selected by minimizing a penalized criterion $\text{crit}(\widehat{f}_\lambda) = \|Y - \widehat{f}_\lambda\|^2 + \text{pen}(\widehat{f}_\lambda)$. Assume that there exists some penalty $\text{pen}_{\min}(\widehat{f}_\lambda)$, such that

- when $\text{pen}(\widehat{f}_\lambda) = K\text{pen}_{\min}(\widehat{f}_\lambda)$ with $K > \sigma^2$, we have an oracle risk bound like Theorem 2.2 for $\widehat{f}_{\widehat{\lambda}}$, and
- when $\text{pen}(\widehat{f}_\lambda) = K\text{pen}_{\min}(\widehat{f}_\lambda)$ with $K < \sigma^2$, we have $\widehat{f}_{\widehat{\lambda}} \approx Y$.

For example, in the setting of Exercise 2.8.1 in Chapter 2, the penalty $\text{pen}_{\min}(\widehat{f}_m) = 2|m|\log(p)$ is a minimal penalty. When such a minimal penalty exists, we then observe the following phenomenon. Define $\widehat{\lambda}_K$ by

$$\widehat{\lambda}_K \in \underset{\lambda \in \Lambda}{\text{argmin}} \left\{ \|Y - \widehat{f}_\lambda\|^2 + K\text{pen}_{\min}(\widehat{f}_\lambda) \right\}.$$

When $K < \sigma^2$, we have $\widehat{f}_{\widehat{\lambda}_K} \approx Y$, whereas when $K > \sigma^2$, we have $\|Y - \widehat{f}_{\widehat{\lambda}_K}\|$ large. The rough idea is then to track a value $\widehat{K}$ where this transition occurs and use this value as an estimator of σ^2. We refer to the original papers [33] and [6] for a more precise description of the slope-heuristic procedure and some non-asymptotic bounds.

Another approach based on pairwise test comparisons on a discretized space has been proposed and analyzed by Baraud and Birgé [31, 16, 18]. The procedure cannot be easily sketched in a couple of lines, so we refer to the original papers. This estimator selection technique has the nice feature of being very flexible, and it enjoys the property to be able to automatically adapt to the distribution of the noise. This last property is extremely desirable, since we usually do not know the distribution of the noise. Unfortunately, the computational complexity of the selection procedure is generally very high and it cannot be directly implemented even in moderate dimensions.

As mentioned above, there is a need for some efforts to strengthen the theory of estimator selection. We refer to Giraud *et al.* [82] for a recent review on this issue in the regression setting.

7.6 Exercises

7.6.1 Expected V-Fold CV ℓ^2-Risk

A natural setting for analyzing the V-fold CV selection procedure is the random design regression setting, where the observations $(Y_1,X_1),\ldots,(Y_n,X_n)$ are i.i.d. with common distribution $\mathbb{P}$. We assume in the following that the variance of Y is finite, and we keep the notations of Section 7.2. Writing $f(x)$ for (a version of) the conditional expectation $f(x)=\mathbb{E}[Y\,|\,X=x]$, we have $Y_i=f(X_i)+\varepsilon_i$ with the $\varepsilon_1,\ldots,\varepsilon_n$ i.i.d. centered and with finite variance.

For any measurable function $g:\mathbb{R}^p\to\mathbb{R}$, we denote by $\|g\|_{L^2(\mathbb{P}^X)}$ the expectation $\|g\|_{L^2(\mathbb{P}^X)}=\mathbb{E}\left[g(X_1)^2\right]^{1/2}$.

1. Prove that $\mathbb{E}[\varepsilon_1|X_1]=0$ and $\mathbb{E}\left[(Y_1-g(X_1))^2\right]=\|g-f\|^2_{L^2(\mathbb{P}^X)}+\text{var}(\varepsilon_1)$.
2. Prove the equality

$$\mathbb{E}\left[(Y_1-F_\lambda(\mathscr{D}_{-I_1},X_1))^2\right]=\mathbb{E}_{-I_1}\left[\|F_\lambda(\mathscr{D}_{-I_1},\cdot)-f\|^2_{L^2(\mathbb{P}^X)}\right]+\text{var}(\varepsilon_1),$$

   where $\mathbb{E}_{-I_1}$ refers to the expectation with respect to $\mathscr{D}_{-I_1}$.
3. Conclude that the expected value of the V-fold CV ℓ^2-risk $\widehat{R}_{CV}[F_\lambda]$ defined by (7.1) is given by

$$\mathbb{E}\left[\widehat{R}_{CV}[F_\lambda]\right]=\mathbb{E}_{-I_1}\left[\|F_\lambda(\mathscr{D}_{-I_1},\cdot)-f\|^2_{L^2(\mathbb{P}^X)}\right]+\text{var}(\varepsilon_1).$$

Remark. Since the variance of ε_1 does not depend on λ, the V-fold CV ℓ^2-risk is equal up to a constant to an unbiased estimator of the integrated risk $\mathbb{E}_{-I_1}\left[\|F_\lambda(\mathscr{D}_{-I_1},\cdot)-f\|^2_{L^2(\mathbb{P}^X)}\right]$. This risk can be viewed as an approximation of the risk $\mathbb{E}\left[\|F_\lambda(\mathscr{D},\cdot)-f\|^2_{L^2(\mathbb{P}^X)}\right]$, which measures the mean L^2-distance between the estimator $F_\lambda(\mathscr{D},\cdot)$ of f and f.

7.6.2 Proof of Corollary 7.5

The proof of Corollary 7.5 builds on the link (7.21) between the Lasso and square-root Lasso estimators. The idea is to bound $\widehat{\sigma}=\|Y-\mathbf{X}\widehat{\beta}\|/\sqrt{n}$ from above and below and apply Theorem 5.1 for the Lasso. To bound $\widehat{\sigma}$, we essentially check that

$$\frac{\|\varepsilon\|-\|\mathbf{X}(\widehat{\beta}-\beta^*)\|}{\sqrt{n}}\leq\widehat{\sigma}\leq\frac{\|\varepsilon\|+\|\mathbf{X}(\widehat{\beta}-\beta^*)\|}{\sqrt{n}} \tag{7.23}$$

and then bound $\|\mathbf{X}(\widehat{\beta}-\beta^*)\|$ with (7.16). In the following, we work on the event

$$\mathscr{A}=\left\{3|\mathbf{X}^T\varepsilon|_\infty\leq\mu\sigma \text{ and } |\sigma-\|\varepsilon\|/\sqrt{n}|\leq(1-2^{-1/2})\sigma\right\}.$$

1. Prove that the event $\mathscr{A}$ has probability at least $1-e^{-L}-(1+e^2)e^{-n/24}$.
2. Check Inequalities (7.23).
3. From Lemma 7.4, prove that under the hypotheses of Corollary 7.5 we have on the event $\mathscr{A}$ for $\delta\leq 1/5$

$$\|\mathbf{X}(\widehat{\beta}-\beta^*)\|^2\leq\frac{18\mu^2|\beta^*|_0}{\delta(1-\delta)\kappa(\beta^*)^2n}\|\varepsilon\|^2\leq\frac{9\delta}{32(1-\delta)}\|\varepsilon\|^2\leq\left(1-2^{-1/2}\right)^2\|\varepsilon\|^2.$$

4. Check that we have $3|\mathbf{X}^T\varepsilon|_\infty\leq 2\mu\widehat{\sigma}$ on the event $\mathscr{A}$.
5. From Theorem 5.1, prove that on the event $\mathscr{A}$ we have

$$\begin{aligned}\|\mathbf{X}(\widehat{\beta}-\beta^*)\|^2&\leq\inf_{\beta\neq0}\left\{\|\mathbf{X}(\beta-\beta^*)\|^2+\frac{4\mu^2\widehat{\sigma}^2}{\kappa(\beta)^2}|\beta|_0\right\}\\&\leq\inf_{\beta\neq0}\left\{\|\mathbf{X}(\beta-\beta^*)\|^2+\frac{202(L+\log(p))\sigma^2}{\kappa(\beta)^2}|\beta|_0\right\}\end{aligned}$$

   and conclude the proof of Corollary 7.5.

7.6.3 Some Properties of Penalty (7.4)

For any positive integers D,N, we denote by X_D and X_N two independent positive random variables, such that X_D^2 and X_N^2 are two independent chi-square random variables with, respectively, D and N degrees of freedom. We write

$$p_D(x)=2^{-D/2}\Gamma(D/2)^{-1}x^{D/2-1}e^{-x/2}\mathbf{1}_{\mathbb{R}^+}(x),\quad x\in\mathbb{R}$$

for the probability distribution function of X_D^2. From (B.4), page 302, in Chapter B we have $\sqrt{D-4}\leq\mathbb{E}[X_D]\leq\sqrt{D}$. In the following, d_m denotes the dimension of the model S_m.

A) Computation of the penalty

We prove in this first part that the penalty $\text{pen}_\pi(m)$ defined by (7.4), page 144, is given by $\text{pen}_\pi(m) = 1.1(n-d_m)x_{n,d_m}$, where x_{n,d_m} is solution in x of

$$\pi_m = (d_m+1)\mathbb{P}\left(F_{d_m+3,n-d_m-1} \geq \frac{n-d_m-1}{d_m+3}x\right) - x(n-d_m-1)\mathbb{P}\left(F_{d_m+1,n-d_m+1} \geq \frac{n-d_m+1}{d_m+1}x\right), \quad (7.24)$$

where $F_{d,r}$ is a Fisher random variable with d and r degrees of freedom.

1. Check that $xp_D(x) = Dp_{D+2}(x)$ for any positive integer D and $x \in \mathbb{R}$.
2. For $\alpha > 0$ prove the equalities

$$\mathbb{E}\left[(X_D^2 - \alpha X_N^2)_+\right] = \int_0^\infty \int_{\alpha y}^\infty xp_D(x)p_N(y)\,dx\,dy - \alpha \int_0^\infty \int_{\alpha y}^\infty yp_D(x)p_N(y)\,dx\,dy$$
$$= D\mathbb{P}\left(X_{D+2}^2 \geq \alpha X_N^2\right) - \alpha N \mathbb{P}\left(X_D^2 \geq \alpha X_{N+2}^2\right).$$

3. Conclude the proof of (7.24).

B) An upper bound on the penalty

In this second part, we prove that for any model m fulfilling

$$\left(\sqrt{d_m+1} + 2\sqrt{\log(8/\pi_m)}\right)^2 \leq n - d_m - 5, \quad (7.25)$$

we have the upper bound

$$\text{pen}_\pi(m) \leq 2.2\frac{n-d_m}{n-d_m-5}\left(\sqrt{d_m+1} + 2\sqrt{\log(8/\pi_m)}\right)^2 \quad (7.26)$$

(we refer to Proposition 4 in Baraud *et al.* [19] for a tighter upper bound).

Let $L > 0$ and N, D be such that

$$t_{D,N,L} = \frac{\sqrt{D} + 2\sqrt{L}}{\sqrt{N-4}} \leq 1.$$

We define the function $F_t : \mathbb{R}^{N+D} \to \mathbb{R}$ by

$$F_t(x_1,\ldots,x_{D+N}) = \sqrt{x_1^2 + \ldots + x_D^2} - t\sqrt{x_{D+1}^2 + \ldots + x_{D+N}^2}\,.$$

1. Prove that $F_{t_{D,N,L}}$ is $\sqrt{2}$-Lipschitz. By considering the variable $F_{t_{D,N,L}}(\varepsilon)$ with ε a standard Gaussian random variable in R^{N+D}, prove with the Gaussian concentration Inequality (B.2) and the Inequalities (B.4) in Appendix B that there exists a standard exponential random variable ξ, such that $X_D - t_{D,N,L}X_N \leq 2\sqrt{\xi} - 2\sqrt{L}$, and thus

$$X_D^2 - 2t_{D,N,L}^2 X_N^2 \leq 8\left(\sqrt{\xi} - \sqrt{L}\right)_+^2.$$

2. Prove that for any $a,b \geq 0$ we have $(a-b)_+^2 \leq (a^2-b^2)_+$ and check the inequality

$$\mathbb{E}\left[\left(X_D^2 - 2t_{D,N,L}^2 X_N^2\right)_+\right] \leq 8\mathbb{E}\left[(\xi - L)_+\right] = 8e^{-L}.$$

3. From Definition (7.4) of $\text{pen}_\pi(m)$ conclude that we have the upper Bound (7.26) when the Condition (7.25) is met.

7.6.4 Selecting the Number of Steps for the Forward Algorithm

We consider the linear regression setting $Y = \mathbf{X}\beta^* + \varepsilon = f^* + \varepsilon$ with $\beta^* \in \mathbb{R}^p$ and the collection of models $\{S_m : m \in \mathcal{M}\}$ defined in Section 2.2, page 32, Chapter 2 for the coordinate-sparse setting. For $\lambda = 1,2,\ldots$, we denote by $\widehat{m}_\lambda = \{j_1,\ldots,j_\lambda\}$ the forward selection algorithm defined recursively by

$$j_\lambda \in \underset{j=1,\ldots,p}{\text{argmin}} \|Y - \text{Proj}_{S_{\{j_1,\ldots,j_{\lambda-1},j\}}} Y\|^2.$$

For $m \subset \{1,\ldots,p\}$, we set $\pi_m = p^{-d_m}$ with $d_m = \dim(S_m)$. In the following, we restrict to models fulfilling

$$\left(\sqrt{2} + 2\sqrt{\log(8p)}\right)^2 d_m \leq n - d_m - 5. \tag{7.27}$$

We define

$$\widehat{\mathcal{M}} = \left\{\widehat{m}_\lambda : \lambda = 1,2,\ldots \text{ such that } d_{\widehat{m}_\lambda} \text{ fulfills (7.27)}\right\},$$

$\widehat{\Lambda} = \left\{\lambda = 1,2,\ldots : \widehat{m}_\lambda \in \widehat{\mathcal{M}}\right\}$, and $\bar{S}_m = S_m + < f^* >$, where $< f^* >$ is the line spanned by f^*.

For $\lambda = 1,2,\ldots$, let us define $\widehat{f}_\lambda = \text{Proj}_{S_{\widehat{m}_\lambda}} Y$. The goal of this exercise is to prove that there exists a numerical constant $C > 1$, such that when $\widehat{f}_{\widehat{\lambda}}$ is a minimizer of Criterion (7.3) over $\left\{\widehat{f}_\lambda : \lambda = 1,2,\ldots\right\}$, we have

$$\mathbb{E}\left[\|\widehat{f}_{\widehat{\lambda}} - f^*\|^2\right] \leq C\mathbb{E}\left[\inf_{\lambda \in \widehat{\Lambda}}\left\{\|\widehat{f}_\lambda - f^*\|^2 + d_{\widehat{m}_\lambda}\log(p)\sigma^2 + \sigma^2\right\}\right]. \tag{7.28}$$

1. Prove that the set $\widehat{m}_{\widehat{\lambda}}$ is a solution of

$$\widehat{m}_{\widehat{\lambda}} \in \underset{m \in \widehat{\mathcal{M}}}{\text{argmin}}\left\{\|Y - \text{Proj}_{S_m} Y\|^2 + \text{pen}_\pi(m)\widehat{\sigma}_m^2\right\}.$$

2. Following lines similar to the beginning of the proof of Theorem 2.2, prove that for any $\lambda \in \widehat{\Lambda}$

$$\begin{aligned}\|\widehat{f}_{\widehat{\lambda}} - f^*\|^2 \leq & \|\widehat{f}_\lambda - f^*\|^2 + 2\text{pen}_\pi(\widehat{m}_\lambda)\widehat{\sigma}_{\widehat{m}_\lambda}^2 + (1.1)^{-1}\|\widehat{f}_\lambda - f^*\|^2 + Z(\lambda) \\ & + (1.1)^{-1}\|\widehat{f}_{\widehat{\lambda}} - f^*\|^2 + Z(\widehat{\lambda}),\end{aligned}$$

where $\quad Z(\lambda) = 1.1\|\text{Proj}_{\bar{S}_{\widehat{m}_\lambda}}\varepsilon\|^2 - \text{pen}_\pi(\widehat{m}_\lambda)\widehat{\sigma}_{\widehat{m}_\lambda}^2.$

3. Check that $\|Y - \mathrm{Proj}_{\bar{S}_m} Y\|^2$ is stochastically larger than $\|\varepsilon - \mathrm{Proj}_{\bar{S}_m} \varepsilon\|^2$, and then

$$\begin{aligned}
\mathbb{E}\left[\sup_{\lambda=1,2,\ldots} Z(\lambda)\right] &\leq \sum_{m\in\mathcal{M}} \mathbb{E}\left[\left(1.1\|\mathrm{Proj}_{\bar{S}_m}\varepsilon\|^2 - \mathrm{pen}_\pi(m)\widehat{\sigma}_m^2\right)_+\right] \\
&\leq \sum_{m\in\mathcal{M}} \mathbb{E}\left[\left(1.1\|\mathrm{Proj}_{\bar{S}_m}\varepsilon\|^2 - \frac{\mathrm{pen}_\pi(m)}{n-d_m}\|\varepsilon - \mathrm{Proj}_{\bar{S}_m}\varepsilon\|^2\right)_+\right] \\
&\leq (1+p^{-1})^p\sigma^2 \leq e\sigma^2.
\end{aligned}$$

4. Prove that when Condition (7.27) is fulfilled, we have $n - d_m \geq 24$ and (7.25) is fulfilled. Conclude that according to (7.26), we then have $\mathrm{pen}_\pi(m) \leq 2.2(n-d_m)$ and

$$\mathrm{pen}_\pi(m) \leq 3\left(\sqrt{2} + 2\sqrt{\log(8p)}\right)^2 d_m.$$

5. Prove that

$$\begin{aligned}
\widehat{\sigma}_m^2 &\leq \frac{2}{n-d_m}\left(\|\varepsilon - \mathrm{Proj}_{S_m}\varepsilon\|^2 + \|f^* - \mathrm{Proj}_{S_m} f^*\|^2\right) \\
&\leq \frac{2}{n-d_m}\left(2n\sigma^2 + \left(\|\varepsilon\|^2 - 2n\sigma^2\right)_+ + \|f^* - \mathrm{Proj}_{S_m} Y\|^2\right).
\end{aligned}$$

6. By combining the two last questions, prove that when $\lambda \in \widehat{\Lambda}$, we have

$$\begin{aligned}
&\mathrm{pen}_\pi(\widehat{m}_\lambda)\widehat{\sigma}_{\widehat{m}_\lambda}^2 \\
&\quad\leq 24\left(\sqrt{2} + 2\sqrt{\log(8p)}\right)^2 d_{\widehat{m}_\lambda}\sigma^2 + 5\left(\|\varepsilon\|^2 - 2n\sigma^2\right)_+ + 5\|f^* - \widehat{f}_\lambda\|^2.
\end{aligned}$$

7. Conclude the proof of (7.28) by combining Questions 2, 3, and 6.

Chapter 8

Multivariate Regression

In the previous chapters, we have focused on a response y that was 1-dimensional. In many cases, we do not focus on a single quantity $y \in \mathbb{R}$, but rather on a T-dimensional vector $y = (y_1, \ldots, y_T) \in \mathbb{R}^T$ of measurements. It is, of course, possible to analyze each coordinate y_t independently, but it is usually wise to analyze simultaneously the T coordinates $y_1, \ldots, y_T$. Actually, when $y_1, \ldots, y_T$ are the outcome of a common process, they usually share some common structures, and handling the T measurements $y_1, \ldots, y_T$ together enables to rely on these structures.

In this chapter, we give a special focus on the case where the measurements lie in the vicinity of an (unknown) low-dimensional space. In a linear model, this kind of structure translates in terms of low-rank of the regression matrix. We present in the next sections a theory for estimating low-rank matrices and then we investigate how we can handle simultaneously low-rank structures with some other sparsity structures.

8.1 Statistical Setting

We now consider the problem of predicting a T-dimensional *vector* y from a p-dimensional vector of covariates. Similarly to the examples discussed in Chapter 2, the linear model is the canonical model and many different situations can be recast as a linear model. Henceforth, we consider the following model

$$y^{(i)} = (A^*)^T x^{(i)} + \varepsilon^{(i)}, \quad i = 1, \ldots, n, \tag{8.1}$$

where $y^{(i)} \in \mathbb{R}^T$, $x^{(i)} \in \mathbb{R}^p$, A^* is a $p \times T$-matrix, and the $\varepsilon^{(i)}$ are i.i.d. with $\mathcal{N}(0, \sigma^2 I_T)$ Gaussian distribution in $\mathbb{R}^T$. Writing Y and E for the $n \times T$ matrices $Y = [y_t^{(i)}]_{i=1,\ldots,n,\ t=1,\ldots,T}$ and $E = [\varepsilon_t^{(i)}]_{i=1,\ldots,n,\ t=1,\ldots,T}$, Model (8.1) can be formulated in a matrix form

$$Y = \mathbf{X}A^* + E, \tag{8.2}$$

where $\mathbf{X}$ is defined as in the previous chapters by $\mathbf{X}_{ij} = x_j^{(i)}$ for $i = 1, \ldots, n$ and $j = 1, \ldots, p$. Let M_k denote the k-th column of a matrix M. If we consider each column of Y independently, then we have T independent linear regressions

$$Y_k = \mathbf{X}A_k^* + E_k, \quad k = 1, \ldots, T.$$

We can consider each regression independently, yet, as mentioned above, these T regressions may share some common structures, and it is wise in this case to analyze the T regressions simultaneously. Assume, for example, that the vectors A_k^* are coordinate sparse. In many cases, the vectors $A_1^*, \ldots, A_T^*$ share the same sparsity pattern, and then it helps to analyze simultaneously the T regression; see Section 8.4.1. The T vectors $A_1^*, \ldots, A_T^*$ may also (approximately) lie in a common (unknown) low-dimensional space. It then means that the rank of A^* is small. We understand in Section 8.3 how we can capitalize on this property. Finally, we investigate in Section 8.4.2 how we can handle simultaneously low-rank properties with coordinate-sparsity.

8.2 Reminder on Singular Values

Singular values play a central role in low-rank estimation. This section is a brief reminder on singular values; we refer to Appendix C for proofs and additional results.

Singular value decomposition

Any $n \times p$ matrix A of rank r can be decomposed as $A = \sum_{j=1}^{r} \sigma_j u_j v_j^T$, where

- $\sigma_1 \geq \ldots \geq \sigma_r > 0$,
- $\{\sigma_1^2, \ldots, \sigma_r^2\}$ are the nonzero eigenvalues of $A^T A$ and AA^T, and
- $\{u_1, \ldots, u_r\}$ and $\{v_1, \ldots, v_r\}$ are two orthonormal families of $\mathbb{R}^n$ and $\mathbb{R}^p$, such that

$$AA^T u_j = \sigma_j^2 u_j \quad \text{and} \quad A^T A v_j = \sigma_j^2 v_j, \quad \text{for } j = 1, \ldots, r.$$

We refer to Theorem C.1, page 311, in Appendix C for a proof. The values $\sigma_1, \sigma_2, \ldots$ are called the singular values of A.

Some matrix norms

In the following, we denote by $\sigma_1(A) \geq \sigma_2(A) \geq \ldots$ the singular values of a matrix A ranked in decreasing order. For $k > \text{rank}(A)$, we define $\sigma_k(A)$ by $\sigma_k(A) = 0$. The Frobenius (or Hilbert–Schmidt) norm of A is defined by

$$\|A\|_F^2 = \sum_{i,j} A_{ij}^2 = \text{trace}(A^T A) = \sum_{k=1}^{\text{rank}(A)} \sigma_k(A)^2.$$

For any integer $q \geq 1$ the Ky–Fan $(2,q)$-norm is defined by

$$\|A\|_{(2,q)}^2 = \sum_{k=1}^{q} \sigma_k(A)^2.$$

For $q = 0$ we set $\|A\|_{(2,0)} = 0$. We observe that $\|A\|_{(2,q)} \leq \|A\|_F$, so $A \to \|A\|_{(2,q)}$ is 1-Lipschitz with respect to the Frobenius norm. For $q = 1$ the Ky–Fan $(2,1)$-norm corresponds to the operator norm

$$\|A\|_{(2,1)} = \sigma_1(A) = |A|_{\text{op}} = \sup_{x:\|x\| \leq 1} \|Ax\|;$$

see Appendix C. In the next sections, we repeatedly use the two following properties (see Theorem C.5, page 315, in Appendix C for a proof):

1. For any matrices $A, B \in \mathbb{R}^{n \times p}$, we have

$$\langle A, B \rangle_F \le \|A\|_{(2,r)} \|B\|_{(2,r)}, \tag{8.3}$$

   where $r = \min(\text{rank}(A), \text{rank}(B))$.

2. For any $q \ge 1$, we have

$$\min_{B:\text{rank}(B) \le q} \|A - B\|_F^2 = \sum_{k=q+1}^{r} \sigma_k(A)^2 \quad \text{for } q < r = \text{rank}(A) \tag{8.4}$$

   and $\min_{B:\text{rank}(B) \le q} \|A - B\|_F^2 = 0$ for $q \ge r$. The minimum is achieved for

$$(A)_{(q)} = \sum_{k=1}^{q \wedge r} \sigma_k(A) u_k v_k^T. \tag{8.5}$$

The matrix $(A)_{(q)}$ is then a "projection" of the matrix A on the set of matrices with rank not larger than q.

8.3 Low-Rank Estimation

8.3.1 When the Rank of A^* is Known

In Model (8.2), with $\{E_{it} : i = 1, \dots, n,\ t = 1, \dots, T\}$ i.i.d. with $\mathcal{N}(0, \sigma^2)$ Gaussian distribution, the negative log-likelihood of a matrix A is

$$-\text{log-likelihood}(A) = \frac{1}{2\sigma^2} \|Y - \mathbf{X}A\|_F^2 + \frac{nT}{2} \log(2\pi\sigma^2),$$

where $\|\cdot\|_F$ is the Frobenius (or Hilbert–Schmidt) norm. If we knew the rank r^* of A^*, then we would estimate A^* by the maximum-likelihood estimator $\widehat{A}_{r^*}$ constrained to have a rank at most r^*, namely

$$\widehat{A}_r \in \underset{\text{rank}(A) \le r}{\text{argmin}} \|Y - \mathbf{X}A\|_F^2, \tag{8.6}$$

with $r = r^*$. This estimator can be computed easily, as explained below.

The next lemma provides a useful formula for $\mathbf{X}\hat{A}_r$ in terms of the singular value decomposition of PY, where P is the orthogonal projector onto the range of $\mathbf{X}$. We refer to Appendix C for a reminder on the Moore–Penrose pseudo-inverse of a matrix.

Lemma 8.1 Computation of $\mathbf{X}\widehat{A}_r$

Write $P = \mathbf{X}(\mathbf{X}^T\mathbf{X})^+\mathbf{X}^T$ *for the projection onto the range of* $\mathbf{X}$, *with* $(\mathbf{X}^T\mathbf{X})^+$ *the Moore–Penrose pseudo-inverse of* $\mathbf{X}^T\mathbf{X}$.

Then, for any $r \geq 1$, *we have* $\mathbf{X}\widehat{A}_r = (PY)_{(r)}$.

As a consequence, denoting by $PY = \sum_{k=1}^{\text{rank}(PY)} \sigma_k u_k v_k^T$ *a singular value decomposition of* PY, *we have for any* $r \geq 1$

$$\mathbf{X}\widehat{A}_r = \sum_{k=1}^{r\wedge\text{rank}(PY)} \sigma_k u_k v_k^T.$$

Proof. Pythagorean formula gives $\|Y - \mathbf{X}M\|_F^2 = \|Y - PY\|_F^2 + \|PY - \mathbf{X}M\|_F^2$ for any $p \times T$-matrix M. Since the rank of $\mathbf{X}\widehat{A}_r$ is at most r, we have $\|PY - \mathbf{X}\widehat{A}_r\|_F^2 \geq \|PY - (PY)_{(r)}\|_F^2$, and hence

$$\begin{aligned} \|Y - (PY)_{(r)}\|_F^2 &= \|Y - PY\|_F^2 + \|PY - (PY)_{(r)}\|_F^2 \\ &\leq \|Y - PY\|_F^2 + \|PY - \mathbf{X}\widehat{A}_r\|_F^2 = \|Y - \mathbf{X}\widehat{A}_r\|_F^2. \end{aligned} \tag{8.7}$$

To conclude the proof of the lemma, we only need to check that we have a decomposition $(PY)_{(r)} = \mathbf{X}\tilde{A}_r$ with $\text{rank}(\tilde{A}_r) \leq r$. From Pythagorean formula, we get

$$\|PY - (PY)_{(r)}\|_F^2 = \|PY - P(PY)_{(r)}\|_F^2 + \|(I-P)(PY)_{(r)}\|_F^2.$$

Since $\text{rank}(P(PY)_{(r)}) \leq r$, the above equality ensures that

$$(PY)_{(r)} = P(PY)_{(r)} = X\underbrace{(X^TX)^+X^T(PY)_{(r)}}_{=:\tilde{A}_r},$$

with $\text{rank}(\tilde{A}_r) \leq \text{rank}((PY)_{(r)}) \leq r$. According to (8.7), the matrix $\tilde{A}_r$ is a minimizer of (8.6) and finally $\mathbf{X}\widehat{A}_r = \mathbf{X}\tilde{A}_r = (PY)_{(r)}$. □

According to Lemma 8.1, for $r \geq \text{rank}(PY)$, we have $\mathbf{X}\widehat{A}_r = \mathbf{X}\widehat{A}_{\text{rank}(PY)}$. Since $\text{rank}(PY) \leq q \wedge T$, with $q = \text{rank}(\mathbf{X}) \leq n \wedge p$, we only need to consider the collection of estimators $\{\widehat{A}_1, \ldots, \widehat{A}_{q\wedge T}\}$. As a nice consequence of Lemma 8.1, the collection of estimators $\{\mathbf{X}\widehat{A}_1, \ldots, \mathbf{X}\widehat{A}_{q\wedge T}\}$ can be computed from a single singular value decomposition of PY. We also observe from the above proof that the matrix $\widehat{A}_r = (\mathbf{X}^T\mathbf{X})^+\mathbf{X}^T(PY)_{(r)}$ is a solution of the minimization problem (8.6). Let us now investigate the quadratic risk $\mathbb{E}\left[\|\mathbf{X}\widehat{A}_r - \mathbf{X}A^*\|_F^2\right]$ of the estimator $\mathbf{X}\widehat{A}_r$.

We remind the reader that $\sigma_k(M)$ denotes the k-th largest singular value of a matrix M, with $\sigma_k(M) = 0$ for $k > \text{rank}(M)$.

Proposition 8.2 Deterministic bound

For any $r \geq 1$ and $\theta > 0$, we have

$$\|\mathbf{X}\widehat{A}_r - \mathbf{X}A^*\|_F^2 \leq c^2(\theta) \sum_{k>r} \sigma_k(\mathbf{X}A^*)^2 + 2c(\theta)(1+\theta)r|PE|_{\text{op}}^2,$$

with $c(\theta) = 1 + 2/\theta$.

Proof. Write $\sum_k \sigma_k(\mathbf{X}A^*)u_k v_k^T$ for a singular value decomposition of $\mathbf{X}A^*$. Following the same lines as in the proof of the previous lemma, we can choose a matrix B of rank at most r, such that

$$\mathbf{X}B = (\mathbf{X}A^*)_{(r)} = \sum_{k=1}^{r} \sigma_k(\mathbf{X}A^*)u_k v_k^T.$$

From the definition of $\widehat{A}_r$, we have that $\|Y - \mathbf{X}\widehat{A}_r\|_F^2 \leq \|Y - \mathbf{X}B\|_F^2$, from which follows that

$$\|\mathbf{X}\widehat{A}_r - \mathbf{X}A^*\|_F^2 \leq \|\mathbf{X}B - \mathbf{X}A^*\|_F^2 + 2\langle E, \mathbf{X}\widehat{A}_r - \mathbf{X}B\rangle_F. \tag{8.8}$$

Since $\mathbf{X}B$ and $\mathbf{X}\widehat{A}_r$ have a rank at most r, the matrix $\mathbf{X}\widehat{A}_r - \mathbf{X}B$ has a rank at most $2r$, and according to (8.3), we obtain

$$\langle E, \mathbf{X}\widehat{A}_r - \mathbf{X}B\rangle_F = \langle PE, \mathbf{X}\widehat{A}_r - \mathbf{X}B\rangle_F \leq \|PE\|_{(2,2r)} \|\mathbf{X}\widehat{A}_r - \mathbf{X}B\|_F. \tag{8.9}$$

We have $\|PE\|_{(2,2r)}^2 \leq 2r|PE|_{\text{op}}^2$ and $\|\mathbf{X}\widehat{A}_r - \mathbf{X}B\|_F \leq \|\mathbf{X}\widehat{A}_r - \mathbf{X}A^*\|_F + \|\mathbf{X}B - \mathbf{X}A^*\|_F$, so using twice the inequality $2xy \leq ax^2 + y^2/a$ for all $a > 0$, $x, y \geq 0$, we obtain by combining (8.8) and (8.9)

$$(1 - 1/a)\|\mathbf{X}\widehat{A}_r - \mathbf{X}A^*\|_F^2 \leq (1 + 1/b)\|\mathbf{X}B - \mathbf{X}A^*\|_F^2 + 2(a+b)r|PE|_{\text{op}}^2.$$

The proposition follows by choosing $a = 1 + \theta/2$ and $b = \theta/2$. □

In order to obtain an upper bound not depending on the noise matrix E, we need a probabilistic upper bound on $|PE|_{\text{op}}$. Investigating the singular spectrum of random matrices is a very active field in mathematics, with applications in statistical physics, data compression, communication networks, statistics, etc. In particular, the operator norm $|PE|_{\text{op}}^2$ is known to be almost surely equivalent to $(\sqrt{q} + \sqrt{T})^2\sigma^2$, as q and T goes to infinity (we remind the reader that q is the rank of $\mathbf{X}$). We also have a non-asymptotic result, which is a nice application of Slepian's lemma [143].

Lemma 8.3 Spectrum of random matrix
The expectated value of the operator norm of PE is upper-bounded by

$$\mathbb{E}\left[|PE|_{\text{op}}\right] \leq \left(\sqrt{q} + \sqrt{T}\right)\sigma, \quad \textit{with } q = \text{rank}(\mathbf{X}). \tag{8.10}$$

We refer to Davidson and Szarek [61] for a proof of this result and to Exercise 12.9.6 (page 288) for a proof of this bound with less tight constants.

The map $E \to |PE|_{op}$ is 1-Lipschitz with respect to the Frobenius norm, since $|PE|_{op} \leq \|PE\|_F \leq \|E\|_F$, so according to the Gaussian concentration Inequality (B.2), page 301, there exists an exponential random variable ξ with parameter 1, such that $|PE|_{op} \leq \mathbb{E}\left[|PE|_{op}\right] + \sigma\sqrt{2\xi}$. Since $(a+b)^2 \leq 2(a^2+b^2)$, we have

$$\mathbb{E}\left[|PE|_{op}^2\right] \leq 2\mathbb{E}\left[|PE|_{op}\right]^2 + 4\sigma^2\mathbb{E}[\xi] \leq 3\left(\sqrt{q}+\sqrt{T}\right)^2\sigma^2. \tag{8.11}$$

According to (8.4), we have

$$\min_{\operatorname{rank}(A)\leq r} \|\mathbf{X}A - \mathbf{X}A^*\|_F^2 \geq \min_{\operatorname{rank}(\mathbf{X}A)\leq r} \|\mathbf{X}A - \mathbf{X}A^*\|_F^2 = \sum_{k>r} \sigma_k(\mathbf{X}A^*)^2,$$

so combining Proposition 8.2 (with $\theta = 1$) with this inequality, and (8.11) gives the following risk bound.

Corollary 8.4 Risk of $\widehat{A}_r$
For any $r \geq 1$, we have the risk bound

$$\begin{aligned}\mathbb{E}\left[\|\mathbf{X}\widehat{A}_r - \mathbf{X}A^*\|_F^2\right] &\leq 9\sum_{k>r}\sigma_k(\mathbf{X}A^*)^2 + 36r\left(\sqrt{q}+\sqrt{T}\right)^2\sigma^2,\\ &\leq 36 \min_{\operatorname{rank}(A)\leq r}\left\{\|\mathbf{X}A - \mathbf{X}A^*\|_F^2 + r\left(\sqrt{q}+\sqrt{T}\right)^2\sigma^2\right\},\end{aligned} \tag{8.12}$$

with $q = \operatorname{rank}(\mathbf{X})$.

Let us comment on this bound. We notice that if $r = r^* = \operatorname{rank}(A^*)$, Bound (8.12) gives

$$\mathbb{E}\left[\|\mathbf{X}\widehat{A}_{r^*} - \mathbf{X}A^*\|_F^2\right] \leq 36r^*\left(\sqrt{q}+\sqrt{T}\right)^2\sigma^2.$$

Conversely, it can be shown (see [80] at the end of Section 3) that there exists a constant $C(\mathbf{X}) > 0$ only depending on the ratio $\sigma_1(\mathbf{X})/\sigma_q(\mathbf{X})$, such that for any $r^* \leq q \wedge T$ with $q = \operatorname{rank}(\mathbf{X})$, the minimax lower bound holds

$$\inf_{\widehat{A}} \sup_{A^*:\,\operatorname{rank}(A^*)=r^*} \mathbb{E}\left[\|\mathbf{X}\widehat{A} - \mathbf{X}A^*\|_F^2\right] \geq C(\mathbf{X})r^*\left(\sqrt{q}+\sqrt{T}\right)^2\sigma^2,$$

where the infimum is over all the estimators $\widehat{A}$. This means that, up to a constant factor, the estimator $\widehat{A}_r$ estimates matrices of rank r at the minimax rate.

Furthermore, according to (8.4), when A^* has a rank larger than r, any estimator $\widehat{A}$ of rank at most r fulfills

$$\|\mathbf{X}A^* - \mathbf{X}\widehat{A}\|_F^2 \geq \|\mathbf{X}A^* - (\mathbf{X}A^*)_{(r)}\|_F^2 = \sum_{k>r}\sigma_k(\mathbf{X}A^*)^2.$$

Therefore, the term $\sum_{k>r}\sigma_k(\mathbf{X}A^*)^2$ is the minimal bias that an estimator of rank r can have.

8.3.2 When the Rank of A^* Is Unknown

When the rank r^* of A^* is unknown, we would like to use the "oracle" estimator $\widehat{A}_{r_o}$, which achieves the best trade-off in (8.12) between the bias term $\sum_{k>r}\sigma_k(\mathbf{X}A^*)^2$ and the variance term $r\left(\sqrt{q}+\sqrt{T}\right)^2\sigma^2$. Of course, the oracle rank r_o is unknown since it depends on the unknown matrix A^*. Similarly to Chapter 2, we select a rank $\widehat{r}$ according to a penalized criterion and show that the risk of $\widehat{A}_{\widehat{r}}$ is almost as small as the risk of $\widehat{A}_{r_o}$.

For a constant $K>1$, we select $\widehat{r}$ by minimizing the criterion

$$\widehat{r}\in\underset{r=1,\dots,q\wedge T}{\operatorname{argmin}}\left\{\|Y-\mathbf{X}\widehat{A}_r\|^2+\sigma^2\mathrm{pen}(r)\right\},\quad \text{with } \mathrm{pen}(r)=Kr\left(\sqrt{T}+\sqrt{q}\right)^2. \tag{8.13}$$

According to (8.6), the estimator $\widehat{A}=\widehat{A}_{\widehat{r}}$ is then solution of

$$\widehat{A}\in\underset{A\in\mathbb{R}^{p\times T}}{\operatorname{argmin}}\left\{\|Y-\mathbf{X}A\|^2+K\,\mathrm{rank}(A)\left(\sqrt{T}+\sqrt{q}\right)^2\sigma^2\right\}, \tag{8.14}$$

since the rank of the solution to (8.14) is not larger than $\mathrm{rank}(PY)\le q\wedge T$. Let us analyze the risk of $\mathbf{X}\widehat{A}$ with $\widehat{A}=\widehat{A}_{\widehat{r}}$.

Theorem 8.5 Oracle risk bound
For any $K>1$, there exists a constant $C_K>1$ depending only on K, such that the estimator $\widehat{A}$ defined by (8.14) fulfills the risk bound (with $q=\mathrm{rank}(\mathbf{X})$)

$$\mathbb{E}\left[\|\mathbf{X}\widehat{A}-\mathbf{X}A^*\|_F^2\right]\le C_K\min_{r=1,\dots,q\wedge T}\left\{\mathbb{E}\left[\|\mathbf{X}\widehat{A}_r-\mathbf{X}A^*\|_F^2\right]+r(\sqrt{T}+\sqrt{q})^2\sigma^2\right\}. \tag{8.15}$$

If we compare Bounds (8.12) and (8.15), we observe that the risk of the estimator $\mathbf{X}\widehat{A}$ is almost as good as the risk of the best of the estimators $\{\mathbf{X}\widehat{A}_r,\ r=1,\dots,q\wedge T\}$. Since the estimator $\widehat{A}$ can be computed with a single singular value decomposition, we can adapt in practice to the rank of the matrix A^*. Combining (8.12) and (8.15), we obtain the upper bound for the risk of $\mathbf{X}\widehat{A}$

$$\mathbb{E}\left[\|\mathbf{X}\widehat{A}-\mathbf{X}A^*\|_F^2\right]\le C'_K\min_{A\in\mathbb{R}^{p\times T}}\left\{\|\mathbf{X}A-\mathbf{X}A^*\|_F^2+\mathrm{rank}(A)(T+q)\sigma^2\right\}, \tag{8.16}$$

with $C'_K>1$ depending only on $K>1$ (again, the minimum in (8.16) is achieved for A with rank not larger than $q\wedge T$).

Proof of Theorem 8.5
1- Deterministic bound

Let us fix $r \in \{1,\ldots,q\wedge T\}$. From the definition of $\widehat{r}$, we have that $\|Y-\mathbf{X}\widehat{A}\|^2 + \sigma^2\text{pen}(\widehat{r}) \le \|Y-\mathbf{X}\widehat{A}_r\|^2+\sigma^2\text{pen}(r)$, from which it follows that

$$\|\mathbf{X}\widehat{A}-\mathbf{X}A^*\|_F^2 \le \|\mathbf{X}\widehat{A}_r-\mathbf{X}A^*\|_F^2 + 2\langle PE, \mathbf{X}\widehat{A}-\mathbf{X}\widehat{A}_r\rangle_F + \sigma^2\text{pen}(r)-\sigma^2\text{pen}(\widehat{r}). \tag{8.17}$$

Similarly as in the proof of Theorem 2.2 in Chapter 2, if we prove that for some $a>1,\, b>0$

$$2\langle PE, \mathbf{X}\widehat{A}-\mathbf{X}\widehat{A}_r\rangle_F - \sigma^2\text{pen}(\widehat{r}) \le a^{-1}\|\mathbf{X}\widehat{A}-\mathbf{X}A^*\|_F^2 + b^{-1}\|\mathbf{X}\widehat{A}_r-\mathbf{X}A^*\|_F^2 + Z_r, \tag{8.18}$$

with Z_r fulfilling $\mathbb{E}[Z_r] \le c\left(\sqrt{q}+\sqrt{T}\right)^2 r\sigma^2$ for some constant $c>0$, then we have from (8.17)

$$\frac{a-1}{a}\mathbb{E}\left[\|\mathbf{X}\widehat{A}-\mathbf{X}A^*\|_F^2\right] \le \frac{b+1}{b}\mathbb{E}\left[\|\mathbf{X}\widehat{A}_r-\mathbf{X}A^*\|_F^2\right] + (K+c)\left(\sqrt{q}+\sqrt{T}\right)^2 r\sigma^2,$$

and (8.15) follows.

Let us prove (8.18). As in the proof of Proposition 8.2, we bound the cross-term in terms of the operator norm of PE. The rank of $\mathbf{X}\widehat{A}-\mathbf{X}\widehat{A}_r$ is at most $\widehat{r}+r$, so according to (8.3), we have for any $a,b>0$

$$\begin{aligned}
&2\langle PE, \mathbf{X}\widehat{A}-\mathbf{X}\widehat{A}_r\rangle_F\\
&\le 2\|PE\|_{(2,\widehat{r}+r)}\, \|\mathbf{X}\widehat{A}-\mathbf{X}\widehat{A}_r\|_F\\
&\le 2|PE|_{\text{op}}\sqrt{\widehat{r}+r}\left(\|\mathbf{X}\widehat{A}-\mathbf{X}A^*\|_F + \|\mathbf{X}\widehat{A}_r-\mathbf{X}A^*\|_F\right)\\
&\le (a+b)\,(\widehat{r}+r)\,|PE|_{\text{op}}^2 + a^{-1}\|\mathbf{X}\widehat{A}-\mathbf{X}A^*\|_F^2 + b^{-1}\|\mathbf{X}\widehat{A}_r-\mathbf{X}A^*\|_F^2,
\end{aligned}$$

where the last line is obtained by applying twice the inequality $2xy \le ax^2+y^2/a$. We then obtain (8.18) with

$$Z_r = (a+b)\,(\widehat{r}+r)\,|PE|_{\text{op}}^2 - \sigma^2\text{pen}(\widehat{r}).$$

It only remains to prove that $\mathbb{E}[Z_r] \le c\left(\sqrt{q}+\sqrt{T}\right)^2 r\sigma^2$ for some $c>0$.

2- Stochastic control

The map $E \to |PE|_{\text{op}}$ is 1-Lipschitz with respect to the Frobenius norm, so according to the Gaussian concentration Inequality (B.2), page 301, and Bound (8.10), there exists an exponential random variable ξ with parameter 1, such that

$$|PE|_{\text{op}} \le \sigma\left(\sqrt{T}+\sqrt{q}\right) + \sigma\sqrt{2\xi}\,.$$

Since $\widehat{r} \le q\wedge T$, taking $a=(3+K)/4$ and $b=(K-1)/4$ we obtain that

$$\begin{aligned}
&\widehat{r}\left((a+b)|PE|_{\text{op}}^2 - K(\sqrt{T}+\sqrt{q})^2\sigma^2\right)\\
&\le \frac{1+K}{2}\,(q\wedge T)\left(|PE|_{\text{op}}^2 - \frac{2K}{1+K}(\sqrt{T}+\sqrt{q})^2\sigma^2\right)_+\\
&\le \frac{1+K}{2}\,(q\wedge T)\left(\left(\sqrt{T}+\sqrt{q}+\sqrt{2\xi}\right)^2 - \frac{2K}{1+K}(\sqrt{T}+\sqrt{q})^2\right)_+\sigma^2.
\end{aligned}$$

From the inequality $(\sqrt{T}+\sqrt{q}+\sqrt{2\xi})^2 \leq (1+\alpha)(\sqrt{T}+\sqrt{q})^2+2\xi(1+1/\alpha)$ with $\alpha = (K-1)/(K+1)$, we obtain

$$\mathbb{E}\left[\widehat{r}\left((a+b)|PE|_{\text{op}}^2 - K(\sqrt{T}+\sqrt{q})^2\sigma^2\right)\right] \leq 2\frac{K(1+K)}{K-1}(q\wedge T)\sigma^2\mathbb{E}[\xi] .$$

Since $q\wedge T \leq \left(\sqrt{T}+\sqrt{q}\right)^2/4$, combining this bound with (8.11), we get (8.18), with

$$\begin{aligned}\mathbb{E}[Z_r] &\leq 3\frac{K+1}{2}\left(\sqrt{q}+\sqrt{T}\right)^2\sigma^2 r + 2\frac{K(1+K)}{K-1}(q\wedge T)\sigma^2 \\ &\leq \left(3\frac{K+1}{2}+\frac{K(1+K)}{2(K-1)}\right)\left(\sqrt{q}+\sqrt{T}\right)^2\sigma^2 r,\end{aligned}$$

and the proof of Theorem 8.5 is complete. □

8.4 Low Rank and Sparsity

As explained at the beginning of the chapter, the matrix A^* in (8.2) is likely not only to have (approximately) a low rank, but also to be sparse in some sense (coordinate sparse, group sparse, etc.). Therefore, we want to exploit simultaneously the low rank structures and the sparse structures to improve our estimation of A^*.

As a first step, we start with the case where the matrix is row sparse.

8.4.1 Row-Sparse Matrices

A natural assumption is that the response $y^{(i)} = \sum_{j=1}^p A^*_{ji} x_j^{(i)} + \varepsilon^{(i)}$ depends from $x^{(i)}$ only through a subset $\{x_j^{(i)} : j \in J^*\}$ of its coordinates. This means that the rows of A^* are zero except those with index in J^*. Again, the difficulty comes from the fact that the set J^* is unknown.

Estimating a row-sparse matrix simply corresponds to a group-sparse regression as described in Chapter 2. Actually, the matrix structure plays no role in this setting, and we can recast the model in a vectorial form. We can stack the columns of A^* into a vector $\text{vect}(A^*)$ of dimension pT and act similarly with Y and E. Then, we end with a simple group-sparse regression

$$\text{vect}(Y) = \widetilde{\mathbf{X}}\,\text{vect}(A^*) + \text{vect}(E),$$

where

$$\widetilde{\mathbf{X}} = \begin{bmatrix} \mathbf{X} & 0 & 0 \\ 0 & \ddots & 0 \\ 0 & 0 & \mathbf{X} \end{bmatrix} \in \mathbb{R}^{nT\times pT},$$

and where the p groups $G_k = \{k+pl : l = 0,\ldots,T-1\}$ for $k = 1,\ldots,p$ gather the indices congruent modulo p. We can estimate $\text{vect}(A^*)$ with the group-Lasso estimator

(see Chapter 5, Section 5.3.1, page 103)

$$\text{vect}(\widehat{A}_\lambda) \in \operatorname*{argmin}_{\beta \in \mathbb{R}^{pT}} \left\{ \|\text{vect}(Y) - \widetilde{\mathbf{X}}\beta\|^2 + \lambda \sum_{j=1}^{p} \|\beta_{G_j}\| \right\}.$$

Writing $A_{j:}$ for the j-th row of A, the above minimization problem is equivalent to

$$\widehat{A}_\lambda \in \operatorname*{argmin}_{A \in \mathbb{R}^{p\times T}} \left\{ \|Y - \mathbf{X}A\|_F^2 + \lambda \sum_{j=1}^{p} \|A_{j:}\| \right\}. \tag{8.19}$$

We assume in the following that the columns $\mathbf{X}_1, \ldots, \mathbf{X}_p$ of $\mathbf{X}$ have unit norm. We observe that in this case the operator norm $|\widetilde{\mathbf{X}}_{G_k}|_{op}$ of

$$\widetilde{\mathbf{X}}_{G_k} = \begin{bmatrix} \mathbf{X}_k & 0 & 0 \\ 0 & \ddots & 0 \\ 0 & 0 & \mathbf{X}_k \end{bmatrix} \in \mathbb{R}^{nT\times T}$$

is 1 for every $k = 1, \ldots, p$. We can then lift the risk Bound (5.20) from Chapter 5, page 105.

Theorem 8.6 Risk bound for row-sparse matrices

For $\lambda = 3\sigma\left(\sqrt{T} + 2\sqrt{\log(p)}\right)$, we have with probability at least $1 - 1/p$

$$\|\mathbf{X}\widehat{A}_\lambda - \mathbf{X}A^*\|_F^2 \le \min_{A\in\mathbb{R}^{p\times T}} \left\{ \|\mathbf{X}A - \mathbf{X}A^*\|_F^2 + \frac{18\sigma^2}{\widetilde{\kappa}_G(A)} |J(A)| \left(T + 4\log(p)\right) \right\}, \tag{8.20}$$

where $J(A) = \{j \in \{1, \ldots, p\} : A_{j:} \neq 0\}$ and

$$\widetilde{\kappa}_G(A) = \min_{B \in \mathscr{C}_G(J(A))} \sqrt{\frac{|J(A)| \|\mathbf{X}B\|_F}{\sum_{j\in J(A)} \|B_{j:}\|}} \quad \text{with } \mathscr{C}_G(J) = \left\{ B : \sum_{j\in J^c} \|B_{j:}\| < 5 \sum_{j\in J} \|B_{j:}\| \right\}.$$

We next investigate whether we can improve upon Theorem 8.5 and Theorem 8.6 by taking simultaneously into account low rank and row sparsity.

8.4.2 Criterion for Row-Sparse and Low-Rank Matrices

We now consider the case where the matrix A^* has a low rank and a small number of nonzero rows. The rank of a matrix is not larger than the number of its nonzero rows, so the estimator (8.19) already has a small rank. But here we have in mind a case where the rank of A^* is much smaller than the number of its nonzero rows, and we want to exploit this feature. The rank of A^* and the location of its nonzero rows are unknown. We first investigate how much we can gain by taking into account simultaneously row sparsity and low rank. In this direction, a model selection estimator will be our benchmark.

Let π be a probability distribution on $\mathscr{P}(\{1,\ldots,p\})$. For example, we can set

$$\pi_J = \left(C_p^{|J|}\right)^{-1} e^{-|J|}(e-1)/(e-e^{-p}) \quad \text{for all} \quad J \subset \{1\ldots,p\}, \tag{8.21}$$

with $C_p^k = p!/(k!(p-k)!)$. For $K > 1$, we define $\widehat{A}$ as a minimizer of the criterion

$$\begin{aligned} \text{Crit}(A) &= \|Y - \mathbf{X}A\|_F^2 + \text{pen}(A)\sigma^2, \\ &\text{with} \quad \text{pen}(A) = K\left(\sqrt{r(A)}\left(\sqrt{T} + \sqrt{|J(A)|}\right) + \sqrt{2\log\left(\pi_{J(A)}^{-1}\right)}\right)^2, \end{aligned} \tag{8.22}$$

where $r(A)$ is the rank of A and where $J(A) = \left\{j \in \{1,\ldots,p\} : A_{j:} \neq 0\right\}$ is the set of nonzero rows of A. Penalty (8.22) is very similar to Penalty (2.9), page 35, for model selection. Actually, the set $V_{J,r} = \left\{A \in \mathbb{R}^{p\times T} : J(A) \subset J,\ r(A) = r\right\}$ is a submanifold of dimension $r(T+J-r)$. Since

$$\sqrt{r}(\sqrt{T} + \sqrt{J}) \geq \sqrt{r(T+J-r)} = \sqrt{\dim(V_{J,r})},$$

Penalty (8.22) can be viewed as an upper bound of the penalty

$$\text{pen}'(A) = K\left(\sqrt{\dim(V_{J,r})} + \sqrt{2\log(\pi_J^{-1})}\right)^2, \quad \text{for } A \in V_{J,r},$$

which has the same form as Penalty (2.9).

The minimization Problem (8.22) has a computational complexity that is prohibitive in high-dimensional settings, since it requires to explore all the subsets $J \subset \{1\ldots,p\}$ in general. Yet, the resulting estimator provides a good benchmark in terms of statistical accuracy. We discuss in the following sections how we can relax (8.22) in order to obtain a convex criterion.

Theorem 8.7 Risk bound for row-sparse and low-rank matrices

For any $K > 1$, there exists a constant $C_K > 1$ depending only on K, such that the estimator $\widehat{A}$ defined by (8.22) fulfills the risk bound

$$\begin{aligned} &\mathbb{E}\left[\|\mathbf{X}\widehat{A} - \mathbf{X}A^*\|_F^2\right] \\ &\leq C_K \min_{A\neq 0}\left\{\|\mathbf{X}A - \mathbf{X}A^*\|_F^2 + r(A)(T + |J(A)|)\sigma^2 + \log\left(\pi_{J(A)}^{-1}\right)\sigma^2\right\}. \end{aligned} \tag{8.23}$$

If we choose π_J as in (8.21), we obtain for some constant $C'_K > 1$ depending only on $K > 1$

$$\begin{aligned} &\mathbb{E}\left[\|\mathbf{X}\widehat{A} - \mathbf{X}A^*\|_F^2\right] \\ &\leq C'_K \min_{A\neq 0}\left\{\|\mathbf{X}A - \mathbf{X}A^*\|_F^2 + r(A)(T + |J(A)|)\sigma^2 + |J(A)|\log\left(ep/|J(A)|\right)\sigma^2\right\}. \end{aligned}$$

We observe that the term $r(A)(T+|J(A)|)$ can be much smaller than $|J(A)|T$ appearing in (8.20) if $r(A)$ is small compared to $\min(|J(A)|,T)$. Similarly, if $|J(A)|$ is small compared to $\text{rank}(\mathbf{X})$, then the above upper bound is small compared to the risk Bound (8.16) for the low-rank case. Thus, with an estimator that takes into account simultaneously row sparsity and low rank, we can get a significant improvement in estimation.

Proof of Theorem 8.7

1- Deterministic bound

We fix a nonzero matrix $A \in \mathbb{R}^{p\times T}$. Starting from $\text{Crit}(\widehat{A}) \leq \text{Crit}(A)$, we obtain

$$\|\mathbf{X}\widehat{A}-\mathbf{X}A^*\|_F^2 \leq \|\mathbf{X}A-\mathbf{X}A^*\|_F^2+2\langle E,\mathbf{X}\widehat{A}-\mathbf{X}A\rangle_F+\text{pen}(A)\sigma^2-\text{pen}(\widehat{A})\sigma^2, \quad (8.24)$$

with

$$\text{pen}(A) \leq 4K\left(r(A)(T+|J(A)|)+\log\left(\pi_{J(A)}^{-1}\right)\right).$$

As in the proofs of Theorem 2.2 and Theorem 8.5, if we prove that there exist some constants $c_1<1$, $c_2,c_3>0$ and a random variable $Z(A)$, such that

$$2\langle E,\mathbf{X}\widehat{A}-\mathbf{X}A\rangle_F-\text{pen}(\widehat{A})\sigma^2 \leq c_1\|\mathbf{X}\widehat{A}-\mathbf{X}A^*\|_F^2+c_2\|\mathbf{X}A-\mathbf{X}A^*\|_F^2+Z(A), \quad (8.25)$$

with $\mathbb{E}[Z(A)] \leq c_3 r(A)T\sigma^2$, then (8.23) follows from (8.24) and (8.25).

Let us prove (8.25). For any $J\subset\{1,\ldots,p\}$, we write $\mathbf{X}_J$ for the matrix obtained from $\mathbf{X}$ by keeping only the columns in J, and we define S_J as the orthogonal of $\text{range}(\mathbf{X}A)$ in $\text{range}(\mathbf{X}A)+\text{range}(\mathbf{X}_J)$, so that

$$\text{range}(\mathbf{X}A)+\text{range}(\mathbf{X}_J)=\text{range}(\mathbf{X}A)\oplus S_J.$$

In particular, the linear span S_J has a dimension at most $|J|$. For notational simplicity, we will write in the following $\widehat{J}$ for $J(\widehat{A})$, $\widehat{r}$ for the rank of $\widehat{A}$, P_J for the orthogonal projector onto S_J and P_A for the orthogonal projector onto the range of $\mathbf{X}A$. Since $\mathbf{X}(\widehat{A}-A)=P_{\widehat{J}}\mathbf{X}\widehat{A}+P_A\mathbf{X}(\widehat{A}-A)$ with $\text{rank}(P_{\widehat{J}}\mathbf{X}\widehat{A})\leq\widehat{r}$, from (8.3) we get

$$\begin{aligned}
&2\langle E,\mathbf{X}\widehat{A}-\mathbf{X}A\rangle_F\\
&=2\langle P_{\widehat{J}}E,P_{\widehat{J}}\mathbf{X}\widehat{A}\rangle_F+2\langle P_AE,P_A\mathbf{X}(\widehat{A}-A)\rangle_F\\
&\leq 2\|P_{\widehat{J}}E\|_{(2,\widehat{r})}\|P_{\widehat{J}}\mathbf{X}\widehat{A}\|_F+2\|P_AE\|_F\,\|P_A\mathbf{X}(\widehat{A}-A)\|_F\\
&\leq \frac{K+1}{2}\left[\|P_{\widehat{J}}E\|_{(2,\widehat{r})}^2+\|P_AE\|_F^2\right]+\frac{2}{K+1}\left[\|P_{\widehat{J}}\mathbf{X}\widehat{A}\|_F^2+\|P_A\mathbf{X}(\widehat{A}-A)\|_F^2\right],
\end{aligned}$$

where we used twice in the last line the inequality $2xy\leq ax^2+a^{-1}y^2$ with $a=(K+1)/2$. According to Pythagorean formula, we have $\|P_{\widehat{J}}\mathbf{X}\widehat{A}\|_F^2+\|P_A\mathbf{X}(\widehat{A}-A)\|_F^2=\|\mathbf{X}(\widehat{A}-A)\|_F^2$. Since $\|\mathbf{X}(\widehat{A}-A)\|_F^2\leq(1+b)\|\mathbf{X}(\widehat{A}-A^*)\|_F^2+(1+b^{-1})\|\mathbf{X}(A-A^*)\|_F^2$, taking $b=(K-1)/4$, we obtain (8.25) with $c_1=a^{-1}(1+b)<1$, $c_2=a^{-1}(1+b^{-1})$ and

$$Z(A)=\frac{K+1}{2}\|P_AE\|_F^2+\widehat{\Delta},\quad\text{where }\widehat{\Delta}=\frac{K+1}{2}\|P_{\widehat{J}}E\|_{(2,\widehat{r})}^2-\text{pen}(\widehat{A})\sigma^2.$$

We observe that $\mathbb{E}\left[\|P_A E\|_F^2\right] = \text{rank}(\mathbf{X}A)\, T\sigma^2 \leq r(A)\, T\sigma^2$, so to conclude the proof of (8.23), it only remains to check that

$$\mathbb{E}\left[\widehat{\Delta}\right] \leq \frac{2K(K+1)}{K-1} T\sigma^2. \tag{8.26}$$

2- Stochastic control

We know from Lemma 8.3 that for any $J \subset \{1,\ldots,p\}$ and $r \leq p \wedge T$,

$$\mathbb{E}\left[\|P_J E\|_{(2,r)}\right] \leq \sqrt{r}\,\mathbb{E}\left[\sigma_1(P_J E)\right] \leq \sigma\sqrt{r}\left(\sqrt{\dim(S_J)} + \sqrt{T}\right) \leq \sigma\sqrt{r}\left(\sqrt{|J|} + \sqrt{T}\right).$$

Since the map $E \to \|P_J E\|_{(2,r)}$ is 1-Lipschitz with respect to the Frobenius norm, the Gaussian concentration Inequality (B.2) ensures for each $J \subset \{1,\ldots,p\}$ and $r \in \{1,\ldots,p\wedge T\}$ the existence of a standard exponential random variable $\xi_{J,r}$, such that

$$\|P_J E\|_{(2,r)} \leq \sigma\left(\sqrt{r}\left(\sqrt{|J|} + \sqrt{T}\right) + \sqrt{2\xi_{J,r}}\right). \tag{8.27}$$

As in the proof of Theorem 2.2 in Chapter 2, we observe that for all $J \subset \{1,\ldots,p\}$ and all $r \in \{1,\ldots,p\wedge T\}$, we have

$$\begin{aligned}
&\left(\sqrt{r}\left(\sqrt{|J|} + \sqrt{T}\right) + \sqrt{2\xi_{J,r}}\right)^2 \\
&\leq \left(\sqrt{r}\left(\sqrt{|J|} + \sqrt{T}\right) + \sqrt{2\log(\pi_J^{-1})} + \sqrt{2\left(\xi_{J,r} - \log(\pi_J^{-1})\right)_+}\right)^2 \\
&\leq \frac{2K}{K+1}\left(\sqrt{r}\left(\sqrt{|J|} + \sqrt{T}\right) + \sqrt{2\log(\pi_J^{-1})}\right)^2 + \frac{4K}{K-1}\left(\xi_{J,r} - \log(\pi_J^{-1})\right)_+.
\end{aligned} \tag{8.28}$$

Since $\widehat{r} \leq p \wedge T$, combining (8.27) with (8.28), we obtain

$$\begin{aligned}
\widehat{\Delta} &= \frac{K+1}{2}\left(\|P_{\widehat{J}} E\|_{(2,\widehat{r})}^2 - \frac{2K}{K+1}\left(\sqrt{\widehat{r}}\left(\sqrt{|\widehat{J}|} + \sqrt{T}\right) + \sqrt{2\log(\pi_{\widehat{J}}^{-1})}\right)^2 \sigma^2\right) \\
&\leq \frac{2K(K+1)}{K-1}\sigma^2\left(\xi_{\widehat{J},\widehat{r}} - \log(\pi_{\widehat{J}}^{-1})\right)_+ \mathbf{1}_{\widehat{r}\geq 1}.
\end{aligned}$$

To conclude the proof of (8.26), we check that

$$\begin{aligned}
\mathbb{E}\left[\left(\xi_{\widehat{J},\widehat{r}} - \log(\pi_{\widehat{J}}^{-1})\right)_+ \mathbf{1}_{\widehat{r}\geq 1}\right] &\leq \sum_{r=1}^{p\wedge T} \sum_{J\subset\{1,\ldots,p\}} \mathbb{E}\left[\left(\xi_{J,r} - \log(\pi_J^{-1})\right)_+\right] \\
&\leq T \sum_{J\subset\{1,\ldots,p\}} \pi_J = T.
\end{aligned}$$

The proof of Theorem 8.7 is complete. □

The above procedure satisfies a nice risk bound, but it is computationally untractable since we cannot explore all the subsets $J \subset \{1,\ldots,p\}$. A natural idea to enforce row sparsity is to use a group penalty as in (8.20). Yet, if we a add a constraint on the rank in (8.20), there is no computationally efficient algorithm for solving exactly this problem. A possible direction for combining sparse and low-rank constraints is to convexify the constraint on the rank. We discuss this issue in the next section.

8.4.3 Convex Criterion for Low-Rank Matrices

We emphasize that for the pure low-rank estimation of Section 8.3, there is no need to convexify Criterion (8.14), since it can be minimized efficiently from a single singular value decomposition. The convexification is only needed when we want to combine low-rank properties with some other structures as row sparsity. Yet, as a first step, we start by analyzing the convexification of (8.14).

The main idea underlying the introduction of the Lasso estimator is to replace the constraint on the number of nonzero coordinates of β by a constraint on the sum of their absolute values. Following the same idea, we can replace the constraint on the rank of A, which is the number of nonzero singular values of A, by a constraint on the nuclear norm of A, which is the sum of the singular values of A. This gives the following convex criterion

$$\widehat{A}_\lambda \in \operatorname*{argmin}_{A\in\mathbb{R}^{p\times T}} \left\{ \|Y - \mathbf{X}A\|_F^2 + \lambda |A|_* \right\}, \tag{8.29}$$

where λ is a positive tuning parameter and $|A|_* = \sum_k \sigma_k(A)$ is the nuclear norm of A. Similarly to the Lasso estimator, we can provide a risk bound for this estimator.

Theorem 8.8 Risk bound for the convex multivariate criterion
Let $K > 1$ and set

$$\lambda = 2K\sigma_1(\mathbf{X})\left(\sqrt{T}+\sqrt{q}\right)\sigma, \quad \textit{with } q = \operatorname{rank}(X).$$

Then, with probability larger than $1 - e^{-(K-1)^2(T+q)/2}$, we have

$$\|\mathbf{X}\widehat{A}_\lambda - \mathbf{X}A^*\|_F^2 \le \inf_A \left\{ \|\mathbf{X}A - \mathbf{X}A^*\|_F^2 + 9K^2 \frac{\sigma_1(\mathbf{X})^2}{\sigma_q(\mathbf{X})^2}\left(\sqrt{T}+\sqrt{q}\right)^2 \sigma^2 \operatorname{rank}(A) \right\}.$$

This risk bound is similar to (8.16), except that there is a constant 1 in front of the bias term (which is good news) and a constant $\sigma_1(\mathbf{X})^2/\sigma_q(\mathbf{X})^2$ in front of the variance term (which is bad news). This last constant can be huge in practice, since the smallest singular values $\sigma_q(\mathbf{X})$ of data matrices $\mathbf{X}$ tend to be very small. When this constant remains of a reasonable size, the estimator $\widehat{A}_\lambda$ has properties similar to those of $\widehat{A}$ defined by (8.14).

Proof. The proof is very similar to the proof of (5.13) for the Lasso estimator.

1- Deterministic bound

We first derive a deterministic bound on $\|\mathbf{X}\widehat{A}_\lambda - \mathbf{X}A^*\|_F^2$.

Lemma 8.9

For $\lambda \geq 2\sigma_1(\mathbf{X}^T E)$, we have

$$\|\mathbf{X}\widehat{A}_\lambda - \mathbf{X}A^*\|_F^2 \leq \inf_{A\in\mathbb{R}^{p\times T}} \left\{ \|\mathbf{X}A - \mathbf{X}A^*\|_F^2 + \frac{9\lambda^2}{4\sigma_q(\mathbf{X})^2}\text{rank}(A) \right\}. \tag{8.30}$$

Proof of Lemma 8.9.

Let us introduce the set $\mathbb{A} := \{A \in \mathbb{R}^{p\times T} : A = P_{\mathbf{X}^T} A\}$, where $P_{\mathbf{X}^T}$ is the orthogonal projector onto the range of $\mathbf{X}^T$. Since we have the orthogonal decomposition $\mathbb{R}^p = \ker(\mathbf{X}) \bigoplus \text{range}(\mathbf{X}^T)$, for all matrices $A \in \mathbb{R}^{p\times T}$, we have

$$\mathbf{X}P_{\mathbf{X}^T}A = \mathbf{X}A \quad \text{and} \quad \text{rank}(P_{\mathbf{X}^T}A) \leq \text{rank}(A).$$

In particular, in (8.30) the infimum over $A \in \mathbb{R}^{p\times T}$ coincides with the infimum over $A \in \mathbb{A}$, and we only need to prove (8.30) with the infimum over $\mathbb{A}$. Similarly, from Inequality (C.6), page 315, in Appendix C, we observe that $|P_{\mathbf{X}^T}\widehat{A}_\lambda|_* \leq |\widehat{A}_\lambda|_*$ with strict inequality if $P_{\mathbf{X}^T}\widehat{A}_\lambda \neq \widehat{A}_\lambda$. Therefore, the estimator $\widehat{A}_\lambda$ belongs to the space $\mathbb{A}$.

The optimality condition (D.3), page 323, in Appendix D for convex functions ensures the existence of a matrix $\widehat{Z} \in \partial|\widehat{A}_\lambda|_*$ such that $-2\mathbf{X}^T(Y - \mathbf{X}\widehat{A}_\lambda) + \lambda\widehat{Z} = 0$. Since $Y = \mathbf{X}A^* + E$, for any $A \in \mathbb{A}$, we have

$$2\langle \mathbf{X}\widehat{A}_\lambda - \mathbf{X}A^*, \mathbf{X}\widehat{A}_\lambda - \mathbf{X}A\rangle_F - 2\langle \mathbf{X}^T E, \widehat{A}_\lambda - A\rangle_F + \lambda\langle \widehat{Z}, \widehat{A}_\lambda - A\rangle_F = 0.$$

The subgradient monotonicity of convex functions (D.2), page 322 ensures that for all $Z \in \partial|A|_*$, we have $\langle \widehat{Z}, \widehat{A}_\lambda - A\rangle_F \geq \langle Z, \widehat{A}_\lambda - A\rangle_F$. As a consequence,

for all $A \in \mathbb{A}$ and for all $Z \in \partial|A|_*$, we have

$$2\langle \mathbf{X}\widehat{A}_\lambda - \mathbf{X}A^*, \mathbf{X}\widehat{A}_\lambda - \mathbf{X}A\rangle_F \leq 2\langle \mathbf{X}^T E, \widehat{A}_\lambda - A\rangle_F - \lambda\langle Z, \widehat{A}_\lambda - A\rangle_F. \tag{8.31}$$

Let us denote by $A = \sum_{k=1}^r \sigma_k u_k v_k^T$ the singular value decomposition of A, with $r = \text{rank}(A)$. We write P_u (respectively, P_v) for the orthogonal projector onto $\text{span}\{u_1, \ldots, u_r\}$ (respectively, onto $\text{span}\{v_1, \ldots, v_r\}$). We also set $P_u^\perp = I - P_u$ and $P_v^\perp = I - P_v$. According to Lemma D.6, page 324, in Appendix D, the subdifferential of $|A|_*$ is given by

$$\partial|A|_* = \left\{ \sum_{k=1}^r u_k v_k^T + P_u^\perp W P_v^\perp \ : \ W \in \mathbb{R}^{p\times T} \text{ with } \sigma_1(W) \leq 1 \right\}.$$

Let us set $W = 2\mathbf{X}^T E/\lambda$. Since $\sigma_1(W) = 2\sigma_1(\mathbf{X}^T E)/\lambda \leq 1$, the matrix $Z = \sum_{k=1}^r u_k v_k^T + P_u^\perp W P_v^\perp$ belongs to $\partial|A|_*$. The decomposition

$$\widehat{A}_\lambda - A = P_u(\widehat{A}_\lambda - A) + P_u^\perp(\widehat{A}_\lambda - A)P_v + P_u^\perp(\widehat{A}_\lambda - A)P_v^\perp$$

gives

$$2\langle \mathbf{X}^T E, \widehat{A}_\lambda - A\rangle_F - \lambda\langle Z, \widehat{A}_\lambda - A\rangle_F$$
$$= 2\langle \mathbf{X}^T E, \widehat{A}_\lambda - A\rangle_F - 2\langle P_u^\perp \mathbf{X}^T E P_v^\perp, \widehat{A}_\lambda - A\rangle_F - \lambda\langle \sum_{k=1}^r u_k v_k^T, \widehat{A}_\lambda - A\rangle_F$$
$$= 2\langle \mathbf{X}^T E, \widehat{A}_\lambda - A\rangle_F - 2\langle \mathbf{X}^T E, P_u^\perp(\widehat{A}_\lambda - A)P_v^\perp\rangle_F - \lambda\langle \sum_{k=1}^r u_k v_k^T, \widehat{A}_\lambda - A\rangle_F$$
$$= 2\langle \mathbf{X}^T E, P_u(\widehat{A}_\lambda - A)\rangle_F + 2\langle \mathbf{X}^T E, P_u^\perp(\widehat{A}_\lambda - A)P_v\rangle_F - \lambda\langle \sum_{k=1}^r u_k v_k^T, P_u(\widehat{A}_\lambda - A)P_v\rangle_F.$$

Lemma C.2, page 313, in Appendix C ensures that $\langle A, B\rangle_F \leq \sigma_1(A)|B|_*$, so

$$2\langle \mathbf{X}^T E, \widehat{A}_\lambda - A\rangle_F - \lambda\langle Z, \widehat{A}_\lambda - A\rangle_F$$
$$\leq 2\sigma_1(\mathbf{X}^T E)\left(|P_u(\widehat{A}_\lambda - A)|_* + |P_u^\perp(\widehat{A}_\lambda - A)P_v|_*\right) + \lambda|P_u(\widehat{A}_\lambda - A)P_v|_*.$$

Since $2\sigma_1(\mathbf{X}^T E) \leq \lambda$, we obtain

$$2\langle \mathbf{X}^T E, \widehat{A}_\lambda - A\rangle_F - \lambda\langle Z, \widehat{A}_\lambda - A\rangle_F$$
$$\leq \lambda|P_u(\widehat{A}_\lambda - A)|_* + \lambda|P_u^\perp(\widehat{A}_\lambda - A)P_v|_* + \lambda|P_u(\widehat{A}_\lambda - A)P_v|_*$$
$$\leq \lambda\sqrt{\text{rank}(A)}\left(\|P_u(\widehat{A}_\lambda - A)\|_F + \|P_u^\perp(\widehat{A}_\lambda - A)P_v\|_F + \|P_u(\widehat{A}_\lambda - A)P_v\|_F\right),$$

where we used in the last line $\text{rank}(P_u) = \text{rank}(P_v) = \text{rank}(A)$ and the inequality $|M|_* \leq \sqrt{\text{rank}(M)}\,\|M\|_F$ from Lemma C.2, page 313, in Appendix C. According to Inequalities (C.6) and (C.7), page 315, in Appendix C, the three above Frobenius norms are upper-bounded by $\|\widehat{A}_\lambda - A\|_F$, so combining the above bound with (8.31) and Al-Kashi formula

$$2\langle \mathbf{X}\widehat{A}_\lambda - \mathbf{X}A^*, \mathbf{X}\widehat{A}_\lambda - \mathbf{X}A\rangle_F = \|\mathbf{X}\widehat{A}_\lambda - \mathbf{X}A^*\|_F^2 + \|\mathbf{X}\widehat{A}_\lambda - \mathbf{X}A\|_F^2 - \|\mathbf{X}A - \mathbf{X}A^*\|_F^2,$$

we obtain

$$\|\mathbf{X}\widehat{A}_\lambda - \mathbf{X}A^*\|_F^2 + \|\mathbf{X}\widehat{A}_\lambda - \mathbf{X}A\|_F^2 \leq \|\mathbf{X}A - \mathbf{X}A^*\|_F^2 + 3\lambda\sqrt{\text{rank}(A)}\,\|\widehat{A}_\lambda - A\|_F.$$

Let us denote by $\mathbf{X}^+$ the Moore–Penrose pseudo-inverse of $\mathbf{X}$ (see Section C.2 in Appendix C). For any matrix $M \in \mathbb{A}$, we have $M = P_{\mathbf{X}^T}M = \mathbf{X}^+\mathbf{X}M$, and thus again according to Lemma C.2, page 313, we have

$$\|M\|_F \leq |\mathbf{X}^+|_{\text{op}}\|\mathbf{X}M\|_F = \sigma_q(\mathbf{X})^{-1}\|\mathbf{X}M\|_F.$$

Since $A - \widehat{A}_\lambda$ belongs to $\mathbb{A}$, we have $\|\widehat{A}_\lambda - A\|_F \leq \sigma_q(\mathbf{X})^{-1}\|\mathbf{X}(\widehat{A}_\lambda - A)\|_F$, and therefore

$$\|\mathbf{X}\widehat{A}_\lambda - \mathbf{X}A^*\|_F^2 \leq \|\mathbf{X}A - \mathbf{X}A^*\|_F^2 - \|\mathbf{X}\widehat{A}_\lambda - \mathbf{X}A\|_F^2 + \frac{3\lambda\sqrt{\text{rank}(A)}}{\sigma_q(\mathbf{X})}\|\mathbf{X}\widehat{A}_\lambda - \mathbf{X}A\|_F$$
$$\leq \|\mathbf{X}A - \mathbf{X}A^*\|_F^2 + \frac{9\lambda^2}{4\sigma_q(\mathbf{X})^2}\text{rank}(A),$$

where we used in the last line the inequality $2ab \leq a^2+b^2$. The proof of Lemma 8.9 is complete. □

2- Stochastic control

To conclude the proof of Theorem 8.8, it remains to check that

$$\mathbb{P}\left(\sigma_1(\mathbf{X}^T E) \geq K\sigma_1(\mathbf{X})(\sqrt{T}+\sqrt{q})\sigma\right) \leq e^{-(K-1)^2(T+q)/2}, \quad \text{for all } K>1. \quad (8.32)$$

Writing $P_{\mathbf{X}}$ for the projection onto the range of $\mathbf{X}$, we have $\mathbf{X}^T E = \mathbf{X}^T P_{\mathbf{X}} E$, so $\sigma_1(\mathbf{X}^T E) \leq \sigma_1(\mathbf{X})\sigma_1(P_{\mathbf{X}} E)$. As in the proof of Theorem 8.5, combining the Gaussian concentration Inequality (B.2), page 301, and the Bound (8.10), we obtain that

$$\sigma_1(P_{\mathbf{X}} E) \leq \sigma\left(\sqrt{T}+\sqrt{q}\right) + \sigma\sqrt{2\xi},$$

for some exponential random variable ξ with parameter 1. Bound (8.32) follows, and the proof of Theorem 8.8 is complete. □

8.4.4 Computationally Efficient Algorithm for Row-Sparse and Low-Rank Matrices

In order to combine the benefits of coordinate sparsity and low rankness, it is natural to penalize the negative log-likelihood by both the group-ℓ^1 and the nuclear norms

$$\widehat{A}_{\lambda,\mu} \in \underset{A\in\mathbb{R}^{p\times T}}{\operatorname{argmin}} \left\{ \|Y-\mathbf{X}A\|_F^2 + \lambda|A|_* + \mu\sum_{j=1}^{p}\|A_{j:}\| \right\},$$

where $A_{j:}$ denotes the j-th row of A and $\lambda, \mu > 0$. The resulting criterion is convex in A.

It is not difficult to combine the analysis for the Group-Lasso estimator and the analysis of Theorem 8.8 in order to get a risk bound for $\widehat{A}_{\lambda,\mu}$. Yet, the resulting risk bound does not improve on the results with the nuclear alone or the group norm alone. To overcome this issue, some iterative algorithms have been proposed, in the spirit of the Iterative Hard Thresholding / Iterative Group Thresholding algorithms of Chapter 6.

The main recipe is to decompose $A = UV$ with $U \in \mathbb{R}^{p\times r}$ and $V \in \mathbb{R}^{r\times T}$, and to notice that
(i) the rank of A is smaller than r by construction; and
(ii) if U is row sparse, then A is also row-sparse.

The target is then to minimize the $\|Y-\mathbf{X}UV\|_F^2$ under the constraint that U is row sparse. It is a hard task as the row-sparse constraint is not convex and in addition the objective function $(U,V) \to \|Y-\mathbf{X}UV\|_F^2$ is not convex. We could try to apply an IGT algorithm on (U,V); yet, this cannot be done directly as, for any $\alpha>0$, we have $(\alpha U)(\alpha^{-1}V) = UV$, so hard thresholding the rows of U is ineffective if the size of the entries of U are not stabilized. To do so, the second recipe is to add to the objective function a penalty proportional to $\|U^TU - V^TV\|_F^2$. Such a penalty ensures that the singular values of U and V are similar, stabilizing the size of the entries of U.

The objective function is then $F(U,V) + \lambda|J(U)|$, where $|J(U)|$ is the number of non-zero rows of U and

$$F(U,V) = \|Y - \mathbf{X}UV\|_F^2 + \frac{1}{2}\|U^TU - V^TV\|_F^2.$$

Similarly as for the IGT algorithm (6.23), page 131, the related proximal algorithm amounts to take a gradient step of F and then to apply a group-thresholding to the rows of U. When initialized with the group-Lasso estimator (8.19), page 168, the iterative algorithm can be shown to enjoy a risk bound similar to (8.23) under a restricted isometry property as (6.10). The proof of this result goes beyond the scope of this chapter, we refer the interested reader to the paper Yu *et al.* [166].

8.5 Discussion and References

8.5.1 Take-Home Message

We can easily take advantage of the low rank of the regression matrix A^* by implementing the estimator (8.14). It is possible to improve significantly the estimator accuracy by taking simultaneously into account low-rank structures with row-sparse structures, as explained in Section 8.4.2. A computationally tractable procedure with optimal estimation rates can be obtained via an iterative algorithm, as described in Yu *et al.* [166].

8.5.2 References

Most of the material presented in this chapter is adapted from Bunea, She, and Wegkamp [44, 43]. Lemma 8.3 comes from Davidson and Szarek [61], and Theorem 8.8 is adapted from Koltchinskii, Lounici, and Tsybakov [99].

We refer to Bach [13] for the convex Criterion (8.29) and examples of applications. Finally, Exercise 8.6.3 is adapted from Giraud [80].

8.6 Exercises

8.6.1 Hard Thresholding of the Singular Values

We consider the estimators $\widehat{A}_r$ defined by (8.6) and for $\lambda > 0$ the selection criterion

$$\widehat{r}_\lambda \in \operatorname*{argmin}_r \left\{ \|Y - \mathbf{X}\widehat{A}_r\|_F^2 + \lambda r \right\}.$$

The selection Criterion (8.13) corresponds to the choice $\lambda = K\left(\sqrt{T} + \sqrt{q}\right)^2 \sigma^2$. With the same notations as those of Lemma 8.1, we write $PY = \sum_k \sigma_k u_k v_k^t$ for the SVD decomposition of PY, with $\sigma_1 \geq \sigma_2 \geq \ldots$.

1. Prove that $\|Y-\mathbf{X}\widehat{A}_r\|_F^2 = \|Y\|_F^2 - \sum_{k=1}^r \sigma_k^2$ for $r \leq \text{rank}(PY)$.
2. Check that $\widehat{r}_\lambda = \max\left\{r : \sigma_r^2 \geq \lambda\right\}$ and conclude that

$$\mathbf{X}\widehat{A}_{\widehat{r}_\lambda} = \sum_k \sigma_k \mathbf{1}_{\sigma_k^2 \geq \lambda}\, u_k v_k^T.$$

8.6.2 Exact Rank Recovery

We denote by r^* the rank of A^*, by $\sigma_1(M) \geq \sigma_2(M) \geq \ldots$ the singular values of M ranked in decreasing order, and we consider the selection procedure (8.13).

1. Prove from the previous exercise that

$$\mathbb{P}(\widehat{r} \neq r^*) = \mathbb{P}\left(\sigma_{r^*+1}(PY) \geq \sqrt{K}(\sqrt{T}+\sqrt{q})\sigma \quad \text{or} \quad \sigma_{r^*}(PY) < \sqrt{K}(\sqrt{T}+\sqrt{q})\sigma\right).$$

2. Deduce from Weyl inequality (Theorem C.6 in Appendix C) that

$$\mathbb{P}(\widehat{r} \neq r^*) \leq \mathbb{P}\left(\sigma_1(PE) \geq \min\left(\sqrt{K}(\sqrt{T}+\sqrt{q})\sigma, \sigma_{r^*}(\mathbf{X}A^*) - \sqrt{K}(\sqrt{T}+\sqrt{q})\sigma\right)\right).$$

3. Assume that $\sigma_{r^*}(\mathbf{X}A^*) \geq 2\sqrt{K}(\sqrt{T}+\sqrt{q})\sigma$. Prove that in this case, the probability to recover the exact rank r^* is lower-bounded by

$$\mathbb{P}(\widehat{r} = r^*) \geq 1 - \exp\left(-\frac{(\sqrt{K}-1)^2}{2}\left(\sqrt{T}+\sqrt{q}\right)^2\right).$$

8.6.3 Rank Selection with Unknown Variance

We consider here the case where both the variance σ^2 and the rank $r^* = \text{rank}(A^*)$ are unknown. A classical selection criterion in this setting is

$$\widehat{r} \in \underset{r<nT/\lambda}{\text{argmin}}\left\{\frac{\|Y-\mathbf{X}\widehat{A}_r\|_F^2}{nT-\lambda r}\right\}, \quad \text{with } \lambda > 0. \tag{8.33}$$

We notice that we can recast this criterion as

$$\widehat{r} \in \underset{r<nT/\lambda}{\text{argmin}}\left\{\|Y-\mathbf{X}\widehat{A}_r\|_F^2 + \lambda r \widehat{\sigma}_r^2\right\} \quad \text{with } \widehat{\sigma}_r^2 = \frac{\|Y-\mathbf{X}\widehat{A}_r\|_F^2}{nT-\lambda r},$$

so it can be viewed as a version of (8.13), with the variance σ^2 replaced by $\widehat{\sigma}_r^2$. In the following, we set $\lambda = K(\sqrt{T}+\sqrt{q})^2$ with $K > 1$, and we assume that $1 \leq r^* = \text{rank}(A^*) \leq nT/(2\lambda)$.

1. Prove that $\|Y - (PY)_{(r^*)}\|_F \leq \|E\|_F$.

2. Deduce from the previous question and Criterion (8.33) that

$$\|Y-(PY)_{(\widehat{r})}\|_F^2 \le \|E\|_F^2 - \frac{\lambda(\widehat{r}-r^*)}{nT-\lambda r_*}\|E\|_F^2.$$

3. Prove from the above inequality and (8.3) that for any $\alpha>0$

$$\begin{aligned}\|\mathbf{X}\widehat{A}_{\widehat{r}}-\mathbf{X}A^*\|_F^2 &\le 2\langle \mathbf{X}\widehat{A}_{\widehat{r}}-\mathbf{X}A^*,E\rangle_F - \frac{\lambda(\widehat{r}-r^*)}{nT-\lambda r_*}\|E\|_F^2 \\ &\le \alpha\|PE\|_{(2,r^*+\widehat{r})}^2+\alpha^{-1}\|\mathbf{X}\widehat{A}_{\widehat{r}}-\mathbf{X}A^*\|_F^2 - \frac{\lambda(\widehat{r}-r^*)}{nT-\lambda r_*}\|E\|_F^2.\end{aligned}$$

4. Check that for $\alpha>1$

$$\frac{\alpha-1}{\alpha}\|\mathbf{X}\widehat{A}_{\widehat{r}}-\mathbf{X}A^*\|_F^2 \le \lambda r^*\frac{\|E\|_F^2}{nT-\lambda r^*}+\alpha r^*|PE|_{\text{op}}^2+\widehat{r}\left(\alpha|PE|_{\text{op}}^2-\lambda\frac{\|E\|_F^2}{nT}\right).$$

5. For $\alpha>1$ and $\delta>0$, such that $K\ge\alpha(1+\delta)/(1-\delta)^2$, combining the bound

$$\begin{aligned}&\widehat{r}\left(\alpha|PE|_{\text{op}}^2-\lambda\frac{\|E\|_F^2}{nT}\right)\le \\ &\quad (q\wedge T)\left(\alpha|PE|_{\text{op}}^2-\lambda(1-\delta)^2\sigma^2\right)_+ +\lambda(q\wedge T)\left((1-\delta)^2\sigma^2-\frac{\|E\|_F^2}{nT}\right)_+\end{aligned}$$

with the Gaussian concentration Inequality (B.2), page 301, prove that for $\delta\ge 1-\sqrt{\frac{nT-4}{nT}}\approx 2/nT$

$$\begin{aligned}\mathbb{E}\left[\widehat{r}\left(\alpha|PE|_{\text{op}}^2-\lambda\frac{\|E\|_F^2}{nT}\right)\right] &\le 2\alpha(1+\delta^{-1})(q\wedge T)\sigma^2+4KnTe^{-\delta^2nT/2+2}\sigma^2 \\ &\le 2\alpha(1+\delta^{-1})(q\wedge T)\sigma^2+8K\delta^{-2}e\sigma^2.\end{aligned}$$

6. Conclude that there exists a constant $C_K>1$ depending only on $K>1$, such that

$$\mathbb{E}\left[\|\mathbf{X}\widehat{A}_{\widehat{r}}-\mathbf{X}A^*\|_F^2\right]\le C_K r^*(T+q)\sigma^2.$$

Compare this bound with (8.16).

Chapter 9

Graphical Models

Graphical modeling is a convenient theory for encoding the conditional dependencies between p random variables $X_1,\dots,X_p$ by a graph g. Graphical models are used in many different frameworks (image analysis, physics, economics, etc.), and they have been proposed for investigating biological regulation networks, brain connections, etc.

The concept of conditional dependence is more suited than the concept of dependence in order to catch "direct links" between variables, as explained in Figure 9.1 below.

When there is a snow storm in Paris, we observe both huge traffic jams and plenty of snowmen in parks. So there is a strong correlation (and thus dependence) between the size of the Parisian traffic jams and the number of snowmen in parks.

Of course, snowmen do not cause traffic jams. Traffic jams and snowmen are correlated only because they are both induced by snow falls. These causality relationships are represented in the side picture by edges.

Conditional dependencies better reflect these relationships. Actually, conditionally in the snow falls, the size of the traffic jams and the number of snowmen are likely to be independent.

Figure 9.1 *Difference between dependence and conditional dependence.*

There are mainly two types of graphical models, which encode conditional dependencies in two different ways. We briefly present these two types based on directed and non-directed graphs in Section 9.2. Our main goal in this chapter will be to learn the graph of conditional dependencies between $(X_1,\dots,X_p)$ from an n-sample of $(X_1,\dots,X_p)$, with a special focus on the case where the graph has few edges and the sample size n is smaller than the number p of variables. As explained below, it is a very hard and non-parametric problem in general, and we will mainly investigate the case where $(X_1,\dots,X_p)$ follows a Gaussian distribution with (unknown) covariance Σ. In this case, the conditional dependencies are encoded in the precision matrix Σ^{-1} and our problem mainly amounts to estimate the locations of the nonzero entries of a sparse precision matrix.

9.1 Reminder on Conditional Independence

We remind the reader that two random variables X and Y are independent conditionally on a variable Z (we write $X \perp\!\!\!\perp Y \mid Z$) if their conditional laws fulfill

$$\text{law}((X,Y)|Z) = \text{law}(X|Z) \otimes \text{law}(Y|Z).$$

In particular, if the distribution of (X,Y,Z) has a positive density f with respect to a σ-finite product measure μ, then

$$\begin{aligned} X \perp\!\!\!\perp Y \mid Z \quad &\Longleftrightarrow \quad f(x,y|z) = f(x|z)f(y|z) \quad \mu\text{-a.e.} \\ &\Longleftrightarrow \quad f(x,y,z) = f(x,z)f(y,z)/f(z) \quad \mu\text{-a.e.} \\ &\Longleftrightarrow \quad f(x|y,z) = f(x|z) \quad \mu\text{-a.e.,} \end{aligned}$$

where $f(x,y|z)$ (resp. $f(x|z)$) represents the conditional density of (x,y) (resp. x) given z.

We also recall that for any measurable function h, we have the property

$$X \perp\!\!\!\perp (Y,W) \mid Z \implies X \perp\!\!\!\perp Y \mid (Z,h(W)). \tag{9.1}$$

In order to avoid unnecessary technicalities, we assume in the remainder of this chapter that the distribution of $(X_1,\dots,X_p)$ has a positive continuous density with respect to σ-finite product measure in $\mathbb{R}^p$.

9.2 Graphical Models

9.2.1 Directed Acyclic Graphical Models

Let us consider a directed graph g (made of nodes and arrows, with the arrows linking some nodes) with p nodes labelled from 1 to p, as in Figure 9.2. We will assume that g is acyclic, which means that no sequence of arrows forms a loop in graph g. We call parents of a nodes b, such that there exists an arrow $b \to a$, and we denote by $\text{pa}(a)$ the set of parents of a. We call descendant of a the nodes that can be reached from a by following some sequence of arrows (a included), and we denote by $\text{de}(a)$ the set of descendants of a. For example, in Figure 9.2, the descendants of 11 are $\{11,12,18\}$ and its parents are $\{5,8,10\}$.

Directed acyclic graphical model

Let g be a Directed Acyclic Graph (DAG). The distribution of the random variable $X = (X_1 \ldots, X_p)$ is a graphical model according to g if it fulfills the property

$$\text{for all } a: \quad X_a \perp\!\!\!\perp \{X_b,\ b \notin \text{de}(a)\} \;\big|\; \{X_c,\ c \in \text{pa}(a)\}.$$

We write $\mathscr{L}(X) \sim g$ when this property is met.

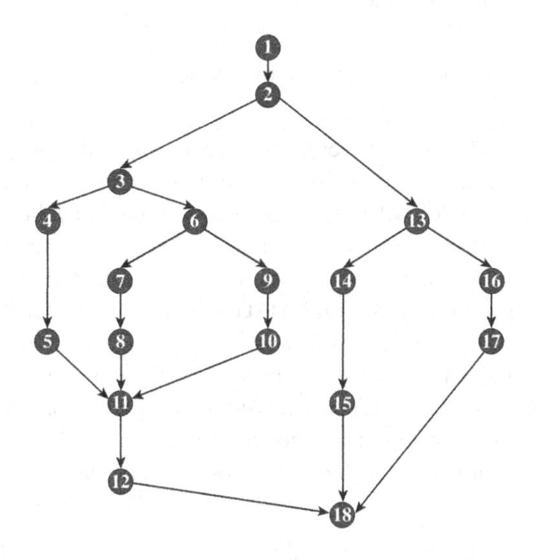

Figure 9.2 *Directed Acyclic Graphical models: $X_a \perp\!\!\!\perp \{X_b,\ b \notin \text{de}(a)\} \mid \{X_c,\ c \in \text{pa}(a)\}$.*

Let us consider two DAGs, g and g', with g a subgraph of g' (denoted by $g \subset g'$) and a random variable X, such that $\mathscr{L}(X) \sim g$. Writing $\text{pa}'(a)$ (respectively, $\text{de}'(a)$) for the parents (respectively, the descendants) of the node a in g', we observe that, due to the acyclicity,

$$\text{pa}(a) \subset \text{pa}'(a) \subset \{1,\ldots,p\} \setminus \text{de}'(a) \subset \{1,\ldots,p\} \setminus \text{de}(a).$$

Accordingly, it follows from $\mathscr{L}(X) \sim g$ and the property (9.1) of the conditional independence, that for any $a = 1,\ldots,p$

$$X_a \perp\!\!\!\perp \left\{X_b : b \notin \text{de}'(a)\right\} \mid \left\{X_c : c \in \text{pa}'(a)\right\}.$$

As a consequence for $g \subset g'$, we have $\mathscr{L}(X) \sim g \implies \mathscr{L}(X) \sim g'$. In particular, there is in general no unique DAG g, such that $\mathscr{L}(X) \sim g$. Yet, we may wonder whether there exists a unique DAG g^* minimal for the inclusion, such that $\mathscr{L}(X) \sim g^*$. It is unfortunately not the case, as can be seen in the following simple example. Consider $X_1,\ldots,X_p$ generated by the autoregressive model

$$X_{i+1} = \alpha X_i + \varepsilon_i \quad \text{with} \quad X_0 = 0, \quad \alpha \neq 0 \quad \text{and} \quad \varepsilon_i \text{ i.i.d.}$$

Since $(X_1,\dots,X_p)$ is a Markov chain, we have

$$(X_1,\dots,X_i) \perp\!\!\!\perp (X_i,\dots,X_p) \mid X_i \quad \text{for all } i=1,\dots,p.$$

As a consequence, the two graphs

$$1 \to 2 \to \dots \to p \quad \text{and} \quad 1 \leftarrow 2 \leftarrow \dots \leftarrow p$$

are minimal graphs for this model.

Be careful with the interpretation of directed graphical models!

Our objective in this chapter is to learn from data a (minimal) graph of conditional dependencies between p variables. Since there is no unique minimal DAG g, such that $\mathscr{L}(X) \sim g$, the problem of estimating "the" minimal acyclic graph of a distribution $\mathbb{P}$ is an ill-posed problem. Yet, it turns out that the minimal DAGs associated with a given distribution only differ by the direction of (some of) their arrows. So instead of trying to estimate a minimal DAG associated to a distribution $\mathbb{P}$, we can try to estimate the locations (but not the directions) of the arrows of the minimal DAGs associated to $\mathbb{P}$. This problem can be solved efficiently under some conditions (see Spirtes *et al.* [144] and Kalisch and Bühlmann [97, 98]). An alternative is to consider another notion of graphical models based on non-directed graphs, which are more suited for our problem. We will follow this alternative in the remainder of this chapter.

Before moving to non-directed graphical models, we emphasize that directed acyclic graphical models are powerful tools for modeling, defining a distribution $\mathbb{P}$ and computing it. These nice features mainly rely on the factorization formula for distributions $\mathbb{P}$ with a positive density f with respect to a σ-finite product measure in $\mathbb{R}^p$. Actually, in such a case, for any DAG g such that $\mathbb{P} \sim g$, we have the factorization formula

$$f(x_1,\dots,x_p) = \prod_{i=1}^{p} f(x_i|x_{\mathrm{pa}(i)}), \tag{9.2}$$

where $f(x_i|x_{\mathrm{pa}(i)})$ is the conditional density of x_i given $x_{\mathrm{pa}(i)}$. This factorization is easily proved by induction; see Exercise 9.6.1. A common way to use (9.2) is to start from a known graph g representing known causality relationships between the variables, and then build the density f from the conditional densities $f(x_i|x_{\mathrm{pa}(i)})$ according to Formula (9.2).

9.2.2 Non-Directed Models

We consider now non-directed graphs g (made of nodes and edges, with the edges linking some nodes) with p nodes labelled from 1 to p, as in Figure 9.3. Graph g induces a symmetric relation on $\{1,\dots,p\}$ by

$$a \overset{g}{\sim} b \iff \text{there is an edge between } a \text{ and } b \text{ in graph } g.$$

We call neighbors of a, the nodes in $\mathrm{ne}(a) = \{b : b \overset{g}{\sim} a\}$, and we set $\mathrm{cl}(a) = \mathrm{ne}(a) \cup \{a\}$.

Non-directed graphical model

The distribution of the random variable $X = (X_1 \dots, X_p)$ is a graphical model according to graph g if it fulfills the property

$$\text{for all } a: \quad X_a \perp\!\!\!\perp \{X_b,\ b \notin \mathrm{cl}(a)\} \mid \{X_c,\ c \in \mathrm{ne}(a)\}.$$

We write $\mathscr{L}(X) \sim g$ when this property is met.

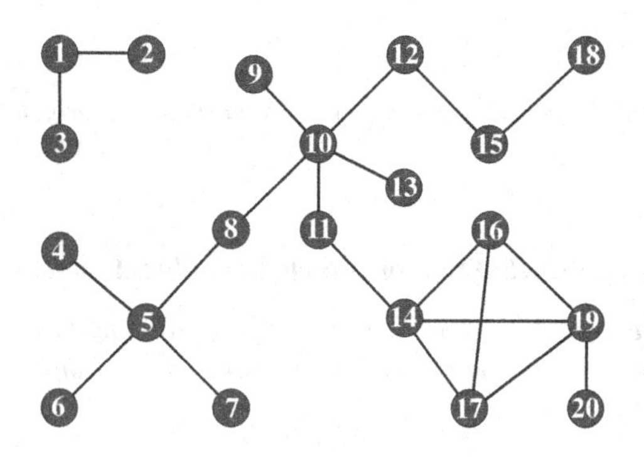

Figure 9.3 *Non-directed graphical models:* $X_a \perp\!\!\!\perp \{X_b : b \nsim a\} \mid \{X_c : c \sim a\}$.

Again, we check that if $\mathscr{L}(X) \sim g$ and $g \subset g'$ then $\mathscr{L}(X) \sim g'$, so there is no unique graph g, such that $\mathscr{L}(X) \sim g$. In particular, if $g_{\#}$ represents the complete graph (where all the nodes are connected together), then $\mathscr{L}(X) \sim g_{\#}$. Yet, when X has a positive continuous density with respect to some σ-finite product measure, there exists a unique minimal graph g_* (for inclusion), such that $\mathscr{L}(X) \sim g_*$. We will prove this result in the Gaussian setting in Section 9.3, and we refer to Lauritzen [101], Chapter 3 for the general case. In the following, we call simply "graph of X" the minimal graph g_*, such that $\mathscr{L}(X) \sim g_*$.

There is a simple connection between directed and non-directed graphical models. Let g be a *directed* graph such that $\mathscr{L}(X) \sim g$. We associate to g the so-called moral graph, which is the *non-directed* graph g_m obtained as follows:

1. For each node, set an edge between its parents in g.
2. Replace all arrows by edges.

We refer to Figure 9.4 for an illustration.

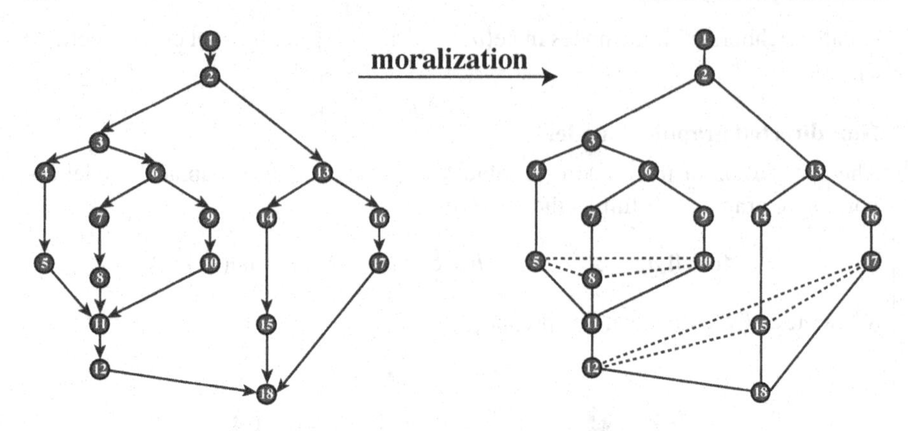

Figure 9.4 *Moralization of a Directed Acyclic Graph. Left: the original graph. Right: the associated moral graph. The extra edges linking the parents are represented with dashed lines.*

> **Lemma 9.1 From directed to non-directed graphical models**
>
> *Let g be a directed graph, and write g_m for its (non-directed) moral graph defined above. Then, if X has a positive density with respect to a σ-finite product measure,*
>
> $$\mathscr{L}(X) \sim g \implies \mathscr{L}(X) \sim g_m.$$

We refer to Exercise 9.6.2 for a proof of this lemma. We emphasize that the moral graph g_m may not coincide with the minimal graph g_* associated to X.

In the following, our goal is to estimate from an n-sample the minimal graph g_*, such that $\mathscr{L}(X) \sim g_*$. In general, it is a very hard problem in high-dimensional settings. Actually, a result due to Hammersley and Clifford [89] ensures that a distribution with positive continuous density f is a graphical model with respect to a non-directed graph g if and only if f fulfills the factorization

$$f(x) = \prod_{c \in \text{cliques}(g)} f_c(x_c), \tag{9.3}$$

where the f_c are positive functions and the cliques of g are the subgraphs of g that are completely connected (all the nodes of a clique are linked together; for example, the nodes 14, 16 ,17, 19 form a clique in Figure 9.3). We refer again to Lauritzen [101] Chapter 3 for a proof of this result. Trying to infer such a minimal decomposition of the density f is a very hard problem in general. Yet, it is tractable in some cases, for example, when $(X_1, \ldots, X_p)$ follows a Gaussian distribution. In the following, we will focus on this issue of estimating g_* from an n-sample of a Gaussian distribution.

9.3 Gaussian Graphical Models

We assume in the remainder of this chapter that X follows a Gaussian $\mathcal{N}(0,\Sigma)$ distribution, with Σ unknown and non-singular. Our goal will be to estimate from an n-sample $X^{(1)},\ldots,X^{(n)}$ of X the non-directed graph g_*, minimal for the inclusion, such that $\mathcal{L}(X) \sim g_*$. We have in mind that the sample size n can be smaller than the dimension p of X.

We will denote by $\mathbf{X}$ the $n \times p$ matrix with rows given by the transpose of $X^{(1)},\ldots,X^{(n)}$ and we write $\widehat{\Sigma}$ for the empirical covariance matrix

$$\widehat{\Sigma} = \frac{1}{n}\sum_{i=1}^{n}(X^{(i)})(X^{(i)})^T = \frac{1}{n}\mathbf{X}^T\mathbf{X}.$$

9.3.1 Connection with the Precision Matrix and the Linear Regression

A nice feature of the Gaussian $\mathcal{N}(0,\Sigma)$ distribution, with Σ non-singular, is that the minimal graph g_* is encoded in the precision matrix $K = \Sigma^{-1}$. Actually, let us define graph g with nodes labelled by $\{1,\ldots,p\}$ according to the symmetric relation for $a \neq b$

$$a \overset{g}{\sim} b \iff K_{a,b} \neq 0. \tag{9.4}$$

The next lemma shows that g is the minimal (non-directed) graph, such that $\mathcal{L}(X) \sim g$.

Lemma 9.2 Gaussian Graphical Models (GGM) and precision matrix

For graph g defined by (9.4), we have

1. *$\mathcal{L}(X) \sim g$ and g is the minimal graph fulfilling this property.*
2. *For any $a \in \{1,\ldots,p\}$, there exists $\varepsilon_a \sim \mathcal{N}(0, K_{aa}^{-1})$ independent of $\{X_b : b \neq a\}$, such that*

$$X_a = -\sum_{b\in \mathrm{ne}(a)} \frac{K_{ab}}{K_{aa}} X_b + \varepsilon_a\,. \tag{9.5}$$

Proof.

1. We write $\mathrm{nn}(a) = \{1,\ldots,p\} \setminus \mathrm{cl}(a)$ for the nodes of g non-neighbor to a. Let us consider the two sets $A = \{a\} \cup \mathrm{nn}(a)$ and $B = \mathrm{ne}(a)$. The precision matrix restricted to A is of the form

$$K_{AA} = \begin{pmatrix} K_{aa} & 0 \\ 0 & K_{\mathrm{nn}(a)\,\mathrm{nn}(a)} \end{pmatrix}.$$

Lemma A.4, page 295, in Appendix A ensures that the distribution of $X_{\{a\}\cup\mathrm{nn}(a)}$ given $X_{\mathrm{ne}(a)}$ is Gaussian with covariance matrix

$$(K_{AA})^{-1} = \begin{pmatrix} K_{aa}^{-1} & 0 \\ 0 & (K_{\mathrm{nn}(a)\,\mathrm{nn}(a)})^{-1} \end{pmatrix}.$$

Since independence is equivalent to zero covariance for Gaussian random variables, the variables X_a and $X_{\text{nn}(a)}$ are independent conditionally on $X_{\text{ne}(a)}$. We note that this property is no longer true if we remove an edge between a and one of its neighbors in g, since in this case the off-diagonal blocks of $(K_{AA})^{-1}$ are nonzero.

2. The second point simply rephrases Formula (A.3), page 295, in Appendix A.

□

In the next subsections, we build on Lemma 9.2 in order to derive procedures for estimating g_*.

9.3.2 Estimating g by Multiple Testing

Corollary A.4 in Appendix A ensures that, for $a \neq b$, the conditional correlation (also called partial correlation) of X_a and X_b given $\{X_c : c \neq a, b\}$ is given by

$$\text{cor}(X_a, X_b | X_c : \ c \neq a, b) = \frac{-K_{ab}}{\sqrt{K_{aa} K_{bb}}}. \tag{9.6}$$

As a consequence, there is an edge in the graph g_* if and only if the conditional correlation $\text{cor}(X_a, X_b | X_c : c \neq a, b)$ is nonzero. A natural idea is then to estimate g_* by testing if these conditional correlations are nonzero. When $n > p - 2$, the conditional correlations can be estimated by replacing K in (9.6) by the inverse of the empirical covariance matrix $\widehat{\Sigma}^{-1}$

$$\widehat{\text{cor}}(X_a, X_b | X_c : c \neq a, b) = \widehat{\rho}_{ab} = \frac{-\left[\widehat{\Sigma}^{-1}\right]_{ab}}{\sqrt{\left[\widehat{\Sigma}^{-1}\right]_{aa} \left[\widehat{\Sigma}^{-1}\right]_{bb}}}.$$

When $\text{cor}(X_a, X_b | X_c : c \neq a, b) = 0$, we have (see Anderson [3], Chapter 4.3)

$$\widehat{t}_{ab} = \sqrt{n - 2 - p} \times \frac{\widehat{\rho}_{ab}}{\sqrt{1 - \widehat{\rho}_{ab}^2}} \sim \text{Student}(n - p - 2).$$

For each $a < b$, let us denote by $\widehat{p}_{ab}$ the p-value associated to the test statistic $|\widehat{t}_{ab}|$. We can then estimate the set of nonzero conditional correlations by applying a multiple testing procedure (see Chapter 10) to the set of p-values $\{\widehat{p}_{ab}, 1 \leq a < b \leq p\}$. This procedure makes perfect sense when n is large, much larger than p. But when $n - p$ is small, the empirical conditional correlation $\widehat{\rho}_{ab}$ is very unstable and the procedure leads to poor results. When $p > n$, the empirical covariance matrix $\widehat{\Sigma}$ is not invertible, so the procedure cannot be implemented.

There have been several propositions to circumvent this issue. One proposition is to work with $\widehat{\text{cor}}(X_a, X_b | X_c : c \in S)$, with S a small subset of $\{1, \ldots, p\} \setminus \{a, b\}$ (like pairs), instead of $\widehat{\text{cor}}(X_a, X_b | X_c : c \neq a, b)$; see, e.g., Wille and Bühlmann [164], and Castelo and Roverato [49]. While the empirical conditional correlation

$\widehat{\text{cor}}(X_a, X_b | X_c : c \in S)$ does not suffer from instability when the cardinality of S is small compared to n, it is unclear what we estimate at the end in general.

If we look carefully at the definition of $\widehat{\text{cor}}(X_a, X_b | X_c : c \neq a, b)$, we observe that the instability for large p comes from the fact that we estimate K by $\widehat{\Sigma}^{-1}$, which is unstable. An idea is then to build a more stable estimator of K.

9.3.3 Sparse Estimation of the Precision Matrix

We have in mind the case where the underlying graph g is sparse (it has a few edges). Since $K_{ab} = 0$ when there is no edge between a and b in g_*, the sparsity of g_* translates into coordinate sparsity for the precision matrix K. Exploiting this sparsity can significantly improve the estimation of K upon $\widehat{\Sigma}^{-1}$. Following the ideas of Chapter 5, we can then estimate K by minimizing the negative log-likelihood penalized by the ℓ^1 norm of K. Let us first derive the likelihood of a $p \times p$ positive symmetric matrix $K \in \mathcal{S}_p^+$

$$\begin{aligned}\text{Likelihood}(K) &= \prod_{i=1}^{n} \sqrt{\frac{\det(K)}{(2\pi)^p}} \exp\left(-\frac{1}{2}(X^{(i)})^T K X^{(i)}\right) \\ &= \left(\frac{\det(K)}{(2\pi)^p}\right)^{n/2} \exp\left(-\frac{n}{2}\langle \widehat{\Sigma}, K\rangle_F\right),\end{aligned}$$

where the last equality follows from $\langle X^{(i)}, KX^{(i)}\rangle = \langle X^{(i)}(X^{(i)})^T, K\rangle_F$. Removing the terms not depending on K, we obtain that the negative-log-likelihood is given (up to a constant) by

$$-\frac{n}{2}\log(\det(K)) + \frac{n}{2}\langle K, \widehat{\Sigma}\rangle_F .$$

Since $-\log(\det(K))$ is convex (Exercise 9.6.3), the above function is convex. Similarly, as for the Lasso estimator, minimizing the negative log-likelihood penalized by the ℓ^1 norm of K will induce a coordinate-sparse estimator. The resulting estimator

$$\widehat{K}_\lambda = \underset{K \in \mathcal{S}_p^+}{\text{argmin}} \left\{ -\frac{n}{2}\log(\det(K)) + \frac{n}{2}\langle K, \widehat{\Sigma}\rangle_F + \lambda \sum_{a \neq b} |K_{ab}| \right\}, \tag{9.7}$$

is usually called the "graphical-Lasso" or simply "glasso" estimator. We point out that we do not penalize the diagonal elements, since they are not expected to be zero. From a theoretical point of view, risk bounds similar to those for the Lasso have been derived for the glasso estimator by Ravikumar *et al.* [130], under some "compatibility conditions" on the true precision matrix, which are hard to interpret. From a practical point of view, minimizing the convex Criterion (9.7) is more challenging than for the Lasso Criterion (5.4), but a powerful numerical scheme has been developed in Friedman *et al.* [73], and the glasso estimator can be computed in quite large dimensions (at least a few thousands). This scheme is implemented in the `R` package `glasso` available at `http://cran.r-project.org/web/packages/glasso/`. Nevertheless, some poor results have been reported in practice by Villers *et al.* [161] when $n \leq p$ compared to some other estimation schemes.

9.3.4 Estimation by Regression

From the second part of Lemma 9.2, we observe that the sparsity of K induces some coordinate-sparsity in the linear regressions (9.5) for $a = 1, \dots, p$. We can then recast our problem in a multivariate regression setting.

Let us define the matrix θ by $\theta_{ab} = -K_{ab}/K_{bb}$ for $b \neq a$ and $\theta_{aa} = 0$. Equation (9.5) ensures that $\mathbb{E}[X_a | X_b : b \neq a] = \sum_b \theta_{ba} X_b$, so the vector $[\theta_{ba}]_{b:b\neq a}$ minimizes

$$\mathbb{E}\left[\left(X_a - \sum_{b:b\neq a} \beta_b X_b\right)^2\right]$$

over the vectors $\beta \in \mathbb{R}^{p-1}$. Writing Θ for the space of $p \times p$ matrices with zero on the diagonal, after summing on a, we obtain

$$\begin{aligned} \theta &\in \underset{\theta\in\Theta}{\operatorname{argmin}}\, \mathbb{E}\left[\sum_{a=1}^{p}\left(X_a - \sum_{b:b\neq a} \theta_{ba} X_b\right)^2\right] \\ &= \underset{\theta\in\Theta}{\operatorname{argmin}}\, \mathbb{E}\left[\|X - \theta^T X\|^2\right] \\ &= \underset{\theta\in\Theta}{\operatorname{argmin}}\, \|\Sigma^{1/2}(I-\theta)\|_F^2, \end{aligned} \tag{9.8}$$

where the last equality comes from $\mathbb{E}[\|AX\|^2] = \langle A, A\Sigma\rangle_F$ for $X \sim \mathcal{N}(0,\Sigma)$. For $a \neq b$, since $K_{ab} \neq 0$ if and only if $\theta_{ab} \neq 0$, an alternative idea to estimate g_* is to build a coordinate-sparse estimator $\widehat{\theta}$ of the matrix θ. If we replace Σ by $\widehat{\Sigma}$ in (9.8), we obtain

$$\|\widehat{\Sigma}^{1/2}(I-\theta)\|_F^2 = \frac{1}{n}\langle (I-\theta), \mathbf{X}^T\mathbf{X}(I-\theta)\rangle_F = \frac{1}{n}\|\mathbf{X}(I-\theta)\|_F^2.$$

Adding a penalty $\Omega(\theta)$ enforcing coordinate sparsity, we end with the estimator

$$\widehat{\theta}_\lambda \in \underset{\theta\in\Theta}{\operatorname{argmin}}\left\{\frac{1}{n}\|\mathbf{X} - \mathbf{X}\theta\|_F^2 + \lambda\Omega(\theta)\right\}.$$

This corresponds to a special case of coordinate-sparse multivariate regression. We discuss below two choices of penalty for enforcing sparsity: the classical ℓ^1 penalty as in the Lasso estimator and a ℓ^1/ℓ^2 penalty as in group-Lasso estimator.

With the ℓ^1 penalty

If we choose for $\Omega(\theta)$ the ℓ^1 norm of θ, we obtain

$$\widehat{\theta}_\lambda^{\ell^1} \in \underset{\theta\in\Theta}{\operatorname{argmin}}\left\{\frac{1}{n}\|\mathbf{X} - \mathbf{X}\theta\|_F^2 + \lambda\sum_{a\neq b}|\theta_{ab}|\right\}. \tag{9.9}$$

Since we have

$$\frac{1}{n}\|\mathbf{X}-\mathbf{X}\theta\|_F^2+\lambda\sum_{a\neq b}|\theta_{ab}|=\sum_{a=1}^{p}\left(\frac{1}{n}\|\mathbf{X}_a-\sum_{b:b\neq a}\theta_{ba}\mathbf{X}_b\|^2+\lambda\sum_{b:b\neq a}|\theta_{ba}|\right),$$

we can split the minimization (9.9) on $\theta\in\Theta$ into p minimization problems in $\mathbb{R}^{p-1}$

$$\left[\widehat{\theta}_\lambda^{\ell^1}\right]_{-a,a}\in\operatorname*{argmin}_{\beta\in\mathbb{R}^{p-1}}\left\{\frac{1}{n}\|\mathbf{X}_a-\sum_b\beta_b\mathbf{X}_b\|^2+\lambda|\beta|_{\ell^1}\right\}\quad\text{for } a=1,\ldots,p,$$

with the notation $\theta_{-a,a}=[\theta_{ba}]_{b:b\neq a}$. We recognize p Lasso estimators that can be computed efficiently; see Section 5.2.4 in Chapter 5.

Unfortunately, there is a difficulty in order to define a non-directed graph $\widehat{g}$ from $\widehat{\theta}^{\ell^1}$. Actually, whereas the zeros of θ are symmetric with respect to the diagonal, no constraint enforces that $\widehat{\theta}_{ab}^{\ell^1}\neq 0$ when $\widehat{\theta}_{ba}^{\ell^1}\neq 0$ and conversely. So we have to choose an arbitrary decision rule in order to build a graph $\widehat{g}$ from $\widehat{\theta}^{\ell^1}$. For example, we may decide to set an edge between a and b in $\widehat{g}$ when either $\widehat{\theta}_{ab}^{\ell^1}\neq 0$ or $\widehat{\theta}_{ba}^{\ell^1}\neq 0$. Another example is to set an edge $a\sim b$ in $\widehat{g}$ when both $\widehat{\theta}_{ab}^{\ell^1}\neq 0$ and $\widehat{\theta}_{ba}^{\ell^1}\neq 0$. In order to avoid these unsatisfactory rules, we can modify the penalty in order to enforce the symmetry of the zeros of $\widehat{\theta}$.

With the ℓ^1/ℓ^2 penalty

The idea is to regroup the non-diagonal indices (a,b) by symmetric pairs $\{(a,b),(b,a)\}$ and apply a group-Lasso penalty, namely

$$\widehat{\theta}_\lambda^{\ell^1/\ell^2}\in\operatorname*{argmin}_{\theta\in\Theta}\left\{\frac{1}{n}\|\mathbf{X}-\mathbf{X}\theta\|_F^2+\lambda\sum_{a<b}\sqrt{\theta_{ab}^2+\theta_{ba}^2}\right\}.\tag{9.10}$$

This estimator has the nice property to be coordinate sparse with symmetric zeros. So there is no ambiguity in order to define a graph $\widehat{g}$ from $\widehat{\theta}_\lambda^{\ell^1/\ell^2}$.

The minimization problem (9.10) is convex in $\mathbb{R}^{p\times p}$; unfortunately, it cannot be split into p subproblems. It is then computationally more intensive to compute $\widehat{\theta}_\lambda^{\ell^1/\ell^2}$ than $\widehat{\theta}_\lambda^{\ell^1}$ in large dimensions. We refer to Exercise 9.6.4 for the description of a block gradient descent algorithm for minimizing (9.10).

Theoretical results

We cannot apply directly the results of Chapter 5 on the Lasso and group-Lasso estimators, since the design matrix $\mathbf{X}$ is random. Nevertheless, we can work conditionally on the design and then integrate the result. For the sake of illustration, we give an example of the kind of results that we can obtain for the estimator $\widehat{\theta}_\lambda^{\ell^1}$.

We introduce the population restricted eigenvalue

$$\phi_*=\max\left\{|\Sigma_{mm}|_{\text{op}}:m\subset\{1,\ldots,p\},\ \text{card}(m)\leq n/(3\log p)\right\}$$

and define the degree of a matrix A by $d(A) = \max\{|A_{:a}|_0 : a = 1, \ldots, p\}$. Note that for $A = \theta$, the degree $d(\theta)$ is equal to the degree of the graph g_*, which is defined by $\text{degree}(g_*) = \max\{\text{ne}(a) : a = 1, \ldots, p\}$.

Theorem 9.3 Risk bound for $\widehat{\theta}^{\ell^1}$

Assume that the diagonal entries of Σ are equal to one and set $\text{cor}(\Sigma) = \max_{a \neq b} |\Sigma_{a,b}|$. *Then there exist some constants $C, C', C_1, C_2, C_3 > 0$, such that when*

$$1 \leq d(\theta) \leq C \frac{n}{\phi_* \log(p)} \bigwedge \frac{1}{\text{cor}(\Sigma)} \tag{9.11}$$

for $\lambda = C'\phi_* \sqrt{\log(p)}$, the estimator $\widehat{\theta}_\lambda^{\ell^1}$ fulfills with probability at least $1 - C_1\left(p^{-C_2} + p^2 e^{-C_2 n}\right)$ the risk bound

$$\|\Sigma^{1/2}(\widehat{\theta}_\lambda^{\ell^1} - \theta)\|_F^2 \leq C_3 \frac{\phi_*^2 \log(p)}{n} |\theta|_0. \tag{9.12}$$

The above theorem provides a control of $\|\Sigma^{1/2}(\widehat{\theta}^{\ell^1} - \theta)\|_F^2$. This quantity does not directly quantify the accuracy of the estimation of g_*, but rather the accuracy of $\widehat{\theta}^{\ell^1}$ for predicting the variables X_a with $(\widehat{\theta}^{\ell^1})^T X$. Actually, assume that we obtain a new observation $X^{\text{new}} \sim \mathcal{N}(0, \Sigma)$ independent of the n-sample used to compute $\widehat{\theta}^{\ell^1}$. Then

$$\sum_{a=1}^{p} \mathbb{E}^{\text{new}} \left[\left(X_a^{\text{new}} - \sum_{b:b \neq a} \widehat{\theta}_{ba}^{\ell^1} X_b^{\text{new}} \right)^2 \right] = \mathbb{E}^{\text{new}} \left[\|X^{\text{new}} - (\widehat{\theta}^{\ell^1})^T X^{\text{new}}\|^2 \right]$$
$$= \|\Sigma^{1/2}(I - \widehat{\theta}^{\ell^1})\|_F^2,$$

where $\mathbb{E}^{\text{new}}$ represents the expectation with respect to X^{new}. Since θ is the orthogonal projection of I onto Θ with respect to the scalar product $\langle \cdot, \Sigma \cdot \rangle_F$, Pythagorean formula gives

$$\sum_{a=1}^{p} \mathbb{E}^{\text{new}} \left[\left(X_a^{\text{new}} - \sum_{b:b \neq a} \widehat{\theta}_{ba}^{\ell^1} X_b^{\text{new}} \right)^2 \right] = \|\Sigma^{1/2}(I - \theta)\|_F^2 + \|\Sigma^{1/2}(\theta - \widehat{\theta}^{\ell^1})\|_F^2.$$

The first term represents the minimal error for predicting the X_a from the $\{X_b : b \neq a\}$ and the second term corresponds to the stochastic error due to the estimation procedure. Theorem 9.3 gives a bound for this last term. Condition (9.11) requires that the degree of the graph g_* remains small compared to $n/\log(p)$. It has been shown that this condition on the degree is unavoidable (see Verzelen [160]), and the degree of the graph seems to be a notion of sparsity well-suited to characterize the statistical difficulty for estimation in the Gaussian graphical model. The second constraint on the product $d(\theta)\text{cor}(\Sigma)$ is due to the use of the lower bound of Exercise 5.5.3 on the restricted eigenvalues, and it is suboptimal.

Proof. The proof mainly relies on Theorem 5.1 in Chapter 5 for the Lasso estimator. Yet, the adaptation to the random design setting is slightly technical and lengthy, and we only sketch the main lines.
The main idea is to start from the formula

$$n^{-1/2}\mathbf{X}_a = \sum_{b\neq a} n^{-1/2}\mathbf{X}_b\theta_{ab} + n^{-1/2}\varepsilon_a \quad \text{for } a=1,\ldots,p,$$

with $\varepsilon_a \sim \mathcal{N}(0, K_{aa}^{-1}I_n)$ and work conditionally on $\{\mathbf{X}_b : b\neq a\}$ for each a. Then, for

$$\lambda \geq 3\max_{a\neq b} |\mathbf{X}_b^T\varepsilon_a/n|,$$

combining the risk Bound (5.13) with Exercise 5.5.3, we obtain (after summing the p bounds)

$$\begin{aligned}\frac{1}{n}\|\mathbf{X}(\widehat{\theta}_\lambda^{\ell^1} - \theta)\|_F^2 &\leq \inf_{A\in\Theta}\left\{\frac{1}{n}\|\mathbf{X}(A-\theta)\|_F^2 + \frac{\lambda^2|A|_0}{\left(1-11d(A)\max_{b\neq a}|\mathbf{X}_a^T\mathbf{X}_b|/n\right)_+}\right\}\\ &\leq \frac{\lambda^2|\theta|_0}{\left(1-11d(\theta)\max_{b\neq a}|\mathbf{X}_a^T\mathbf{X}_b|/n\right)_+}. \end{aligned} \tag{9.13}$$

To conclude, all we need is to be able to replace $n^{-1/2}\mathbf{X}$ by $\Sigma^{1/2}$ in the above bound. In a high-dimensional setting, where $p>n$, we do not have $\|\mathbf{X}v\|^2/n \approx \|\Sigma^{1/2}v\|^2$ for all vector $v\in\mathbb{R}^p$, yet, the next lemma shows that it is true with high probability simultaneously for all sparse vectors v. More precisely, there exist some constants $0<c_-<c_+<+\infty$, such that we have with high probability

$$c_-\|\Sigma^{1/2}v\|^2 \leq \frac{1}{n}\|\mathbf{X}v\|^2 \leq c_+\|\Sigma^{1/2}v\|^2$$

for all v fulfilling $|v|_0 \leq n/(3\log(p))$.

Lemma 9.4 Restricted isometry constants

For $1\leq d < n\wedge p$, there exists an exponential random variable ξ with parameter 1, such that

$$\inf_{\beta:|\beta|_0\leq d}\frac{n^{-1/2}\|\mathbf{X}\beta\|}{\|\Sigma^{1/2}\beta\|} \geq 1 - \frac{\sqrt{d}+\sqrt{2\log(C_p^d)+\delta_d}+\sqrt{2\xi}}{\sqrt{n}}, \tag{9.14}$$

where $\delta_d = \log\left(4\pi\log(C_p^d)\right)/\sqrt{8\log(C_p^d)}$.

Similarly, there exists an exponential random variable ξ' with parameter 1, such that

$$\sup_{\beta:|\beta|_0\leq d}\frac{n^{-1/2}\|\mathbf{X}\beta\|}{\|\Sigma^{1/2}\beta\|} \leq 1 + \frac{\sqrt{d}+\sqrt{2\log(C_p^d)+\delta_d}+\sqrt{2\xi'}}{\sqrt{n}}. \tag{9.15}$$

Proof. We refer to Exercise 9.6.8 for a proof of this lemma. □

To conclude the proof of Theorem 9.3, it remains to check that under the hypotheses of Theorem 9.3, with probability at least $1 - C_1(p^{-C_2} + p^2 e^{-C_2 n})$, we have

1. $\max_{a \neq b} |\mathbf{X}_b^T \varepsilon_a / n| \leq C\sqrt{\log(p)/n}$ and $\max_{a \neq b} |\mathbf{X}_a^T \mathbf{X}_b|/n \leq C_4 \mathrm{cor}(\Sigma)$, and
2. $d(\widehat{\theta}_\lambda^{\ell^1}) \leq n/(6\log(p))$ when $\lambda \geq C'\phi^* \sqrt{\log(p)/n}$,
3. $n^{-1}\|\mathbf{X}(\widehat{\theta}_\lambda^{\ell^1} - \theta)\|_F^2 \geq c_- \|\Sigma^{1/2}(\widehat{\theta}_\lambda^{\ell^1} - \theta)\|_F^2$.

The first point ensures that under the assumptions of the theorem, we have with large probability $\lambda \geq 3\max_{a \neq b} |\mathbf{X}_b^T \varepsilon_a / n|$ and $d(\theta)\max_{b \neq a} |\mathbf{X}_a^T \mathbf{X}_b|/n \leq 1/22$. The two last points allows to bound $\|\Sigma^{1/2}(\widehat{\theta}_\lambda^{\ell^1} - \theta)\|^2$ by a constant times $n^{-1}\|\mathbf{X}(\widehat{\theta}_\lambda^{\ell^1} - \theta)\|^2$. Plugging this bounds in (9.13), we get the risk Bound (9.12).

Checking the abovethree points is somewhat lengthy; we only point out the main arguments. The first point can be proved with the Gaussian concentration Inequality (B.2) and by noticing that $\mathbf{X}_b^T \varepsilon_a = \|\mathbf{X}_b\| N_{a,b}$, where $N_{a,b}$ is independent of $\|\mathbf{X}_b\|$ and follows a Gaussian distribution with variance at most 1. The second point can be proved by combining Lemma 3.2 in the Appendix of Giraud *et al.* [82], Exercise 5.5.3, and Lemma 9.4. The last point is obtained by combining Lemma 9.4 with $d(\widehat{\theta}_\lambda^{\ell^1} - \theta) \leq d(\widehat{\theta}_\lambda^{\ell^1}) + d(\theta)$ and the second point. □

Theorem 9.3 describes the prediction performance of the estimator $\widehat{\theta}^{\ell^1}$ but gives little information on the recovery of g_*. Various conditions have been proposed to ensure the recovery of graph g_* by the estimator $\widehat{\theta}^{\ell^1}$; see, for example, Wainwright [162] or Meinshausen and Bühlmann [120]. Unfortunately, these conditions are not likely to be met in high-dimensional settings, and the best we can expect in practice is a partial recovery of graph g_*.

9.4 Practical Issues

Hidden variables

It may happen that we do not observe some variables that have a strong impact on conditional dependencies. For example, if we investigate the regulation between genes, we may lack the measurements for a key gene that regulates many genes of interest. In such a case, even if the graph of conditional dependencies between all the variables (observed and non-observed) is sparse, the graph of conditional dependencies between the sole observed variables will not be sparse in general; see Exercise 9.6.5. Therefore, the presence of hidden variables can strongly impact the inference in graphical models. We refer to Exercise 9.6.5 for details on this issue and an estimation procedure taking into account hidden variables.

Non-Gaussian variables

Our data may not have a Gaussian distribution. As explained above, inferring the conditional dependencies of a general distribution is unrealistic in high-dimensional settings, even when this distribution is a graphical model with respect to a sparse graph.

In particular, trying to infer graph g and the functions f_c in the decomposition (9.3) is hopeless in a high-dimensional setting without additional structural assumptions. We need to restrict either to some simple classes of graph g or to some special classes of densities f. A possible approach is to assume that some transformations of the data is Gaussian. For example, write F_j for the cumulative distribution function of X_j and Φ for the cumulative distribution of the standard Gaussian distribution. Then, the variable $Z_j = \Phi^{-1}(F_j(X_j))$ follows a standard Gaussian distribution. A structural hypothesis is to *assume* that in addition $(Z_1,\ldots,Z_p)$ has a Gaussian distribution with non-singular covariance matrix. Then, since the minimal graphs of $(X_1,\ldots,X_p)$ and $(Z_1,\ldots,Z_p)$ coincide, we are back to the Gaussian setting. The point is that in practice we do not know F_j. A first approach is to estimate F_j with some non-parametric estimator $\widehat{F}_j$ and work with the transformed data $(\Phi^{-1}(\widehat{F}_1(X_1)),\ldots,\Phi^{-1}(\widehat{F}_p(X_p)))$ as if they were distributed according to a Gaussian distribution.

A more powerful approach amounts to estimate the correlation matrix of $(Z_1,\ldots,Z_p)$ from the ranked statistics of $X_1,\ldots,X_p$ and then replace in the procedures described above the empirical covariance matrix $\widehat{\Sigma}$ by this estimated correlation matrix; see Exercise 9.6.7 for details and Section 9.5.2 for references. This procedure is implemented in the `R` package `huge` available at `http://cran.r-project.org/web/packages/huge/`.

9.5 Discussion and References

9.5.1 Take-Home Message

Graphical modeling is a nice theory for encoding the conditional dependencies between random variables. Estimating the minimal graph depicting the conditional dependencies is in general out of reach in high-dimensional settings. Yet, it is possible for Gaussian random variables, since in this case the minimal graph simply corresponds to the zero entries of the inverse of the covariance matrix of the random variables. The estimation of the graph can even be achieved with a sample size smaller than the number of variables, as long as the degree of the graph remains small enough.

In practice, we should not expect more than a partial recovery of the graph when the sample size is smaller than the number of variables. Furthermore, the variables are not likely to be Gaussian, and we may also have to deal with some hidden variables. Handling these two issues requires some additional work, as explained in Section 9.4.

9.5.2 References

We refer to Lauritzen [101] for an exhaustive book on graphical models. The grapical-Lasso procedure has been proposed by Banerjee, El Ghaoui, and d'Aspremont [15] and the numerical optimization has been improved by Friedman, Hastie, and Tibshirani [73]. The regression procedure with the ℓ^1 penalty has been proposed and analyzed by Meinshausen and Bühlmann [120]. Theorem 9.3 is a simple combination of Proposition 5.1 in Chapter 5, Lemma 3.2 in Appendix of Giraud *et al.* [82], and Lemma 1 in Giraud [78].

Similarly to Chapter 7, we have at our disposal many different estimation procedures, which all depend on at least one tuning parameter. Therefore, we have a large family $\widehat{\mathscr{G}}$ of estimated graphs, and we need a criterion in order to select "at best" a graph among this family. GGMselect [81] is an `R` package that has been designed for this issue `http://cran.r-project.org/web/packages/GGMselect/`. It selects a graph $\widehat{g}$ among $\widehat{\mathscr{G}}$ by minimizing a penalized criterion closely linked to Criterion (7.3) for estimator selection. The resulting estimator satisfies some oracle-like inequality similar to (2.12).

Exercise 9.6.5 on graphical models with hidden variables is mainly based on Chandrasekaran, Parrilo and Willsky [54]. Exercise 9.6.6 is derived from Cai *et al.* [46], and Exercise 9.6.7 on Gaussian copula graphical models is adapted from Lafferty *et al.* [100], Liu *et al.* [112], and Xue and Zou [165].

9.6 Exercises

9.6.1 Factorization in Directed Models

We will prove the factorization formula (9.2) by induction. With no loss of generality, we can assume that node p is a leaf in graph g (which means that it has no descendant).

1. Prove that $f(x_1,\ldots,x_p) = f(x_p|x_{\mathrm{pa}(p)})f(x_1,\ldots,x_{p-1})$.
2. Prove the factorization formula (9.2).

9.6.2 Moralization of a Directed Graph

Let us consider a directed graph g. The moral graph g^m of the directed graph g is the non-directed graph obtained by linking all the parents of each node together and by replacing directed edges by non-directed edges. Assume that $X = (X_1,\ldots,X_p)$ has a positive density with respect to a σ-finite product measure. We will prove that if $\mathscr{L}(X) \sim g$ then $\mathscr{L}(X) \sim g^m$.

1. Starting from the factorization formula (9.2), prove that there exists two functions g_1 and g_2, such that

$$f(x) = g_1(x_a, x_{\mathrm{ne}^m(a)})g_2(x_{\mathrm{nn}^m(a)}, x_{\mathrm{ne}^m(a)}),$$

   where $\mathrm{ne}^m(a)$ represents the nodes neighbor to a in g^m and $\mathrm{nn}^m(a) = \{1,\ldots,p\} \setminus \mathrm{cl}^m(a)$, with $\mathrm{cl}^m(a) = \mathrm{ne}^m(a) \cup \{a\}$.
2. Prove that $f(x|x_{\mathrm{ne}^m(a)}) = f(x_a|x_{\mathrm{ne}^m(a)})f(x_{\mathrm{nn}^m(a)}|x_{\mathrm{ne}^m(a)})$.
3. Conclude that $\mathscr{L}(X) \sim g^m$.

9.6.3 Convexity of $-\log(\det(\mathbf{K}))$

We will prove the convexity of $K \to -\log(\det(K))$ on the set of symmetric positive definite matrices. Let K and S be two symmetric positive definite matrices. Since $K^{-1/2}SK^{-1/2}$ is symmetric, there exist an orthogonal matrix U and a diagonal matrix D, such that $K^{-1/2}SK^{-1/2} = UDU^T$. We set $Q = K^{1/2}U$.

1. Check that $K = QQ^T$ and $S = QDQ^T$.
2. For $\lambda \in [0,1]$, prove that

$$-\log(\det(\lambda S + (1-\lambda)K) = -\log(\det(K)) - \log(\det(\lambda D + (1-\lambda)I)).$$

3. From the convexity of $x \to -\log(x)$, conclude that

$$\begin{aligned}-\log(\det(\lambda S + (1-\lambda)K) &\leq -\log(\det(K)) - \lambda \log(\det(D)) \\ &= -(1-\lambda)\log(\det(K)) - \lambda \log(\det(S)).\end{aligned}$$

9.6.4 Block Gradient Descent with the ℓ^1/ℓ^2 Penalty

We fix $\lambda > 0$, and we consider the estimator $\widehat{\theta}$ defined by (9.10). We will assume that the columns $\mathbf{X}_a$ of $\mathbf{X}$ fulfill $n^{-1}\mathbf{X}_a^T\mathbf{X}_a = 1$.

1. When $\|(\theta_{ab}, \theta_{ba})\| \neq 0$, check that the partial gradient ∇_{ab} of Criterion (9.10) according to $(\theta_{ab}, \theta_{ba})$ is

$$\nabla_{ab} = -\frac{2}{n}\begin{pmatrix}\mathbf{X}_a^T(\mathbf{X}_b - \sum_{k\neq b}\theta_{kb}\mathbf{X}_k) \\ \mathbf{X}_b^T(\mathbf{X}_a - \sum_{k\neq a}\theta_{ka}\mathbf{X}_k)\end{pmatrix} + \frac{\lambda}{\|(\theta_{ab}, \theta_{ba})\|}\begin{pmatrix}\theta_{ab} \\ \theta_{ba}\end{pmatrix}$$

2. We define $\Delta = \begin{pmatrix}\Delta_{ab} \\ \Delta_{ba}\end{pmatrix}$ with

$$\Delta_{ab} = \frac{1}{n}\mathbf{X}_a^T(\mathbf{X}_b - \sum_{k\neq a,b}\theta_{kb}\mathbf{X}_k).$$

   Prove that minimizing (9.10) in the variables $(\theta_{ab}, \theta_{ba})$ gives

$$\begin{pmatrix}\widehat{\theta}_{ab} \\ \widehat{\theta}_{ba}\end{pmatrix} = \left(1 - \frac{\lambda}{2\|\Delta\|}\right)_+ \begin{pmatrix}\Delta_{ab} \\ \Delta_{ba}\end{pmatrix}.$$

3. Propose an algorithm in order to compute the solution $\widehat{\theta}$ of (9.10).
4. Show that if we add a second penalty $\gamma\|\theta\|_F^2$ to Criterion (9.10), the only change in the above algorithm is that the update is divided by $(1+\gamma)$.

9.6.5 Gaussian Graphical Models with Hidden Variables

We assume that $\{1,\ldots,p\} = O \cup H$, with $O \cap H = \emptyset$. We consider a Gaussian random variable $X = \begin{pmatrix}X_O \\ X_H\end{pmatrix} \sim \mathcal{N}(0,\Sigma)$ with $\Sigma = \begin{pmatrix}\Sigma_{OO} & \Sigma_{OH} \\ \Sigma_{HO} & \Sigma_{HH}\end{pmatrix}$. In particular, the variable X_O follows a $\mathcal{N}(0,\Sigma_{OO})$ distribution. We set $\tilde{K}_O = (\Sigma_{OO})^{-1}$.

1. Prove that $\tilde{K}_O = K_{OO} - K_{OH}(K_{HH})^{-1}K_{HO}$, where $K := \Sigma^{-1} = \begin{pmatrix} K_{OO} & K_{OH} \\ K_{HO} & K_{HH} \end{pmatrix}$.
2. Let g_O be the minimal (non-directed) graph such that $\mathscr{L}(X_O) \sim g_O$. Is the graph g_O a subgraph of the minimal graph g, such that $\mathscr{L}(X) \sim g$?
3. We assume that $\mathscr{L}(X) \sim g$ with g sparse. What can we say about the sparsity of K_{OO}? What can we say about the rank of $K_{OH}(K_{HH})^{-1}K_{HO}$ compared to $h = \text{card}(H)$?
4. Very often, we have variables $(X_1, \ldots, X_p)$, with conditional dependencies depicted by a sparse graph g, but we cannot observe all the variables. Furthermore, we even do not know the actual number of unobserved variables. In other words, we only observe an n-sample of $X_O \sim \mathcal{N}(0, \Sigma_{OO})$ and we have no information on X_H and the size of H. Nevertheless, we want to reconstruct on the basis of these observations, the graph g_O^* with nodes labelled by O and defined by

$$a \overset{g_O^*}{\sim} b \Longleftrightarrow a \overset{g}{\sim} b.$$

   Explain why it is equivalent to estimate the location of the nonzero coordinates of K_{OO}.
5. We have in mind that h is small. We have seen that $\tilde{K}_O = K_{OO} - L$, with K_{OO} sparse and $\text{rank}(L) \leq h$. Propose an estimation procedure inspired by the glasso.
6. Propose another procedure inspired by the regression approach.

9.6.6 Dantzig Estimation of Sparse Gaussian Graphical Models

Let $X^{(1)}, \ldots, X^{(n)}$ be a i.i.d. sample of a Gaussian distribution $\mathcal{N}(0, \Sigma)$ in $\mathbb{R}^p$, with Σ non-singular and $K = \Sigma^{-1}$ sparse. In this exercise, we investigate the estimation of K with a matrix version of the Dantzig selector (see Exercise 5.5.5, page 113)

$$\widehat{K} \in \underset{B \in \mathbb{R}^{p \times p} : |\widehat{\Sigma} B - I|_\infty \leq \lambda}{\text{argmin}} |B|_{1,\infty}, \tag{9.16}$$

where $\widehat{\Sigma} = n^{-1} \sum_{i=1}^n X^{(i)} (X^{(i)})^T$ is the empirical covariance matrix, λ is non-negative and

$$|A|_\infty = \max_{i,j} |A_{ij}| \quad \text{and} \quad |A|_{1,\infty} = \max_j \sum_i |A_{ij}|.$$

Henceforth, A_j will refer to the j-th column of the matrix A.

1. Check that for any matrices of appropriate size, we have $|AB|_\infty \leq |A|_\infty |B|_{1,\infty}$ and also $|AB|_\infty \leq |A|_{1,\infty} |B|_\infty$ when A is symmetric.
2. We define the matrix $\widetilde{K}$ by $\widetilde{K}_{ij} = \widehat{\beta}_i^{(j)}$, where $\widehat{\beta}^{(j)}$ is solution of

$$\widehat{\beta}^{(j)} \in \underset{\beta \in \mathbb{R}^p : |\widehat{\Sigma} \beta - e_j|_\infty \leq \lambda}{\text{argmin}} |\beta|_1,$$

   with e_j the j-th vector of the canonical basis of $\mathbb{R}^p$. Prove that $\widetilde{K}$ is a solution of (9.16).

In the following, $\widehat{K}$ refers to the solution $\widetilde{K}$ defined above. This solution can be computed very efficiently, since it simply amounts to compute p Dantzig selectors.

A) Deterministic bound on $|\widehat{\mathbf{K}} - \mathbf{K}|_\infty$

In this part, we consider λ fulfilling

$$\lambda \geq |K|_{1,\infty}|\widehat{\Sigma} - \Sigma|_\infty. \tag{9.17}$$

1. Prove that when (9.17) is met, we have $|\widehat{\Sigma}K - I|_\infty \leq \lambda$, and therefore $|\widehat{K}_j|_1 \leq |K_j|_1$ for all $j = 1, \ldots, p$.
2. Prove the inequalities

$$\begin{aligned}|\widehat{K} - K|_\infty &\leq |K|_{1,\infty}\left(|\widehat{\Sigma}\widehat{K} - I|_\infty + |(\Sigma - \widehat{\Sigma})\widehat{K}|_\infty\right) \\ &\leq |K|_{1,\infty}\left(\lambda + |K|_{1,\infty}|\widehat{\Sigma} - \Sigma|_\infty\right).\end{aligned}$$

3. Conclude that when (9.17) is met, we have

$$|\widehat{K} - K|_\infty \leq 2\lambda|K|_{1,\infty}. \tag{9.18}$$

B) Probabilistic bound on $|\widehat{\Sigma} - \Sigma|_\infty$

We assume henceforth that $\Sigma_{aa} = 1$ for $a = 1, \ldots, p$. Since Σ is non-singular, we have $|\Sigma_{ab}| < 1$ for all $a \neq b$.

1. Let X be a $\mathcal{N}(0, \Sigma)$ Gaussian random variable. We set $Z_1 = (2(1+\Sigma_{12}))^{-1/2}(X_1 + X_2)$ and $Z_2 = (2(1-\Sigma_{12}))^{-1/2}(X_1 - X_2)$. Prove that Z_1 and Z_2 are i.i.d., with $\mathcal{N}(0,1)$ Gaussian distribution and

$$X_1X_2 = \frac{1}{4}\left(2(1+\Sigma_{12})Z_1^2 - 2(1-\Sigma_{12})Z_2^2\right).$$

2. Check that for $0 \leq x \leq 1/2$, we have $-\log(1-x) \leq x + x^2$ and $-\log(1+x) \leq -x + x^2/2$ and prove the bound for $0 \leq s \leq 1/4$

$$\mathbb{E}\left[e^{sX_1X_2}\right] = ((1-(1+\Sigma_{12})s)(1+(1-\Sigma_{12})s))^{-1/2} \leq \exp(\Sigma_{12}s + 2s^2).$$

   Check that this bound still holds when $|\Sigma_{12}| = 1$.
3. For any $a, b \in \{1, \ldots, p\}$, $0 \leq s \leq 1/4$ and $t > 0$, prove that

$$\mathbb{P}\left(\widehat{\Sigma}_{ab} - \Sigma_{ab} > t\right) \leq e^{-snt}e^{2s^2n}.$$

4. For $0 < t \leq 1$, setting $s = t/4$, prove that

$$\mathbb{P}\left(|\widehat{\Sigma} - \Sigma|_\infty > t\right) \leq p(p+1)e^{-nt^2/8}.$$

5. For $\log(p) \leq n/32$, prove that

$$\mathbb{P}\left(|\widehat{\Sigma} - \Sigma|_\infty > 4\sqrt{\frac{2\log(p)}{n}}\right) \leq \frac{2}{p^2}.$$

C) Bounds in sup norm and Frobenius norm

We define $d = \max_{j=1,\dots,p} |K_j|_0$, which corresponds to the degree of the minimal graph associated to the Gaussian distribution $\mathcal{N}(0,\Sigma)$.

1. For $\lambda = 4|K|_{1,\infty}\sqrt{2\log(p)/n}$, by combining the results of Parts A and B, prove that with probability at least $1-2/p^2$, we have

$$|\widehat{K}-K|_\infty \leq 8|K|_{1,\infty}^2\sqrt{\frac{2\log(p)}{n}}\,. \tag{9.19}$$

2. For any $J \subset \{1,\dots,p\}^2$, we define the matrix K^J by $K^J_{ij} = K_{ij}\mathbf{1}_{(i,j)\in J}$. In the following, we set $J = \left\{(i,j) : |\widehat{K}_{ij}| > |\widehat{K}-K|_\infty\right\}$. Prove that for all $j \in \{1,\dots,p\}$, we have

$$|\widehat{K}^{J^c}_j|_1 = |\widehat{K}_j|_1 - |\widehat{K}^J_j|_1 \leq |K_j|_1 - |\widehat{K}^J_j|_1 \leq |K_j - \widehat{K}^J_j|_1\,,$$

   and hence $|\widehat{K}^{J^c}|_{1,\infty} \leq |K-\widehat{K}^J|_{1,\infty}$.

3. Prove that $K_{ij} \neq 0$ for all $(i,j) \in J$, and then

$$|\widehat{K}-K|_{1,\infty} \leq 2|\widehat{K}^J-K|_{1,\infty} \leq 2d|\widehat{K}^J-K|_\infty \leq 4d|\widehat{K}-K|_\infty\,.$$

4. Conclude that

$$\|\widehat{K}-K\|_F^2 \leq p|\widehat{K}-K|_{1,\infty}|\widehat{K}-K|_\infty \leq 4pd|\widehat{K}-K|_\infty^2.$$

5. Prove that when $\lambda = 4|K|_{1,\infty}\sqrt{2\log(p)/n}$, with probability at least $1-2/p^2$, we have

$$\|\widehat{K}-K\|_F^2 \leq 512pd|K|_{1,\infty}^4\frac{\log(p)}{n}\,.$$

The estimation procedure (9.16) is implemented in the R package `flare` available at `http://cran.r-project.org/web/packages/flare/`.

9.6.7 Gaussian Copula Graphical Models

We consider here a case where $X = (X_1,\dots,X_p)$ is not a Gaussian random variable. We assume, as in Section 9.4, that there exists a random variable $Z = (Z_1,\dots,Z_p)$, with $\mathcal{N}(0,\Sigma^Z)$ Gaussian distribution and p increasing differentiable functions $f_a : \mathbb{R} \to \mathbb{R}$, for $a = 1,\dots,p$, such that $(X_1,\dots,X_p) = (f_1(Z_1),\dots,f_p(Z_p))$. We also assume that Σ^Z is non-singular, and $\Sigma^Z_{aa} = 1$ for $a = 1,\dots,p$. We define $K^Z = (\Sigma^Z)^{-1}$. We remind the reader of the classical equality when $\Sigma^Z_{aa} = 1$ for $a = 1,\dots,p$

$$\mathbb{P}(Z_a > 0, Z_b > 0) = \frac{1}{4}\left(1 + \frac{2}{\pi}\arcsin(\Sigma^Z_{ab})\right) \quad \text{for any } a,b \in \{1,\dots,p\}\,.$$

1. Prove that the minimal graph g^* associated to $(X_1,\dots,X_p)$ coincides with the minimal graph associated to $(Z_1,\dots,Z_p)$. In particular, for $a \neq b$ there is an edge between a and b in g^* if and only if $K^Z_{ab} \neq 0$.

2. Let $\widetilde{Z}$ be an independent copy of Z and define $\widetilde{X} = (f_1(\widetilde{Z}_1), \ldots, f_p(\widetilde{Z}_p))$. Prove that for any $a, b \in \{1, \ldots, p\}$

$$\begin{aligned}\tau_{ab} &= \mathbb{E}\left[\text{sign}\left((X_a - \widetilde{X}_a)(X_b - \widetilde{X}_b)\right)\right] \\ &= \mathbb{E}\left[\text{sign}\left((Z_a - \widetilde{Z}_a)(Z_b - \widetilde{Z}_b)\right)\right] = \frac{2}{\pi}\arcsin(\Sigma^Z_{ab}),\end{aligned} \tag{9.20}$$

where $\text{sign}(x) = \mathbf{1}_{x>0} - \mathbf{1}_{x\leq 0}$.

A) Kendall's tau

We assume now that we only observe an i.i.d. sample $X^{(1)}, \ldots, X^{(n)}$ of X. In particular, the functions $f_1, \ldots, f_p$ and the covariance Σ^Z are unknown. Since the minimal graph g^* associated to $(X_1, \ldots, X_p)$ can be read on the precision matrix $K^Z = (\Sigma^Z)^{-1}$, our goal here is to estimate K^Z from the observations $X^{(1)}, \ldots, X^{(n)}$. The main idea is to build on the equation (9.20) in order to construct an estimator $\widehat{\Sigma}^Z$ of Σ^Z based on $X^{(1)}, \ldots, X^{(n)}$, and then apply the procedure (9.16) with $\widehat{\Sigma}$ replaced by $\widehat{\Sigma}^Z$.

For any $a, b \in \{1, \ldots, p\}$, we define $\widehat{\Sigma}^Z_{ab} = \sin(\pi\widehat{\tau}_{ab}/2)$ where

$$\widehat{\tau}_{ab} = \frac{2}{n(n-1)}\sum_{i<j}\text{sign}\left((X_a^{(i)} - X_a^{(j)})(X_b^{(i)} - X_b^{(j)})\right).$$

1. Check that the function $F : \mathbb{R}^{2n} \to [-1, 1]$ defined by

$$F\left((x_1, y_1), \ldots, (x_n, y_n)\right) = \frac{2}{n(n-1)}\sum_{i<j}\text{sign}\left((x_i - x_j)(y_i - y_j)\right)$$

   fulfills

$$|F\left((x_1, y_1), \ldots, (x_i, y_i), \ldots, (x_n, y_n)\right) - F\left((x_1, y_1), \ldots, (x_i', y_i'), \ldots, (x_n, y_n)\right)| \leq \frac{4}{n}$$

   for any $x_1, \ldots, x_n, y_1, \ldots, y_n, x_i', y_i' \in \mathbb{R}$.

2. From McDiarmid concentration inequality (Theorem B.5, page 299, in Appendix B), prove that for $a < b$ and $t > 0$

$$\mathbb{P}(|\widehat{\tau}_{ab} - \tau_{ab}| > t) \leq 2e^{-nt^2/8} \quad \text{and} \quad \mathbb{P}\left(|\widehat{\Sigma}^Z_{ab} - \Sigma^Z_{ab}| > t\right) \leq 2e^{-nt^2/(2\pi^2)}.$$

3. Conclude that for any $t > 0$, we have

$$\mathbb{P}\left(|\widehat{\Sigma}^Z - \Sigma^Z|_\infty > t\right) \leq p^2 e^{-nt^2/(2\pi^2)} \text{ and } \mathbb{P}\left(|\widehat{\Sigma}^Z - \Sigma^Z|_\infty > 2\pi\sqrt{\frac{2\log(p)}{n}}\right) \leq \frac{1}{p^2}.$$

B) Graph estimation

Let $\widehat{K}^Z$ be the matrix obtained by solving the minimization problem (9.16), with $\widehat{\Sigma}$ replaced by $\widehat{\Sigma}^Z$. From $\widehat{K}^Z$, we build the graph $\widehat{g}$ by setting an edge between a and b if and only if both $\widehat{K}^Z_{ab}$ and $\widehat{K}^Z_{ba}$ are nonzero. We use below the same notations as in Exercise 9.6.6.

1. From (9.18), prove that when $\lambda = 2\pi |K^Z|_{1,\infty}\sqrt{2\log(p)/n}$, with probability at least $1-1/p^2$, we have

$$|\widehat{K}^Z - K^Z|_\infty \leq 4\pi |K^Z|_{1,\infty}^2 \sqrt{\frac{2\log(p)}{n}}.$$

2. Assume that all the entries K^Z_{ab} of K^Z are either 0 or larger in absolute value than $4\pi|K^Z|^2_{1,\infty}\sqrt{2\log(p)/n}$. Prove that $g^* \subset \widehat{g}$, with probability at least $1-1/p^2$.

9.6.8 Restricted Isometry Constant for Gaussian Matrices

We prove in the following the upper Bound (9.15). The proof of the lower bound follows exactly the same lines. For a linear span $V \subset \mathbb{R}^p$ and an $n \times p$ matrix Z, we introduce the notation

$$\lambda_V(Z) = \sup_{v \in V \setminus \{0\}} \frac{n^{-1/2}\|Zv\|}{\|v\|}.$$

1. Define the matrix Z by $Z = \mathbf{X}\Sigma^{-1/2}$. Check that the Z_{ij} are i.i.d. with $\mathcal{N}(0,1)$ Gaussian distribution.
2. We write $\mathcal{M}$ for the set gathering all the subsets of $\{1,\ldots,p\}$. For $m \in \mathcal{M}$, we define S_m as the linear space spanned by $\{e_j : j \in m\}$, where $e_1,\ldots,e_p$ is the canonical basis of $\mathbb{R}^p$. To each linear span S_m, we associate the linear span $V_m = \Sigma^{1/2}S_m$. Check that for $d \leq n \wedge p$

$$\sup_{\beta : |\beta|_0 \leq d} \frac{n^{-1/2}\|\mathbf{X}\beta\|}{\|\Sigma^{1/2}\beta\|} = \sup_{m \in \mathcal{M} : |m| = d} \lambda_{V_m}(Z). \tag{9.21}$$

In the following, we prove that for any collection $V_1,\ldots,V_N$ of d dimensional linear spaces, we have

$$\sup_{i=1,\ldots,N} \lambda_{V_i}(Z) \leq 1 + \frac{\sqrt{d} + \sqrt{2\log(N)} + \delta_N + \sqrt{2\xi}}{\sqrt{n}}, \tag{9.22}$$

where ξ is an exponential random variable and

$$\delta_N = \frac{\log\left(\frac{1}{N} + \sqrt{4\pi\log(N)}\right)}{\sqrt{2\log(N)}}.$$

The upper Bound (9.15) then follows by simply combining (9.21) with (9.22).

3. Let P_{V_i} be the orthogonal projector onto V_i. From Lemma 8.3 in Chapter 8, we know that the largest singular value $\sigma_1(ZP_{V_i})$ of ZP_{V_i} fulfills the inequality $\mathbb{E}[\sigma_1(ZP_{V_i})] \leq \sqrt{n} + \sqrt{d}$. Applying the Gaussian concentration inequality and noticing that $\sqrt{n}\lambda_{V_i}(Z) \leq \sigma_1(ZP_{V_i})$, prove that there exist some exponential random variables $\xi_1,\ldots,\xi_N$, such that for all $i = 1,\ldots,N$

$$\lambda_{V_i}(Z) \leq \mathbb{E}[\lambda_{V_i}(Z)] + \sqrt{2\xi_i/n} \leq 1 + \sqrt{d/n} + \sqrt{2\xi_i/n}.$$

4. Applying again the Gaussian concentration inequality, prove that there exists an exponential random variable ξ, such that

$$\sup_{i=1,\dots,N} \lambda_{V_i}(Z) \leq 1 + \sqrt{d/n} + \mathbb{E}\left[\max_{i=1,\dots,N} \sqrt{2\xi_i/n}\right] + \sqrt{2\xi/n}.$$

5. Prove that for any $s > 0$,

$$\mathbb{E}\left[\max_{i=1,\dots,N} \sqrt{2\xi_i}\right] \leq s^{-1} \log\left(\sum_{i=1}^{N} \mathbb{E}\left[e^{s\sqrt{2\xi_i}}\right]\right).$$

6. Check that

$$\mathbb{E}\left[e^{s\sqrt{2\xi_i}}\right] = \int_0^\infty e^{su-u^2/2} u\, du \leq s\sqrt{2\pi}\, e^{s^2/2} + 1.$$

7. Choosing $s = \sqrt{2\log(N)}$, prove that $\mathbb{E}\left[\max_{i=1,\dots,N} \sqrt{2\xi_i}\right] \leq \sqrt{2\log(N)} + \delta_N$ and conclude the proof of (9.22).

Chapter 10

Multiple Testing

In this chapter, we switch from the estimation problem to the test problem, and we explain some possible ways to handle the impact of high dimensionality in this context. More precisely, we will focus on the problem of performing simultaneously a large number of tests. This issue is of major importance in practice: Many scientific experiments seek to determine if a given factor has an impact on various quantities of interest. For example, we can seek for the possible side effects (headache, stomach pain, drowsiness, etc.) induced by a new drug. From a statistical perspective, this amounts to test *simultaneously* for each quantity of interest the hypothesis "the factor has no impact on this quantity" against "the factor has an impact on this quantity." As we have seen in Chapter 1, considering simultaneously many different tests induces a loss in our ability to discriminate between the two hypotheses. We present in this chapter the theoretical basis for reducing at best this deleterious effect. We start by illustrating the issue on a simple example, and then we introduce the bases of False Discovery Rate control.

10.1 Introductory Example

10.1.1 Differential Expression of a Single Gene

Let us assume that we have r measurements for the expression of a gene g in two different conditions A and B (corresponding, for example, to some normal cells and some cancerous cells).

Conditions	Measurements
A	$X_1^A,\ldots,X_r^A$
B	$X_1^B,\ldots,X_r^B$

We want to know if there is a difference in the expression of this gene between these two conditions A and B. In the formalism of test theory, we want to discriminate between the two hypotheses:

- $\mathscr{H}_0$: "the means of the X_i^A and the X_i^B are the same"
- $\mathscr{H}_1$: "the means of the X_i^A and the X_i^B are different"

A classical test statistic

Setting $Z_i = X_i^A - X_i^B$ for $i = 1, \ldots, r$ we can reject $\mathscr{H}_0$ when

$$\widehat{S} := \frac{|\overline{Z}|}{\sqrt{\widehat{\sigma}^2/r}} \geq s = \text{some threshold,} \tag{10.1}$$

with $\overline{Z}$ the empirical mean of the Z_i and $\widehat{\sigma}^2$ the empirical variance of the Z_i. The threshold s is chosen, such that the probability to wrongly reject $\mathscr{H}_0$ is not larger than α.

Case of Gaussian measurements

In the special case where

$$X_i^A \overset{\text{i.i.d.}}{\sim} \mathscr{N}(\mu_A, \sigma_A^2) \quad \text{and} \quad X_i^B \overset{\text{i.i.d.}}{\sim} \mathscr{N}(\mu_B, \sigma_B^2),$$

the hypothesis $\mathscr{H}_0$ corresponds to "$\mu_A = \mu_B$" and the hypothesis $\mathscr{H}_1$ corresponds to "$\mu_A \neq \mu_B$." In addition, if the variables $X_1^A, \ldots, X_r^A, X_1^B, \ldots, X_r^B$ are independent, then the $Z_i = X_i^A - X_i^B$ are i.i.d. with $\mathscr{N}(\mu_A - \mu_B, \sigma_A^2 + \sigma_B^2)$-Gaussian distribution. In this case, the statistic $\widehat{S}$ defined by (10.1) is distributed under the null hypothesis as the absolute value of a student random variable $\mathscr{T}(r-1)$, with $r-1$ degrees of freedom. Let us define the non-increasing function

$$T_0(s) = \mathbb{P}(|\mathscr{T}(r-1)| \geq s), \quad \text{for} \quad s \in \mathbb{R}.$$

We can associate to the test statistic $\widehat{S}$, the p-value

$$\widehat{p} = T_0(\widehat{S}).$$

Then, the test $\psi_\alpha = \mathbf{1}_{\widehat{p} \leq \alpha}$ has level α (the probability to wrongly reject $\mathscr{H}_0$ is α) and we have:

- If the p-value $\widehat{p}$ is larger than α, then the hypothesis $\mathscr{H}_0$ is not rejected.
- If the p-value $\widehat{p}$ is not larger than α, then the hypothesis $\mathscr{H}_0$ is rejected.

10.1.2 Differential Expression of Multiple Genes

DNA microarrays and the Next Generation Sequencing (NGS) technologies allow us to measure the expression level of thousands of genes simultaneously. Our statistical objective is then to test *simultaneously* for all genes $g \in \{1, \ldots, m\}$:

- $\mathscr{H}_{0,g}$: "the mean expression levels of the gene g in conditions A and B are the same"
- $\mathscr{H}_{1,g}$: "the mean expression levels of the gene g in conditions A and B are different"

If we reject $\mathscr{H}_{0,g}$ when the p-value $\widehat{p}_g$ is not larger than α, then for each *individual* gene g, the probability to reject wrongly $\mathscr{H}_{0,g}$ is at most α. Nevertheless, if we consider the m genes *simultaneously* the number of hypotheses $\mathscr{H}_{0,g}$ wrongly rejected (called false positives) can be high. Actually, the mean number of false positives is

$$\mathbb{E}\left[\text{False Positives}\right] = \sum_{g:\mathscr{H}_{0,g}\text{ true}} \mathbb{P}_{\mathscr{H}_{0,g}}(\widehat{p}_g \leq \alpha) = \text{card}\left\{g : \mathscr{H}_{0,g} \text{ is true}\right\} \times \alpha$$

since the p-values are such that $\mathbb{P}_{\mathscr{H}_{0,g}}(\widehat{p}_g \leq \alpha) = \alpha$ for every g. For example, for typical values like $\alpha = 5\%$ and $\text{card}\left\{g : \mathscr{H}_{0,g} \text{ is true}\right\} = 10000$, we obtain on average 500 false positives.

From a biological point of view, a gene that presents a differential expression between the two conditions A and B is a gene that is suspected to be involved in the response to the change of "environment" between A and B. Further experiments must be carried out in order to validate (or not) this "discovery." If in our list of genes suspected to present a differential expression there are 500 false positives, it means that we have 500 false discoveries, and a lot of time and money will be spent in useless experiments. Therefore, biologists ask for a list of genes that contains as few false positives as possible. Of course, the best way to avoid false positives is to declare no gene positive, but, in this case, there is no discovery and the data are useless. In this chapter, we will present a couple of procedures designed to control the number of false discoveries in multiple testing settings while not losing too much in terms of power.

10.2 Statistical Setting

In the remainder of this chapter, we consider the following setting. We have m families of probability distribution $\{\mathbb{P}_\theta : \theta \in \Theta_i\}$ with $i = 1, \ldots, m$, and we consider simultaneously the m tests

$$\mathscr{H}_{0,i} : \theta \in \Theta_{0,i} \quad \text{against} \quad \mathscr{H}_{1,i} : \theta \in \Theta_{1,i} \qquad \text{for } i = 1, \ldots, m,$$

where $\Theta_{0,i}$ and $\Theta_{1,i}$ are two disjointed subsets of Θ_i.

10.2.1 p-Values

For each test i, we have access to some data X_i. A p-value $\widehat{p}_i$ for the test indexed by i, is any random variable which is $\sigma(X_i)$-measurable (it can be computed from the data), taking values in $[0, 1]$ and fulfilling the distributional property

$$\sup_{\theta \in \Theta_{0,i}} \mathbb{P}_\theta(\widehat{p}_i \leq u) \leq u, \quad \text{for all } u \in [0, 1]. \tag{10.2}$$

We say that, under the null hypotheses, the p-values are stochastically larger than a uniform random variable.
Let us explain on an example how we can define some p-values.

Example.

Let $\widehat{S}_i$ be any real random variable, which is $\sigma(X_i)$-measurable (it can be computed from the data). For $\theta \in \Theta_i$, let us denote by $T_\theta(s) = \mathbb{P}_\theta(\widehat{S}_i \geq s)$ the tail distribution of the statistic $\widehat{S}_i$ under $\mathbb{P}_\theta$. We can associate to the statistic $\widehat{S}_i$, the p-value for the test i

$$\widehat{p}_i = \sup_{\theta \in \Theta_{0,i}} T_\theta(\widehat{S}_i). \tag{10.3}$$

It corresponds to the maximum probability under the null hypothesis to observe a value for our statistic not smaller than the value $\widehat{S}_i$ that we have actually observed.

The next proposition ensures that (10.3) is indeed a p-value.

Proposition 10.1 $\widehat{p}_i$ defined by (10.3) is a p-value

The random variable $\widehat{p}_i$ defined by (10.3) is distributed as a p-value:

$$\sup_{\theta \in \Theta_{0,i}} \mathbb{P}_\theta(\widehat{p}_i \leq u) \leq u, \quad \textit{for all } u \in [0,1].$$

Proof. For any $\theta \in \Theta_{0,i}$ and $u \in [0,1]$, we have

$$\mathbb{P}_\theta(\widehat{p}_i \leq u) = \mathbb{P}_\theta\left(\sup_{\theta \in \Theta_{0,i}} T_\theta(\widehat{S}_i) \leq u\right) \leq \mathbb{P}_\theta\left(T_\theta(\widehat{S}_i) \leq u\right). \tag{10.4}$$

For $u \in [0,1]$, we define $T_\theta^{-1}(u) = \inf\{s \in \mathbb{R} : T_\theta(s) \leq u\}$. Since T_θ is non-increasing, we have

$$]T_\theta^{-1}(u), +\infty[\subset \{s \in \mathbb{R} : T_\theta(s) \leq u\} \subset [T_\theta^{-1}(u), +\infty[\,.$$

Let us consider apart the cases where $T_\theta(T_\theta^{-1}(u)) \leq u$ and $T_\theta(T_\theta^{-1}(u)) > u$.

- When $T_\theta(T_\theta^{-1}(u)) \leq u$, we have $\{s \in \mathbb{R} : T_\theta(s) \leq u\} = [T_\theta^{-1}(u), +\infty[$ and then

$$\mathbb{P}_\theta\left(T_\theta(\widehat{S}_i) \leq u\right) = \mathbb{P}_\theta\left(\widehat{S}_i \geq T_\theta^{-1}(u)\right) = T_\theta(T_\theta^{-1}(u)) \leq u.$$

- When $T_\theta(T_\theta^{-1}(u)) > u$, we have $\{s \in \mathbb{R} : T_\theta(s) \leq u\} =]T_\theta^{-1}(u), +\infty[$, and therefore

$$\mathbb{P}_\theta\left(T_\theta(\widehat{S}_i) \leq u\right) = \mathbb{P}_\theta\left(\widehat{S}_i > T_\theta^{-1}(u)\right) = \lim_{\varepsilon \searrow 0} T_\theta(T_\theta^{-1}(u) + \varepsilon) \leq u,$$

  where the last inequality comes from $T_\theta^{-1}(u) + \varepsilon \in \{s \in \mathbb{R} : T_\theta(s) \leq u\}$ for all $\varepsilon > 0$.

Combining the two last displays with (10.4), we have proved that $\mathbb{P}_\theta(\widehat{p}_i \leq u) \leq u$ for all $\theta \in \Theta_{0,i}$ and $u \in [0,1]$. □

10.2.2 Multiple-Testing Setting

We assume that for each test $i \in \{1,\ldots,m\}$, we have access to a p-value $\widehat{p}_i$. A multiple-testing procedure is a procedure that takes as input the vector of p-values $(\widehat{p}_1,\ldots,\widehat{p}_m)$ corresponding to the m tests and returns a set of indices $\widehat{R} = R(\widehat{p}_1,\ldots,\widehat{p}_m) \subset I = \{1,\ldots,m\}$, which gives the set of the null hypotheses $\{\mathscr{H}_{0,i} : i \in \widehat{R}\}$ that are rejected. Writing I_0 for the set

$$I_0 = \{i \in \{1,\ldots,m\} : \mathscr{H}_{0,i} \text{ is true}\}, \tag{10.5}$$

we call *false positive* (FP) the indices $i \in \widehat{R} \cap I_0$ and *true positive* (TP) the indices $i \in \widehat{R} \setminus I_0$. In the following, we will use the notations

$$\mathsf{FP} = \text{card}(\widehat{R} \cap I_0) \quad \text{and} \quad \mathsf{TP} = \text{card}(\widehat{R} \setminus I_0).$$

Ideally, we would like a procedure that selects $\widehat{R}$ in such a way that FP is small and TP is large. Of course, there is a balance to find between these two terms, since a severe control of the number of false positives usually induces a small number of true positives.

10.2.3 Bonferroni Correction

The Bonferroni correction provides a severe control of the number of false positives. It is designed in order to control the probability of existence of false positives $\mathbb{P}(\mathsf{FP} > 0)$. It is defined by

$$\widehat{R}_{\text{Bonf}} = \{i : \widehat{p}_i \leq \alpha/m\}.$$

Let us denote by m_0 the cardinality of I_0. According to Proposition 10.1, we have

$$\mathbb{P}(\mathsf{FP} > 0) = \mathbb{P}(\exists i \in I_0 : \widehat{p}_i \leq \alpha/m) \leq \sum_{i \in I_0} \sup_{\theta \in \Theta_{0,i}} \mathbb{P}_\theta(\widehat{p}_i \leq \alpha/m) \leq m_0\alpha/m \leq \alpha.$$

The probability of existence of false positives is thus smaller than α. The Bonferroni procedure avoids false positives but produces only a few true positives in general. Actually, it amounts to using the level α/m for each test, which can be very conservative when m is large.

In the next section, we will describe some procedures that control the (mean) proportion of false positives among $\widehat{R}$ instead of the absolute number of false positives.

10.3 Controlling the False Discovery Rate

The False Discovery Proportion (FDP) corresponds to the proportion $\mathsf{FP}/(\mathsf{FP}+\mathsf{TP})$ of false positives among the positives (with the convention $0/0 = 0$). The False Discovery Rate (FDR) is defined as the mean False Discovery Proportion

$$\mathsf{FDR} = \mathbb{E}\left[\frac{\mathsf{FP}}{\mathsf{FP}+\mathsf{TP}} \mathbf{1}_{\{\mathsf{FP}+\mathsf{TP} \geq 1\}}\right].$$

This quantity was introduced in the '90s, and it is now widely used in science, especially in biostatistics.

10.3.1 Heuristics

Let us try to guess what could be a procedure that controls the FDR. To start, we notice that if we want to have FP as small as possible and TP as large as possible, then the rejected p-values should correspond to the smallest p-values. So the only issue is to determine how many p-values can be rejected while keeping the FDR lower than α. Therefore, we will focus on rejection sets $\widehat{R}$ of the form

$$\widehat{R} = \{i \in I : \widehat{p}_i \leq t(\widehat{p}_1, \ldots, \widehat{p}_m)\}, \tag{10.6}$$

with $t : [0,1]^m \to [0,1]$. In the following, we will seek some functions t that prevent an FDR larger than α, while maximizing the size of $\widehat{R}$.

Let us investigate informally this point. According to (10.2), for a given threshold $\tau > 0$, the number FP of false positives in the rejection set $\widehat{R} = \{i : \widehat{p}_i \leq \tau\}$ fulfills

$$\mathbb{E}[\mathsf{FP}] = \mathbb{E}\left[\sum_{i \in I_0} \mathbf{1}_{\{\widehat{p}_i \leq \tau\}}\right] \leq \sum_{i \in I_0} \sup_{\theta \in \Theta_{0,i}} \mathbb{P}_\theta(\widehat{p}_i \leq \tau) \leq \mathrm{card}(I_0)\, \tau \leq m\tau. \tag{10.7}$$

Let us denote by $\widehat{p}_{(1)} \leq \ldots \leq \widehat{p}_{(m)}$ the p-values ranked in a non-decreasing order. For any $\tau \in [\widehat{p}_{(k)}, \widehat{p}_{(k+1)}[$, we have $\{i : \widehat{p}_i \leq \tau\} = \{i : \widehat{p}_i \leq \widehat{p}_{(k)}\}$. Therefore, we only have to focus on a threshold $t(\widehat{p}_1, \ldots, \widehat{p}_m)$ in (10.6) of the form $t(\widehat{p}_1, \ldots, \widehat{p}_m) = \widehat{p}_{(\widehat{k})}$, with $\widehat{k} = \widehat{k}(\widehat{p}_1, \ldots, \widehat{p}_m)$. With this choice, we notice that $\mathrm{card}(\widehat{R}) = \widehat{k}$, so according to (10.7), we expect to have for $t(\widehat{p}_1, \ldots, \widehat{p}_m) = \widehat{p}_{(\widehat{k})}$

$$\frac{\mathsf{FP}}{\mathsf{FP} + \mathsf{TP}} = \frac{\mathsf{FP}}{\mathrm{card}(\widehat{R})} \overset{?}{\leq} \frac{m\widehat{p}_{(\widehat{k})}}{\widehat{k}} \quad \text{“on average.”}$$

This non-rigorous computation suggests that we should have an FDR upper-bounded by $\alpha > 0$, as soon as the integer $\widehat{k}$ fulfills

$$\widehat{p}_{(\widehat{k})} \leq \alpha \widehat{k}/m. \tag{10.8}$$

Since we want to have the cardinality of $\widehat{R}$ as large as possible, and since $\mathrm{card}(\widehat{R}) = \widehat{k}$, we then choose $\widehat{k} = \max\{k : \widehat{p}_{(k)} \leq \alpha k/m\}$. The rejection set suggested by the above discussion is then $\widehat{R} = \{i \in I : \widehat{p}_i \leq \widehat{p}_{(\widehat{k})}\}$ or equivalently

$$\widehat{R} = \left\{i \in I : \widehat{p}_i \leq \alpha \widehat{k}/m\right\}, \quad \text{with } \widehat{k} = \max\{k : \widehat{p}_{(k)} \leq \alpha k/m\}.$$

Such a rejection set is illustrated in Figure 10.1.

10.3.2 Step-Up Procedures

The previous informal discussion suggests to choose a set of rejected hypotheses $\widehat{R}$ in the form

$$\widehat{R} = \left\{i \in I : \widehat{p}_i \leq \alpha\beta(\widehat{k})/m\right\} \quad \text{with} \quad \widehat{k} = \max\{k \in I : \widehat{p}_{(k)} \leq \alpha\beta(k)/m\} \tag{10.9}$$

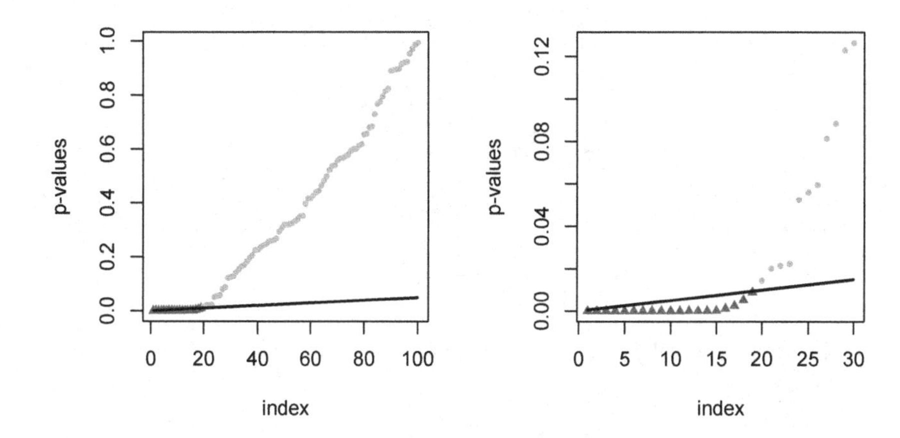

Figure 10.1 *Some p-values $\widehat{p}_{(1)} \leq \ldots \leq \widehat{p}_{(m)}$ ranked in increasing order and their position relative to the line $k \to \alpha k/m$ (in black). The right picture is an enlargement of the left one. Dark gray triangles: p-values corresponding to rejected hypotheses. Light gray dots: p-values corresponding to non-rejected hypotheses.*

where $\beta : \{1,\ldots,m\} \to \mathbb{R}^+$ is some non-decreasing function, and $\widehat{p}_{(1)} \leq \ldots \leq \widehat{p}_{(m)}$ are the p-values ranked in non-decreasing order. When $\{k \in I : \widehat{p}_{(k)} \leq \alpha\beta(k)/m\} = \emptyset$, we set $\widehat{R} = \emptyset$. Figure 10.1 gives an illustration of the choice of $\widehat{k}$ for the function $\beta(k) = k$ suggested by the previous discussion. The next theorem provides a bound on the FDR of the procedure (10.9).

Theorem 10.2 General control of the FDR

Let $\beta : \{1,\ldots,m\} \to \mathbb{R}^+$ be a non-decreasing function and for $\alpha > 0$ define $\widehat{R}$ by (10.9), with the convention that $\widehat{R} = \emptyset$ when $\{k \in I : \widehat{p}_{(k)} \leq \alpha\beta(k)/m\} = \emptyset$.

Writing $m_0 = \text{card}(I_0)$ with I_0 defined by (10.5), we have the following upper bound on the FDR *of the procedure*

$$\text{FDR}(\widehat{R}) \leq \alpha \, \frac{m_0}{m} \sum_{j \geq 1} \frac{\beta(j \wedge m)}{j(j+1)} . \qquad (10.10)$$

Proof. We use below the convention that $\widehat{k} = 0$ when $\widehat{R} = \emptyset$. We observe from the

Definition (10.9) that $\widehat{k} = \text{card}(\widehat{R})$. From the definition of the FDR, we have

$$\begin{aligned}
\text{FDR}(\widehat{R}) &= \mathbb{E}\left[\frac{\text{card}\left\{i \in I_0 : \widehat{p}_i \leq \alpha\beta(\widehat{k})/m\right\}}{\widehat{k}} \mathbf{1}_{\widehat{k}\geq 1}\right] \\
&= \sum_{i\in I_0} \mathbb{E}\left[\mathbf{1}_{\widehat{p}_i\leq \alpha\beta(\widehat{k})/m} \frac{\mathbf{1}_{\widehat{k}\geq 1}}{\widehat{k}}\right].
\end{aligned}$$

For $\widehat{k} \geq 1$, we have

$$\frac{1}{\widehat{k}} = \sum_{j\geq 1} \frac{\mathbf{1}_{j\geq \widehat{k}}}{j(j+1)},$$

so applying first Fubini, then $\beta(\widehat{k}) \leq \beta(j \wedge m)$ for $j \geq \widehat{k}$, and finally Proposition 10.1, we obtain

$$\begin{aligned}
\text{FDR}(\widehat{R}) &\overset{\text{Fubini}}{=} \sum_{i\in I_0}\sum_{j\geq 1} \frac{1}{j(j+1)} \underbrace{\mathbb{E}\left[\mathbf{1}_{j\geq \widehat{k}} \mathbf{1}_{\widehat{p}_i\leq \alpha\beta(\widehat{k})/m} \mathbf{1}_{\widehat{k}\geq 1}\right]}_{\leq \mathbb{P}\left(\widehat{p}_i\leq \alpha\beta(j\wedge m)/m\right)} \\
&\overset{\text{Prop. 10.1}}{\leq} \sum_{i\in I_0}\sum_{j\geq 1} \frac{1}{j(j+1)} \times \alpha\beta(j\wedge m)/m \\
&\leq \alpha\,\frac{m_0}{m} \sum_{j\geq 1} \frac{\beta(j\wedge m)}{j(j+1)}.
\end{aligned}$$

The proof of Theorem 10.2 is complete. □

We point out that the upper Bound (10.10) is sharp in the sense that for α small enough, there exist some distributions of the p-values $(\widehat{p}_1, \ldots, \widehat{p}_m)$, such that (10.10) is an equality; see Theorem 5.2 in Guo and Rao [88].

If we choose β in such a way that the sum fulfills the condition

$$\sum_{j\geq 1} \frac{\beta(j\wedge m)}{j(j+1)} \leq 1, \tag{10.11}$$

then the FDR of the procedure is less than α. We notice that for our guess $\beta(k) = k$ suggested by the informal discussion, the upper Bound (10.10) is equal to $\alpha m_0 H_m/m$ with $H_m = 1 + 1/2 + \ldots + 1/m$. In particular, this choice of β does not meet Condition (10.11). A popular choice of function β fulfilling Condition (10.11) is the linear function $\beta(k) = k/H_m$. The procedure (10.9) with this choice of β is called the Benjamini–Yekutieli procedure.

Corollary 10.3 FDR control of Benjamini–Yekutieli procedure

We set $H_m = 1 + 1/2 + \ldots + 1/m$. The multiple-testing procedure defined by $\widehat{R} = \emptyset$ when $\{k \in I : \widehat{p}_{(k)} \leq \alpha k/(mH_m)\} = \emptyset$ and otherwise

$$\widehat{R} = \left\{ i \in I : \widehat{p}_i \leq \alpha \widehat{k}/(mH_m) \right\}, \quad \text{with} \quad \widehat{k} = \max\left\{k \in I : \widehat{p}_{(k)} \leq \alpha k/(mH_m)\right\} \tag{10.12}$$

has an FDR *upper-bounded by α.*

Another example of function β fulfilling (10.11) is

$$\beta(k) = \alpha \frac{k(k+1)}{2m}, \quad \text{for } k = 1, \ldots, m.$$

For this choice of β, we have $\beta(k) \geq k/H_m$ when $k + 1 \geq 2m/H_m$, so the resulting procedure tends to reject more hypotheses than the Benjamini–Yekutieli procedure when there are many positives.

10.3.3 FDR Control under the WPRD Property

In the Benjamini–Yekutieli procedure, function β is linear with a slope $1/H_m \sim 1/\log(m)$. The smaller the slope, the less true positives we have. Since (10.10) is sharp for some distributions of the p-values, we know that we cannot choose a larger slope in general. Yet, we may wonder if in some cases we can choose a linear function β with a larger slope while keeping the FDR smaller than α.

The discussion in Section 10.3.1 suggests that for some distributions of the p-values, we may expect some FDR control with the choice $\beta(k) = k$ instead of $\beta(k) = k/H_m$ in Procedure (10.9). The choice $\beta(k) = k$ corresponds to the Benjamini–Hochberg procedure defined by

$$\widehat{R} = \left\{ i \in I : \widehat{p}_i \leq \alpha \widehat{k}/m \right\} \quad \text{where} \quad \widehat{k} = \max\left\{k \in I : \widehat{p}_{(k)} \leq \alpha k/m\right\}, \tag{10.13}$$

with the convention $\widehat{R} = \emptyset$ when $\left\{k \in I : \widehat{p}_{(k)} \leq \alpha k/m\right\} = \emptyset$. We give below a simple distributional condition on the p-values which ensures an FDR control at level α for the Benjamini–Hochberg procedure.

In the next definition, we will say that a function $g : [0,1]^m \to \mathbb{R}^+$ is non-decreasing if for any $p, q \in [0,1]^m$ such that $p_i \geq q_i$ for all $i = 1, \ldots, m$, we have $g(p) \geq g(q)$.

Weak Positive Regression Dependency

The distribution of the p-values $(\widehat{p}_1, \ldots, \widehat{p}_m)$ is said to fulfill the Weak Positive Regression Dependency Property (WPRD) if for any bounded measurable non-decreasing function $g : [0,1]^m \to \mathbb{R}^+$ and for all $i \in I_0$, the function

$$u \to \mathbb{E}\left[g(\widehat{p}_1, \ldots, \widehat{p}_m) \,|\, \widehat{p}_i \leq u\right] \quad \text{is non-decreasing} \tag{10.14}$$

on the interval $\{u \in [0,1] : \mathbb{P}(\widehat{p}_i \leq u) > 0\}$.

The set of distributions fulfilling the WPRD property includes the independent distributions.

Lemma 10.4 Independent p-values fulfills the WPRD
Assume that the $(\widehat{p}_i)_{i\in I_0}$ are independent random variables and that the $(\widehat{p}_i)_{i\in I\setminus I_0}$ are independent from the $(\widehat{p}_i)_{i\in I_0}$. Then, the distribution of $(\widehat{p}_1,\ldots,\widehat{p}_m)$ fulfills the WPRD property.

Proof. Let us consider some $i \in I_0$ and some bounded measurable non-decreasing function $g:[0,1]^m \to \mathbb{R}^+$. With no loss of generality, we can assume (for notational simplicity) that $i=1$. The random variable $\widehat{p}_1$ is independent of $(\widehat{p}_2,\ldots,\widehat{p}_m)$, so for any u such that $\mathbb{P}(\widehat{p}_1 \le u) > 0$, we have

$$\mathbb{E}\left[g(\widehat{p}_1,\ldots,\widehat{p}_m)\,|\,\widehat{p}_1 \le u\right] = \int_{(x_2,\ldots,x_m)\in[0,1]^{m-1}} \mathbb{E}\left[g(\widehat{p}_1,x_2,\ldots,x_m)\,|\,\widehat{p}_1 \le u\right] \mathbb{P}(\widehat{p}_2\in dx_2,\ldots,\widehat{p}_m \in dx_m)\,.$$

To prove the lemma, we only need to check that $u \to \mathbb{E}\left[g(\widehat{p}_1,x_2,\ldots,x_m)\,|\,\widehat{p}_1 \le u\right]$ is non-decreasing for all $x_2,\ldots,x_m \in [0,1]$. Since the function g is non-decreasing, the function $g_1 : x_1 \to g(x_1,x_2,\ldots,x_m)$ is also non-decreasing. Writing $g_1^{-1}(t) = \inf\{x\in[0,1] : g_1(x) \ge t\}$, with $\inf\{\emptyset\} = +\infty$, we have

$$\begin{aligned}
\mathbb{E}\left[g(\widehat{p}_1,x_2,\ldots,x_m)\,|\,\widehat{p}_1 \le u\right] &= \mathbb{E}\left[g_1(\widehat{p}_1)\,|\,\widehat{p}_1 \le u\right] \\
&= \int_0^{+\infty} \mathbb{P}(g_1(\widehat{p}_1) \ge t\,|\,\widehat{p}_1 \le u)\,dt \\
&= \int_0^{+\infty} \mathbb{P}(\widehat{p}_1 \ge \text{or} > g_1^{-1}(t)\,|\,\widehat{p}_1 \le u)\,dt,
\end{aligned}$$

since

$$]g_1^{-1}(t),1] \subset \{p\in[0,1] : g_1(p) \ge t\} \subset [g_1^{-1}(t),1].$$

To conclude, we simply notice that both

$$u \to \mathbb{P}(\widehat{p}_1 \ge g_1^{-1}(t)\,|\,\widehat{p}_1 \le u) = \left(1 - \frac{\mathbb{P}(\widehat{p}_1 < g_1^{-1}(t))}{\mathbb{P}(\widehat{p}_1 \le u)}\right)_+,$$

and

$$u \to \mathbb{P}(\widehat{p}_1 > g_1^{-1}(t)\,|\,\widehat{p}_1 \le u) = \left(1 - \frac{\mathbb{P}(\widehat{p}_1 \le g_1^{-1}(t))}{\mathbb{P}(\widehat{p}_1 \le u)}\right)_+$$

are non-decreasing for all $t \in \mathbb{R}^+$. □

Another example of p-values fulfilling the WPRD property is the p-values associated to some $(S_1,\ldots,S_m)$ distributed according to a $\mathcal{N}(0,\Sigma)$ Gaussian distribution, with $\Sigma_{ij} \geq 0$ for all $i, j = 1, \ldots, m$; see Exercise 10.6.3. We refer to Benjamini and Yekutieli [27] for some other examples.

The next theorem ensures that the FDR of the Benjamini–Hochberg procedure (10.13) is upper-bounded by α when the WPRD property is met.

Theorem 10.5 FDR control of Benjamini–Hochberg procedure

When the distribution of the p-values fulfills the WPRD property, the multiple-testing procedure defined by $\widehat{R} = \emptyset$ when $\left\{k \in I : \widehat{p}_{(k)} \leq \alpha k/m\right\} = \emptyset$, and otherwise

$$\widehat{R} = \left\{i \in I : \widehat{p}_i \leq \alpha \widehat{k}/m\right\}, \quad \textit{with} \quad \widehat{k} = \max\left\{k \in I : \widehat{p}_{(k)} \leq \alpha k/m\right\},$$

has an FDR *upper-bounded by* α.

Proof. We use again the convention $\widehat{k} = 0$ when $\widehat{R} = \emptyset$. Since $\widehat{k} = \text{card}(\widehat{R})$, we have

$$\begin{aligned}
\text{FDR}(\widehat{R}) &= \mathbb{E}\left[\frac{\text{card}\left\{i \in I_0 : \widehat{p}_i \leq \alpha\widehat{k}/m\right\}}{\widehat{k}} \mathbf{1}_{\widehat{k}\geq 1}\right] \\
&= \sum_{i\in I_0} \mathbb{E}\left[\mathbf{1}_{\widehat{p}_i \leq \alpha\widehat{k}/m} \frac{\mathbf{1}_{\widehat{k}\geq 1}}{\widehat{k}}\right] \\
&= \sum_{i\in I_0} \sum_{k=1}^{m} \frac{1}{k} \mathbb{P}\left(\widehat{k} = k \text{ and } \widehat{p}_i \leq \alpha k/m\right) \\
&= \sum_{i\in I_0} \sum_{k=k_i^*}^{m} \frac{1}{k} \mathbb{P}\left(\widehat{k} = k \,|\, \widehat{p}_i \leq \alpha k/m\right) \mathbb{P}(\widehat{p}_i \leq \alpha k/m),
\end{aligned}$$

where $k_i^* = \inf\{k \in \mathbb{N} : \mathbb{P}(\widehat{p}_i \leq \alpha k/m) > 0\}$ and with the convention that the sum from k_i^* to m is zero if $m < k_i^*$. By Proposition 10.1, we have $\mathbb{P}(\widehat{p}_i \leq \alpha k/m) \leq \alpha k/m$, so we obtain

$$\begin{aligned}
\text{FDR}(\widehat{R}) &\leq \sum_{i\in I_0} \sum_{k=k_i^*}^{m} \frac{\alpha}{m} \mathbb{P}\left(\widehat{k} = k \,|\, \widehat{p}_i \leq \alpha k/m\right) \\
&\leq \frac{\alpha}{m} \sum_{i\in I_0} \sum_{k=k_i^*}^{m} \left[\mathbb{P}\left(\widehat{k} \leq k \,|\, \widehat{p}_i \leq \alpha k/m\right) - \mathbb{P}\left(\widehat{k} \leq k-1 \,|\, \widehat{p}_i \leq \alpha k/m\right)\right].
\end{aligned}$$

The function

$$g(\widehat{p}_1, \ldots, \widehat{p}_m) = \mathbf{1}_{\left\{\max\left\{j : \widehat{p}_{(j)} \leq \alpha j/m\right\} \leq k\right\}} = \mathbf{1}_{\{\widehat{k}\leq k\}}$$

is non-decreasing with respect to $(\widehat{p}_1, \ldots, \widehat{p}_m)$, so the WPRD property ensures that for $k \geq k_i^*$

$$\mathbb{P}\left(\widehat{k} \leq k \,|\, \widehat{p}_i \leq \alpha k/m\right) \leq \mathbb{P}\left(\widehat{k} \leq k \,|\, \widehat{p}_i \leq \alpha(k+1)/m\right).$$

We then obtain a telescopic sum and finally

$$\mathsf{FDR}(\widehat{R}) \le \frac{\alpha}{m} \sum_{i\in I_0} \mathbf{1}_{k_i^*\le m} \mathbb{P}\left(\widehat{k} \le m \,\middle|\, \widehat{p}_i \le \frac{\alpha(m+1)}{m}\right) \le \frac{m_0}{m}\alpha \le \alpha.$$

The proof of Theorem 10.5 is complete. □

10.4 Illustration

We illustrate the implementation of the Benjamini–Hochberg procedure on a microarray data set from Golub *et al.* [86]. We first load the data that are in the `multtest` package available on the website `http://www.bioconductor.org`.

```
library(multtest)
library(stats)
data(golub)      # load the data
```

The Golub data set is a 3051×38 matrix. Each row reports the expression level for $m = 3051$ genes. The first 27 columns correspond to patients with leukemia of type "ALL," the last 11 columns correspond to patients with leukemia of type "AML." Our goal is to find genes that have a differential expression between these two conditions. Therefore, for each gene we perform a t-test and we record the corresponding p-value.

```
golub1<-golub[,1:27]    # data for Leukemia ALL
golub2<-golub[,28:38]   # data for Leukemia AML
m<-3051
p<-rep(0,m)
# compute the p-values with a two-sample t-test
for (i in 1:m)  p[i]<-t.test(golub1[i,],golub2[i,])$p.value
```

We then compute the number $\widehat{k}$ of p-values rejected according to Formula (10.13) and the rejection set $\widehat{R}$. We also print the names of the genes for which a differential expression has been detected.

```
k<-sum(sort(p)<=0.05*(1:m)/m) # number of p-values rejected
R<-(1:m)[p<=0.05*k/m]         # rejection set
print(golub.gnames[R,2])      # print the names of the genes
```

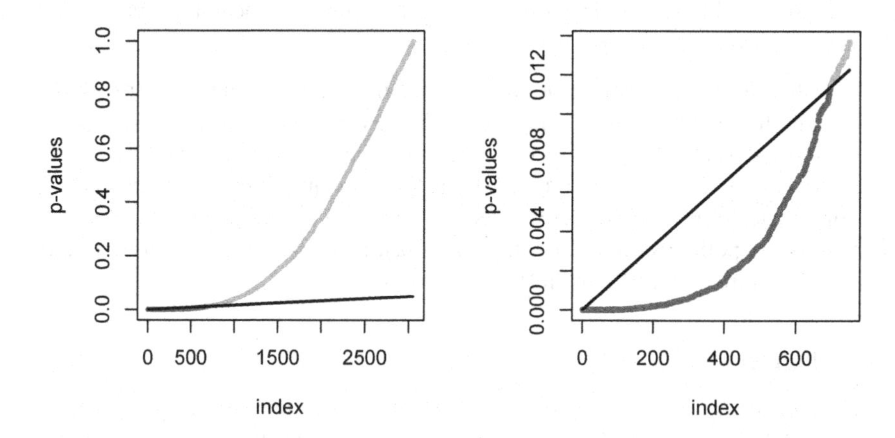

Figure 10.2 *The p-values ranked in increasing order. The dark-gray p-values correspond to rejected hypotheses, and the light-gray ones correspond to those not rejected. The black line represents the map* $k \to \alpha k/m$. *The right picture is an enlargement of the left one.*

Finally, we plot the rejected p-values; see Figure 10.2.

```
par(mfrow=c(1,2))
plot(1:m,sort(p),col=c(rep(2,k),rep(3,m-k)),type="p",pch=20,cex=0.7,
xlab="index",ylab="p-values")
points(1:m,0.05*(1:m)/m,col=1,type="l",lwd=2)
plot(1:750,sort(p)[1:750],type="p",col=c(rep(2,k),rep(3,750-k)),
pch=20,cex=0.7,xlab="index",ylab="p-values")
points(1:750,0.05*(1:750)/m,col=1,type="l",lwd=2)
```

We refer the interested reader to the paper by Dudoit, Fridlyand, and Speed [68] for a careful analysis of this data set.

10.5 Discussion and References

10.5.1 Take-Home Message

When we perform simultaneously m tests of level α the average number of false positives (tests for which $\mathcal{H}_0$ is wrongly rejected) is αm_0, where m_0 is the number of tests for which $\mathcal{H}_0$ is true. When m_0 is large, this number of false positives can be large, even larger than the number of true positives (tests for which $\mathcal{H}_0$ has been correctly rejected). In order to avoid this deleterious effect of multiple testing, we must apply a multiple-testing selection procedure on the p-values of the m tests.

A first possibility is to control the probability of existence of false positives. This can be performed by the Bonferroni procedure which simply rejects the $\mathscr{H}_0$-hypotheses for which the p-values are not larger than α/m. The main drawback of this procedure is its lack of power for m large.

An alternative is to control the FDR, which is the mean ratio of the number of false positives by the total number of rejected hypotheses $\mathscr{H}_0$. The Benjamini–Yekutieli procedure offers such a control, but it can be more conservative than the Bonferroni procedure in some cases; see Exercise 10.6.1. The Benjamini–Hochberg procedure is a more powerful procedure, but it offers a control on the FDR only under some distributional hypotheses on the p-values. This last procedure is widely used in the scientific literature, especially in biology and medicine.

10.5.2 References

The FDR concept and the procedure (10.13) have been introduced in the seminal paper of Benjamini and Hochberg [26], while the procedure (10.12) has been introduced and analyzed by Benjamini and Yekutieli [27]. The proofs of Theorem 10.2 and Theorem 10.5 presented in these notes are adapted from Blanchard and Roquain [34].

It turns out that the controls of the FDR obtained in Theorem 10.2 and Theorem 10.5 are of level $m_0\alpha/m$ instead of α. Since m_0 is unknown, we cannot directly correct this level. A lot of work has been done in order to achieve a better level, mainly by trying to estimate m_0. We refer the interested reader to the survey by Roquain [136] for references on this topic and many other issues related to FDR control. We finally refer to Goeman and Solari [84], Dudoit and van der Laan [69], and Dickhaus [64] for a recent review and two detailed books on multiple testing for genomics and life sciences.

10.6 Exercises

10.6.1 FDR versus FWER

The Family Wise Error Rate (FWER) is defined as $\mathsf{FWER} = \mathbb{P}(\mathsf{FP} > 0)$.

1. Prove that $\mathsf{FDR} \leq \mathsf{FWER}$.
2. What control of the FDR offers the Bonferroni procedure?
3. Prove that the number of hypotheses rejected by the Bonferroni procedure is smaller than the number of hypotheses rejected by the Benjamini–Hochberg procedure (10.13).
4. Check that the Benjamini–Yekutieli procedure (10.12) rejects more hypothesis than the Bonferroni procedure only when $\widehat{k} \geq H_m$, with $\widehat{k}$ and H_m defined in Corollary 10.3.

10.6.2 WPRD Property

We prove in this exercise that if the distribution of the p-values $(\widehat{p}_1,\dots,\widehat{p}_m)$ fulfills Property (10.14) for any non-decreasing indicator function $g = \mathbf{1}_\Gamma$, then it fulfills the WPRD property. Below, $g : [0,1]^m \to \mathbb{R}^+$ denotes a bounded measurable non-decreasing function.

1. Prove that for any $i \in \{1,\dots,m\}$ and u, such that $\mathbb{P}(\widehat{p}_i \le u) > 0$, we have

$$\mathbb{E}\left[g(\widehat{p}_1,\dots,\widehat{p}_m)\,|\,\widehat{p}_i \le u\right] = \int_0^{+\infty} \mathbb{P}\left(g(\widehat{p}_1,\dots,\widehat{p}_m) \ge t\,|\,\widehat{p}_i \le u\right) dt\,.$$

2. Check that the indicator function $(\widehat{p}_1,\dots,\widehat{p}_m) \to \mathbf{1}_{g^{-1}([t,+\infty[)}(\widehat{p}_1,\dots,\widehat{p}_m)$ is non-decreasing for all $t \ge 0$, and conclude that $(\widehat{p}_1,\dots,\widehat{p}_m)$ fulfills the WPRD property.

10.6.3 Positively Correlated Normal Test Statistics

Assume that $(\widehat{S}_1,\dots,\widehat{S}_m)$ is distributed according to a $\mathcal{N}(\mu,\Sigma)$-Gaussian distribution, with $\Sigma_{ij} \ge 0$ for all $i,j = 1,\dots,m$. We want to test $\mathcal{H}_{0,i}$: "$\mu_i = 0$" against $\mathcal{H}_{1,i}$: "$\mu_i > 0$." We consider the tests $\mathbf{1}_{\widehat{S}_i \ge s_i}$ for $i = 1,\dots,m$. The associated p-values are $\widehat{p}_i = T_i(\widehat{S}_i)$ with $T_i(s) = \mathbb{P}(\varepsilon_i \ge s)$, where ε_i has a $\mathcal{N}(0,\Sigma_{ii})$-Gaussian distribution.

For any vector $v \in \mathbb{R}^m$ and any $1 \le i \le m$, we define $v_{-i} = (v_1,\dots,v_{i-1},v_{i+1},\dots,v_m)$. For two subsets A,B of $\{1,\dots,p\}$, we denote by $\Sigma_{A,B}$ the matrix $[\Sigma_{ij}]_{i\in A, j\in B}$.

1. Check with Lemma A.4 in Appendix A that the conditional distribution of $\widehat{S}_{-i}$ given $\widehat{S}_i = x$ is the Gaussian distribution with mean $\mu_{-i} + \Sigma_{-i,i}(x-\mu_i)/\Sigma_{i,i}$ and covariance matrix $\Sigma_{-i,-i} - \Sigma_{-i,i}\Sigma_{i,-i}/\Sigma_{i,i}$.
2. For $u > 0$, prove that to any bounded measurable non-decreasing function $g : [0,1]^m \to \mathbb{R}^+$, we can associate a bounded measurable non-increasing function $f : \mathbb{R}^m \to \mathbb{R}^+$, such that

$$\mathbb{E}\left[g(\widehat{p}_1,\dots,\widehat{p}_m)\,|\,\widehat{p}_i \le u\right] = \mathbb{E}\left[f(\widehat{S}_i,\widehat{S}_{-i})\,|\,\widehat{S}_i \ge T_i^{-1}(u)\right],$$

   with $T_i^{-1}(u) = \inf\{s \in \mathbb{R} : T_i(s) \le u\}$.
3. We define $\phi(x) = \mathbb{E}\left[f(x,\mu_{-i} + \Sigma_{-i,i}(x-\mu_i)/\Sigma_{i,i} + \varepsilon_{-i})\right]$, where the random variable ε_{-i} follows a $\mathcal{N}(0,\Sigma_{-i,-i} - \Sigma_{-i,i}\Sigma_{i,-i}/\Sigma_{i,i})$ Gaussian distribution. Prove the equalities

$$\begin{aligned}\mathbb{E}\left[g(\widehat{p}_1,\dots,\widehat{p}_m)\,|\,\widehat{p}_i \le u\right] &= \mathbb{E}\left[\phi(\widehat{S}_i)\,|\,\widehat{S}_i \ge T_i^{-1}(u)\right]\\ &= \int_0^{+\infty} \mathbb{P}\left(\phi(\widehat{S}_i) \ge t\,|\,\widehat{S}_i \ge T_i^{-1}(u)\right) dt\,.\end{aligned}$$

4. Check that ϕ is non-increasing, and prove that the p-values associated with the tests $\mathbf{1}_{\widehat{S}_i \ge s_i}$ fulfill the WPRD property.

Chapter 11

Supervised Classification

The goal of automatic classification is to predict at best the class y of an object x from some observations. A typical example is the spam filter of our mailbox, which predicts (more or less fairly) whether a mail is a spam or not. It is omnipresent in our daily life, by filtering the spams in our mailbox, reading automatically the post code on our postal letters, or recognizing faces in photos that we post on social networks. It is also very important in sciences, e.g., in medicine for early diagnosis of diseases from high-throughput data and in the industry, e.g., for detecting potential customers from their profiles.

In this chapter, we consider the setting of supervised classification: We have a data set recording the label (or class) y of n observed points (or objects) $x \in \mathscr{X}$, and we want to build from these data a function $h(x)$ that predicts the label of any point $x \in \mathscr{X}$. The function h is called a *classifier*. When we want to predict a label y with a function $h(x)$, we fall into the regression setting. Yet, the discrete nature of the labels allows us to strongly weaken the statistical modeling of the pair (x,y) by merely requiring that the observations $(X_i, Y_i)_{i=1,\ldots,n}$ are i.i.d.

We focus in this chapter on binary classification (only two classes). In Section 11.2, we present Vapnik's theory, which can be viewed as an analog of the model selection problem of Chapter 2 in the context of supervised classification. Due to the discrete nature of the labels y, the theory relies heavily on some combinatorial arguments. Similarly as for model selection, the classifier of Section 11.2 enjoys some good statistical properties, but suffers from a prohibitive computational cost. As in Chapter 5, some practical procedures can be built from a convex relaxation of the procedure of Section 11.2. We describe two convex relaxations in Section 11.3 leading to the popular AdaBoost and Support Vector Machine (SVM) algorithms.

11.1 Statistical Modeling

11.1.1 Bayes Classifier

For the sake of simplicity, we restrict ourselves in this chapter to the case where we only have two classes (as for the spam filter) labelled by -1 and $+1$. The problem of automatic classification can then be modeled as follows. Let $\mathscr{X}$ be some measurable space. Each outcome $X \in \mathscr{X}$ has a label $Y \in \{-1, +1\}$. We only observe the points

$X \in \mathscr{X}$, and our aim is to find a (measurable) function $h : \mathscr{X} \to \{-1,+1\}$, called *classifier*, such that $h(X)$ predicts at best the label Y.

Let us first quantify the prediction accuracy of a classifier h. Assume that the couple $(X,Y) \in \mathscr{X} \times \{-1,+1\}$ is sampled from a distribution $\mathbb{P}$. For a classifier $h : \mathscr{X} \to \{-1,+1\}$, the probability of misclassification is

$$L(h) = \mathbb{P}(Y \neq h(X)).$$

In the following, we will quantify the quality of a classifier h by its probability $L(h)$ of misclassification. This measure of quality is natural, yet we point out that some other measures can be more suited in some specific contexts. For example, while we do not care if a spam ends from time to time in our main mailbox, we definitely want to avoid that a personal mail ends in our spam box. In such a case, there is an asymmetry between the two different types of error, and another measure of quality should be considered.

Ideally, we would like to classify the data according to the classifier h_* minimizing the probability $L(h)$ of misclassification. Since $|Y - h(X)| \in \{0,2\}$, we have

$$L(h) = \frac{1}{4}\mathbb{E}\left[(Y - h(X))^2\right] = \frac{1}{4}\mathbb{E}\left[(Y - \mathbb{E}[Y|X])^2\right] + \frac{1}{4}\mathbb{E}\left[(\mathbb{E}[Y|X] - h(X))^2\right].$$

Therefore, $L(h)$ is minimal for the Bayes classifier

$$h_*(X) = \text{sign}(\mathbb{E}[Y|X]) \quad \text{where} \quad \text{sign(x)} = \mathbf{1}_{x>0} - \mathbf{1}_{x\leq 0} \quad \text{for } x \in \mathbb{R}. \tag{11.1}$$

If the distribution $\mathbb{P}$ were known, we would simply use the Bayes classifier h_* in order to have the smallest possible probability of misclassification. Unfortunately, the distribution $\mathbb{P}$ is usually unknown, so we cannot compute the Bayes classifier h_*.

In practice, we only have access to some training data $(X_i,Y_i)_{i=1,\ldots,n}$ i.i.d. with distribution $\mathbb{P}$, and our goal is to build from this training data a classifier $\widehat{h} : \mathscr{X} \to \{-1,+1\}$, such that $L(\widehat{h}) - L(h_*)$ is as small as possible.

11.1.2 Parametric Modeling

A first approach is to assume that the distribution $\mathbb{P}$ belongs to a parametric family of distributions. Conditioning on Y, we have

$$\mathbb{P}(X \in dx, Y = k) = \pi_k \mathbb{P}(X \in dx|Y = k), \quad \text{with } \pi_k = \mathbb{P}(Y = k) \text{ for } k \in \{-1,1\}.$$

So in order to parametrize the distribution $\mathbb{P}$, we only need to parametrize the two conditional distributions $\mathbb{P}(X \in dx|Y = 1)$ and $\mathbb{P}(X \in dx|Y = -1)$.

A popular model when $\mathscr{X} = \mathbb{R}^d$ is to assume that the conditional distributions $\mathbb{P}(X \in dx|Y = k)$ are Gaussian with mean μ_k and covariance Σ_k. When we have $\Sigma_1 = \Sigma_{-1} = \Sigma$, the Bayes classifier is given by

$$h_*(x) = \text{sign}\left(\left\langle \Sigma^{-1}(\mu_1 - \mu_{-1}), x - \frac{\mu_1 + \mu_{-1}}{2}\right\rangle + \log(\pi_1/\pi_{-1})\right);$$

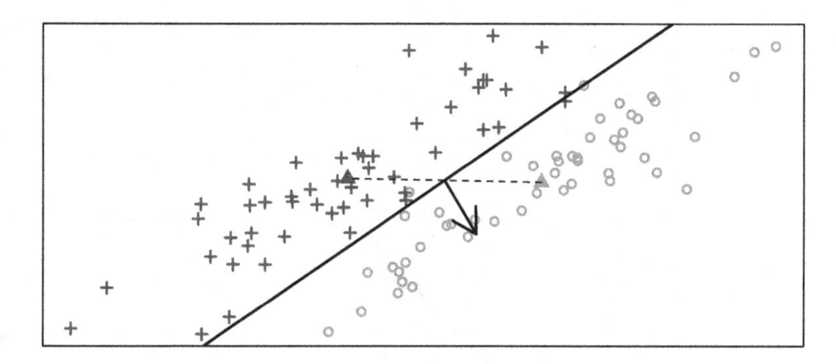

Figure 11.1 *Linear discriminant analysis. The triangles represent the means* μ_1 *and* μ_{-1}*, the arrow represents the vector* $\widehat{\Sigma}^{-1}(\widehat{\mu}_1 - \widehat{\mu}_{-1})$*, and the black line represents the frontier between* $\{\widehat{h}_{LDA} = 1\}$ *and* $\{\widehat{h}_{LDA} = -1\}$.

see Exercise 11.5.1. A point x is then classified according to its position relative to the affine hyperplane orthogonal to $\Sigma^{-1}(\mu_1 - \mu_{-1})$ and with offset $\log(\pi_1/\pi_{-1}) - (\mu_1 - \mu_{-1})^T\Sigma^{-1}(\mu_1 + \mu_{-1})/2$. In practice, we can classify the data with

$$\widehat{h}_{\text{LDA}}(x) = \text{sign}\left(\left\langle \widehat{\Sigma}^{-1}(\widehat{\mu}_1 - \widehat{\mu}_{-1}), x - \frac{\widehat{\mu}_1 + \widehat{\mu}_{-1}}{2} \right\rangle + \log(\widehat{\pi}_1/\widehat{\pi}_{-1})\right), \tag{11.2}$$

where $\widehat{\pi}_k$ is the empirical proportion of the label k, the mean $\widehat{\mu}_k$ is the empirical mean of the X_i, such that $Y_i = k$, and $\widehat{\Sigma}$ is the empirical covariance of the data. This leads to the *Linear Discriminant Analysis* (LDA). We refer to Exercise 11.5.1 for more details, and to Figure 11.1 for an illustration.

The parametric modeling is powerful when the model is correct, but in many cases, we do not know the distribution of the points X given the labels Y, and an incorrect modeling can lead to poor results; see Figure 11.2 for an illustration where the LDA fails.

11.1.3 Semi-Parametric Modeling

Since the Bayes classifier (11.1) only depends on the conditional distribution of Y given X, we can avoid to model the distribution of X as above. From the formula

$$\mathbb{E}[Y|X] = \mathbb{P}(Y = 1|X) - \mathbb{P}(Y = -1|X) = 2\mathbb{P}(Y = 1|X) - 1,$$

we obtain that the Bayes classifier $h_*(x)$ is given by

$$h_*(x) = \text{sign}\big(\mathbb{P}(Y = 1|X = x) - 1/2\big), \quad \text{for } x \in \mathcal{X}.$$

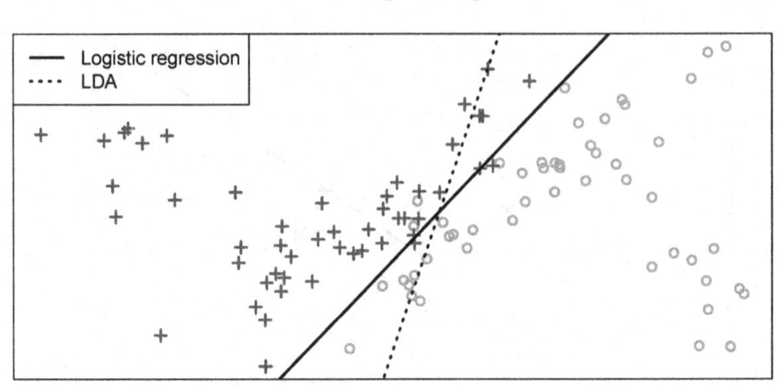

Figure 11.2 *The black line represents the separating hyperplane of the logistic regression. The dashed line represents the separating hyperplane of the LDA.*

A classical approach is to assume a parametric model for the conditional probability $\mathbb{P}(Y=1|X=x)$. The most popular model in $\mathbb{R}^d$ is probably the *logistic model*, where

$$\mathbb{P}(Y=1|X=x) = \frac{\exp(\alpha + \langle \beta, x\rangle)}{1+\exp(\alpha + \langle \beta, x\rangle)} \quad \text{for all } x \in \mathscr{X}, \tag{11.3}$$

with $\alpha \in \mathbb{R}$ and $\beta \in \mathbb{R}^d$. In this case, we have $\mathbb{P}(Y=1|X=x) > 1/2$ if and only if $\exp(\alpha + \langle \beta, x\rangle) > 1$, so the Bayes classifier h_* has the simple form

$$h_*(x) = \text{sign}\big(\alpha + \langle \beta, x\rangle\big) \quad \text{for all } x \in \mathscr{X}.$$

We observe that the frontier between $\{h_* = 1\}$ and $\{h_* = -1\}$ is again an affine hyperplane, with orthogonal direction β and offset α.

We can estimate the parameters (α, β) by maximizing the conditional likelihood of Y given X

$$(\widehat{\alpha}, \widehat{\beta}) \in \underset{(\alpha,\beta)\in\mathbb{R}^{d+1}}{\text{argmax}} \left\{ \prod_{i:Y_i=1} \frac{\exp(\alpha + \langle \beta, X_i\rangle)}{1+\exp(\alpha + \langle \beta, X_i\rangle)} \prod_{i:Y_i=-1} \frac{1}{1+\exp(\alpha + \langle \beta, X_i\rangle)} \right\},$$

and compute the classifier $\widehat{h}_{\text{logistic}}(x) = \text{sign}\big(\widehat{\alpha} + \langle \widehat{\beta}, x\rangle\big)$ for all $x \in \mathscr{X}$. We emphasize that even if the Bayes classifiers have the same shape in the LDA and in the logistic modeling, the two procedures do not lead to the same classifier in general. In particular, if the conditional distribution of the X given Y is far from Gaussian, the LDA can produce some very poor results, while the logistic model will work as long as the modeling (11.3) remains valid; see Figure 11.2 for such a case.

11.1.4 Non-Parametric Modeling

We may wish to weaken further our hypotheses on the distribution of (X,Y) and adopt a non-parametric point of view. Instead of assuming that the distribution of (X,Y) belongs to some parametric or semi parametric set of distributions (as in Sections 11.1.2 and 11.1.3), we will rather assume that h_* is "smooth" in some sense (suited to the classification setting).

A classical approach in non-parametric estimation is to replace the ideal risk minimization by some constrained empirical risk minimization. In our case, we cannot minimize $h \to L(h)$, since $L(h)$ is unknown, but we can minimize instead the empirical probability of misclassification

$$\widehat{L}_n(h) := \frac{1}{n}\sum_{i=1}^{n} \mathbf{1}_{Y_i \neq h(X_i)} = \widehat{\mathbb{P}}_n(Y \neq h(X)), \tag{11.4}$$

where $\widehat{\mathbb{P}}_n = \frac{1}{n}\sum_{i=1}^n \delta_{(X_i,Y_i)}$. There is in general no unique minimizer of $\widehat{L}_n$, and even if an unconstrained minimizer of the empirical risk perfectly classifies the labels in the data set, it produces in general a very poor prediction for a new point. We must then restrict the minimization of $h \to \widehat{L}_n(h)$ to a set $\mathscr{H}$ of classifiers with limited "flexibility"

$$\widehat{h}_{\mathscr{H}} \in \operatorname*{argmin}_{h\in\mathscr{H}} \widehat{L}_n(h), \quad \text{with } \widehat{L}_n \text{ defined by (11.4).} \tag{11.5}$$

As we will see in the next section, the appropriate notion of "flexibility" in this context corresponds to some combinatorial complexity of the set $\mathscr{H}$ measuring the classification flexibility offered by the classifiers in $\mathscr{H}$. We stress that the set $\mathscr{H}$ of classifiers, usually called *dictionary*, plays the same role as the model S in the model selection setting of Chapter 2. In particular, we face the same issues as in Chapter 2: How does $\widehat{h}_{\mathscr{H}}$ behave compared to h_*? Which dictionary $\mathscr{H}$ should be chosen? We investigate these two issues in the next section.

11.2 Empirical Risk Minimization

We analyze in this section the classifier $\widehat{h}_{\mathscr{H}}$ defined by (11.5), and we explain how we can handle the problem of the choice of $\mathscr{H}$. Decomposing the difference between the misclassification probabilities $L(\widehat{h}_{\mathscr{H}})$ and $L(h_*)$, we find

$$0 \leq L(\widehat{h}_{\mathscr{H}}) - L(h_*) = \underbrace{\min_{h\in\mathscr{H}} L(h) - L(h_*)}_{\text{approximation error}} + \underbrace{L(\widehat{h}_{\mathscr{H}}) - \min_{h\in\mathscr{H}} L(h)}_{\text{stochastic error}}.$$

The first term is a bias term that measures the ability of the classifiers $h \in \mathscr{H}$ to produce a classification as good as the Bayes classifier h_*. This approximation error is purely deterministic, and enlarging the dictionary $\mathscr{H}$ can only reduce it. The second term measures the error made by minimizing over $h \in \mathscr{H}$ the empirical misclassification probability $\widehat{L}_n(h)$ instead of the true misclassification probability $L(h)$. This term is stochastic, and it tends to increase when $\mathscr{H}$

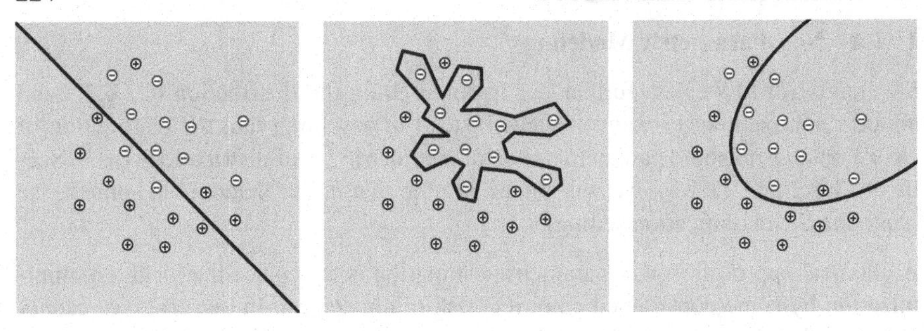

Figure 11.3 *Examples of classification produced by different dictionaries. Left: dictionary of linear classifiers $\mathcal{H}_{\text{lin}}$. Center: dictionary of polygon classifiers $\mathcal{H}_{\text{poly}}$. Right: dictionary of quadratic forms.*

increases. This phenomenon is illustrated in Figure 11.3. In this illustration, with $\mathcal{X} = \mathbb{R}^2$, the classifiers of the dictionary $\mathcal{H}_{\text{lin}} = \{h(x) = \text{sign}(\langle w,x\rangle) : \|w\| = 1\}$ are not flexible enough and they produce a poor classification. In this case, the approximation error is large. On the other hand, the classifiers of the dictionary $\mathcal{H}_{\text{poly}} = \{h(x) = 2\mathbf{1}_A(x) - 1 : A \text{ polygon in } \mathcal{X}\}$ are very flexible and can always classify exactly the data $(X_i, Y_i)_{i=1,\ldots,n}$ when the X_i are distinct. The empirical error $\widehat{L}_n(\widehat{h}_{\mathcal{H}_{\text{poly}}})$ is then 0, but $\widehat{h}_{\mathcal{H}_{\text{poly}}}$ tends to produce a poor classification of new data (X,Y), and the stochastic term $L(\widehat{h}_{\mathcal{H}}) - \min_{h\in\mathcal{H}} L(h)$ is large. The last example, based on a less flexible set of quadratic classifiers, produces a better result, even though its empirical error is larger than the one of $\widehat{h}_{\mathcal{H}_{\text{poly}}}$.

To choose a good dictionary $\mathcal{H}$, we shall then find a good balance between the approximation properties of $\mathcal{H}$ and its size. The first step toward a procedure for selecting the dictionary $\mathcal{H}$ is to assess the misclassification probability of the empirical risk minimizer $\widehat{h}_{\mathcal{H}}$.

11.2.1 Misclassification Probability of the Empirical Risk Minimizer

As mentioned above, increasing the size of $\mathcal{H}$ tends to increase the stochastic error $L(\widehat{h}_{\mathcal{H}}) - \min_{h\in\mathcal{H}} L(h)$. Actually, it is not really the size of the dictionary that matters, but rather its flexibility in terms of classification. For example, we cannot classify correctly the three labelled points $\{((0,1),+1),((1,1),-1),((1,0),+1)\}$ with a classifier in $\mathcal{H}_{\text{lin}}$. Conversely, for any set of labelled points $(x_i,y_i)_{i=1,\ldots,n}$ with distinct $x_1,\ldots,x_n$, there exists $h \in \mathcal{H}_{\text{poly}}$ such that $h(x_i) = y_i$.

In order to capture this classification flexibility, we introduce the shattering coefficient

$$\mathbb{S}_n(\mathcal{H}) = \max_{(x_1,\ldots,x_n)\in\mathcal{X}^n} \text{card}\{(h(x_1),\ldots,h(x_n)) : h \in \mathcal{H}\}, \tag{11.6}$$

which gives the maximal number of different labelling of n points that the classifiers in $\mathcal{H}$ can produce. For example, since n distinct points can be arbitrarily labelled with classifiers in $\mathscr{H}_{\text{poly}}$, we have $\mathbb{S}_n(\mathscr{H}_{\text{poly}}) = 2^n$. On the contrary, the number of possible labelling of n points with classifiers in $\mathscr{H}_{\text{lin}}$ is more limited. Actually, Proposition 11.6 in Section 11.2.2 ensures that $\mathbb{S}_n(\mathscr{H}_{\text{lin}}) \leq (n+1)^2$ in dimension 2. The next theorem provides an upper bound on the stochastic error and a confidence interval for the misclassification probability $L(\widehat{h}_{\mathscr{H}})$ in terms of the shattering coefficient (11.6). Again, the proof relies on a concentration inequality.

Theorem 11.1 Control of the stochastic error

For any $t > 0$, with probability at least $1 - e^{-t}$, we have

$$L(\widehat{h}_{\mathscr{H}}) - \min_{h \in \mathscr{H}} L(h) \leq 4\sqrt{\frac{2\log(2\mathbb{S}_{\mathscr{H}}(n))}{n}} + \sqrt{\frac{2t}{n}} \tag{11.7}$$

and

$$\left|L(\widehat{h}_{\mathscr{H}}) - \widehat{L}_n(\widehat{h}_{\mathscr{H}})\right| \leq 2\sqrt{\frac{2\log(2\mathbb{S}_{\mathscr{H}}(n))}{n}} + \sqrt{\frac{t}{2n}}. \tag{11.8}$$

Proof. We split the proof of Theorem 11.1 into three lemmas. The first lemma shows that the left-hand terms in (11.7) and (11.8) can be upper-bounded in terms of the maximum difference over $\mathscr{H}$ between the empirical misclassification probability and the true misclassification probability

$$\widehat{\Delta}_n(\mathscr{H}) = \sup_{h \in \mathscr{H}} \left|\widehat{L}_n(h) - L(h)\right|. \tag{11.9}$$

The second lemma ensures that the difference between $\widehat{\Delta}_n(\mathscr{H})$ and its expectation is smaller than $\sqrt{\xi/(2n)}$, with ξ a standard exponential random variable. Finally, the third lemma upper-bounds the expectation of $\widehat{\Delta}_n(\mathscr{H})$ by $2\sqrt{2\log(2\mathbb{S}_{\mathscr{H}}(n))/n}$.

Lemma 11.2

We have the upper bounds

$$L(\widehat{h}_{\mathscr{H}}) - \min_{h \in \mathscr{H}} L(h) \leq 2\widehat{\Delta}_n(\mathscr{H}) \quad \text{and} \quad \left|L(\widehat{h}_{\mathscr{H}}) - \widehat{L}_n(\widehat{h}_{\mathscr{H}})\right| \leq \widehat{\Delta}_n(\mathscr{H}).$$

Proof of Lemma 5.12. For any $h \in \mathscr{H}$, we have $\widehat{L}_n(\widehat{h}_{\mathscr{H}}) \leq \widehat{L}_n(h)$, and therefore

$$\begin{aligned} L(\widehat{h}_{\mathscr{H}}) - L(h) &= L(\widehat{h}_{\mathscr{H}}) - \widehat{L}_n(\widehat{h}_{\mathscr{H}}) + \widehat{L}_n(\widehat{h}_{\mathscr{H}}) - L(h) \\ &\leq L(\widehat{h}_{\mathscr{H}}) - \widehat{L}_n(\widehat{h}_{\mathscr{H}}) + \widehat{L}_n(h) - L(h) \\ &\leq 2\widehat{\Delta}_n(\mathscr{H}). \end{aligned}$$

Since this inequality is true for any $h \in \mathscr{H}$, the first bound of Lemma 11.2 follows. The second bound is obvious. □

In order to prove Theorem 11.1, it remains to prove that

$$\widehat{\Delta}_n(\mathscr{H}) \leq 2\sqrt{\frac{2\log(2\mathbb{S}_{\mathscr{H}}(n))}{n}} + \sqrt{\frac{t}{2n}},$$

with probability at least $1-e^{-t}$. The first step is a concentration inequality for $\widehat{\Delta}_n(\mathscr{H})$.

Lemma 11.3

With probability at least $1-e^{-t}$, *we have*

$$\widehat{\Delta}_n(\mathscr{H}) \leq \mathbb{E}\left[\widehat{\Delta}_n(\mathscr{H})\right] + \sqrt{\frac{t}{2n}}.$$

Proof of Lemma 11.3. We have $\widehat{\Delta}_n(\mathscr{H}) = F((X_1,Y_1),\ldots,(X_n,Y_n))$, with

$$\begin{aligned} F \ : \quad & (\mathscr{X} \times \{-1,+1\})^n \quad \to \quad \mathbb{R} \\ & ((x_1,y_1),\ldots,(x_n,y_n)) \mapsto \frac{1}{n} \sup_{h\in\mathscr{H}} \left| \sum_{i=1}^{n} \mathbf{1}_{y_i \neq h(x_i)} - L(h) \right|. \end{aligned}$$

For any $(x_1,y_1),\ldots,(x_n,y_n),(x_i',y_i') \in \mathscr{X} \times \{-1,+1\}$, we have

$$\left|F((x_1,y_1),\ldots,(x_i',y_i'),\ldots,(x_n,y_n)) - F((x_1,y_1),\ldots,(x_i,y_i),\ldots,(x_n,y_n))\right| \leq \frac{1}{n},$$

so according to McDiarmid concentration inequality (Theorem B.5, page 299, in Appendix B), with probability at least $1-e^{-2ns^2}$, we have $\widehat{\Delta}_n(\mathscr{H}) \leq \mathbb{E}\left[\widehat{\Delta}_n(\mathscr{H})\right] + s$. Lemma 11.3 follows by setting $s = \sqrt{t/(2n)}$. □

It remains to bound the expectation of $\widehat{\Delta}_n(\mathscr{H})$ in terms of $\mathbb{S}_{\mathscr{H}}(n)$.

Lemma 11.4

For any dictionary $\mathscr{H}$, *we have the upper bound*

$$\mathbb{E}\left[\widehat{\Delta}_n(\mathscr{H})\right] \leq 2\sqrt{\frac{2\log(2\mathbb{S}_{\mathscr{H}}(n))}{n}}.$$

Proof of Lemma 11.4. The proof of Lemma 11.4 is divided into two parts. Part **(i)** is based on a classical and elegant symmetrization argument. Part **(ii)** is a classical application of Jensen inequality.

(i) The first step is to upper-bound $\mathbb{E}\left[\widehat{\Delta}_n(\mathscr{H})\right]$ with the symmetrization lemma

(Theorem B.10, page 305, in Appendix B). Let $\sigma_1,\dots,\sigma_n$ be n i.i.d. random variables uniformly distributed on $\{-1,+1\}$ and independent of $(X_i,Y_i)_{i=1,\dots,n}$. According to (B.7), with $Z_i=(X_i,Y_i)$ for $i=1,\dots,n$ and $f(Z_i)=\mathbf{1}_{Y_i\neq h(X_i)}$, we obtain

$$\mathbb{E}\left[\widehat{\Delta}_n(\mathscr{H})\right]\leq 2\mathbb{E}\mathbb{E}_\sigma\left[\sup_{h\in\mathscr{H}}\left|\frac{1}{n}\sum_{i=1}^n\sigma_i\mathbf{1}_{Y_i\neq h(X_i)}\right|\right],$$

where $\mathbb{P}_\sigma$ refers to the expectation with respect to $\sigma_1,\dots,\sigma_n$. This term can be simply upper-bounded by

$$\mathbb{E}\left[\widehat{\Delta}_n(\mathscr{H})\right]\leq 2\max_{y\in\{-1,+1\}^n}\max_{x\in\mathscr{X}^n}\mathbb{E}_\sigma\left[\sup_{h\in\mathscr{H}}\left|\frac{1}{n}\sum_{i=1}^n\sigma_i\mathbf{1}_{y_i\neq h(x_i)}\right|\right].$$

At this point, we notice that we have replaced an expectation with respect to the unknown probability distribution $\mathbb{P}$ by an expectation with respect to the known probability distribution $\mathbb{P}_\sigma$.

(ii) For any $(x,y)\in\mathscr{X}^n\times\{-1,+1\}^n$, let us define the set

$$\mathscr{V}_{\mathscr{H}}(x,y)=\left\{\left(\mathbf{1}_{y_1\neq h(x_1)},\dots,\mathbf{1}_{y_n\neq h(x_n)}\right):\ h\in\mathscr{H}\right\}.$$

The last upper bound on $\mathbb{E}\left[\widehat{\Delta}_n(\mathscr{H})\right]$ can be written as

$$\mathbb{E}\left[\widehat{\Delta}_n(\mathscr{H})\right]\leq\frac{2}{n}\times\max_{y\in\{-1,+1\}^n}\max_{x\in\mathscr{X}^n}\mathbb{E}_\sigma\left[\sup_{v\in\mathscr{V}_{\mathscr{H}}(x,y)}|\langle\sigma,v\rangle|\right],$$

where $\langle x,y\rangle$ is the canonical scalar product on $\mathbb{R}^n$. We notice that for any $y\in\{-1,1\}^n$ there is a bijection between $\mathscr{V}_{\mathscr{H}}(x,y)$ and the set $\{(h(x_1),\dots,h(x_n)):h\in\mathscr{H}\}$. As a consequence, we have the upper bound

$$\max_{y\in\{-1,+1\}^n}\max_{x\in\mathscr{X}^n}\mathrm{card}(\mathscr{V}_{\mathscr{H}}(x,y))\leq\mathbb{S}_n(\mathscr{H}).$$

In view of the last two inequalities, in order to conclude the proof of Lemma 11.4, it simply remains to prove the following result.

Lemma 11.5

For $\sigma_1,\dots,\sigma_n$ i.i.d. with $\mathbb{P}_\sigma(\sigma_i=1)=\mathbb{P}_\sigma(\sigma_i=-1)=1/2$, we have

$$\mathbb{E}_\sigma\left[\sup_{v\in\mathscr{V}}|\langle\sigma,v\rangle|\right]\leq\sqrt{2n\log(2\mathrm{card}(\mathscr{V}))},\quad\textit{for any finite }\mathscr{V}\subset[-1,1]^n.\quad(11.10)$$

Proof of Lemma 11.5. Writing $\mathscr{V}^\#=\mathscr{V}\cup-\mathscr{V}$, Jensen inequality ensures that for any $s > 0$

$$\mathbb{E}_\sigma\left[\sup_{v\in\mathcal{V}}|\langle\sigma,v\rangle|\right] = \mathbb{E}_\sigma\left[\sup_{v\in\mathcal{V}^\#}\langle\sigma,v\rangle\right] \leq \frac{1}{s}\log\mathbb{E}_\sigma\left[\sup_{v\in\mathcal{V}^\#}e^{s\langle\sigma,v\rangle}\right]$$
$$\leq \frac{1}{s}\log\left(\sum_{v\in\mathcal{V}^\#}\mathbb{E}_\sigma\left[e^{s\langle\sigma,v\rangle}\right]\right). \tag{11.11}$$

Combining the facts that the σ_i are independent, $(e^x+e^{-x}) \leq 2e^{x^2/2}$ for all $x\in\mathbb{R}$ and $v_i^2 \leq 1$ for all $v\in\mathcal{V}^\#$, we have

$$\mathbb{E}_\sigma\left[e^{s\langle\sigma,v\rangle}\right] = \prod_{i=1}^n\mathbb{E}_\sigma\left[e^{sv_i\sigma_i}\right] = \prod_{i=1}^n\frac{1}{2}(e^{sv_i}+e^{-sv_i}) \leq \prod_{i=1}^n e^{s^2v_i^2/2} \leq e^{ns^2/2}.$$

Plugging this inequality in (11.11), we obtain

$$\mathbb{E}_\sigma\left[\sup_{v\in\mathcal{V}}|\langle\sigma,v\rangle|\right] \leq \frac{\log(\mathrm{card}(\mathcal{V}^\#))}{s}+\frac{ns}{2} \qquad \text{for any } s>0.$$

The right-hand side is minimal for $s = \sqrt{2\log(\mathrm{card}(\mathcal{V}^\#))/n}$, which gives the upper bound

$$\mathbb{E}_\sigma\left[\sup_{v\in\mathcal{V}}|\langle\sigma,v\rangle|\right] \leq \sqrt{2n\log(\mathrm{card}(\mathcal{V}^\#))}\,.$$

We finally obtain (11.10) by noticing that $\mathrm{card}(\mathcal{V}^\#) \leq 2\,\mathrm{card}(\mathcal{V})$. □

The proof of Lemma 11.4 is complete, and Bounds (11.7) and (11.8) are obtained by combining Lemma 11.2, Lemma 11.3, and Lemma 11.4. □

Theorem 11.1 provides a control of the misclassification probability in terms of the shattering coefficient (11.6). The shattering coefficient offers a good notion of complexity for a set $\mathcal{H}$ of classifiers, but its computation can be tricky in practice. In the next section, we prove that a nice combinatorial property of the shattering coefficients provides a simple upper bound on $\mathbb{S}_n(\mathcal{H})$, depending on $\mathcal{H}$ only through a single quantity, the Vapnik–Chervonenkis dimension of $\mathcal{H}$.

11.2.2 Vapnik–Chervonenkis Dimension

By convention, we set $\mathbb{S}_0(\mathcal{H}) = 1$. From the definition (11.6) of $\mathbb{S}_n(\mathcal{H})$, we have $\mathbb{S}_n(\mathcal{H}) \leq 2^n$ for all $n\in\mathbb{N}$. We call Vapnik–Chervonenkis (VC) dimension of $\mathcal{H}$ the integer $d_{\mathcal{H}}$ defined by

$$d_{\mathcal{H}} = \sup\left\{d\in\mathbb{N} : \mathbb{S}_d(\mathcal{H}) = 2^d\right\} \in \mathbb{N}\cup\{+\infty\}. \tag{11.12}$$

It corresponds to the maximum number of points in $\mathcal{X}$ that can be arbitrarily classified by the classifiers in $\mathcal{H}$. The next proposition gives an upper bound on the shattering coefficient $\mathbb{S}_n(\mathcal{H})$ in terms of the VC dimension $d_{\mathcal{H}}$.

Proposition 11.6 Sauer's lemma

Let $\mathscr{H}$ be a set of classifiers with finite VC dimension $d_{\mathscr{H}}$. For any $n \in \mathbb{N}$, we have

$$\mathbb{S}_n(\mathscr{H}) \leq \sum_{i=0}^{d_{\mathscr{H}}} C_n^i \leq (n+1)^{d_{\mathscr{H}}} \quad \text{with} \quad C_n^i = \begin{cases} \frac{n!}{i!(n-i)!} & \text{for } n \geq i \\ 0 & \text{for } n < i. \end{cases}$$

Proof. We first prove by induction on k the inequality

$$\mathbb{S}_k(\mathscr{H}) \leq \sum_{i=0}^{d_{\mathscr{H}}} C_k^i \tag{11.13}$$

for any $\mathscr{H}$ with finite VC dimension $d_{\mathscr{H}}$.

Let us consider the case $k = 1$. If $d_{\mathscr{H}} = 0$, then $\mathbb{S}_1(\mathscr{H}) < 2$ and so $\mathbb{S}_1(\mathscr{H}) = 1 = C_1^0$. If $d_{\mathscr{H}} \geq 1$, we have $\mathbb{S}_1(\mathscr{H}) = 2$, which is also equal to $C_1^0 + C_1^1$.

Assume now that (11.13) is true for all $k \leq n-1$. Let us consider $\mathscr{H}$, with finite VC dimension $d_{\mathscr{H}}$. When $d_{\mathscr{H}} = 0$, no set of point can be shattered, so all points can only be labelled in one way. Therefore, $\mathbb{S}_k(\mathscr{H}) = 1$ and (11.13) is true for all k. We assume now that $d_{\mathscr{H}} \geq 1$. Let $x_1, \ldots, x_n$ be n points in $\mathscr{X}$ and define

$$\mathscr{H}(x_1, \ldots, x_n) = \{(h(x_1), \ldots, h(x_n)) : h \in \mathscr{H}\}.$$

We want to prove that

$$\text{card}(\mathscr{H}(x_1, \ldots, x_n)) \leq \sum_{i=0}^{d_{\mathscr{H}}} C_k^i. \tag{11.14}$$

The set $\mathscr{H}(x_1, \ldots, x_n)$ depends only on the values of $h \in \mathscr{H}$ on $\{x_1, \ldots, x_n\}$, so we can replace $\mathscr{H}$ by $\mathscr{F} = \{h|_{\{x_1,\ldots,x_n\}} : h \in \mathscr{H}\}$ in the definition of $\mathscr{H}(x_1, \ldots, x_n)$. Since $d_{\mathscr{F}}$ is not larger than $d_{\mathscr{H}}$, it is enough to prove (11.14) for $\mathscr{F}$. Therefore, we assume (with no loss of generality) that $\mathscr{X} = \{x_1, \ldots, x_n\}$ and $\mathscr{H} = \mathscr{F}$. Let us consider the set

$$\mathscr{H}' = \{h \in \mathscr{H} : h(x_n) = 1 \text{ and } h' = h - 2 \times \mathbf{1}_{\{x_n\}} \in \mathscr{H}\}.$$

Since $\mathscr{H}(x_1, \ldots, x_n) = \mathscr{H}'(x_1, \ldots, x_n) \cup (\mathscr{H} \setminus \mathscr{H}')(x_1, \ldots, x_n)$, we have

$$\text{card}(\mathscr{H}(x_1, \ldots, x_n)) \leq \text{card}(\mathscr{H}'(x_1, \ldots, x_n)) + \text{card}((\mathscr{H} \setminus \mathscr{H}')(x_1, \ldots, x_n)). \tag{11.15}$$

Let us bound apart the cardinality of $\mathscr{H}'(x_1, \ldots, x_n)$ and the cardinality of $(\mathscr{H} \setminus \mathscr{H}')(x_1, \ldots, x_n)$.

1. We note that $\text{card}(\mathscr{H}'(x_1, \ldots, x_n)) = \text{card}(\mathscr{H}'(x_1, \ldots, x_{n-1}))$ since $h(x_n) = 1$ for all $h \in \mathscr{H}'$. Let us check that the VC dimension $d_{\mathscr{H}'}$ of $\mathscr{H}'$ is at most $d_{\mathscr{H}} - 1$. Actually, if d points $x_{i_1}, \ldots, x_{i_d}$ of $\mathscr{X} = \{x_1, \ldots, x_n\}$ are shattered by $\mathscr{H}'$, then $x_n \notin \{x_{i_1}, \ldots, x_{i_d}\}$ since $h(x_n) = 1$ for all $h \in \mathscr{H}'$. Furthermore, the set

$\{x_{i_1},\dots,x_{i_d},x_n\}$ is shattered by $\mathscr{H}'\cup\{h'=h-2\times\mathbf{1}_{\{x_n\}}:h\in\mathscr{H}\}$, which is included in $\mathscr{H}$ according to the definition of $\mathscr{H}'$. So, $d+1\leq d_{\mathscr{H}}$, which implies $d_{\mathscr{H}'}\leq d_{\mathscr{H}}-1$. Applying (11.13) with $k=n-1$, we obtain

$$\operatorname{card}\left(\mathscr{H}'(x_1,\dots,x_n)\right)=\operatorname{card}\left(\mathscr{H}'(x_1,\dots,x_{n-1})\right)\leq\sum_{i=0}^{d_{\mathscr{H}}-1}C_{n-1}^i. \tag{11.16}$$

2. When $h,h'\in\mathscr{H}\setminus\mathscr{H}'$ fulfill $h(x_i)=h'(x_i)$ for $i=1,\dots,n-1$, they also fulfill $h(x_n)=h'(x_n)$; otherwise, either h or h' would belong to $\mathscr{H}'$. Therefore, we have as above $\operatorname{card}((\mathscr{H}\setminus\mathscr{H}')(x_1,\dots,x_n))=\operatorname{card}((\mathscr{H}\setminus\mathscr{H}')(x_1,\dots,x_{n-1}))$. Furthermore, $d_{\mathscr{H}\setminus\mathscr{H}'}$ is not larger than $d_{\mathscr{H}}$, since $\mathscr{H}\setminus\mathscr{H}'\subset\mathscr{H}$, so Equation (11.13) with $k=n-1$ gives

$$\operatorname{card}\left((\mathscr{H}\setminus\mathscr{H}')(x_1,\dots,x_n)\right)=\operatorname{card}\left((\mathscr{H}\setminus\mathscr{H}')(x_1,\dots,x_{n-1})\right)\leq\sum_{i=0}^{d_{\mathscr{H}}}C_{n-1}^i. \tag{11.17}$$

Combining (11.15), (11.16), and (11.17), we obtain

$$\operatorname{card}\left(\mathscr{H}(x_1,\dots,x_n)\right)\leq\sum_{i=1}^{d_{\mathscr{H}}}C_{n-1}^{i-1}+\sum_{i=0}^{d_{\mathscr{H}}}C_{n-1}^i=\sum_{i=0}^{d_{\mathscr{H}}}C_n^i,$$

since $C_{n-1}^i+C_{n-1}^{i-1}=C_n^i$ for $i\geq 1$. As a consequence, (11.13) is true for $k=n$, and the induction is complete.

The second upper bound of the proposition is obtained by

$$\sum_{i=0}^{d}C_n^i\leq\sum_{i=0}^{d}\frac{n^i}{i!}\leq\sum_{i=0}^{d}C_d^i n^i=(1+n)^d.$$

The proof of Proposition 11.6 is complete. □

Remark. The reader may check (by induction) that for $d\leq n$, we also have

$$\sum_1^d C_n^i\leq\left(\frac{en}{d}\right)^d,$$

which improves the bound $\sum_1^d C_n^i\leq(n+1)^d$ when $3\leq d\leq n$.

Let us give some examples of VC dimension for some simple dictionaries on $\mathscr{X}=\mathbb{R}^d$. The proofs are left as exercises.

Example 1: Linear classifiers.
The VC dimension of the set $\mathscr{H}=\{h(x)=\operatorname{sign}(\langle w,x\rangle):\|w\|=1\}$ of linear classifiers is d (see Exercise 11.5.2).

Example 2: Affine classifiers.
The VC dimension of the set $\mathscr{H}=\{h(x)=\operatorname{sign}(\langle w,x\rangle+b):\|w\|=1,\ b\in\mathbb{R}\}$ of affine classifiers is $d+1$ (see Exercise 11.5.2).

Example 3: Hyper-rectangle classifiers.
The VC dimension of the set $\mathscr{H} = \{h(x) = 2\mathbf{1}_A(x) - 1 : A \text{ hyper-rectangle of } \mathbb{R}^d\}$ of hyper-rectangle classifiers is $2d$.

Example 4: Convex polygon classifiers.
The VC dimension of the set $\mathscr{H} = \{h(x) = 2\mathbf{1}_A(x) - 1 : A \text{ convex polygon of } \mathbb{R}^d\}$ of convex polygon classifiers is $+\infty$ (consider n points on the unit sphere: For any subset of these points, you can choose their convex hull as convex polygon).

Finally, we can state the following corollary of Theorem 11.1.

Corollary 11.7 Control of the stochastic error for Vapnick dictionaries

For $\mathscr{H}$ with VC dimension $1 \leq d_{\mathscr{H}} < +\infty$, for any $t > 0$, we have the upper bound

$$L(\widehat{h}_{\mathscr{H}}) \leq \min_{h \in \mathscr{H}} L(h) + 4\sqrt{\frac{2d_{\mathscr{H}} \log(2n+2))}{n}} + \sqrt{\frac{2t}{n}}$$

with probability at least $1 - e^{-t}$.

Let us now investigate the problem of the choice of the dictionary $\mathscr{H}$.

11.2.3 Dictionary Selection

Let us consider a collection $\{\mathscr{H}_1, \ldots, \mathscr{H}_M\}$ of dictionaries. Similarly to Chapter 2, we would like to select among this collection, the dictionary $\mathscr{H}_o$ with the smallest misclassification probability $L(\widehat{h}_{\mathscr{H}_o})$. The so-called *oracle* dictionary $\mathscr{H}_o$ depends on the unknown distribution $\mathbb{P}$, so it is not accessible to the statistician. In the following, we will build on Theorem 11.1 in order to design a data-driven procedure for selecting a dictionary $\mathscr{H}_{\widehat{m}}$ among the collection $\{\mathscr{H}_1, \ldots, \mathscr{H}_M\}$, with performances similar to those of $\mathscr{H}_o$.

The oracle dictionary $\mathscr{H}_o$ is obtained by minimizing the misclassification probability $L(\widehat{h}_{\mathscr{H}})$ over $\mathscr{H} \in \{\mathscr{H}_1, \ldots, \mathscr{H}_M\}$. A first idea is to select $\mathscr{H}_{\widehat{m}}$ by minimizing over the collection $\{\mathscr{H}_1, \ldots, \mathscr{H}_M\}$ the empirical misclassification probability $\widehat{L}_n(\widehat{h}_{\mathscr{H}})$. This selection procedure will not give good results, since for any $\mathscr{H} \subset \mathscr{H}'$ we always have $\widehat{L}_n(\widehat{h}_{\mathscr{H}'}) \leq \widehat{L}_n(\widehat{h}_{\mathscr{H}})$ by Definition (11.5). So the procedure will tend to select the largest dictionary. For designing a good selection procedure, we have to take into account the fluctuations of $\widehat{L}_n(\widehat{h}_{\mathscr{H}})$ around $L(\widehat{h}_{\mathscr{H}})$, as in Chapter 2. The Bound (11.8) in Theorem 11.1 gives us a control of these fluctuations. Building on this bound, we have the following result.

Theorem 11.8 Dictionary selection

Let us consider the dictionary selection procedure

$$\widehat{m} \in \underset{m=1,\ldots,M}{\operatorname{argmin}} \left\{ \widehat{L}_n(\widehat{h}_{\mathscr{H}_m}) + \operatorname{pen}(\mathscr{H}_m) \right\}, \quad \text{with} \quad \operatorname{pen}(\mathscr{H}) \geq 2\sqrt{\frac{2\log(2\mathbb{S}_n(\mathscr{H}))}{n}}.$$

Then, for any $t > 0$, with probability at least $1 - e^{-t}$, we have

$$L(\widehat{h}_{\mathscr{H}_{\widehat{m}}}) \leq \min_{m=1,\ldots,M} \left\{ \inf_{h \in \mathscr{H}_m} L(h) + 2\operatorname{pen}(\mathscr{H}_m) \right\} + \sqrt{\frac{2\log(M) + 2t}{n}}. \tag{11.18}$$

Before proving Theorem 11.8, let us comment on Bound (11.18). Since $\min_{h\in\mathscr{H}} L(h) \leq L(\widehat{h}_{\mathscr{H}})$, we obtain with probability $1 - e^{-t}$

$$L(\widehat{h}_{\mathscr{H}_{\widehat{m}}}) \leq L(\widehat{h}_{\mathscr{H}_o}) + 2\operatorname{pen}(\mathscr{H}_o) + \sqrt{\frac{2\log(M) + 2t}{n}}.$$

In particular, we can compare the misclassification probability of the selected classifier with the misclassification probability of the best classifier among the collection $\{\widehat{h}_{\mathscr{H}_1}, \ldots, \widehat{h}_{\mathscr{H}_M}\}$.

We also notice that the second term of Bound (11.18) increases as $\sqrt{2\log(M)/n}$ with the number M of candidate dictionaries.

Proof of Theorem 11.8. We recall the notation $\widehat{\Delta}_n(\mathscr{H}) = \sup_{h\in\mathscr{H}} \left|\widehat{L}_n(h) - L(h)\right|$. The Lemma 11.2 ensures that

$$L(\widehat{h}_{\mathscr{H}_{\widehat{m}}}) \leq \widehat{L}_n(\widehat{h}_{\mathscr{H}_{\widehat{m}}}) + \widehat{\Delta}_n(\mathscr{H}_{\widehat{m}}).$$

According to Lemma 11.3 and Lemma 11.4, we have for $s > 0$ and $m \in \{1, \ldots, M\}$

$$\mathbb{P}\left(\widehat{\Delta}_n(\mathscr{H}_m) > \operatorname{pen}(\mathscr{H}_m) + \sqrt{s/(2n)}\right) \leq e^{-s}.$$

For $s = \log(M) + t$, the union bound ensures that

$$\mathbb{P}\left(\widehat{\Delta}_n(\mathscr{H}_m) \leq \operatorname{pen}(\mathscr{H}_m) + \sqrt{\frac{\log(M) + t}{2n}}, \quad \text{for all } m = 1, \ldots, M\right) \geq 1 - e^{-t} \tag{11.19}$$

Therefore, according to the definition of the selection criterion, we have with probability at least $1 - e^{-t}$

$$\begin{aligned} L(\widehat{h}_{\mathscr{H}_{\widehat{m}}}) &\leq \widehat{L}_n(\widehat{h}_{\mathscr{H}_{\widehat{m}}}) + \operatorname{pen}(\mathscr{H}_{\widehat{m}}) + \sqrt{\frac{\log(M) + t}{2n}} \\ &\leq \min_{m=1,\ldots,M} \left\{ \widehat{L}_n(\widehat{h}_{\mathscr{H}_m}) + \operatorname{pen}(\mathscr{H}_m) \right\} + \sqrt{\frac{\log(M) + t}{2n}}. \end{aligned} \tag{11.20}$$

To conclude, we only need to control the size of $\widehat{L}_n(\widehat{h}_{\mathscr{H}_m})$ in terms of $\inf_{h\in\mathscr{H}_m} L(h)$. This can be done directly by combining (11.7) and (11.8), but the resulting bound is not tight.

In order to compare $\widehat{L}_n(\widehat{h}_{\mathscr{H}_m})$ to $\inf_{h\in\mathscr{H}_m} L(h)$, let us notice that for any $h \in \mathscr{H}_m$, we have

$$\widehat{L}_n(\widehat{h}_{\mathscr{H}_m}) \leq \widehat{L}_n(h) \leq L(h) + \widehat{\Delta}_n(\mathscr{H}_m),$$

so taking the infimum over $h \in \mathscr{H}_m$, we obtain for all $m = 1, \ldots, M$

$$\widehat{L}_n(\widehat{h}_{\mathscr{H}_m}) \leq \inf_{h\in\mathscr{H}_m} L(h) + \widehat{\Delta}_n(\mathscr{H}_m).$$

Combining this bound with (11.19) and (11.20), we obtain with probability at least $1 - e^{-t}$

$$L(\widehat{h}_{\mathscr{H}_{\widehat{m}}}) \leq \min_{m=1,\ldots,M} \left\{ \inf_{h\in\mathscr{H}_m} L(h) + 2\,\mathrm{pen}(\mathscr{H}_m) \right\} + 2\sqrt{\frac{\log(M)+t}{2n}}.$$

The proof of Theorem 11.8 is complete. □

Remark. Combining (11.19) and Lemma 11.2, we obtain the confidence interval for the misclassification probability

$$\mathbb{P}\left(L(\widehat{h}_{\mathscr{H}_{\widehat{m}}}) \in \left[\widehat{L}_n(\widehat{h}_{\mathscr{H}_{\widehat{m}}}) - \delta(\widehat{m},t), \widehat{L}_n(\widehat{h}_{\mathscr{H}_{\widehat{m}}}) + \delta(\widehat{m},t)\right]\right) \geq 1 - e^{-t},$$

$$\text{with} \quad \delta(\widehat{m},t) = \mathrm{pen}(\mathscr{H}_{\widehat{m}}) + \sqrt{\frac{\log(M)+t}{2n}}.$$

11.3 From Theoretical to Practical Classifiers

11.3.1 Empirical Risk Convexification

The empirical risk minimization classifier analyzed in the previous section has some very nice statistical properties, but it cannot be used in practice because of its computational cost. Actually, there is no efficient way to minimize (11.5), since neither $\mathscr{H}$ nor $\widehat{L}_n$ are convex. The situation is very similar to the situation met in Chapter 2 for the model selection procedure (2.9). As in Chapter 5, we will derive some practical classifiers from (11.5) by a convex relaxation of the minimization (11.5). Some of the most popular classification algorithms are obtained by following this principle. The empirical misclassification probability $\widehat{L}_n$ will be replaced by some convex surrogate and the set of classifiers $\mathscr{H}$ will be replaced by some convex functional set $\mathscr{F} \subset \mathbb{R}^{\mathscr{X}}$.

Let us consider some convex set $\mathscr{F}$ of functions from $\mathscr{X}$ to $\mathbb{R}$. A function $f \in \mathscr{F}$ is not a classifier, but we can use it for classification by classifying the data points according to the sign of f. In other words, we can associate to f the classifier $\mathrm{sign}(f)$. The empirical misclassification probability of this classifier can be written as

$$\widehat{L}_n(\mathrm{sign}(f)) = \frac{1}{n}\sum_{i=1}^{n} \mathbf{1}_{\{Y_i\,\mathrm{sign}(f)(X_i)<0\}} = \frac{1}{n}\sum_{i=1}^{n} \mathbf{1}_{\{Y_i f(X_i)<0\}}.$$

Let us replace this empirical misclassification probability $\widehat{L}_n$ by some convex surrogate, which is more amenable to numerical computations. A simple and efficient way to obtain a convex criterion is to replace the loss function $z \to \mathbf{1}_{z<0}$ by some convex function $z \to \ell(z)$. Building on this simple idea, we will focus in the following on classifiers obtained by the procedure

$$\widehat{h}_{\mathscr{F}} = \text{sign}(\widehat{f}_{\mathscr{F}}),$$

$$\text{where} \quad \widehat{f}_{\mathscr{F}} \in \underset{f \in \mathscr{F}}{\text{argmin}} \widehat{L}_n^{\ell}(f), \quad \text{with} \quad \widehat{L}_n^{\ell}(f) = \frac{1}{n} \sum_{i=1}^{n} \ell(Y_i f(X_i)). \quad (11.21)$$

This classifier can be computed efficiently, since both $\mathscr{F}$ and $\widehat{L}_n^{\ell}$ are convex. Many classical classifiers are obtained by solving (11.21), with some specific choices of $\mathscr{F}$ and ℓ; see Sections 11.3.3 and 11.3.4 for some examples.

Some popular convex loss ℓ

It is natural to consider a convex loss function ℓ, which is non-increasing and non-negative. Usually, we also ask that $\ell(z) \geq \mathbf{1}_{z<0}$ for all $z \in \mathbb{R}$, since in this case we can give an upper bound on the misclassification probability; see Theorem 11.10. Some classical loss functions are

- the exponential loss $\ell(z) = e^{-z}$,
- the logit loss $\ell(z) = \log_2(1 + e^{-z})$, and
- the hinge loss $\ell(z) = (1 - z)_+$, with $(x)_+ = \max(0, x)$.

A plot of these three functions is given in Figure 11.4.

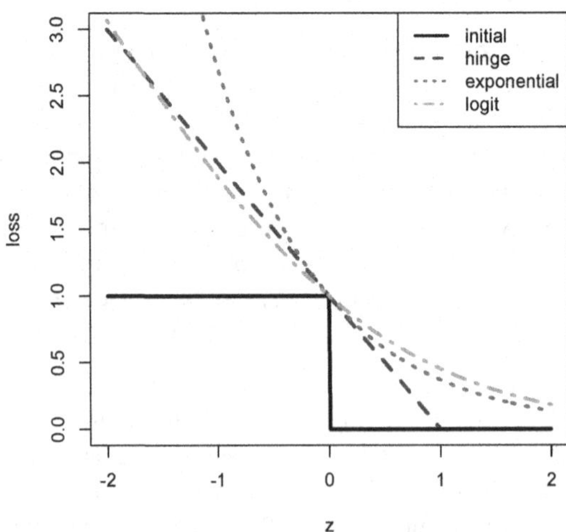

Figure 11.4 *Plot of the exponential, hinge, and logit losses.*

Some classical functional sets $\mathscr{F}$

The main popular convex functional sets $\mathscr{F}$ can be grouped into two classes.

A first popular class of sets $\mathscr{F}$ is obtained by taking a linear combination of a finite family $\mathscr{H} = \{h_1, \dots, h_p\}$ of classifiers

$$\mathscr{F} = \left\{ f : f(x) = \sum_{j=1}^{p} \beta_j h_j(x) \text{ with } \beta_j \in \mathscr{C} \right\}, \tag{11.22}$$

where $\mathscr{C}$ is a convex subset of $\mathbb{R}^p$. Typical choices for $\mathscr{C}$ are the ℓ^1-ball $\{\beta \in \mathbb{R}^p : |\beta|_1 \leq R\}$, the simplex $\left\{\beta \in \mathbb{R}^p : \beta_j \geq 0, \sum_{j=1}^{p} \beta_j \leq 1\right\}$, or the whole space $\mathbb{R}^p$. This choice appears, for example, in boosting methods; see Section 11.3.4. The basic classifiers $\{h_1, \dots, h_p\}$ are often called *weak learners*. A popular choice of weak learners in $\mathbb{R}^d$ is $h_j(x) = \text{sign}(x_j - t_j)$ with $t_j \in \mathbb{R}$.

A second popular class of sets $\mathscr{F}$ is obtained by taking a ball of a Reproducing Kernel Hilbert Space (RKHS). We refer to Appendix E for a brief introduction to RKHS. Since the smoothness of a function in an RKHS is driven by its norm (see, e.g., Formula (E.5), page 327, in Appendix E), a ball of an RKHS corresponds to a set of smooth functions. Let $\mathscr{F}_k$ be an RKHS with reproducing kernel k, and write $\|f\|_{\mathscr{F}}$ for the Hilbert norm of $f \in \mathscr{F}_k$. For notational simplicity, in the following, we simply write $\mathscr{F}$ for $\mathscr{F}_k$. Minimizing $\widehat{L}_n^{\ell}$ over the ball $\{f \in \mathscr{F} : \|f\|_{\mathscr{F}} \leq R\}$ is equivalent to minimizing over $\mathscr{F}$ the dual Lagrangian problem

$$\widehat{f}_{\mathscr{F}} \in \operatorname*{argmin}_{f \in \mathscr{F}} \widetilde{L}_n^{\ell}(f), \quad \text{with} \quad \widetilde{L}_n^{\ell}(f) = \frac{1}{n} \sum_{i=1}^{n} \ell(Y_i f(X_i)) + \lambda \|f\|_{\mathscr{F}}^2, \tag{11.23}$$

for some $\lambda > 0$. This kind of classifier appears, for example, in Support Vector Machine algorithms, presented in Section 11.3.3. The solution of (11.23) fulfills the following representation formula.

Proposition 11.9 Representation formula

The minimization problem (11.23) is equivalent to $\widehat{f}_{\mathscr{F}} = \sum_{j=1}^{n} \widehat{\beta}_j k(X_j, .)$, *with*

$$\widehat{\beta} \in \operatorname*{argmin}_{\beta \in \mathbb{R}^n} \left\{ \frac{1}{n} \sum_{i=1}^{n} \ell\left(\sum_{j=1}^{n} \beta_j Y_i k(X_j, X_i) \right) + \lambda \sum_{i,j=1}^{n} \beta_i \beta_j k(X_i, X_j) \right\}. \tag{11.24}$$

Proof. Let V be the linear space spanned by $k(X_1, .), \dots, k(X_n, .)$, where $k(x, .)$ refers to the map $y \to k(x, y)$. Decomposing $f = f_V + f_{V^\perp}$ according to the orthogonal decomposition $\mathscr{F} = V \bigoplus V^\perp$, we have by the reproducing property (E.2), page 325,

$$f(X_i) = \langle f, k(X_i, .) \rangle_{\mathscr{F}} = \langle f_V, k(X_i, .) \rangle_{\mathscr{F}} = f_V(X_i).$$

Combining this formula with the Pythagorean formula, we obtain

$$\widetilde{L}_n^\ell(f_V + f_{V^\perp}) = \frac{1}{n}\sum_{i=1}^n \ell(Y_i f_V(X_i)) + \lambda\|f_V\|_{\mathscr{F}}^2 + \lambda\|f_{V^\perp}\|_{\mathscr{F}}^2.$$

Since λ is positive, any minimizer $\widehat{f}$ of $\widetilde{L}_n^\ell$ must fulfill $\widehat{f}_{V^\perp} = 0$, so it is of the form

$$\widehat{f}_{\mathscr{F}} = \sum_{i=1}^n \widehat{\beta}_i k(X_i, .).$$

Furthermore, the reproducing property (E.2) ensures again that $\langle k(X_i,.), k(X_j,.)\rangle_{\mathscr{F}} = k(X_i, X_j)$, so

$$\left\| \sum_{j=1}^n \beta_j k(X_j,.) \right\|_{\mathscr{F}}^2 = \sum_{i,j=1}^n \beta_i\beta_j k(X_i, X_j).$$

The proof of Proposition 11.9 is complete. □

The representation formula is of major importance in practice, since it reduces the infinite-dimensional minimization problem (11.23) into an n-dimensional convex minimization problem (11.24) that can be solved efficiently. In Section 11.3.3 on Support Vector Machines, we will give a more precise description of the solution of this problem when ℓ is the hinge loss.

Another important feature of the representation formula is that we only need to know the positive definite kernel k in order to compute the classifier $\widehat{h}_{\mathscr{F}}$. In particular, we do not need to identify the RKHS associated to k in order to define and compute the estimator (11.24). The RKHS $\mathscr{F}$ is only used implicitly in order to understand the nature of the classifier $\widehat{h}_{\mathscr{F}}$.

11.3.2 Statistical Properties

The classifier $\widehat{h}_{\mathscr{F}}$ given by (11.21) with $\mathscr{F}$ and ℓ convex has the nice feature of a low computational cost, but does it have some good statistical properties?

Link with the Bayes classifier

The empirical risk minimizer $\widehat{h}_{\mathscr{H}}$ of Section 11.2 minimizes the empirical version $\widehat{L}_n(h)$ of the misclassification probability $L(h) = \mathbb{P}(Y \neq h(X))$ over some set $\mathscr{H}$ of classifiers. The function $\widehat{f}_{\mathscr{F}}$ minimizes instead the empirical version $\widehat{L}_n^\ell(h)$ of $\widehat{L}^\ell(h) = \mathbb{E}\left[\ell(Yf(X))\right]$ over some functional set $\mathscr{F}$. The classifier $\widehat{h}_{\mathscr{H}}$ can then be viewed as an empirical version of the Bayes classifier h_* which minimizes $\mathbb{P}(Y \neq h(X))$ over the set of measurable functions $h : \mathscr{X} \to \{-1, +1\}$, whereas the function $\widehat{f}_{\mathscr{F}}$ is an empirical version of the function f_*^ℓ, which minimizes $\mathbb{E}\left[\ell(Yf(X))\right]$ over the set of measurable functions $f : \mathscr{X} \to \mathbb{R}$. A first point is to understand the link between the Bayes classifier h_* and the sign of the function f_*^ℓ. It turns out that under some weak assumptions on ℓ, the sign of f_*^ℓ exactly coincides with the Bayes classifier h_*, so

$\text{sign}(f_*^\ell)$ minimizes the misclassification probability $\mathbb{P}(Y \neq h(X))$. Let us check this point.

Conditioning on X, we have

$$\begin{aligned}\mathbb{E}\left[\ell(Yf(X)\right] &= \mathbb{E}\left[\mathbb{E}\left[\ell(Yf(X)|X\right]\right] \\ &= \mathbb{E}\left[\ell(f(X))\mathbb{P}(Y=1|X) + \ell(-f(X))(1-\mathbb{P}(Y=1|X))\right].\end{aligned}$$

Assume that ℓ is decreasing, differentiable, and strictly convex (e.g., exponential or logit loss). The above expression is minimum for $f_*^\ell(X)$ solution of

$$\frac{\ell'(-f(X))}{\ell'(f(X))} = \frac{\mathbb{P}(Y=1|X)}{1-\mathbb{P}(Y=1|X)},$$

when such a solution exists. Since ℓ is strictly convex, we then have $f(X) > 0$ if and only if $\ell'(-f(X))/\ell'(f(X)) > 1$, so

$$f_*^\ell(X) > 0 \iff \mathbb{P}(Y=1|X) > 1/2 \iff \mathbb{E}[Y|X] = 2\mathbb{P}(Y=1|X) - 1 > 0.$$

Since $h_*(X) = \text{sign}(\mathbb{E}\left[Y|X\right])$ (see Section 11.1), we obtain $\text{sign}(f_*^\ell) = h_*$. This equality also holds true for the hinge loss ℓ (check it!).

To sum up the above discussion, the target function f_*^ℓ approximated by $\widehat{f}_{\mathscr{F}}$ does perfectly make sense for the classification problem, since, under some weak assumptions, its sign coincides with the best possible classifier h_* (the Bayes classifier).

Upper-bound on the misclassification probability

We focus now on the misclassification probability $L(\widehat{h}_{\mathscr{F}})$ of the classifier $\widehat{h}_{\mathscr{F}} = \text{sign}(\widehat{f}_{\mathscr{F}})$ given by (11.21). In practice, it is important to have an upper bound on the misclassification probability $L(\widehat{h}_{\mathscr{F}})$, which can be computed from the data. The next theorem provides such an upper bound for some typical examples of set $\mathscr{F}$.

Theorem 11.10 Confidence bound on $L(\widehat{h}_{\mathscr{F}})$

For any $R > 0$, we set $\Delta\ell(R) = |\ell(R) - \ell(-R)|$. We assume here that the loss-function ℓ is convex, non-increasing, non-negative, α-Lipschitz on $[-R, R]$ and fulfills $\ell(z) \geq \mathbf{1}_{z<0}$ for all z in $\mathbb{R}$. We consider the classifier $\widehat{h}_{\mathscr{F}}$ given by (11.21).

(a) *When $\mathscr{F}$ is of the form (11.22), with $\mathscr{C} = \{\beta \in \mathbb{R}^p : |\beta|_1 \leq R\}$, we have with probability at least $1 - e^{-t}$*

$$L(\widehat{h}_{\mathscr{F}}) \leq \widehat{L}_n^\ell(\widehat{f}_{\mathscr{F}}) + 4\alpha R\sqrt{\frac{2\log(2p)}{n}} + \Delta\ell(R)\sqrt{\frac{t}{2n}}. \tag{11.25}$$

(b) *Let $\mathscr{F}$ be the ball of radius R of an RKHS with kernel k fulfilling $k(x,x) \leq 1$ for all $x \in \mathscr{X}$. Then, we have with probability at least $1 - e^{-t}$*

$$L(\widehat{h}_{\mathscr{F}}) \leq \widehat{L}_n^\ell(\widehat{f}_{\mathscr{F}}) + \frac{4\alpha R}{\sqrt{n}} + \Delta\ell(R)\sqrt{\frac{t}{2n}}. \tag{11.26}$$

Proof. We first prove a general upper bound for $L(\widehat{h}_{\mathscr{F}})$, similar to Theorem 11.1.

Lemma 11.11

Assume that $\sup_{f\in\mathscr{F}}|f(x)|\leq R<+\infty$. *For any loss* ℓ *fulfilling the hypotheses of Theorem 11.10, we have with probability at least* $1-e^{-t}$

$$L(\widehat{h}_{\mathscr{F}})\leq \widehat{L}_n^{\ell}(\widehat{f}_{\mathscr{F}})+\frac{4\alpha}{n}\max_{x\in\mathscr{X}^n}\mathbb{E}_{\sigma}\left[\sup_{f\in\mathscr{F}}\left|\sum_{i=1}^n\sigma_i f(x_i)\right|\right]+\Delta\ell(R)\sqrt{\frac{t}{2n}}, \tag{11.27}$$

where $\sigma_1,\ldots,\sigma_n$ *are i.i.d. random variables with distribution* $\mathbb{P}_{\sigma}(\sigma_i=1)=\mathbb{P}_{\sigma}(\sigma_i=-1)=1/2$.

Proof of Lemma 11.11. The proof of this lemma relies on the same arguments as the proof of Theorem 11.1. We set

$$\widehat{\Delta}_n^{\ell}(\mathscr{F})=\sup_{f\in\mathscr{F}}\left|\widehat{L}_n^{\ell}(f)-L^{\ell}(f)\right|\quad \text{with } L^{\ell}(f)=\mathbb{E}\left[\ell(Yf(X))\right].$$

The first point is to notice that since $\ell(z)\geq \mathbf{1}_{z<0}$ for all real z, we have

$$\begin{aligned} L(\widehat{h}_{\mathscr{F}})=\mathbb{P}(Y\widehat{f}_{\mathscr{F}}(X)<0)&\leq \mathbb{E}\left[\ell(Y\widehat{f}_{\mathscr{F}}(X))\right]\\ &\leq \widehat{L}_n^{\ell}(\widehat{f}_{\mathscr{F}})+\widehat{\Delta}_n^{\ell}(\mathscr{F}). \end{aligned}$$

As in Lemma 11.3, the McDiarmid concentration inequality (Theorem B.5, page 299, in Appendix B) ensures that with probability at least $1-e^{-t}$, we have

$$\widehat{\Delta}_n^{\ell}(\mathscr{F})\leq \mathbb{E}\left[\widehat{\Delta}_n^{\ell}(\mathscr{F})\right]+\Delta\ell(R)\sqrt{\frac{t}{2n}}.$$

To conclude the proof of the lemma, it only remains to prove that

$$\mathbb{E}\left[\widehat{\Delta}_n^{\ell}(\mathscr{F})\right]\leq \frac{4\alpha}{n}\max_{x\in\mathscr{X}^n}\mathbb{E}_{\sigma}\left[\sup_{f\in\mathscr{F}}\left|\sum_{i=1}^n\sigma_i f(x_i)\right|\right]. \tag{11.28}$$

Following exactly the same lines as in the proof of Lemma 11.4 (replacing $\mathbf{1}_{Y_i\neq h(X_i)}$ by $\ell(Y_i f(X_i))-\ell(0)$), we obtain

$$\mathbb{E}\left[\widehat{\Delta}_n^{\ell}(\mathscr{F})\right]\leq \frac{2}{n}\max_{y\in\{-1,+1\}^n}\max_{x\in\mathscr{X}^n}\mathbb{E}_{\sigma}\left[\sup_{f\in\mathscr{F}}\left|\sum_{i=1}^n\sigma_i(\ell(y_i f(x_i))-\ell(0))\right|\right].$$

We finally use the α-Lipschitz property of ℓ to conclude: According to the Contraction principle (Proposition B.11, page 306, in Appendix B, with $\varphi(z)=\ell(z)-\ell(0)$ and $\mathscr{Z}=\{[y_i f(x_i)]_{i=1,\ldots,n}: f\in\mathscr{F}\}$), we have

$$\begin{aligned} \mathbb{E}_{\sigma}\left[\sup_{f\in\mathscr{F}}\left|\sum_{i=1}^n\sigma_i(\ell(y_i f(x_i))-\ell(0))\right|\right] &\leq 2\alpha\mathbb{E}_{\sigma}\left[\sup_{f\in\mathscr{F}}\left|\sum_{i=1}^n\sigma_i y_i f(x_i)\right|\right]\\ &= 2\alpha\mathbb{E}_{\sigma}\left[\sup_{f\in\mathscr{F}}\left|\sum_{i=1}^n\sigma_i f(x_i)\right|\right], \end{aligned}$$

where we used in the last line that $(\sigma_1,\ldots,\sigma_n)$ has the same distribution as $(y_1\sigma_1,\ldots,y_n\sigma_n)$ for any $y\in\{-1,1\}^n$. Combining the last two bounds gives (11.28), and the proof of Lemma 11.11 is complete. □

(a) Let us prove now Bound (11.25). The map $\beta \to \sum_{i=1}^n \sigma_i \sum_{j=1}^p \beta_j h_j(x_i)$ is linear, so it reaches its maximum and minimum on the ℓ^1-ball $\mathscr{C}$ at one of the vertices of $\mathscr{C}$. Therefore, we have

$$\mathbb{E}_\sigma\left[\sup_{f\in\mathscr{F}}\left|\sum_{i=1}^n \sigma_i f(x_i)\right|\right] = R\,\mathbb{E}_\sigma\left[\max_{j=1,\ldots,p}\left|\sum_{i=1}^n \sigma_i h_j(x_i)\right|\right].$$

It remains to apply Inequality (11.10) with $\mathscr{V}=\{(h_j(x_1),\ldots,h_j(x_n)) : j=1,\ldots,p\}$, whose cardinality is at most p in order to obtain

$$\mathbb{E}_\sigma\left[\sup_{f\in\mathscr{F}}\left|\sum_{i=1}^n \sigma_i f(x_i)\right|\right] \leq R\sqrt{2n\log(2p)}.$$

Bound (11.25) then follows from Lemma 11.11.

(b) We now turn to the second Bound (11.26) and write $\|.\|_{\mathscr{F}}$ for the norm in the RKHS. According to the reproducing formula (E.2), page 325, and the Cauchy–Schwartz inequality, we have

$$\left|\sum_{i=1}^n \sigma_i f(x_i)\right| = \left|\left\langle f, \sum_{i=1}^n \sigma_i k(x_i,.)\right\rangle_{\mathscr{F}}\right| \leq \|f\|_{\mathscr{F}}\left\|\sum_{i=1}^n \sigma_i k(x_i,.)\right\|_{\mathscr{F}}.$$

Since $\|f\|_{\mathscr{F}}\leq R$, applying Jensen inequality, we obtain

$$\begin{aligned}\mathbb{E}_\sigma\left[\sup_{\|f\|_{\mathscr{F}}\leq R}\left|\sum_{i=1}^n \sigma_i f(x_i)\right|\right] &\leq R\,\mathbb{E}_\sigma\left[\left\|\sum_{i=1}^n \sigma_i k(x_i,.)\right\|_{\mathscr{F}}\right]\\ &\leq R\sqrt{\mathbb{E}_\sigma\left[\left\|\sum_{i=1}^n \sigma_i k(x_i,.)\right\|_{\mathscr{F}}^2\right]}\\ &= R\sqrt{\sum_{i,j=1}^n k(x_i,x_j)\mathbb{E}_\sigma[\sigma_i\sigma_j]},\end{aligned}$$

where we have used $\langle k(x_i,.),k(x_j,.)\rangle_{\mathscr{F}} = k(x_i,x_j)$. Since $k(x,x)\leq 1$ and $\mathbb{E}[\sigma_i\sigma_j]=0$ for $i\neq j$, we get

$$\mathbb{E}_\sigma\left[\sup_{\|f\|_{\mathscr{F}}\leq R}\left|\sum_{i=1}^n \sigma_i f(x_i)\right|\right] \leq R\sqrt{\sum_{i=1}^n k(x_i,x_i)\mathbb{E}_\sigma[\sigma_i^2]} \leq R\sqrt{n}.$$

Combining again the reproducing property with the Cauchy–Schwartz inequality, we obtain

$$|f(x)| = |\langle f, k(x,.)\rangle_{\mathscr{F}}| \leq R\sqrt{k(x,x)} \leq R.$$

So $\mathscr{F}$ fulfills the hypotheses of the Lemma 11.11, which gives

$$L(\widehat{h}_{\mathscr{F}}) \leq \widehat{L}_n^{\ell}(\widehat{f}_{\mathscr{F}}) + \frac{4\alpha R}{\sqrt{n}} + \Delta\ell(R)\sqrt{\frac{t}{2n}} .$$

The proof of Theorem 11.10 is complete. □

It is possible to derive risk bounds similar to (11.7) for $L(\widehat{h}_{\mathscr{H}})$; we refer to Boucheron, Bousquet, and Lugosi [37] for a review of such results. In the remainder of this chapter, we will describe two very popular classification algorithms: the Support Vector Machines and AdaBoost.

11.3.3 Support Vector Machines

The Support Vector Machine (SVM) algorithm corresponds to the estimator (11.23) with the hinge loss $\ell(z) = (1-z)_+$. The final classification is performed according to $\widehat{h}_{\mathscr{F}}(x) = \text{sign}(\widehat{f}_{\mathscr{F}}(x))$. We stress that there is a unique solution to (11.23) when ℓ is the convex loss since (11.23) is strictly convex when $\lambda > 0$. It turns out that there is a very nice geometrical interpretation of the solution $\widehat{f}_{\mathscr{F}}$, from which originates the name "Support Vector Machines."

Proposition 11.12 Support Vectors

The solution of (11.23) is of the form $\widehat{f}_{\mathscr{F}}(x) = \sum_{i=1}^n \widehat{\beta}_i k(X_i, x)$, with

$$\begin{cases} \widehat{\beta}_i = 0 & \text{if } \; Y_i \widehat{f}_{\mathscr{F}}(X_i) > 1 \\ \widehat{\beta}_i = Y_i/(2\lambda n) & \text{if } \; Y_i \widehat{f}_{\mathscr{F}}(X_i) < 1 \\ 0 \leq Y_i \widehat{\beta}_i \leq 1/(2\lambda n) & \text{if } \; Y_i \widehat{f}_{\mathscr{F}}(X_i) = 1 . \end{cases}$$

The vectors X_i with index i, such that $\widehat{\beta}_i \neq 0$, are called support vectors.

Proof. Writing K for the matrix $[k(X_i, X_j)]_{i,j=1,\ldots,n}$, we know from the representation Formula (11.24) that the solution of (11.23) is of the form $\widehat{f}_{\mathscr{F}} = \sum_{j=1}^n \widehat{\beta}_j k(X_j, .)$, with

$$\widehat{\beta} \in \underset{\beta \in \mathbb{R}^n}{\text{argmin}} \left\{ \frac{1}{n} \sum_{i=1}^n \left(1 - Y_i[K\beta]_i\right)_+ + \lambda \beta^T K \beta \right\}.$$

The above minimization problem is not smooth, so we introduce some slack variables $\widehat{\xi}_i = (1 - Y_i[K\widehat{\beta}]_i)_+$ and rewrite the minimization problem as

$$(\widehat{\beta}, \widehat{\xi}) \in \underset{\substack{\beta, \xi \in \mathbb{R}^n \text{ such that} \\ \xi_i \geq 1 - Y_i[K\beta]_i \\ \xi_i \geq 0}}{\text{argmin}} \left\{ \frac{1}{n} \sum_{i=1}^n \xi_i + \lambda \beta^T K \beta \right\}. \tag{11.29}$$

This problem is now smooth and convex, and the Karush–Kuhn–Tucker conditions for the Lagrangian dual problem

$$(\widehat{\beta},\widehat{\xi}) \in \underset{\beta,\xi\in\mathbb{R}^n}{\operatorname{argmin}} \left\{ \frac{1}{n}\sum_{i=1}^{n} \xi_i + \lambda \beta^T K \beta - \sum_{i=1}^{n} \left(\alpha_i(\xi_i - 1 + Y_i[K\beta]_i) + \gamma_i \xi_i \right) \right\} \quad (11.30)$$

gives the formulas for $i,j = 1,\dots,n$

$$\text{first-order conditions:} \quad 2\lambda [K\widehat{\beta}]_j = \sum_{i=1}^{n} K_{ij}\alpha_i Y_i \quad \text{and} \quad \alpha_j + \gamma_j = \frac{1}{n},$$

$$\text{slackness conditions:} \quad \min(\alpha_i, \widehat{\xi}_i - 1 + Y_i[K\widehat{\beta}]_i) = 0 \quad \text{and} \quad \min(\gamma_i, \widehat{\xi}_i) = 0.$$

The first first-order condition is fulfilled with $\widehat{\beta}_i = \alpha_i Y_i/(2\lambda)$. Since $\widehat{f}_{\mathscr{F}}(X_i) = [K\widehat{\beta}]_i$, the first slackness condition enforces that $\widehat{\beta}_i = 0$ if $Y_i \widehat{f}_{\mathscr{F}}(X_i) > 1$. The second slackness condition, together with the second first-order optimality condition, enforces that $\widehat{\beta}_i = Y_i/(2\lambda n)$ if $\widehat{\xi}_i > 0$ and $0 \le Y_i\widehat{\beta}_i \le 1/(2\lambda n)$ otherwise. To conclude the proof of the proposition, we notice that when $\widehat{\xi}_i > 0$, we have $\widehat{\beta}_i$ and α_i nonzero, and therefore $Y_i \widehat{f}_{\mathscr{F}}(X_i) = 1 - \widehat{\xi}_i < 1$ according to the first slackness condition. □

We observe that when the matrix $K = [k(X_i,X_j)]_{i,j=1,\dots,n}$ is non-singular, there is a unique solution $\widehat{\beta}$ to (11.24). We refer to Exercise 11.5.5 for the computation of $\widehat{\beta}$. An implementation of the SVM is available, e.g., in the `R` package `kernlab` at `http://cran.r-project.org/web/packages/kernlab/`.

Let us now interpret geometrically Proposition 11.12.

Geometrical interpretation: linear kernel

We start with the simplest kernel $k(x,y) = \langle x,y \rangle$ for all $x,y \in \mathbb{R}^d$. The associated RKHS is the space of linear forms $\mathscr{F} = \{\langle w,.\rangle : w \in \mathbb{R}^d\}$. In this case,

$$\widehat{f}_{\mathscr{F}}(x) = \sum_{i=1}^{n} \widehat{\beta}_i \langle X_i, x\rangle = \langle \widehat{w}, x\rangle \quad \text{with} \quad \widehat{w} = \sum_{i=1}^{n} \widehat{\beta}_i X_i,$$

so the classifier $\widehat{h}_{\mathscr{F}}(x) = \text{sign}(\langle \widehat{w}, x\rangle)$ assigns labels to points according to their position relative to the hyperplane $\{x \in \mathbb{R}^d : \langle \widehat{w}, x\rangle = 0\}$. The normal $\widehat{w}$ to the hyperplane is a linear combination of the support vectors, which are the data points X_i, such that $Y_i\langle \widehat{w}, X_i\rangle \le 1$. They are represented by squares in Figure 11.5. The hyperplanes $\{x \in \mathbb{R}^d : \langle \widehat{w}, x\rangle = +1\}$ and $\{x \in \mathbb{R}^d : \langle \widehat{w}, x\rangle = -1\}$ are usually called margin hyperplanes.

We notice the following important property of the SVM. If we add to the learning data set a point X_{n+1}, which fulfills $Y_{n+1}\langle \widehat{w}, X_{n+1}\rangle > 1$, then the vector $\widehat{w}$ and the classifier $\widehat{h}_{\mathscr{F}}$ do not change. In other words, only data points that are wrongly classified or classified with not enough margin (i.e., $Y_i\langle \widehat{w}, X_i\rangle \le 1$) do influence the separating hyperplane $\{x \in \mathbb{R}^d : \langle \widehat{w}, x\rangle = 0\}$. This property is in contrast with the LDA classifier (11.2), where all the points have an equal weight in the definition of the $\widehat{\mu}_k$ and $\widehat{\Sigma}$.

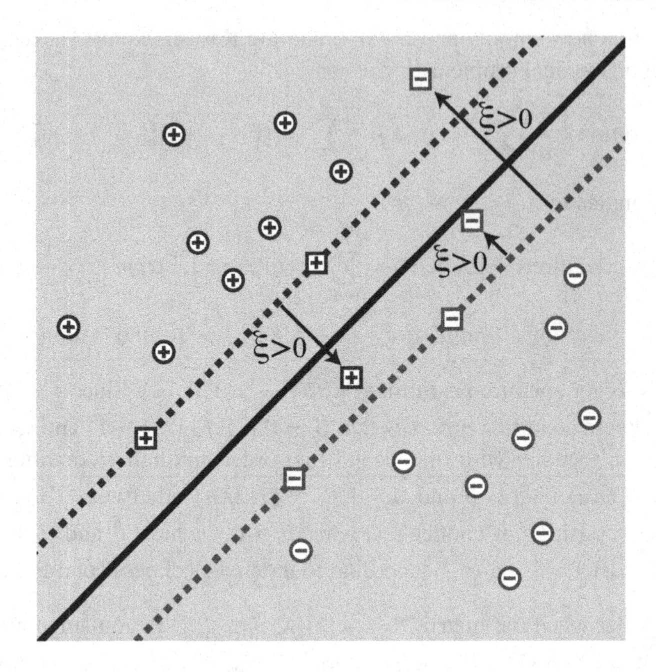

Figure 11.5 *Classification with a linear SVM: The separating hyperplane* $\{x \in \mathbb{R}^d : \langle \widehat{w}, x \rangle = 0\}$ *is represented in black, and the two margin hyperplanes* $\{x \in \mathbb{R}^d : \langle \widehat{w}, x \rangle = +1\}$ *and* $\{x \in \mathbb{R}^d : \langle \widehat{w}, x \rangle = -1\}$ *are represented in dotted blue and red, respectively. The support vectors are represented by squares.*

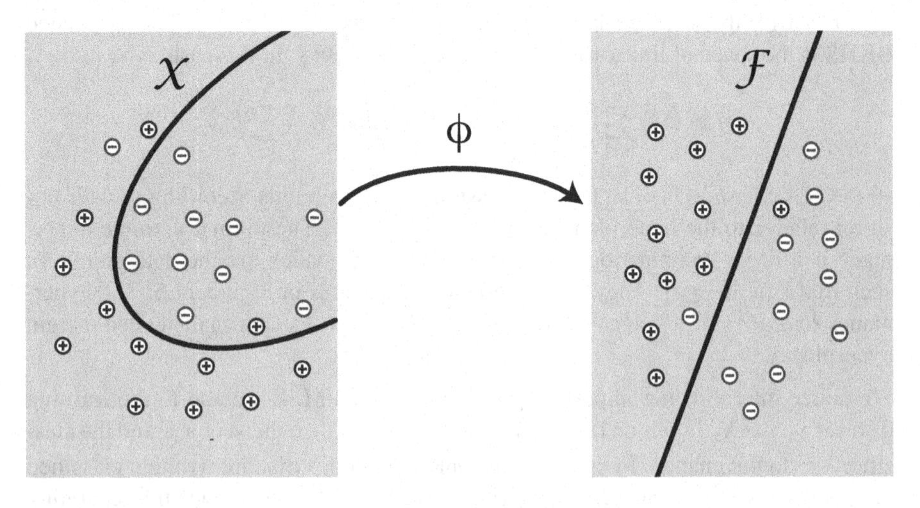

Figure 11.6 *Classification with a non-linear kernel: The linear classification in* $\mathscr{F}$ *produces a non-linear classification in* $\mathscr{X}$ *via the reciprocal image of* ϕ.

Geometrical interpretation: arbitrary positive definite kernels

Let us denote by $\phi : \mathscr{X} \to \mathscr{F}$ the map $\phi(x) = k(x,.)$, usually called the "feature map." According to the reproducing property (E.2), page 325, and Proposition 11.12, we have

$$\widehat{f}_{\mathscr{F}}(x) = \langle \widehat{f}_{\mathscr{F}}, \phi(x)\rangle_{\mathscr{F}} = \left\langle \sum_{i=1}^{n} \widehat{\beta}_i \phi(X_i), \phi(x) \right\rangle_{\mathscr{F}} .$$

A point $x \in \mathscr{X}$ is classified by $\widehat{h}_{\mathscr{F}}$ according to the sign of the above scalar product. Therefore, the points $\phi(x) \in \mathscr{F}$ are classified according to the linear classifier on $\mathscr{F}$

$$f \mapsto \text{sign}\left(\langle \widehat{w}_{\phi}, f\rangle_{\mathscr{F}}\right), \quad \text{where } \widehat{w}_{\phi} = \sum_{i=1}^{n} \widehat{\beta}_i \phi(X_i).$$

The separating frontier $\left\{x \in \mathscr{X} : \widehat{f}_{\mathscr{F}}(x) = 0\right\}$ of the classifier $\widehat{h}_{\mathscr{F}}$ is therefore the reciprocal image by ϕ of the intersection of the hyperplane $\left\{f \in \mathscr{F} : \langle \widehat{w}_{\phi}, f\rangle_{\mathscr{F}} = 0\right\}$ in $\mathscr{F}$ with the range of ϕ, as represented in Figure 11.6. We observe that the kernel k then delinearizes the SVM, in the sense that it produces a non-linear classifier $\widehat{h}_{\mathscr{F}}$ with almost the same computational cost as a linear one in $\mathbb{R}^n$.

You can observe SVM in action with the following recreative applet:
`http://cs.stanford.edu/people/karpathy/svmjs/demo/`.

Why are RKHS useful?

There are mainly two major reasons for using RKHS. The first reason is that using RKHS allows to delinearize some algorithms by mapping $\mathscr{X}$ in $\mathscr{F}$ with $\phi : x \to k(x,.)$, as represented in Figure 11.6. It then provides non-linear algorithms with almost the same computational complexity as a linear one.

The second reason is that it allows us to apply to any set $\mathscr{X}$ some algorithms that are defined for vectors. Assume, for example, that we want to classify some proteins or molecules according to their therapeutic properties. Let $\mathscr{X}$ represents our set of molecules. For any $x, y \in \mathscr{X}$, let us represent by $k(x,y)$ some measure of similarity between x and y. If the kernel $k : \mathscr{X} \times \mathscr{X} \to \mathbb{R}$ is positive definite, then we can directly apply the SVM algorithm in order to classify them; see Figure 11.7. Of course, the key point in this case is to properly design the kernel k. Usually, the kernel $k(x,y)$ is designed according to some properties of x, y that are known to be relevant for the classification problem. For example, the number of common short sequences is a useful index of similarity between two proteins. The computational complexity for evaluating $k(x,y)$ is also an issue that is crucial in many applications with complex data.

We point out that RKHS can be used in many other statistical settings in order to either delinearize an algorithm or to apply a vectorial algorithm to non-vectorial data. In principle, it can be used with any algorithm relying only on scalar products $\langle x, y\rangle$ by replacing these scalar products by the kernel evaluation $k(x,y)$. Some popular examples are the kernel-PCA (for a PCA with non-vectorial data; see Exercise 11.5.6) or the kernel-smoothing in regression.

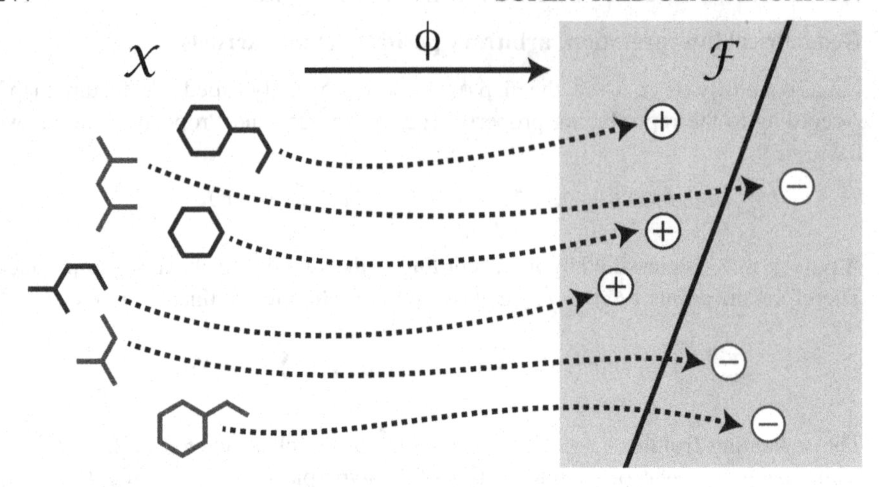

Figure 11.7 *Classification of molecules with an SVM.*

Finally, we stress again that as long as we have a representation formula as (11.24), we do not need to identify the RKHS associated to a positive definite kernel k in order to implement an algorithm based on RKHS. This property is usually referred to as the "kernel trick."

11.3.4 AdaBoost

AdaBoost is an algorithm that computes an approximation of the estimator (11.21) with the exponential loss $\ell(z) = e^{-z}$ and the functional space $\mathscr{F} = \text{span}\{h_1, \ldots, h_p\}$, where $h_1, \ldots, h_p$ are p arbitrary classifiers.

The principle of the AdaBoost algorithm is to perform a greedy minimization of

$$\widehat{f}_{\mathscr{F}} \in \underset{f \in \text{span}\{h_1,\ldots,h_p\}}{\text{argmin}} \left\{ \frac{1}{n} \sum_{i=1}^{n} \exp(-Y_i f(X_i)) \right\}.$$

More precisely, it computes a sequence of functions $\widehat{f}_m$ for $m = 0, \ldots, M$ by starting from $\widehat{f}_0 = 0$ and then solving for $m = 1, \ldots, M$

$$\widehat{f}_m = \widehat{f}_{m-1} + \beta_m h_{j_m},$$
$$\text{where} \quad (\beta_m, j_m) \in \underset{\substack{j=1,\ldots,p \\ \beta \in \mathbb{R}}}{\text{argmin}} \ \frac{1}{n} \sum_{i=1}^{n} \exp\left(-Y_i\left(\widehat{f}_{m-1}(X_i) + \beta h_j(X_i)\right)\right).$$

The final classification is performed according to $\widehat{h}_M(x) = \text{sign}(\widehat{f}_M(x))$, which is an approximation of $\widehat{h}_{\mathscr{H}}$ defined by (11.21).

The exponential loss allows us to compute (β_m, j_m) very efficiently. Actually, setting $w_i^{(m)} = n^{-1}\exp(-Y_i\widehat{f}_{m-1}(X_i))$, we have

$$\frac{1}{n}\sum_{i=1}^{n}\exp\left(-Y_i(\widehat{f}_{m-1}(X_i)+\beta h_j(X_i))\right) = (e^{\beta}-e^{-\beta})\sum_{i=1}^{n} w_i^{(m)}\mathbf{1}_{h_j(X_i)\neq Y_i} + e^{-\beta}\sum_{i=1}^{n} w_i^{(m)}.$$

When no initial classifier h_j perfectly classifies the data $(X_i, Y_i)_{i=1,\ldots,n}$, so that the condition

$$\mathrm{err}_m(j) = \frac{\sum_{i=1}^n w_i^{(m)}\mathbf{1}_{h_j(X_i)\neq Y_i}}{\sum_{i=1}^n w_i^{(m)}} < 1 \quad \text{for all } j = 1,\ldots,p$$

is met, the minimizers (β_m, j_m) are given by

$$j_m = \underset{j=1,\ldots,p}{\mathrm{argmin}}\,\mathrm{err}_m(j) \quad \text{and} \quad \beta_m = \frac{1}{2}\log\left(\frac{1-\mathrm{err}_m(j_m)}{\mathrm{err}_m(j_m)}\right).$$

Noticing that $-Y_ih(X_i) = 2\mathbf{1}_{Y_i\neq h(X_i)} - 1$, we obtain the standard formulation of the AdaBoost algorithm.

AdaBoost

Init: $w_i^{(1)} = 1/n$, for $i = 1,\ldots,n$

Iterate: For $m = 1,\ldots,M$ do

$$\begin{aligned} j_m &= \underset{j=1,\ldots,p}{\mathrm{argmin}}\,\mathrm{err}_m(j) \\ 2\beta_m &= \log(1-\mathrm{err}_m(j_m)) - \log(\mathrm{err}_m(j_m)) \\ w_i^{(m+1)} &= w_i^{(m)}\exp(2\beta_m\mathbf{1}_{h_{j_m}(X_i)\neq Y_i} - \beta_m), \quad \text{for } i = 1,\ldots,n \end{aligned}$$

Output: $\widehat{f}_M(x) = \sum_{m=1}^M \beta_m h_{j_m}(x)$.

We notice that the AdaBoost algorithm gives more and more weight in $\mathrm{err}_m(j)$ to the data points X_i, which are wrongly classified at the stage m.

You can observe AdaBoost in action (with half-plane weak-learners h_j) with the following recreative applet: `http://cseweb.ucsd.edu/~yfreund/adaboost/`.

11.3.5 Classifier Selection

The practical implementation of the SVM classifier or the AdaBoost classifier requires us to choose different quantities: the kernel k and the tuning parameter λ for the SVM, the weak learners $\{h_1,\ldots,h_p\}$ and the integer M for AdaBoost. The most popular technique for choosing these quantities is to apply a V-fold cross-validation scheme (see Chapter 7). Cross-validation can also be used to compare the SVM to the boosting, the Linear Discriminant Analysis, the logistic regression, etc.

11.4 Discussion and References

11.4.1 Take-Home Message

The discrete nature of the labels y enables us to relax the modeling of the data, by merely assuming that the data are i.i.d. For this reason, some may believe that the theory of classification presented in Section 11.2 is "model-free." It is not really the case, since Estimator (11.5) relies on a dictionary $\mathscr{H}$, which must provide a good approximation of the Bayes classifier h_* in the sense that $\min_{h\in\mathscr{H}} L(h) - L(h_*)$ must be small. Otherwise, the misclassification probability of Estimator (11.5) will be large and the theory is useless. There are then some implicit assumptions on the distribution of the data that are hidden in the choice of the dictionary $\mathscr{H}$ (or of the collection of dictionaries in Section 11.2.3). Yet, Estimator (11.5) enjoys some very nice statistical properties under very weak assumptions. Its main drawback is its high computational complexity that is prohibitive when n is larger than a few tens. As in Chapter 5, a powerful strategy is to relax the minimization problem (11.5) in order to obtain classifiers that are both computationally efficient and statistically grounded. Combining this strategy with the use of kernels provides a very flexible theory of classification that can handle data such as text, graph, images, etc. As usual, the resulting classifiers depend on some tuning parameters and, again, cross-validation is a useful technique for selecting them.

11.4.2 References

The mathematical foundations of supervised classification date back to the seminal work of Vapnik and Chervonenkis [155, 156] in the '70s, and this topic has since attracted a lot of effort. For the reader interested in going beyond the basic concepts presented in this chapter, we refer to the book by Devroye, Györfi, and Lugosi [62] and the survey by Boucheron, Bousquet, and Lugosi [37] for recent developments on the topic and a comprehensive bibliography. For more practical consideration, we refer to the book by Hastie, Tibshirani, and Friedman [91], where many practical algorithms are described and discussed. Finally, we point out that the concepts introduced here arise also for the ranking problem (rank at best some data, as your favorite search engine does); see, e.g., Clémençon, Lugosi and Vayatis [58].

11.5 Exercises

11.5.1 Linear Discriminant Analysis

Let us consider a couple of random variables (X,Y) with values in $\mathbb{R}^d \times \{-1,1\}$, and distribution

$$\mathbb{P}(Y=k) = \pi_k > 0 \quad \text{et} \quad \mathbb{P}(X \in dx | Y=k) = g_k(x)\,dx, \;\; k \in \{-1,1\},\; x \in \mathbb{R}^d,$$

where $\pi_{-1} + \pi_1 = 1$ and g_{-1}, g_1 are two densities in $\mathbb{R}^d$.

1. What is the distribution of X?

2. Check that the Bayes classifier is given by $h_*(x) = \text{sign}(\pi_1 g_1(x) - \pi_{-1} g_{-1}(x))$ for $x \in \mathbb{R}^d$.

We assume henceforth that

$$g_k(x) = (2\pi)^{-d/2} \sqrt{\det(\Sigma_k^{-1})} \exp\left(-\frac{1}{2}(x-\mu_k)^T \Sigma_k^{-1}(x-\mu_k)\right), \qquad k = -1, 1,$$

for two non-singular matrices Σ_{-1}, Σ_1 and $\mu_{-1}, \mu_1 \in \mathbb{R}^d$, with $\mu_{-1} \neq \mu_1$.

3. Prove that when $\Sigma_{-1} = \Sigma_1 = \Sigma$, the condition $\pi_1 g_1(x) > \pi_{-1} g_{-1}(x)$ is equivalent to
$$(\mu_1 - \mu_{-1})^T \Sigma^{-1} \left(x - \frac{\mu_1 + \mu_{-1}}{2}\right) > \log(\pi_{-1}/\pi_1).$$
4. What is the nature of the frontier between $\{h_* = 1\}$ and $\{h_* = -1\}$ in this case?
5. We assume in addition that $\pi_1 = \pi_{-1}$. Prove that
$$\mathbb{P}(h_*(X) = 1 | Y = -1) = \Phi(-d(\mu_1, \mu_{-1})/2),$$
where Φ is the standard Gaussian cumulative function and $d(\mu_1, \mu_{-1})$ is the Mahalanobis distance associated to Σ defined by $d(\mu_1, \mu_{-1}) = \|\Sigma^{-1/2}(\mu_1 - \mu_{-1})\|$.
6. When $\Sigma_1 \neq \Sigma_{-1}$, what is the nature of the frontier between $\{h_* = 1\}$ and $\{h_* = -1\}$?

In this exercise, we have analyzed the risk of the Bayes classifier h^*, which has a full knowledge of the parameters $\pi_1, \mu_1, \mu_{-1}, \Sigma$ of the distribution. We refer to Section 12.7.1, page 270, for an analysis of the LDA classifier (11.2), when $\Sigma = \sigma^2 I_n$.

11.5.2 VC Dimension of Linear Classifiers in $\mathbb{R}^d$

For any $w \in \mathbb{R}^d$, we denote by $h_w : \mathbb{R}^d \to \{-1, 1\}$, the classifier

$$h_w(x) = \text{sign}(\langle w, x \rangle), \quad \text{for } x \in \mathbb{R}^d.$$

We compute below the VC dimension of $\mathscr{H} = \{h_w : w \in \mathbb{R}^d\}$.

1. We write $e_1, \ldots, e_d$ for the canonical basis of $\mathbb{R}^d$. Prove that for any $\delta \in \{-1, 1\}^d$ there exists $w_\delta \in \mathbb{R}^d$, such that $h_{w_\delta}(e_i) = \delta_i$ for $i = 1, \ldots, d$. Give a lower bound on $d_{\mathscr{H}}$.
2. For any $x_1, \ldots, x_{d+1} \in \mathbb{R}^d$, there exists $\lambda \in \mathbb{R}^{d+1}$ nonzero, such that $\sum_{i=1}^{d+1} \lambda_i x_i = 0$. We can assume that there exists j, such that $\lambda_j > 0$, by changing λ in $-\lambda$ if necessary. We define δ by $\delta_i = \text{sign}(\lambda_i)$. By considering the sum $\sum_{i=1}^{d+1} \lambda_i \langle w_\delta, x_i \rangle$, prove that there exists no $w_\delta \in \mathbb{R}^d$, such that $h_{w_\delta}(x_i) = \delta_i$, for $i = 1, \ldots, d+1$. Conclude that $d_{\mathscr{H}} = d$.
3. We define $\tilde{\mathscr{H}} = \{h_{w,b} : w \in \mathbb{R}^d,\ b \in \mathbb{R}\}$, where $h_{w,b}(x) = \text{sign}(\langle w, x \rangle - b)$. Prove that the VC dimension of $\tilde{\mathscr{H}}$ is $d + 1$.

11.5.3 Linear Classifiers with Margin Constraints

Assume that $\mathscr{F}$ is a Hilbert space, and consider $\{x_1, \ldots, x_M\} \subset \mathscr{F}$ with $\|x_i\|_{\mathscr{F}} \leq A$ for any $i = 1, \ldots, M$. For $r, R > 0$ we set

$$\mathscr{H} = \left\{h_w : w \in \mathscr{F},\ \|w\|_{\mathscr{F}} \leq R \text{ and } |\langle w, x_i \rangle_{\mathscr{F}}| \geq r \text{ for } i = 1, \ldots, M\right\},$$

where $h_w : \{x_1, \ldots, x_M\} \to \{-1, 1\}$ is defined by $h_w(x) = \text{sign}\left(\langle w, x \rangle_{\mathscr{F}}\right)$ for any $x \in \{x_1, \ldots, x_M\}$. We assume henceforth that $M > A^2R^2/r^2$. We will prove that $d_{\mathscr{H}} \leq A^2R^2/r^2$.

1. For $n \leq M$ and $\sigma_1, \ldots, \sigma_n$ i.i.d. uniform on $\{-1, 1\}$, prove that

$$\mathbb{E}\left[\Big\|\sum_{i=1}^{n} \sigma_i x_i\Big\|_{\mathscr{F}}^2\right] = \sum_{i=1}^{n} \mathbb{E}\left[\|\sigma_i x_i\|_{\mathscr{F}}^2\right] \leq nA^2.$$

2. Conclude that there exists $y \in \{-1, 1\}^n$, such that $\left\|\sum_{i=1}^n y_i x_i\right\|_{\mathscr{F}}^2 \leq nA^2$.
3. Assume there exists $w \in \mathscr{F}$, such that $h_w \in \mathscr{H}$ and $y_i \langle w, x_i \rangle_{\mathscr{F}} \geq r$ for $i = 1, \ldots, n$. Prove that

$$nr \leq \left\langle w, \sum_{i=1}^{n} y_i x_i \right\rangle_{\mathscr{F}} \leq RA\sqrt{n}.$$

4. Show that $\mathbb{S}_n(\mathscr{H}) < 2^n$ for $R^2A^2/r^2 < n \leq M$ and conclude.

11.5.4 Spectral Kernel

The spectral kernel is a classical kernel for classifying "words" from a finite alphabet $\mathscr{A}$. For $x \in \bigcup_{n \geq q} \mathscr{A}^n$ and $s \in \mathscr{A}^q$, we set

$$N_s(x) = \text{number of occurence of } s \text{ in } x.$$

The spectral kernel is then defined by

$$k(x, y) = \sum_{s \in \mathscr{A}^q} N_s(x) N_s(y)$$

for all $x, y \in \bigcup_{n \geq q} \mathscr{A}^n$. It counts the number of common sequences of length q in x and y.

1. Prove that k is a positive definite kernel on $\bigcup_{n \geq q} \mathscr{A}^n$.
2. Check that the computational complexity for computing $k(x, y)$ is at most of order $\ell(x) + \ell(y)$, where $\ell(x)$ is the length of x.

11.5.5 Computation of the SVM Classifier

We consider the SVM classifier $\widehat{h}_{\text{SVM}} = \text{sign}\left(\sum_{j=1}^n \widehat{\beta}_j k(X_j, .)\right)$, with $\widehat{\beta}$ solution of (11.29). We assume that $K = [k(X_i, X_j)]_{i,j=1,\ldots,n}$ is non-singular.

1. From the Lagrangian problem (11.30) and the Karush–Kuhn–Tucker conditions prove that

$$\widehat{\beta} \in \underset{\beta \in \mathbb{R}^n}{\operatorname{argmin}} \left\{ \lambda \beta^T K \beta - \sum_{i=1}^{n} \widehat{\alpha}_i (Y_i (K\beta)_i - 1) \right\}$$

for some $\widehat{\alpha} \in \mathbb{R}^n$, fulfilling $0 \leq \widehat{\alpha}_i \leq 1/n$ for all $i = 1, \ldots, n$.

2. By Lagrangian duality, we know that $\widehat{\alpha}$ is a solution of

$$\widehat{\alpha} \in \underset{0 \leq \alpha_i \leq 1/n}{\operatorname{argmax}} \min_{\beta \in \mathbb{R}^n} \left\{ \lambda \beta^T K \beta - \sum_{i=1}^{n} \alpha_i (Y_i (K\beta)_i - 1) \right\}.$$

Proves that $\widehat{\beta}_i = \widehat{\alpha}_i Y_i / (2\lambda)$, where

$$\widehat{\alpha} \in \underset{0 \leq \alpha_i \leq 1/n}{\operatorname{argmax}} \left\{ \sum_{i=1}^{n} \alpha_i - \frac{1}{4\lambda} \sum_{i,j=1}^{n} K_{ij} y_i y_j \alpha_i \alpha_j \right\}.$$

In particular, the SVM classifier $\widehat{h}_{\text{SVM}}$ can be computed from a simple constrained quadratic maximization.

11.5.6 Kernel Principal Component Analysis

As discussed above, RKHS allows us to "delinearize" some linear algorithms. We give an example here with the Principal Component Analysis (see Exercise 1.6.4, page 22). Assume that we have n points $X^{(1)}, \ldots, X^{(n)} \in \mathcal{X}$, and let us consider an RKHS $\mathcal{F}$ associated to a positive definite kernel k on $\mathcal{X}$. We denote by ϕ the map from $\mathcal{X}$ to $\mathcal{F}$ defined by $\phi : x \to k(x, .)$. The principle of Kernel Principal Component Analysis (KPCA) is to perform a PCA on the points $\phi(X^{(1)}), \ldots, \phi(X^{(n)})$ mapped by ϕ in the RKHS. We then seek for the space $\mathcal{V}_d \subset \mathcal{F}$ fulfilling

$$\mathcal{V}_d \in \underset{\dim(\mathcal{V}) \leq d}{\operatorname{argmin}} \sum_{i=1}^{n} \|\phi(X^{(i)}) - \mathcal{P}_{\mathcal{V}} \phi(X^{(i)})\|_{\mathcal{F}}^2,$$

where the infimum is taken over all the subspaces $\mathcal{V} \subset \mathcal{F}$ with dimension not larger than d, and $\mathcal{P}_{\mathcal{V}}$ denotes the orthogonal projection onto $\mathcal{V}$ with respect to the Hilbert norm $\|.\|_{\mathcal{F}}$ on $\mathcal{F}$. In the following, we denote by $\mathcal{L}$ the linear map

$$\mathcal{L} : \mathbb{R}^n \to \mathcal{F}$$

$$\alpha \to \mathcal{L}\alpha = \sum_{i=1}^{n} \alpha_i \phi(X^{(i)}).$$

1. By adapting the arguments of Proposition 11.9, prove that $\mathcal{V}_d = \mathcal{L} V_d$, with V_d fulfilling

$$V_d \in \underset{\dim(V) \leq d}{\operatorname{argmin}} \sum_{i=1}^{n} \|\phi(X^{(i)}) - \mathcal{P}_{\mathcal{L}V} \phi(X^{(i)})\|_{\mathcal{F}}^2,$$

where the infimum is taken over all the subspaces $V \subset \mathbb{R}^n$ with dimension not larger than d.

2. We denote by K the $n \times n$ matrix with entries $K_{i,j} = k(X^{(i)}, X^{(j)})$ for $i, j = 1, \ldots, n$. We assume in the following that K is non-singular. Prove that $\|\mathscr{L}K^{-1/2}\alpha\|^2_{\mathscr{F}} = \|\alpha\|^2$ for any $\alpha \in \mathbb{R}^n$.
3. Let V be subspace of $\mathbb{R}^n$ of dimension d and denote by $(b_1, \ldots, b_d)$ an orthonormal basis of the linear span $K^{1/2}V$. Prove that $(\mathscr{L}K^{-1/2}b_1, \ldots, \mathscr{L}K^{-1/2}b_d)$ is an orthonormal basis of $\mathscr{L}V$.
4. Prove the identities

$$\begin{aligned}\mathscr{P}_{\mathscr{L}V}\mathscr{L}\alpha &= \sum_{k=1}^{d} \langle \mathscr{L}K^{-1/2}b_k, \mathscr{L}\alpha\rangle_{\mathscr{F}} \mathscr{L}K^{-1/2}b_k \\ &= \mathscr{L}K^{-1/2}\mathrm{Proj}_{K^{1/2}V}K^{1/2}\alpha,\end{aligned}$$

where $\mathrm{Proj}_{K^{1/2}V}$ denotes the orthogonal projector onto $K^{1/2}V$ in $\mathbb{R}^n$.
5. Let us denote by $(e_1, \ldots, e_n)$ the canonical basis of $\mathbb{R}^n$. Check that

$$\begin{aligned}\sum_{i=1}^{n} \|\phi(X^{(i)}) - \mathscr{P}_{\mathscr{L}V}\phi(X^{(i)})\|^2_{\mathscr{F}} &= \sum_{i=1}^{n} \|\mathscr{L}e_i - \mathscr{L}K^{-1/2}\mathrm{Proj}_{K^{1/2}V}K^{1/2}e_i\|^2_{\mathscr{F}} \\ &= \sum_{i=1}^{n} \|K^{1/2}e_i - \mathrm{Proj}_{K^{1/2}V}K^{1/2}e_i\|^2 \\ &= \|K^{1/2} - \mathrm{Proj}_{K^{1/2}V}K^{1/2}\|^2_F,\end{aligned}$$

where $\|.\|_F$ denotes the Frobenius norm $\|A\|^2_F = \sum_{i,j} A^2_{ij}$.
6. With Theorem C.5, page 315, in Appendix C, prove that $V_d = \mathrm{span}\{v_1, \ldots, v_d\}$, where $v_1, \ldots, v_d$ are eigenvectors of K associated to the d largest eigenvalues of K.
7. We set $f_k = \mathscr{L}K^{-1/2}v_k$ for $k = 1, \ldots, d$. Check that $(f_1, \ldots, f_d)$ is an orthonormal basis of $\mathscr{V}_d$ and

$$\mathscr{P}_{\mathscr{V}}\phi(X^{(i)}) = \sum_{k=1}^{n} \langle v_k, K^{1/2}e_i\rangle f_k.$$

So, in the basis $(f_1, \ldots, f_d)$, the coordinates of the orthogonal projection of the point $\phi(X^{(i)})$ onto $\mathscr{V}$ are $(\langle v_1, K^{1/2}e_i\rangle, \ldots, \langle v_d, K^{1/2}e_i\rangle)$.

Chapter 12

Clustering

In the previous chapters, the data were considered homogeneous: all the observations were distributed according to a common model. Such an assumption is valid for data coming from small-scale controlled experiments, but it is highly unrealistic in the era of "Big Data", where data come from multiple sources. A recipe for dealing with such inhomogeneous data is to consider them as an assemblage of several homogeneous data sets, corresponding to homogeneous "subpopulations". Then each subpopulation can be treated either independently or jointly. The main hurdle in this approach is to recover the unknown subpopulations, which is the main goal of clustering algorithms.

Clustering algorithms can also be used for some other purposes. Two important motivations are scientific understanding and data quantization. In many fields, finding groups of items with similar behavior is of primary interest, as finding structures is a first step towards the scientific understanding of complex systems. For example, finding groups of genes with similar expression level profiles is important in biology, as these genes are likely to be involved in a common regulatory mechanism. Summarizing a cloud of n data points by a smaller cloud of K points is another important motivation for clustering. For example, we may wish to summarize the expression level profiles of tens of thousands of genes by a small number of representative profiles (templates) and then only work with these templates which are lighter to handle.

The methodology for clustering can be based on some proximity-separation paradigms, or on some statistical models. The proximity-separation paradigm is model free and offers some easy-to-understand algorithms. It is yet difficult to define a clear "ground truth" objective in this perspective, and to evaluate the performance of a given algorithm. The statistical paradigm, based on a probabilistic modeling, is more easily amenable to interpretation and statistical analysis. Most of this chapter's focus is on this approach.

12.1 Proximity-Separation-Based Clustering

12.1.1 Clustering According to Proximity and Separation

Assume that we have n data points $X_1,\dots,X_n$ in a metric space $(\mathscr{X},d)$. Informally, the goal of clustering is to find a partition $G=\{G_1,\dots,G_K\}$ of the indices $\{1,\dots,n\}$, such that, data points with indices within a group are similar, and those with indices in different groups are different. Can we make this statement more formal?

The answer is no, in general, as the two objectives "group similar points" and "separate different points" can be contradictory. For example, consider the cells of your two legs, while standing. The cells of one leg can be considered similar, as they are in the same connected components. And the cells of different legs can be considered different, as they are in different connected components. Yet, when standing, the cells of your right foot are much closer to the cells of your left foot, than to the cells of your right thigh. Shall we group together cells of feet, or cells of a common leg? From this toy example, we observe that we have no "ground truth". Is it better to split cells in terms of "left / right legs" or in terms of "foot / thigh"? Hence, there are as many notions of proximity-separation-based clustering as there are notions of proximity and separation.

In this section, we briefly present the two most widely used algorithms for proximity-separation clustering: Kmeans algorithm and hierarchical clustering algorithm(s).

12.1.2 Kmeans Paradigm

The principle of Kmeans clustering is to represent the n data points $X_1,\dots,X_n \in \mathscr{X}$ by K representative points $\theta_1,\dots,\theta_K \in \mathscr{X}$ such that the cumulative residual square distance

$$\sum_{i=1}^{n} \min_{k=1,\dots,K} d(x_i,\theta_k)^2,$$

is minimal. Hence, Kmeans clustering seeks to solve the minimization problem

$$\left(\widehat{\theta}_1,\dots,\widehat{\theta}_K\right) \in \operatorname*{argmin}_{(\theta_1,\dots,\theta_K)\in\mathscr{X}^K} \sum_{i=1}^{n} \min_{k=1,\dots,K} d(x_i,\theta_k)^2. \tag{12.1}$$

Then, the partition $\widehat{G}^{Kmeans}$ is defined by

$$\widehat{G}_k^{Kmeans} = \left\{ i \in \{1,\dots,n\} : d(x_i,\theta_k) = \min_{k'=1,\dots,K} d(x_i,\theta_{k'}) \right\},$$

with ties broken randomly.

In general, solving the minimization problem (12.1) is NP-hard, and even hard to approximate [11]. We refer to Section 12.3 for an analysis of Kmeans clustering.

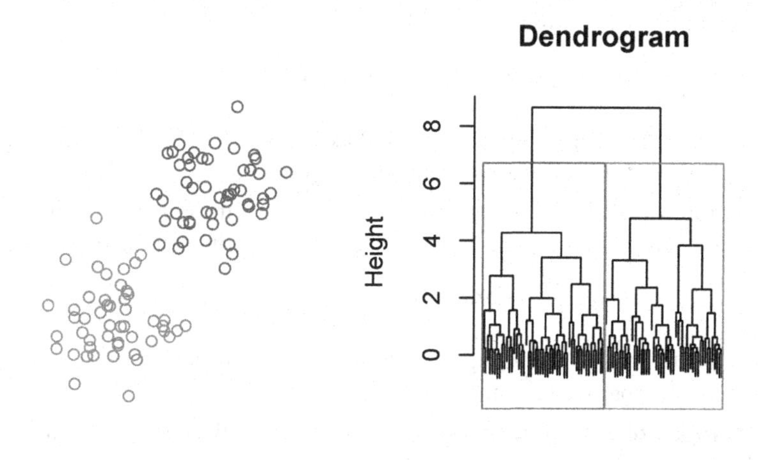

Figure 12.1 *Left: data points in $\mathbb{R}^2$. Right: dendrogram of hierarchical clustering with Euclidean distance d and complete linkage ℓ. The colors correspond to the clustering output when selecting $K = 2$ clusters.*

12.1.3 Hierarchical Clustering Algorithms

The principle of Kmeans clustering is to try to minimize the global criterion (12.1). The strategy in hierarchical clustering is different. The principle is to merge data points step-by-step, by merging at each step the two closest groups of points.

More precisely, the hierarchical clustering algorithms cluster data points sequentially, starting from a trivial partition with n singletons (each data point is a cluster on its own) and then merging them step-by-step until eventually getting a single cluster with all the data points. At the end of the process, we obtain a hierarchical family of nested clusterings and the data scientist can choose her favorite one.

Linkage

In hierarchical clustering, the recipe for merging points is quite simple: at each step the algorithm merges the two closest clusters (in a sense to be defined) of the current clustering, letting the other clusters unchanged. This requires the definition of a "distance" $\ell(G,G')$ between clusters G and G', usually called "linkage". Let $d(x,y)$ be any distance on $\mathcal{X}$; typically $d(x,y) = \|x-y\|$ or $d(x,y) = |x-y|_1$ when $\mathcal{X} = \mathbb{R}^p$. Some classical examples of linkage are:

- **Single linkage:** single linkage corresponds to the smallest distance between the points of the two clusters

$$\ell_{single}(G,G') = \min\left\{d(x_i,x_j) : i \in G,\ j \in G'\right\}.$$

Single linkage clustering tends to produce clusters looking like "chains", and we can have within a cluster two data points x,y with $d(x,y)$ very large.

- **Complete linkage:** complete linkage is somewhat the opposite of single linkage. It corresponds to the largest distance between the points of the two clusters

$$\ell_{complete}(G,G') = \max\left\{d(x_i,x_j) : x_i \in G,\ x_j \in G'\right\}.$$

Complete linkage clustering tends to produce "compact" clusters where all data points are close to each other.

- **Average linkage:** average linkage corresponds to the average distance between the points of the clusters G,G'

$$\ell_{average}(G,G') = \frac{1}{|G||G'|} \sum_{i\in G,\ j\in G'} d(x_i,x_j).$$

The clustering produced by average linkage is less "chainy" than those produced by single linkage and less compact than those produced by complete linkage.

Hierarchical Clustering Algorithm

Hierarchical clustering algorithms start from the trivial partition $G^{(n)} = \{\{1\},\ldots,\{n\}\}$ with n clusters, and then sequentially merge clusters two-by-two. At each step, the algorithm merges the two clusters G,G' available at this step with the smallest linkage $\ell(G,G')$. The output is a sequence of clustering $G^{(1)},\ldots,G^{(n)}$ with $K = 1,\ldots,n$ clusters. These clusterings are nested, in the sense that for $j \leq k$ the partition $G^{(k)}$ is a sub-partition of $G^{(j)}$.

Hierarchical clustering

- Input: data points $X_1,\ldots,X_n$ and a linkage ℓ
- Initialization: $G^{(n)} = \{\{1\},\ldots,\{n\}\}$
- Iterations: for $t = n,\ldots,2$
  - find $(\widehat{a},\widehat{b}) \in \text{argmin}_{(a,b)}\,\ell(G_a^{(t)},G_b^{(t)})$
  - build $G^{(t-1)}$ from $G^{(t)}$ by merging $G_{\widehat{a}}^{(t)}$ and $G_{\widehat{b}}^{(t)}$. The other clusters are left unchanged.
- Output: the n partitions $G^{(1)},\ldots,G^{(n)}$ of $\{1,\ldots,n\}$.

Dendrogram

It is popular to represent the sequence of clustering $G^{(1)},\dots,G^{(n)}$ with a dendrogram, which is a tree, rooted in $G^{(1)}$, and whose leaves correspond to $G^{(n)}$. The dendrogram depicts how the merging is performed. The partition $G^{(k)}$ can be read on the dendrogram as follows, see Figure 12.1:

1. locate the level where there are exactly k branches in the dendrogram;
2. cut the dendrogram at this level in order to get k subtrees; and
3. read the clustering as follows: Each subtree corresponds to one cluster, gathering the points corresponding to its leaves.

The height in the tree represents the distance between two clusters. A classical recipe for choosing the number k of clusters is to look for a level k where the height between two successive merges increases abruptly.

Hierarchical clustering algorithms are popular, as they are simple to understand and to visualize. When the clusters are well separated, they succeed to recover the hidden partition G^*, see Exercise 12.9.1. Yet, hierarchical clustering is based on local information, and does not take into account global information on the distribution of the cloud of points, especially at the first steps. As the mistakes in the first steps cannot be repaired in the following steps, it is a strong limitation for clustering in a less separated case. It is also not clear which global criteria hierarchical clustering seek to minimize and for which data distributions it has an optimal behavior.

In the remaining of this chapter, we change the point of view. Starting from a data distribution modeling, we seek algorithms achieving the best possible performance, with computational complexity constraints.

12.2 Model-Based Clustering

12.2.1 Gaussian Sub-Population Model

We assume from now on that the n observations $X_1,\dots,X_n \in \mathbb{R}^p$ are independent, but not identically distributed. We denote by $\mu_i = \mathbb{E}[X_i]$ the mean of X_i and by $\Sigma_i = \mathrm{Cov}(X_i)$ the covariance of X_i. As discussed in introduction, we assume that the distribution of the X_i is homogeneous across some subpopulations. This means that there exists an *unknown* partition $G^* = \{G_1^*,\dots,G_K^*\}$ of $\{1,\dots,n\}$, such that, within a group G_k^* all the means and covariances are equal. In this chapter, we restrict ourselves to the Gaussian setting and consider the following model.

Definition Gaussian cluster model

We assume that

1. *the observations $X_1,\dots,X_n \in \mathbb{R}^p$, are independent,*
2. *there exists a partition $G^* = \{G_1^*,\dots,G_K^*\}$ of $\{1,\dots,n\}$, and $\theta_1,\dots,\theta_K \in \mathbb{R}^p$, $\Lambda_1,\dots,\Lambda_K \in \mathbb{R}^{p\times p}$ such that*

$$\text{for all } i \in G_k^*: \quad X_i \sim \mathcal{N}(\theta_k,\Lambda_k). \tag{12.2}$$

As in the previous chapter, we denote by $\mathbf{X} \in \mathbb{R}^{n\times p}$ the matrix whose i-th row is given by X_i. We define similarly the matrices

- $M \in \mathbb{R}^{n\times p}$ the matrix whose i-th row is given by $\mathbb{E}[X_i] = \mu_i$;
- $\mathbf{E} \in \mathbb{R}^{n\times p}$ the matrix whose i-th row is given by $E_i = X_i - \mu_i$;
- $\Theta \in \mathbb{R}^{K\times p}$ the matrix whose k-th row is given by θ_k;
- and $A \in \mathbb{R}^{n\times K}$ the membership matrix defined by $A_{ik} = \mathbf{1}_{i\in G_k^*}$, for $i = 1,\ldots,n$, and $k = 1,\ldots,K$.

Then we have the compact formula

$$\mathbf{X} = M + \mathbf{E} = A\Theta + \mathbf{E}.$$

Remark: a popular variant of the Gaussian cluster model is the Gaussian mixture model. This model has an additional generating feature compared to the Gaussian cluster model: Instead of being arbitrary, the partition G^* is generated by sampling for each observation i the label of the group according to a probability distribution π on $\{1,\ldots,K\}$. We refer to Exercise 12.9.9 for properties of this model and the associated EM algorithm.

12.2.2 MLE Estimator

The negative log-likelihood of the distribution $\mathcal{N}(\theta_k, \Lambda_k)$ with respect to the observation X_i is

$$\frac{1}{2}(X_i - \theta_k)^T \Lambda_k^{-1}(X_i - \theta_k) + \frac{1}{2}\log(\det(\Lambda_k)) + \frac{p}{2}\log(2\pi),$$

so the maximum likelihood estimator of the partition G^* is given by

$$\widehat{G}^{MLE} \in \operatorname*{argmin}_G \sum_{k=1}^K \min_{\Lambda_k \in \mathcal{S}_p^+} \min_{\theta_k\in\mathbb{R}^p} \sum_{i\in G_k} \left((X_i - \theta_k)^T \Lambda_k^{-1}(X_i - \theta_k) + \log(\det(\Lambda_k))\right) \tag{12.3}$$

where $\mathcal{S}_p^+$ is the set of $p\times p$ positive semi-definite matrices and where the first minimum is over all partitions G of $\{1,\ldots,n\}$ into K groups.

The MLE estimator $\widehat{G}^{MLE}$ suffers from several drawbacks.

1. A first drawback is related to the cardinality of the set of partitions of $\{1,\ldots,n\}$ into K groups. This cardinality, known as Sterling number of the second kind, grows exponentially fast with n like $K^n/K!$, see Exercise 12.9.8. Hence, the computational cost of scanning the set of partitions of $\{1,\ldots,n\}$ into K groups is prohibitive and $\widehat{G}^{MLE}$ cannot be computed except for very small sample sizes.
2. From a more statistical point of view, the estimation of Λ_k is unstable in high-dimensional settings, and even degenerated when p is larger than n.

To avoid the second issue, a classical solution is to consider Criterion (12.3) with all Λ_k set to $\sigma^2 I_n$. This leads to the already met Kmeans criterion.

12.3 Kmeans

12.3.1 Kmeans Algorithm

When setting all Λ_k to $\sigma^2 I_n$, Criterion (12.3) reduces to the Kmeans criterion (12.1). Let us observe that

$$\min_{\theta_k \in \mathbb{R}^p} \sum_{i \in G_k} \|X_i - \theta_k\|^2 = \sum_{i \in G_k} \|X_i - \overline{X}_{G_k}\|^2, \quad \text{where} \quad \overline{X}_{G_k} = \frac{1}{|G_k|} \sum_{i \in G_k} X_i$$

is the mean of the observations in G_k.
Then, the minimization of (12.3) when all Λ_k have been set to $\sigma^2 I_n$, is given by the Kmeans clustering.

$$\widehat{G}^{Kmeans} \in \underset{G}{\operatorname{argmin}} \operatorname{Crit}_K(G), \quad \text{with } \operatorname{Crit}_K(G) = \sum_{k=1}^{K} \sum_{i \in G_k} \|X_i - \overline{X}_{G_k}\|^2. \tag{12.4}$$

We have the alternative formulas for Kmeans

Lemma 12.1 Alternative formulas for Kmeans.

Symmetrization: we have

$$\operatorname{Crit}_K(G) = \frac{1}{2} \sum_{k=1}^{K} \frac{1}{|G_k|} \sum_{i,j \in G_k} \|X_i - X_j\|^2. \tag{12.5}$$

Matrix form: the Kmeans minimization problem can be written as

$$\widehat{G}^{Kmeans} \in \underset{G}{\operatorname{argmax}} \langle \mathbf{X}\mathbf{X}^T, B(G) \rangle_F, \tag{12.6}$$

where the $p \times p$ matrix $B(G)$ is defined by

$$B_{ij}(G) = \frac{1}{|G_k|} \quad \text{if } i, j \in G_k,\ k = 1, \ldots, K,$$

and $B_{ij}(G) = 0$ if i and j are in different groups.

Proof of Lemma 12.1.
Symmetrization: Let us observe that

$$\frac{1}{|G_k|} \sum_{i \in G_k} \|X_i - \overline{X}_{G_k}\|^2 = \frac{1}{|G_k|} \sum_{i \in G_k} \langle X_i - \overline{X}_{G_k}, X_i \rangle - \frac{1}{|G_k|} \sum_{i \in G_k} \langle X_i - \overline{X}_{G_k}, \overline{X}_{G_k} \rangle.$$

with the second sum equal to 0 by bilinearity of the scalar product. Hence, by symmetry

$$\sum_{i\in G_k} \|X_i - \overline{X}_{G_k}\|^2 = \frac{1}{|G_k|}\sum_{i,j\in G_k} \langle X_i - X_j, X_i\rangle$$
$$= \frac{1}{2}\left(\frac{1}{|G_k|}\sum_{i,j\in G_k} \langle X_i - X_j, X_i\rangle + \frac{1}{|G_k|}\sum_{i,j\in G_k} \langle X_j - X_i, X_j\rangle\right)$$
$$= \frac{1}{2}\frac{1}{|G_k|}\sum_{i,j\in G_k} \|X_i - X_j\|^2.$$

Summing over k gives Equation (12.5).

Matrix form: We observe that

$$\sum_{k=1}^{K} \frac{1}{|G_k|}\sum_{i,j\in G_k} \|X_i\|^2 = \sum_{i=1}^{n} \|X_i\|^2 = \|\mathbf{X}\|_F^2,$$

so expanding the squares in (12.5) gives

$$\mathrm{Crit}_K(G) = \|\mathbf{X}\|_F^2 - \sum_{k=1}^{K}\sum_{i,j\in G_k} \frac{1}{|G_k|}\langle X_i, X_j\rangle.$$

The first term in the right-hand side is independent of G, and the second term can be written as $\langle \mathbf{X}\mathbf{X}^T, B(G)\rangle_F$, since $[\mathbf{X}\mathbf{X}^T]_{ij} = \langle X_i, X_j\rangle$. Equation (12.6) follows. □

12.3.2 Bias of Kmeans

Let us denote by Γ the diagonal matrix, with entries $\Gamma_{ii} = \mathrm{Tr}(\mathrm{Cov}(X_i)) = \mathrm{Tr}(\mathrm{Cov}(E_i))$. Since, for any $i \neq j$

$$\mathbb{E}\left[\|X_i - X_j\|^2\right] = \|\mathbb{E}[X_i - X_j]\|^2 + \mathrm{Tr}(\mathrm{Cov}(X_i - X_j)) = \|\mu_i - \mu_j\|^2 + \Gamma_{ii} + \Gamma_{jj},$$

we get that for any partition G of $\{1,\ldots,n\}$ into K groups

$$\mathbb{E}[\mathrm{Crit}_K(G)] = \frac{1}{2}\sum_{k=1}^{K}\frac{1}{|G_k|}\sum_{i,j\in G_k}\left(\|\mu_i - \mu_j\|^2 + \Gamma_{ii} + \Gamma_{jj}\right)\mathbf{1}_{i\neq j}$$
$$= \frac{1}{2}\sum_{k=1}^{K}\frac{1}{|G_k|}\sum_{i,j\in G_k}\|\mu_i - \mu_j\|^2 + \mathrm{Tr}(\Gamma) - \sum_{k=1}^{K}\frac{1}{|G_k|}\sum_{i\in G_k}\Gamma_{ii}. \quad (12.7)$$

When all the covariances are the same, we have $\Gamma_{ii} = \mathrm{Tr}(\Gamma)/p$ so

$$\mathbb{E}[\mathrm{Crit}_K(G)] = \frac{1}{2}\sum_{k=1}^{K}\frac{1}{|G_k|}\sum_{i,j\in G_k}\|\mu_i - \mu_j\|^2 + \left(1 - \frac{K}{p}\right)\mathrm{Tr}(\Gamma).$$

Since the second term in the right-hand side of the equality is independent of G, we observe that $\mathbb{E}[\mathrm{Crit}_K(G)]$ is minimum for the target partition G^*, as it is the only partition for which $\mu_i = \mu_j$ for any i, j in the same group.

Yet, when the Γ_{ii} do not all have the same value, the minimum of (12.7) may not be G^*. Actually, smaller values than $\mathbb{E}[\text{Crit}_K(G^*)]$ can be achieved by splitting groups with large Γ_{ii} in order to increase the last sum in (12.7). We refer to Exercise 12.9.2 for such an example.

The origin of the bias is easily seen on the formulation (12.6) of the Kmeans criterion. Indeed, since

$$(\mathbf{X}\mathbf{X}^T)_{ij} = (A\Theta\Theta^T A^T)_{ij} + E_i^T E_j + E_i^T(\Theta^T A^T)_j + (A\Theta)_i^T E_j,$$

with $\mathbb{E}\left[E_i^T E_j\right] = \mathbf{1}_{i=j}\Gamma_{ii}$, we have

$$\mathbb{E}\left[\mathbf{X}\mathbf{X}^T\right] = A\Theta\Theta^T A^T + \Gamma,$$

which is the sum of a block structured matrix $A\Theta\Theta^T A^T$ and a diagonal matrix Γ. When the diagonal matrix Γ is not proportional to the idendity, it can blur the signal $A\Theta\Theta^T A^T$ as in Exercise 12.9.2. To overcome this issue, a solution is to "debias" the Gram matrix $\mathbf{X}\mathbf{X}^T$ by removing an estimator $\widehat{\Gamma}$ of Γ.

12.3.3 Debiasing Kmeans

For a given estimator $\widehat{\Gamma}$ of Γ, the debiased Kmeans criterion is defined by

$$\widehat{G}^{Debiased-Kmeans} \in \underset{G}{\operatorname{argmax}} \langle \mathbf{X}\mathbf{X}^T - \widehat{\Gamma}, B(G)\rangle_F, \tag{12.8}$$

where the $p \times p$ matrix $B(G)$ is defined by

$$B_{ij}(G) = \frac{1}{|G_k|} \quad \text{if } i,j \in G_k,\ k = 1,\ldots,K,$$

and $B_{ij}(G) = 0$ if i and j are in different groups.

A first approach to handle the bias induced by $\widetilde{\Gamma}_{ii} := \|E_i\|^2$ is to remove the diagonal of $\mathbf{X}\mathbf{X}^T$, which amounts to take

$$\widehat{\Gamma}^{(0)} = \operatorname{diag}(\mathbf{X}\mathbf{X}^T).$$

Let us define the quantities

$$\|\Theta\|_{2\infty} = \max_k \|\theta_k\|, \quad |\Lambda|_{\text{op}} = \max_k |\Lambda_k|_{\text{op}} \quad \text{and} \quad \|\Lambda\|_F = \max_k \|\Lambda_k\|_F. \tag{12.9}$$

Then, since

$$\widehat{\Gamma}_{ii}^{(0)} - \widetilde{\Gamma}_{ii} = \|X_i\|^2 - \|E_i\|^2 = \|\mu_i\|^2 + 2\langle \mu_i, E_i\rangle,$$

with $\langle \mu_i, E_i\rangle$ distributed as a $\mathcal{N}(0, \mu_i^T\Sigma_i\mu_i)$ Gaussian random variable, and since

$$\mu_i^T\Sigma_i\mu_i \leq |\Sigma_i|_{\text{op}}\|\mu_i\|^2 \leq |\Lambda|_{\text{op}}\|\Theta\|_{2\infty},$$

we get for any $L \geq 0$

$$\mathbb{P}\left[|\widehat{\Gamma}^{(0)} - \widetilde{\Gamma}|_\infty \leq \|\Theta\|_{2\infty}^2 + \|\Theta\|_{2\infty}\sqrt{2|\Lambda|_{\text{op}}(\log(n) + L)}\right] \leq 1 - e^{-L}.$$

We observe that this bound involves $\|\Theta\|_{2\infty}^2$, which can be much larger than the mean differences $\|\mu_i - \mu_j\|^2$. Worst, this error is systematic, in the sense that $\mathbb{E}\left[\widehat{\Gamma}_{ii}^{(0)} - \widetilde{\Gamma}_{ii}\right] = \|\mu_i\|^2$ for $i = 1, \ldots, n$. Since debiasing with $\widehat{\Gamma}^{(0)}$ amounts to replace Γ_{ii} by $\Gamma_{ii} - \mathbb{E}\left[\widehat{\Gamma}_{ii}^{(0)}\right] = -\|\mu_i\|^2$ in (12.7), removing the diagonal amounts to replace the bias

$$-\sum_{k=1}^{K} \frac{1}{|G_k|} \sum_{i \in G_k} \Gamma_{ii} \quad \text{by the bias} \quad \sum_{k=1}^{K} \frac{1}{|G_k|} \sum_{i \in G_k} \|\mu_i\|^2.$$

In the case where the means in the different groups have quite different norms, this bias tends to enforce a splitting of the groups with small means, and a merging of the groups with large means.

In order to avoid this unwanted phenomenon, it is possible to derive some more subtle estimators $\widehat{\Gamma}$, with error bound independent of $\|\Theta\|_{2\infty}$. Let us explain a first step in this direction. A more complex estimator completely removing the dependence in $\|\Theta\|_{2\infty}$ is described in Exercise 12.9.3.

If we knew the partition G^*, "estimating[1]" the random quantity $\widetilde{\Gamma}_{ii}$ (or the parameter Γ_{ii}) would be a simple task. Actually, if $i' \neq i$ belongs to the same group as i, then

$$\langle X_i - X_{i'}, X_i \rangle = \|E_i\|^2 - \langle E_i, E_{i'} \rangle + \langle \mu_i, E_i - E_{i'} \rangle, \tag{12.10}$$

with $\mathbb{E}\left[\langle E_i, E_{i'} \rangle\right] = \mathbb{E}\left[\langle \mu_i, E_i - E_{i'} \rangle\right] = 0$. So $\langle X_i - X_{i'}, X_i \rangle$ is an unbiased "estimator" of $\widetilde{\Gamma}_{ii}$ (and Γ_{ii}).
The difficulty is that we do not know G^*, and we need to estimate Γ_{ii} to estimate G^*. We can build yet on (12.10), by replacing i' by a data-driven choice $\widehat{i}$. We observe that we have the decomposition

$$\langle X_i - X_{\widehat{i}}, X_i \rangle - \langle X_i - X_{i'}, X_i \rangle = \langle X_i - X_{\widehat{i}}, X_{i'} \rangle - \langle X_i - X_{i'}, X_{\widehat{i}} \rangle,$$

and we will be able to control the size of the right-hand side if we are able to control $\max_k |\langle X_i - X_{\widehat{i}}, X_k \rangle|$. This observation motivates the definition of the following estimator

$$\widehat{\Gamma}_{ii}^{(1)} = \langle X_i - X_{\widehat{i}}, X_i \rangle, \quad \text{with} \quad \widehat{i} \in \underset{j}{\operatorname{argmin}} \max_{k: k \neq i, j} |\langle X_i - X_j, X_k \rangle|. \tag{12.11}$$

This estimator fulfills the following bound.

> **Lemma 12.2** *Assume that each group G_k has a cardinality as least 2. Then, with probability at least $1 - 2e^{-L}$, we have*
>
> $$|\widehat{\Gamma}^{(1)} - \widetilde{\Gamma}|_\infty$$
> $$\leq 6\|\Theta\|_{2\infty}\sqrt{|\Lambda|_{op}(3\log(n) + L)} + 10(\|\Lambda\|_F\sqrt{2\log(n) + L}) \vee (|\Lambda|_{op}(4\log(n) + 2L)).$$

[1]we use the word "estimation" even if $\widetilde{\Gamma}_{ii}$ is a random quantity, not a parameter

Proof of Lemma 12.2. Take $i' \neq i$ in the same group as i. Starting from the decomposition

$$\langle X_i - X_{\widehat{i}}, X_i\rangle = \langle X_i - X_{i'}, X_i\rangle + \langle X_i - X_{\widehat{i}}, X_{i'}\rangle - \langle X_i - X_{i'}, X_{\widehat{i}}\rangle$$

and using $\mu_i = \mu_{i'}$, as well as the definition of $\widehat{i}$, we get

$$\begin{aligned}
|\widehat{\Gamma}^{(1)}_{ii} - \widetilde{\Gamma}_{ii}| \leq & |\langle \mu_i, E_i - E_{i'}\rangle| + |\langle E_{i'}, E_i\rangle| + \max_{k\neq i,\widehat{i}} |\langle X_i - X_{\widehat{i}}, X_k\rangle| + \max_{k\neq i,i'} |\langle X_i - X_{i'}, X_k\rangle| \\
\leq & |\langle \mu_i, E_i - E_{i'}\rangle| + |\langle E_{i'}, E_i\rangle| + 2\max_{k\neq i,i'} |\langle X_i - X_{i'}, X_k\rangle| \\
\leq & 3\max_{k,i,i'} |\langle \mu_k, E_i - E_{i'}\rangle| + 5\max_{k,i} |\langle E_i, E_k\rangle|.
\end{aligned}$$

Since $\langle \mu_k, E_i - E_{i'}\rangle$ follows a $\mathcal{N}(0, \mu_k^T(\Sigma_i + \Sigma_{i'})\mu_k)$ Gaussian distribution with $\mu_k^T(\Sigma_i + \Sigma_{i'})\mu_k \leq 2|\Lambda|_{\text{op}}\|\Theta\|_{2\infty}$, a simple union bound gives

$$\begin{aligned}
&\mathbb{P}\left[\max_{k,i,i'} |\langle \mu_k, E_i - E_{i'}\rangle| > 2\|\Theta\|_{2\infty}\sqrt{|\Lambda|_{\text{op}}(3\log(n) + L)}\right] \\
&\leq \sum_{i,i',k=1}^{n} \mathbb{P}\left[|\langle \mu_k, E_i - E_{i'}\rangle| > 2\|\Theta\|_{2\infty}\sqrt{|\Lambda|_{\text{op}}(3\log(n) + L)}\right] \leq e^{-L}.
\end{aligned}$$

We can write the scalar product $\langle E_i, E_k\rangle$ as $\varepsilon_i^T \Sigma_i^{1/2}\Sigma_k^{1/2}\varepsilon_k$, with $\varepsilon_i, \varepsilon_k$, two independent standard Gaussian random variables in $\mathbb{R}^p$. Hence, Hanson-Wright inequality (B.6) on page 303 ensures that

$$\mathbb{P}\left[|\langle E_i, E_k\rangle| > (2\|\Sigma_i^{1/2}\Sigma_k^{1/2}\|_F\sqrt{L}) \vee (4|\Sigma_i^{1/2}|_{\text{op}}|\Sigma_k^{1/2}|_{\text{op}}L)\right] \leq e^{-L}.$$

So, a union bound gives

$$\mathbb{P}\left[\max_{i\neq k} |\langle E_i, E_k\rangle| > (2\|\Lambda\|_F\sqrt{2\log(n) + L}) \vee (4|\Lambda|_{\text{op}}(2\log(n) + L))\right] \leq e^{-L}. \tag{12.12}$$

The proof of Lemma 12.2 is complete. □

We observe that the upper bound in Lemma 12.2 still depends on $\|\Theta\|_{2\infty}$. The improvement upon $\widehat{\Gamma}^{(0)}$ is that the dependence is linear, instead of quadratic. In order to get rid of this dependence, we must go one step further and consider an estimator of the form $\langle X_i - X_{\widehat{i}_1}, X_i - X_{\widehat{i}_2}\rangle$ for two well-chosen indices $\widehat{i}_1, \widehat{i}_2$. We refer to the Exercise 12.9.3 for such an estimator.

12.4 Semi-Definite Programming Relaxation

While Kmeans has the nice feature of not requiring to estimate all the covariance matrices Λ_k, it suffers from the same computational issue as the MLE, since we need to explore the set of partitions in order to minimize it.

In the spirit of Chapter 5, an approach to overcome this issue is to convexify the optimization problem. In this direction, the formulation (12.6) is useful as the set $\{B(G) : G \text{ partition into } K \text{ groups}\}$ is easily amenable to a convexification. Next lemma gives a helpful representation of this set.

Lemma 12.3 *Let us denote by* $\mathbf{1}$ *the n-dimension vector with all entries equal to 1, and by* $\mathcal{S}_n^+$ *the set of* $n \times n$ *symmetric positive semi-definite matrices. Then, for any* $1 \le K \le n$, *we have*

$$\{B(G) : G \text{ partition of } \{1,\ldots,n\} \text{ into } K \text{ groups}\} \\ = \left\{B \in \mathcal{S}_n^+ : B^2 = B,\ B\mathbf{1} = \mathbf{1},\ \mathrm{Tr}(B) = K,\ B \ge 0\right\} \tag{12.13}$$

where the matrix $B(G)$ *has been defined in Lemma 12.1, page 257 and where* $B \ge 0$ *denotes the set of constraints* $B_{ij} \ge 0$, *for* $i, j = 1, \ldots, n$.

Proof of Lemma 12.3. First, we observe that all matrices $B(G)$ fulfill the properties $B \in \mathcal{S}_n^+,\ B^2 = B,\ B\mathbf{1} = \mathbf{1},\ \mathrm{Tr}(B) = K,\ B \ge 0$, so the left-hand side set is included in the right-hand side set.
Let us consider a matrix B in the right-hand side set. The properties $B \in \mathcal{S}_n^+$, $B^2 = B$, and $\mathrm{Tr}(B) = K$ enforce that B is an orthogonal projector of rank K. Let us consider an index a corresponding to the maximum value on the diagonal $B_{aa} = \max_b B_{bb}$. The three properties $B_{aa} = [B^2]_{aa}$ (since $B^2 = B$), $B_{ab} \le \sqrt{B_{aa}B_{bb}} \le B_{aa}$ (Cauchy–Schwartz), and $\sum_b B_{ab} = 1$ (since $B\mathbf{1} = \mathbf{1}$) give

$$B_{aa} = \sum_{b=1}^{n} B_{ab}B_{ab} \le \sum_{b=1}^{n} B_{ab}B_{aa} = B_{aa}.$$

So the inequality must be an equality in the above display. As a consequence, B_{ab} is either equal to 0 or B_{aa}. Let us define $G = \{b : B_{ab} = B_{aa}\}$. Since

$$1 = \sum_{b=1}^{p} B_{ab} = \sum_{b \in G} B_{ab} = |G| B_{aa},$$

we get that $B_{ab} = 1/|G|$ for all $b \in G$, and $B_{ab} = 0$ otherwise.
Furthermore, since for any $b \in G$ we have

$$B_{aa} = B_{ab} \le \sqrt{B_{aa}B_{bb}} \le B_{aa},$$

we obtain that $B_{bb} = B_{aa}$ for all $b \in G$. Following the same reasoning as for a, we get that $B_{bc} = 1/|G|$ for all $c \in G$, and $B_{bc} = 0$ otherwise. So we have the block-decomposition

$$B = \begin{bmatrix} B_{GG} & B_{GG^c} \\ B_{G^cG} & B_{G^cG^c} \end{bmatrix} = \begin{bmatrix} \frac{1}{|G|} & 0 \\ 0 & B_{G^cG^c} \end{bmatrix}.$$

We observe that the matrix $B_{G^cG^c}$ belongs to the set

$$\left\{B \in \mathcal{S}_{n-|G|}^+ : B^2 = B,\ B\mathbf{1} = \mathbf{1},\ \mathrm{Tr}(B) = K-1,\ B \ge 0\right\},$$

so we can apply the same reasoning as before and by induction we get that there exists $G_1,\dots,G_K$ disjoints such that

$$B = \begin{bmatrix} \frac{1}{|G_1|} & 0 & \dots & 0 \\ \ddots & \ddots & \ddots & 0 \\ 0 & \dots & \frac{1}{|G_K|} & 0 \\ 0 & \dots & \dots & 0 \end{bmatrix},$$

where the $(K+1)$-th group with only 0, may be empty. To conclude, we observe that since $B\mathbf{1} = \mathbf{1}$, the $(K+1)$-th group is actually empty, so $\{G_1,\dots,G_K\}$ is a partition of $\{1,\dots,n\}$ and $B = B(\{G_1,\dots,G_K\})$. □

A consequence of Lemmas 12.1 and 12.3, is that Kmeans algorithm (12.4) is equivalent to the maximization problem

$$\widehat{B}^{Kmeans} \in \operatorname*{argmax}_{B \text{ in } (12.13)} \langle \mathbf{X}\mathbf{X}^T, B\rangle,$$

with the partition $\widehat{G}^{Kmeans}$ obtained by grouping together indices i,j such that $\widehat{B}^{Kmeans}_{ij} \neq 0$.

From an optimization point of view, we observe that the objective function $B \to \langle \mathbf{X}\mathbf{X}^T, B\rangle$ is linear, the constraints $B\mathbf{1} = \mathbf{1}$, $\text{Tr}(B) = K$, $B \geq 0$ are linear and the constraint $B \in \mathscr{S}_n^+$ is conic. Hence, the difficulty of the optimization problem is only due to the projector constraint $B^2 = B$. Hence, a simple way to convexify the problem is to drop out this constraint $B^2 = B$ and to compute the solution of the Semi-Definite Programming (SDP) problem

$$\widehat{B}^{SDP} \in \operatorname*{argmax}_{B \in \mathscr{C}} \langle \mathbf{X}\mathbf{X}^T, B\rangle, \quad \text{with} \quad \mathscr{C} = \left\{B \in \mathscr{S}_n^+ : \, B\mathbf{1} = \mathbf{1}, \, \text{Tr}(B) = K, \, B \geq 0\right\}, \tag{12.14}$$

which can be solved numerically in polynomial time.

Since we have dropped the condition $B^2 = B$ in $\mathscr{C}$, we may not have $\widehat{B}^{SDP} \in \{B(G) : G \text{ partition into } K \text{ groups}\}$, so in general, the matrix $\widehat{B}^{SDP}$ is not linked to a specific $\widehat{G}^{SDP}$ by $\widehat{B}^{SDP} = B(\widehat{G}^{SDP})$. In order to get a partition, we need to apply some "rounding" to the entries of $\widehat{B}^{SDP}$. For example, we can run a basic clustering algorithm on the entries of $\widehat{B}^{SDP}$, like hierarchical clustering algorithm to get an estimated partition $\widehat{G}^{SDP}$.

For the same reason as before, when Γ is not proportional to the identity, it is wise to replace the Gram matrix $\mathbf{X}\mathbf{X}^T$ in (12.14) by a debiased version $\mathbf{X}\mathbf{X}^T - \widehat{\Gamma}$.

This SDP-Kmeans clustering algorithm has some nice theoretical properties, and can be computed in polynomial time. Yet, in practice, it suffers of a quite high computational cost since $\mathscr{C}$ gathers $O(n^2)$ linear constraints together with a semi-definite positive constraint. Computationally more efficient alternatives are needed for n larger than a few hundreds. Similarly as in Chapter 6, instead of following the convexification approach of Chapter 5, we may try a more greedy approach, seeking to approximately minimize the Kmeans criterion (12.4). This approach is implemented in the Lloyd algorithm.

12.5 Lloyd Algorithm

The principle of the Lloyd algorithm is to decouple the estimation of the partition G and the estimation of the centers θ_k in (12.4), and to alternate between the two estimation problems.

Lloyd algorithm

Initialization: start from $\widehat{G}^{(0)}$ and $t=0$.

Iterate: until convergence

1. $\widehat{\theta}_k^{(t)} = \overline{X}_{G_k^{(t)}}$, for $k=1,\ldots,K$;
2. $G_k^{(t+1)} = \left\{i : k \in \operatorname*{argmin}_{j=1,\ldots,K} \|X_i - \widehat{\theta}_j^{(t)}\|^2\right\}$, for $k=1,\ldots,K$;
3. $t \leftarrow t+1$.

Output: $G^{(t)} = \left\{G_1^{(t)},\ldots,G_K^{(t)}\right\}$

Each iteration of the Lloyd algorithm is cheap in terms of computations, and the running cost of the Lloyd algorithm is low. Yet, it suffers from two drawbacks. First, as Kmeans, it suffers from a bias, and this issue is even stronger in this case as it happens even when the covariances are all equal (see below). Second, the output of the algorithm is very sensitive to the initialization, so it must be initialized with care. Let us detail these two points.

12.5.1 Bias of Lloyd Algorithm

Let us consider two groups G_j and G_k. At the second stage of Lloyd iterations, we ideally would like to have $\|X_i - \overline{X}_{G_j}\|^2 < \|X_i - \overline{X}_{G_k}\|^2$ as soon as $\|\mu_i - \overline{\mu}_{G_j}\|^2 < \|\mu_i - \overline{\mu}_{G_k}\|^2$. Let us investigate the population values of these quantities. Writing $\tilde{\mathbf{1}}_G$ for the vector with entries equal to $[\tilde{\mathbf{1}}_G]_i = 1/|G|$ if $i \in G$ and $[\tilde{\mathbf{1}}_G]_i = 0$ if $i \notin G$, and writing $\overline{\Gamma}_G$ for the average of the values $\{\Gamma_{ii} : i \in G\}$ and e_i for the i-th vector of the canonical basis, we have

$$\begin{aligned}
&\mathbb{E}\left[\|X_i - \overline{X}_{G_j}\|^2 - \|X_i - \overline{X}_{G_k}\|^2\right] \\
&= (e_i - \tilde{\mathbf{1}}_{G_j})^T \mathbb{E}\left[\mathbf{X}\mathbf{X}^T\right](e_i - \tilde{\mathbf{1}}_{G_j}) - (e_i - \tilde{\mathbf{1}}_{G_k})^T \mathbb{E}\left[\mathbf{X}\mathbf{X}^T\right](e_i - \tilde{\mathbf{1}}_{G_k}) \\
&= \|\mu_i - \overline{\mu}_{G_j}\|^2 - \|\mu_i - \overline{\mu}_{G_k}\|^2 + (e_i - \tilde{\mathbf{1}}_{G_j})^T \Gamma (e_i - \tilde{\mathbf{1}}_{G_j}) - (e_i - \tilde{\mathbf{1}}_{G_k})^T \Gamma (e_i - \tilde{\mathbf{1}}_{G_k}) \\
&= \|\mu_i - \overline{\mu}_{G_j}\|^2 - \|\mu_i - \overline{\mu}_{G_k}\|^2 + \left(\frac{\overline{\Gamma}_{G_j}}{|G_j|} - \frac{\overline{\Gamma}_{G_k}}{|G_k|}\right) - 2\Gamma_{ii}\left(\frac{\mathbf{1}_{i\in G_j}}{|G_j|} - \frac{\mathbf{1}_{i\in G_k}}{|G_k|}\right).
\end{aligned}$$

We observe that $\mathbb{E}\left[\|X_i - \overline{X}_{G_j}\|^2 - \|X_i - \overline{X}_{G_k}\|^2\right]$ is equal to $\|\mu_i - \overline{\mu}_{G_j}\|^2 - \|\mu_i - \overline{\mu}_{G_k}\|^2$ plus two additional terms. These two terms have different impact on the population version of the Lloyd algorithm. The first term tends to favor groups with small sizes, and as for the Kmeans criterion it tends to split groups with large variances. The second term is of a different nature: it tends to favor the group $G_j^{(t)}$ to which i belongs to at step t. As a consequence, this term can prevent from reclassifying a misclassified X_i.

We emphasize that, contrary to the Kmeans criterion (12.4), this bias exists even in the case where Γ is proportional to the identity matrix $\Gamma = |\Gamma|_\infty I_n$. Indeed, in this case we have

$$\mathbb{E}\left[\|X_i - \overline{X}_{G_j}\|^2 - \|X_i - \overline{X}_{G_k}\|^2\right] = \|\mu_i - \overline{\mu}_{G_j}\|^2 - \|\mu_i - \overline{\mu}_{G_k}\|^2 + |\Gamma|_\infty \left(\frac{1 - 2\mathbf{1}_{i\in G_j}}{|G_j|} - \frac{1 - 2\mathbf{1}_{i\in G_k}}{|G_k|}\right).$$

So, even when all the covariances Σ_i are the same, we need to remove the bias from the Lloyd updates. Similarly to Kmeans, we reduce the bias by subtracting to $\mathbf{X}\mathbf{X}^T$ an estimator $\widehat{\Gamma}$ of Γ.

Debiased Lloyd algorithm

Input: $\widehat{\Gamma}$, $\widehat{G}^{(0)}$ and $t = 0$.

Iterate: until convergence

1. $\widehat{\theta}_k^{(t)} = \overline{X}_{G_k^{(t)}}$, for $k = 1, \ldots, K$;
2. $G_k^{(t+1)} = \left\{i : k \in \underset{j=1,\ldots,K}{\operatorname{argmin}} (e_i - \tilde{\mathbf{1}}_{G_j^{(t)}})^T (\mathbf{X}\mathbf{X}^T - \widehat{\Gamma})(e_i - \tilde{\mathbf{1}}_{G_j^{(t)}})\right\}$, for $k = 1, \ldots, K$;
3. $t \leftarrow t + 1$.

Output: $G^{(t)} = \left\{G_1^{(t)}, \ldots, G_K^{(t)}\right\}$

For the special case where all the covariances are equal to a common matrix Σ, so that $\Gamma = \text{Tr}(\Sigma) I_n$, we can use the following simple estimator

$$\widehat{\Gamma} = \frac{1}{2} \min_{i \neq j} \|X_i - X_j\|^2 I_n \tag{12.15}$$

of Γ.

Lemma 12.4 *When all the covariances are equal to a common matrix Σ, so that $\Gamma = \text{Tr}(\Sigma) I_n$, and when $2 \leq K < n$, the Estimator (12.15) fulfills the bound*

$$\mathbb{P}\left[\|\widehat{\Gamma} - \Gamma\|_\infty \leq (3\sqrt{2\|\Sigma\|_F^2 (2\log(n) + L)}) \vee (10|\Sigma|_{\text{op}} (2\log(n) + L))\right] \geq 1 - 3e^{-L}.$$

Proof of Lemma 12.4. For any $i \neq j$, we have

$$\begin{aligned}
\|X_i - X_j\|^2 &= \|\mu_i - \mu_j\|^2 + 2\langle \mu_i - \mu_j, E_i - E_j\rangle + \|E_i - E_j\|^2 \\
&= \left(\|\mu_i - \mu_j\| + \left\langle \frac{\mu_i - \mu_j}{\|\mu_i - \mu_j\|}, E_i - E_j \right\rangle\right)^2 - \left\langle \frac{\mu_i - \mu_j}{\|\mu_i - \mu_j\|}, E_i - E_j \right\rangle^2 \\
&\quad + \|E_i - E_j\|^2 \\
&\geq - \left\langle \frac{\mu_i - \mu_j}{\|\mu_i - \mu_j\|}, E_i - E_j \right\rangle^2 + \|E_i - E_j\|^2.
\end{aligned}$$

The random variable $\varepsilon = (2\Sigma)^{-1/2}(E_i - E_j)$ follows a standard Gaussian distribution, so from (B.5), with $S = -2\Sigma$ we have

$$\mathbb{P}\left[\|E_i - E_j\|^2 \geq 2\text{Tr}(\Sigma) - (4\sqrt{2\|\Sigma\|_F^2 L}) \vee (16|\Sigma|_{\text{op}} L)\right] \geq 1 - e^{-L}.$$

In addition, the random variable $\left\langle \frac{\mu_i - \mu_j}{\|\mu_i - \mu_j\|}, E_i - E_j \right\rangle$ follows a Gaussian distribution with variance bounded by $2|\Sigma|_{\text{op}}$, so

$$\mathbb{P}\left[\left\langle \frac{\mu_i - \mu_j}{\|\mu_i - \mu_j\|}, E_i - E_j \right\rangle^2 \geq 4|\Sigma|_{\text{op}} L\right] \leq e^{-L}.$$

Hence, we have

$$\mathbb{P}\left[\frac{1}{2}\|X_i - X_j\|^2 \geq \text{Tr}(\Sigma) - 2|\Sigma|_{\text{op}} L - (2\sqrt{2\|\Sigma\|_F^2 L}) \vee (8|\Sigma|_{\text{op}} L)\right] \geq 1 - 2e^{-L}.$$

With $2a + (2b) \vee (8a) \leq (3b) \vee (10a)$ and a union bound, we obtain that

$$\mathbb{P}\left[\frac{1}{2}\min_{i \neq j}\|X_i - X_j\|^2 \geq \text{Tr}(\Sigma) - (3\sqrt{2\|\Sigma\|_F^2(2\log(n) + L)}) \vee (10|\Sigma|_{\text{op}}(2\log(n) + L))\right] \geq 1 - 2e^{-L}.$$

Let us turn now to the upper bound. Taking any $i \neq j$ in the same group (always possible as $K < n$), we have

$$\|X_i - X_j\|^2 = \|E_i - E_j\|^2,$$

so applying (B.5) with $S = 2\Sigma$, we get

$$\mathbb{P}\left[\frac{1}{2}\|X_i - X_j\|^2 \leq \text{Tr}(\Sigma) + (2\sqrt{2\|\Sigma\|_F^2 L}) \vee (8|\Sigma|_{\text{op}} L)\right] \geq 1 - e^{-L}.$$

Since this upper bound still holds when replacing $\|X_i - X_j\|^2$ by $\min_{i \neq j}\|X_i - X_j\|^2$, this concludes the proof of Lemma 12.4. □

Now that the bias of Lloyd algorithm is fixed, it remains to give a good initialization $\widehat{G}^{(0)}$ to the algorithm.

12.6 Spectral Algorithm

Many state-of-the-art procedures in clustering have two stages. The first step produces a primary estimation of the groups, typically with spectral algorithms, while the second step is a refinement where each observation is reclassified according to a more specialized algorithm. In our Gaussian cluster model (12.2), the spectral algorithm is a good and simple algorithm in order to get a primary estimation of the groups, and these groups can then be refined by running the (debiased) Lloyd algorithm.

The principle of spectral algorithm is to compute the K leading eigenvectors of $\mathbf{X}\mathbf{X}^T$ and apply a rounding procedure on it, in order to get a clustering. Let us explain why this approach makes sense.

We remind the reader that $\mathbb{E}\left[\mathbf{X}\mathbf{X}^T\right] = A\Theta\Theta^T A^T + \Gamma$, where $A \in \mathbb{R}^{n\times K}$ is the membership matrix defined by $A_{ik} = \mathbf{1}_{i\in G_k}$. Hence, $\mathbb{E}\left[\mathbf{X}\mathbf{X}^T\right]$ is the sum of a rank K matrix, which is structured by the partition G, and a diagonal matrix Γ. Hence it makes sense to look at the best rank K approximation of $\mathbf{X}\mathbf{X}^T$, in order to estimate the partition G. Writing $\mathbf{X}\mathbf{X}^T = \sum_k \lambda_k v_k v_k^T$ for an eigenvalue decomposition of $\mathbf{X}\mathbf{X}^T$, with eigenvalues ranked in decreasing order, the best rank K approximation of $\mathbf{X}\mathbf{X}^T$ is $\left(\mathbf{X}\mathbf{X}^T\right)_{(K)} = \sum_{k=1}^K \lambda_k v_k v_k^T$. At this stage, we can either apply some rounding procedure on the rows/columns of $\left(\mathbf{X}\mathbf{X}^T\right)_{(K)}$, or directly to the eigenvectors $v_1,\ldots,v_k$, as suggested by the following lemma.

Lemma 12.5 *Let*

$$A\Theta\Theta^T A^T = \sum_{k=1}^K d_k u_k u_k^T$$

be a spectral decomposition of $A\Theta\Theta^T A^T$, *and set* $U = [u_1,\ldots,u_k] \in \mathbb{R}^{n\times K}$. *Then, there exist* $Z_1,\ldots,Z_K \in \mathbb{R}^K$, *such that*

$$U_{i:} = Z_k \text{ for all } i \in G_k, \quad \text{and} \quad \|Z_k - Z_\ell\|^2 = \frac{1}{|G_k|} + \frac{1}{|G_\ell|}, \quad \text{for } k \neq \ell.$$

Proof of Lemma 12.5. Let us define $\Delta = \text{diag}\left(|G_1|^{-1/2},\ldots,|G_K|^{-1/2}\right)$. We notice that the columns of $A\Delta$

$$[A\Delta]_{:k} = \left[\frac{\mathbf{1}_{i\in G_k^*}}{|G_k^*|^{1/2}}\right]_{i=1,\ldots,n}$$

are orthonormal. Let us consider VDV^T an eigenvalue decomposition of $\Delta^{-1}\Theta\Theta^T\Delta^{-1}$. Since

$$A\Theta\Theta^T A^T = A\Delta VDV^T(A\Delta)^T,$$

and since the columns of $A\Delta$ are orthonormal, we get that $A\Theta\Theta^T A^T = UDU^T$ with $U = A\Delta V$ orthogonal. Hence, for $k = 1,\ldots,K$, we have

$$U_{i:} = [\Delta V]_{k:} = |G_k|^{-1/2} V_{k:}, \quad \text{for } i \in G_k.$$

The vectors $Z_k = [\Delta V]_{k:}$ are orthogonal with norm $\|Z_k\|^2 = 1/|G_k|$, as $\|V_{k:}\|^2 = 1$. Hence, for $k \neq \ell$

$$\|Z_k - Z_\ell\|^2 = \frac{1}{|G_k|} + \frac{1}{|G_\ell|}.$$

The proof of Lemma 12.5 is complete. □

When Γ is proportional to the identity matrix, the eigenvectors of $A\Theta\Theta^T A^T + \Gamma$ and $A\Theta\Theta^T A^T$ are the same, so we can directly work on the eigenvalues of $\mathbf{X}\mathbf{X}^T$. When Γ is not proportional to the identity matrix, similarly to the Kmeans criterion, it is wise to reduce the bias of $\mathbf{X}\mathbf{X}^T$ by considering the eigenvalue decomposition of $\mathbf{X}\mathbf{X}^T - \widehat{\Gamma}$, for some estimator $\widehat{\Gamma}$ of Γ. Let us sum up the spectral algorithm.

(Debiased) Spectral clustering algorithm

1. Compute the eigenvalue decomposition $\mathbf{X}\mathbf{X}^T - \widehat{\Gamma} = \sum_{k=1}^{n} d_k v_k v_k^T$, with eigenvalues ranked in decreasing order; and
2. Apply a rounding procedure on the rows of $v = [v_1, \ldots, v_K]$ in order to get a partition of $\{1, \ldots, n\}$.

There are many possible choices of rounding procedures, for example hierarchical clustering. In the two clusters problem investigated in Section 12.7.2, the rounding procedure will simply be based on the sign of the entries.
It is worth mentioning that the spectral algorithm corresponds to a relaxed version of the SDP algorithm (12.14), where we have removed in $\mathscr{C}$ the constraints $B\mathbf{1} = \mathbf{1}$ and $B \geq 0$.

Lemma 12.6

Let $\mathbf{X}\mathbf{X}^T - \widehat{\Gamma} = \sum_{k=1}^{n} d_k v_k v_k^T$ *be an eigenvalue decomposition of* $\mathbf{X}\mathbf{X}^T - \widehat{\Gamma}$, *with eigenvalues ranked in decreasing order, and set* $v = [v_1, \ldots, v_K]$. *Then*

$$vv^T \in \underset{B \in \mathscr{C}_{sp}}{\operatorname{argmax}} \langle \mathbf{X}\mathbf{X}^T - \widehat{\Gamma}, B \rangle, \quad \text{where } \mathscr{C}_{sp} = \left\{ B \in \mathscr{S}_n^+ : \operatorname{Tr}(B) = K,\ I - B \in \mathscr{S}_n^+ \right\}. \tag{12.16}$$

You may have noticed that the condition $I - B \in \mathscr{S}_n^+$, which was not present in $\mathscr{C}$ has appeared in $\mathscr{C}_{sp}$. It turns out that this condition was actually implicit in $\mathscr{C}$, since any matrix $B \in \mathscr{C}$ has eigenvalues between 0 and 1. Indeed, let v be an eigenvector of $B \in \mathscr{C}$, with eigenvalue $\lambda \geq 0$. Since $\sum_b B_{ab} v_b = \lambda v_a$ and $\sum_b B_{ab} = 1$, with $B_{ab} \geq 0$, we obtain that all (λv_a) belong to the convex hull of the coordinates $\{v_a : a = 1, \ldots, n\}$. This is in particular true for an entry v_a with largest absolute value, which enforces that $0 \leq \lambda \leq 1$.

Proof of Lemma 12.6. Let $\Pi_K = \operatorname{diag}(1, \ldots, 1, 0, \ldots, 0)$ with K ones and $n - K$ zeros.

We observe that for any diagonal matrix $D = \mathrm{diag}(d_1, \ldots, d_n)$, with $d_1 \geq \ldots \geq d_n$,

$$\max_{B:\, 0 \leq B_{kk} \leq 1,\, \sum_k B_{kk} = K} \langle D, B \rangle = \sum_{k=1}^{K} d_k = \langle D, \Pi_K \rangle.$$

In addition, $\Pi_K \in \mathscr{C}_{sp}$ and $\mathscr{C}_{sp} \subset \{B : 0 \leq B_{kk} \leq 1, \sum_k B_{kk} = K\}$. This inclusion is due to the two constraints $B \in \mathscr{S}_n^+$ and $I - B \in \mathscr{S}_n^+$. So,

$$\Pi_K = \underset{B \in \mathscr{C}_{sp}}{\mathrm{argmax}} \langle D, B \rangle.$$

Let VDV^T be an eigenvalue decomposition of $\mathbf{X}\mathbf{X}^T - \widehat{\Gamma}$, with $D = \mathrm{diag}(d_1, \ldots, d_n)$, such that $d_1 \geq \ldots \geq d_n$. Since $V^T BV$ and B have the same spectral properties, $\mathscr{C}_{sp} = \{V^T BV : B \in \mathscr{C}_{sp}\}$. Hence,

$$vv^T = V\Pi_K V^T \in \underset{B \in \mathscr{C}_{sp}}{\mathrm{argmin}} \langle D, V^T BV \rangle = \underset{B \in \mathscr{C}_{sp}}{\mathrm{argmin}} \langle VDV^T, B \rangle.$$

The proof of Lemma 12.6 is complete. □

We now have at hand a practical two-steps clustering strategy: first, a global localization with spectral clustering, and then a refinement with Lloyd algorithm. Let us analyze theoretically these two steps.

12.7 Recovery Bounds

In this section, we investigate the ability of spectral clustering and (corrected) Lloyd algorithm to recover the groups. In order to avoid an inflation of technicalities, we focus on the most simple setting where there are only two groups with means symmetric with respect to 0. More precisely, we assume that there exists an unobserved sequence $z_1, \ldots, z_n \in \{-1, +1\}$ of binary labels such that the observations $X_1, \ldots, X_n$ are independent, and the distribution of X_i is a Gaussian distribution $\mathcal{N}(z_i\theta, \sigma^2 I_p)$ for $i = 1, \ldots, n$.

Stacking as before the observations $X_1, \ldots, X_n$ into a $n \times p$ matrix $\mathbf{X}$, we then observe

$$\mathbf{X} = z\theta^T + E, \tag{12.17}$$

where $z \in \{-1, +1\}^n$ and the E_{ij} are i.i.d. with a $\mathcal{N}(0, \sigma^2)$ distribution. The underlying partition is $G^* = \{\{i : z_i = 1\}, \{i : z_i = -1\}\}$.

A good clustering algorithm is an algorithm that recovers the vector z, up to a sign change. Hence, if $\widehat{z} \in \{-1, +1\}^n$ encodes the clustering output by this algorithm, ($\widehat{z}_i = 1$ if $i \in \widehat{G}_1$ and $\widehat{z}_i = -1$ if $i \in \widehat{G}_2$), we measure the quality of the clustering by the metric

$$\mathrm{recov}(\widehat{z}) := \frac{1}{n} \min_{\delta \in \{-1, +1\}} |z - \delta \widehat{z}|_0 \tag{12.18}$$

which counts the proportion of mismatches between $\widehat{G}$ and G^*.

12.7.1 Benchmark: Supervised Case

As a warm-up, let us consider the arguably simpler problem, where the labels $z_1,\ldots,z_n$ are observed and we want to predict the unobserved label z_{new} of a new data point X_{new}. This situation would happen in the clustering problem if we were able to cluster correctly n data points and wanted to cluster a remaining one. This situation exactly corresponds to a supervised classification problem as in Chapter 11, with learning data set $\mathscr{L} = (X_a, Z_a)_{a=1,\ldots,n}$. The Bayes misclassification error has been quantified in Exercise 11.5.1 (page 246) and $\mathbb{P}[Y_{\text{new}} \neq h^*(X_{\text{new}}]$ was shown to decrease as $\exp(-2\theta^2/\sigma^2)$.

Yet, the Bayes classifier in Exercise 11.5.1 uses the knowledge of θ, so this rate does not take into account the price to pay for not knowing θ. In order to derive optimal classification rates, we focus on a fully Bayesian scenario where the labels $Z_1,\ldots,Z_n$ have been generated as an i.i.d. sequence with uniform distribution on $\{-1,1\}$, and the mean θ has been sampled uniformly over the sphere $\partial B(0,\Delta/2)$ independently of $Z_1,\ldots,Z_n$. As above, conditionally on $Z_1,\ldots,Z_n,\mu$, the X_i are independent Gaussian random variables with mean $Z_i\theta$ and covariance $\sigma^2 I_p$.
In this Bayesian setting, the misclassification error $\mathbb{P}[Z_{\text{new}} \neq \widehat{h}(X_{\text{new}})]$ for a classifier h, plays a similar role as the proportion of mismatches (12.18) in the clustering problem. The classifier $\widehat{h}$ minimizing the misclassification probability $\mathbb{P}\left[Z_{\text{new}} \neq \widehat{h}(X_{\text{new}})\right]$ over all the $\sigma(\mathscr{L})$-measurable classifiers $\widehat{h}$ is the Bayes classifier given by

$$\widehat{h}(x) = \text{sign}\big(\mathbb{P}[Z=1|X=x,\mathscr{L}] - \mathbb{P}[Z=-1|X=x,\mathscr{L}]\big).$$

This classifier $\widehat{h}$ can be computed explicitly and is given by

$$\widehat{h}(x) = \text{sign}\left(\left\langle \frac{1}{n}\sum_{i=1}^{n} Z_i X_i, x\right\rangle\right),$$

see Exercise 12.9.4 for a proof of this formula.
Denoting by γ the uniform distribution on $\partial B(0,\Delta/2)$, the probability of misclassification of the Bayes classifier is given by

$$\begin{aligned}\mathbb{P}\left[Z_{\text{new}} \neq \widehat{h}(X_{\text{new}})\right] &= \int_{\partial B(0,\Delta/2)} \mathbb{P}\left[Z\widehat{h}(X) < 0 \middle| \theta\right] d\gamma(\theta)\\ &= \int_{\partial B(0,\Delta/2)} \mathbb{P}\left[\langle \theta + \frac{\sigma}{\sqrt{n}}\varepsilon, \theta + \sigma\varepsilon'\rangle < 0 \middle| \theta\right] d\gamma(\theta),\end{aligned}$$

where ε and ε' are two independent standard Gaussian random variables in $\mathbb{R}^p$. The above conditional probability is invariant over $\partial B(0,\Delta/2)$, hence we only need to evaluate it for a fixed $\theta \in \partial B(0,\Delta/2)$, say $\theta_\Delta = [\Delta/2, 0, \ldots, 0]$. Let us set $W = -2\left(\Delta\sqrt{1+1/n}\right)^{-1} \langle \theta_\Delta, \frac{1}{\sqrt{n}}\varepsilon + \varepsilon'\rangle$ which follows a standard Gaussian distribution in $\mathbb{R}$ and $Q = -\langle \varepsilon, \varepsilon' \rangle$. Then, we have

$$\mathbb{P}\left[Z_{\text{new}} \neq \widehat{h}(X_{\text{new}})\right] = \mathbb{P}\left[\langle \theta_\Delta + \frac{\sigma}{\sqrt{n}}\varepsilon, \theta_\Delta + \sigma\varepsilon' \rangle < 0\right]$$
$$= \mathbb{P}\left[\frac{\Delta^2}{4\sigma^2} < \frac{\Delta}{2\sigma}\sqrt{1+\frac{1}{n}}W + \frac{1}{\sqrt{n}}Q\right]$$
$$\leq \mathbb{P}\left[\frac{\Delta^2}{8\sigma^2} < \frac{\Delta}{2\sigma}\sqrt{1+\frac{1}{n}}W\right] + \mathbb{P}\left[\frac{\Delta^2}{8\sigma^2} < \frac{1}{\sqrt{n}}Q\right]$$

Since $\Delta = 2\|\theta\|$, the first probability is smaller than $\exp(-\|\theta\|^2/(8\sigma^2))$. According to Hanson-Wright concentration inequality (B.6) page 303 with $A = -I_p$, the second one is upper-bounded by

$$\mathbb{P}\left[Q > \frac{\sqrt{n}\Delta^2}{8\sigma^2}\right] \leq \exp\left(-\frac{\sqrt{n}\Delta^2}{32\sigma^2} \wedge \frac{n\Delta^4}{256p\sigma^4}\right) = \exp\left(-\frac{\sqrt{n}\|\theta\|^2}{8\sigma^2} \wedge \frac{n\|\theta\|^4}{16p\sigma^4}\right).$$

Hence, $\mathbb{P}\left[Z_{\text{new}} \neq \widehat{h}(X_{\text{new}})\right] \leq 2e^{-s^2/16}$ with

$$s^2 = \frac{\|\theta\|^4}{\|\theta\|^2\sigma^2 + \frac{p}{n}\sigma^4}. \tag{12.19}$$

We observe two interesting regimes in (12.19). When n is large, much larger than $p\sigma^2/\|\theta\|^2$, the SNR s^2 essentially reduces to $s^2 \approx \|\theta\|^2/\sigma^2$ as for the Bayes classifier, see Exercise 11.5.1 (page 246). Conversely, in a high-dimensional regime, with p much larger than $n\|\theta\|^2/\sigma^2$, the SNR is approximately equal to $s^2 \approx n\|\theta\|^4/(p\sigma^4)$. This new regime compared to Exercise 11.5.1 is due to the fact that, contrary to the Bayes classifier, we are estimating the mean θ and $p\sigma^2/n$ corresponds to the variance induced by this estimation.

12.7.2 Analysis of Spectral Clustering

When $\mathbf{X}$ follows Model (12.17), we have

$$\mathbb{E}\left[\mathbf{X}\mathbf{X}^T\right] = \|\theta\|^2 zz^T + \Gamma,$$

with $\Gamma_{ii} = \text{Tr}(\text{cov}(X_i))$. As all the covariances are assumed to be equal to $\sigma^2 I_p$, the matrix $\Gamma = p\sigma^2 I_n$ is proportional to the identity, and hence, we do not need to debias $\mathbf{X}\mathbf{X}^T$ in the spectral clustering algorithm. Hence we set $\widehat{\Gamma} = 0$. Since zz^T is of rank one, we only focus on the first eigenvector $\hat{v}_1$ of $\mathbf{X}\mathbf{X}^T$.

The first eigenvector $\hat{v}_1$ of $\mathbf{X}\mathbf{X}^T$ does not provide a clustering of $\{1,\ldots,n\}$ into two groups and a rounding procedure is needed (second step of spectral algorithm). One of the nice features of Model (12.17) is that we can choose a very simple rounding procedure. Actually, as, hopefully, $\hat{v}_1 \approx \pm z/\|z\|$, we can simply take the sign of $\hat{v}_1$ in order to get a partition of $\{1,\ldots,n\}$ into two groups, corresponding to positive and negative entries of $\hat{v}_1$.

We consider then the following spectral algorithm

$$\widehat{z} = \text{sign}(\hat{v}_1), \quad \text{with } \hat{v}_1 \text{ a leading eigenvector of } \frac{1}{n}\mathbf{X}\mathbf{X}^T. \tag{12.20}$$

Theorem 12.7 *Assume that* $\mathbf{X}$ *follows the model (12.17). Then, there exists a numerical constant* $c \geq 1$*, such that, with probability at least* $1 - 2e^{-n/2}$*, the spectral clustering (12.20) fulfills the recovery bound*

$$recov(\widehat{z}) \leq 1 \wedge \frac{c}{s^2}, \tag{12.21}$$

with s^2 *defined in (12.19).*

We observe that the upper bound (12.21) is decreasing with s^2, yet not exponentially fast as in the supervised case investigated above. Yet, when combining spectral clustering with Lloyd algorithm, we obtain an exponential decay with s^2, see Theorem 12.14.

The remaining of this subsection is devoted to the proof of Theorem 12.7.

Proof of Theorem 12.7.

Let us first connect the Hamming distance $|z - \delta\widehat{z}|_0$ to the square norm $\|z - \delta\sqrt{n}\hat{v}_1\|^2$.

Lemma 12.8

For any $x \in \{-1,1\}^n$ *and* $y \in \mathbb{R}^n$*, we have*

$$|x - \text{sign}(y)|_0 \leq \min_{\alpha>0} \|x - \alpha y\|^2.$$

This lemma simply follows from the inequality

$$\mathbf{1}_{x_i \neq \text{sign}(y_i)} = \mathbf{1}_{x_i \neq \text{sign}(\alpha y_i)} \leq |x_i - \alpha y_i|^2,$$

for any $\alpha > 0$ and $i = 1,\ldots,n$.

From Lemma 12.8 with $\alpha = \sqrt{n}$ and $\|z\|^2 = n$, we get

$$\begin{aligned}\min_{\delta=-1,+1} |z - \delta\text{sign}(\hat{v}_1)|_0 &\leq \min_{\delta=-1,+1} \|z - \delta\sqrt{n}\hat{v}_1\|^2 = 2n(1 - |\langle z/\sqrt{n}, \hat{v}_1\rangle|) \\ &\leq 2n(1 - \langle z/\sqrt{n}, \hat{v}_1\rangle^2),\end{aligned}$$

where we used $|\langle z/\sqrt{n}, \hat{v}_1\rangle| \leq \|z/\sqrt{n}\|\|\hat{v}_1\| = 1$ in the last inequality.

Notice that $z/\sqrt{n}$ is a unit-norm leading eigenvector of $\frac{1}{n}\|\theta\|^2 zz^T$, associated to the eigenvalue $\|\theta\|^2$. Notice also that the second eigenvalue of $\frac{1}{n}\|\theta\|^2 zz^T$ is 0, since

$\frac{1}{n}\|\theta\|^2 zz^T$ is a rank one matrix. Combining the previous bound with Davis-Kahan inequality (C.11), page 318, with $A = \frac{1}{n}\|\theta\|^2 zz^T$ and $B = \frac{1}{n}\mathbf{X}\mathbf{X}^T = A + W$, we get

$$\text{recov}(\widehat{z}) = \min_{\delta=-1,+1} \frac{1}{n}|z - \delta\widehat{z}|_0 \le 8 \inf_{\lambda\in\mathbb{R}} \frac{|\lambda I_n + W|_{\text{op}}^2}{\|\theta\|^4}. \tag{12.22}$$

It remains to upper bound $|W + \lambda I_n|_{\text{op}}$.

Proposition 12.9 *Let D be any diagonal matrix. Then, there exists two exponential random variables ξ, ξ' with parameter 1, such that the operator norm of*

$$W' = \frac{1}{n}\mathbf{X}\mathbf{X}^T - \frac{\|\theta\|^2}{n} zz^T - D$$

is upper-bounded by

$$\frac{|W'|_{\text{op}}}{\sigma^2} \le 4\sqrt{\frac{p}{n}\left(6 + 2\frac{\xi}{n}\right)} + \left(48 + \frac{16\xi}{n}\right) + 2\frac{\|\theta\|}{\sigma}\left(1 + \sqrt{\frac{2\xi'}{n}}\right) + \left|\frac{p}{n}I_n - \frac{D}{\sigma^2}\right|_\infty \tag{12.23}$$

Let us explain how Theorem 12.7 follows from this bound. According to (12.22) with $\lambda = -p\sigma^2/n$ and (12.23) with $D = p\sigma^2 I_n/n$, we have with probability at least $1 - 2e^{-n/2}$, the upper bound

$$\min_{\delta=-1,+1} \frac{1}{n}|z - \delta\widehat{z}|_0 \le 1 \wedge \left(\frac{30\sqrt{p/n} + 159 + 16\|\theta\|/\sigma}{\|\theta\|^2/\sigma^2}\right)^2.$$

The right-hand side is smaller than 1 only if $16 \le \|\theta\|/\sigma$, so $159 \le 10\|\theta\|/\sigma$, from which follows

$$\min_{\delta=-1,+1} \frac{1}{n}|z - \delta\widehat{z}|_0 \le 1 \wedge \left(\frac{30\sqrt{p/n} + 26\|\theta\|/\sigma}{\|\theta\|^2/\sigma^2}\right)^2$$
$$\le 1 \wedge \left(1800\frac{p/n + \|\theta\|^2/\sigma^2}{\|\theta\|^4/\sigma^4}\right) = 1 \wedge \frac{1800}{s^2},$$

which gives (12.21). It remains to prove Proposition 12.9.

Proof of Proposition 12.9.
We have $nW' = (EE^T - p\sigma^2 I_n) + (p\sigma^2 I_n - nD) + E\theta z^T + z\theta^T E^T$. Let us first control the quadratic term.

Lemma 12.10
There exists a random variable ξ with exponential distribution with parameter 1 such that

$$|EE^T - p\sigma^2 I_n|_{\text{op}} \le 4\sigma^2\sqrt{p(6n + 2\xi)} + (48n + 16\xi)\sigma^2. \tag{12.24}$$

Proof of Lemma 12.10.
The proof proceeds into three steps: first, a discretization, then Gaussian concentration, and finally a union bound. Dividing both sides of (12.24) by σ^2, we can assume with no loss of generality that $\sigma^2 = 1$.

Step 1: Discretization. Let $\partial B_{\mathbb{R}^n}(0,1)$ denote the unit sphere in $\mathbb{R}^n$. For any symmetric matrix A, the operator norm of A is equal to

$$|A|_{op} = \sup_{x \in \partial B_{\mathbb{R}^n}(0,1)} |\langle Ax, x\rangle|.$$

When $A = EE^T - p\sigma^2 I_n$, since $\partial B_{\mathbb{R}^n}(0,1)$ is an infinite set, we cannot directly use a union bound in order to control the fluctuation of the supremum. Yet, we notice that for two close x and y in $\partial B_{\mathbb{R}^n}(0,1)$, the values $\langle Ax, x\rangle$ and $\langle Ay, y\rangle$ are also close. Hence, the recipe is to discretize the ball $\partial B_{\mathbb{R}^n}(0,1)$ and to control the supremum over $\partial B_{\mathbb{R}^n}(0,1)$ by a supremum over the discretization of the ball plus the error made when replacing $\partial B_{\mathbb{R}^n}(0,1)$ by its discretization.
A set $\mathcal{N}_\varepsilon \subset \partial B_{\mathbb{R}^n}(0,1)$ is called an ε-net of $\partial B_{\mathbb{R}^n}(0,1)$, if for any $x \in \partial B_{\mathbb{R}^n}(0,1)$, there exists $y \in \mathcal{N}_\varepsilon$ such that $\|x-y\| \leq \varepsilon$. Next lemma links the operator norm of a matrix to a supremum over an ε-net.

Lemma 12.11
For any symmetric matrix $A \in \mathbb{R}^{n\times n}$ and any ε-net of $\partial B_{\mathbb{R}^n}(0,1)$, we have

$$|A|_{op} \leq \frac{1}{1-2\varepsilon} \sup_{x\in\mathcal{N}_\varepsilon} |\langle Ax, x\rangle|. \tag{12.25}$$

Proof of Lemma 12.11.
Let $x^* \in \partial B_{\mathbb{R}^n}(0,1)$ be such that $|A|_{op} = |\langle Ax^*, x^*\rangle|$ and let $y \in \mathcal{N}_\varepsilon$ fulfilling $\|x^* - y\| \leq \varepsilon$. According to the decomposition

$$\langle Ax^*, x^*\rangle = \langle Ay, y\rangle + \langle A(x^*-y), y\rangle + \langle Ax^*, x^*-y\rangle,$$

and the triangular inequality, we have

$$\begin{aligned} |A|_{op} = |\langle Ax^*, x^*\rangle| &\leq |\langle Ay, y\rangle| + |\langle A(x^*-y), y\rangle| + |\langle Ax^*, x^*-y\rangle| \\ &\leq \sup_{y\in\mathcal{N}_\varepsilon} |\langle Ay, y\rangle| + 2|A|_{op}\varepsilon. \end{aligned}$$

Bound (12.25) then follows. □

Next lemma provides an upper bound on the cardinality of a minimal ε-net of $\partial B_{\mathbb{R}^n}(0,1)$.

Lemma 12.12
For any $n \in \mathbb{N}$ and $\varepsilon > 0$, there exists an ε-net of $\partial B_{\mathbb{R}^n}(0,1)$ with cardinality upper-bounded by

$$|\mathcal{N}_\varepsilon| \leq \left(1 + \frac{2}{\varepsilon}\right)^n$$

We refer to Exercise 12.9.5 for a proof of this lemma based on volumetric arguments. Choosing $\varepsilon = 1/4$, we get the existence of an $1/4$-net $\mathcal{N}_{1/4}$ of $\partial B_{\mathbb{R}^n}(0,1)$ with cardinality at most 9^n and such that

$$|EE^T - pI_n|_{\text{op}} \leq 2 \max_{x \in \mathcal{N}_{1/4}} |\langle (EE^T - pI_n)x, x\rangle| = 2 \max_{x \in \mathcal{N}_{1/4}} \left| \|E^T x\|^2 - p \right|. \quad (12.26)$$

Step 2: Gaussian concentration. Let $E_{:i}$ denote the ith column of E. We observe that $x^T E_{:i} \sim \mathcal{N}(0, x^T x)$ and that the $(x^T E_{:i})_{i=1,\dots,p}$ are independent since the columns $E_{:i}$ are independent. Hence, since $\mathcal{N}_{1/4} \subset \partial B_{\mathbb{R}^n}(0,1)$, the coordinates $[E^T x]_i = E_{:i}^T x$ are i.i.d. $\mathcal{N}(0,1)$, and the random vector $E^T x$ follows a standard Gaussian distribution $\mathcal{N}(0, I_p)$ in $\mathbb{R}^p$. Hanson-Wright inequality $(B.5)$ with $S = I_p$ and $S = -I_p$ ensures that there exist two exponential random variables ξ_x, ξ_x', such that

$$\|E^T x\|^2 - p \geq \sqrt{8p\xi_x} \vee 8\xi_x \quad \text{and} \quad p - \|E^T x\|^2 \geq \sqrt{8p\xi_x'} \vee 8\xi_x'.$$

Therefore, combining with (12.26), we obtain the concentration bound

$$|EE^T - pI_n|_{\text{op}} \leq 2\left(2\sqrt{2p \max_{x \in \mathcal{N}_{1/4}} (\xi_x \vee \xi_x')} + 8 \max_{x \in \mathcal{N}_{1/4}} (\xi_x \vee \xi_x')\right). \quad (12.27)$$

Step 3: Union bound. A union bound gives

$$\mathbb{P}\left[\max_{x \in \mathcal{N}_{1/4}} (\xi_x \vee \xi_x') > \log(2|\mathcal{N}_{1/4}|) + t\right] \leq 2 \sum_{x \in \mathcal{N}_{1/4}} \exp(-\log(2|\mathcal{N}_{1/4}|) - t) = e^{-t},$$

so there exists an exponential random variable ξ with parameter 1 such that

$$\max_{x \in \mathcal{N}_{1/4}} (\xi_x \vee \xi_x') \leq \log(2|\mathcal{N}_{1/4}|) + \xi \leq 3n + \xi.$$

Combining this bound with (12.27), we obtain (12.24). The proof of Lemma 12.10 is complete. □

Let us now control the cross-terms in W'.

Lemma 12.13

There exists an exponential random variable ξ' with parameter 1, such that

$$\frac{1}{n}|z\theta^T E^T|_{\text{op}} = \frac{1}{n}|E\theta z^T|_{\text{op}} \leq \|\theta\|\sigma\left(1 + \sqrt{\frac{2\xi'}{n}}\right). \quad (12.28)$$

Proof of Lemma 12.13. Again, we can assume with no loss of generality that $\sigma = 1$. Let us set $u = \theta/\|\theta\|$ and $v = z/\sqrt{n}$.
We observe that for x with norm 1,

$$\|Euv^T x\| = |v^T x| \|Eu\| \leq \|Eu\|,$$

with equality for $x = v$. Hence $|Euv^T|_{op} = \|Eu\|$.
For the same reasons as in Step 2 of the proof of Lemma 12.10, the random variable Eu follows a standard Gaussian $\mathcal{N}(0, I_n)$ distribution. Hence, according to the Gaussian concentration inequality, there exists an exponential random variable ξ' with parameter 1, such that

$$|Euv^T|_{op} = \|Eu\| \leq \mathbb{E}[\|Eu\|] + \sqrt{2\xi'} \leq \sqrt{\mathbb{E}[\|Eu\|^2]} + \sqrt{2\xi'} = \sqrt{n} + \sqrt{2\xi'}.$$

Since $\frac{1}{n}|E\theta z^T|_{op} = \frac{\|\theta\|}{\sqrt{n}}|Euv^T|$, Bound (12.28) follows. □

Combining Lemma 12.10, Lemma 12.13, the decomposition

$$nW' = (EE^T - p\sigma^2 I_n) + (p\sigma^2 I_n - nD) + E\theta z^T + z\theta^T E^T,$$

and the equality $|p\sigma^2 I_n - nD|_{op} = |p\sigma^2 I_n - nD|_\infty$, we get (12.23). The proof of Proposition 12.9 is complete. □

12.7.3 Analysis of Lloyd Algorithm

In the setting (12.17), page 269, Kmeans criterion and Lloyd algorithm take a very simple form. Let $\widetilde{G}$ be a partition with two groups: one group with mean $\tilde{\theta}$ and one group with mean $-\tilde{\theta}$. Let $\tilde{z} \in \{-1, +1\}^n$ encodes the partition $\widetilde{G}$ as follows: $\widetilde{G}_1 = \{i : \tilde{z}_i = 1\}$ is the group with mean $\tilde{\theta}$ and $\widetilde{G}_2 = \{i : \tilde{z}_i = -1\}$ is the group with mean $-\tilde{\theta}$. Then, the Kmeans criterion associated to $\widetilde{G}$ is

$$\begin{aligned} \text{Crit}_K(\widetilde{G}) &= \min_{\tilde{\theta} \in \mathbb{R}^p} \left(\sum_{i:\tilde{z}_i=1} \|X_i - \tilde{\theta}\|^2 + \sum_{i:\tilde{z}_i=-1} \|X_i + \tilde{\theta}\|^2 \right) \\ &= \min_{\tilde{\theta} \in \mathbb{R}^p} \sum_{i=1}^n \|X_i - \tilde{z}_i \tilde{\theta}\|^2. \end{aligned}$$

Since $\|X_i - \tilde{z}_i\tilde{\theta}\|^2 = \|\tilde{z}_i X_i - \tilde{\theta}\|^2$, we have

$$\min_{\tilde{\theta} \in \mathbb{R}^p} \sum_{i=1}^n \|X_i - \tilde{z}_i \tilde{\theta}\|^2 = \sum_{i=1}^n \|\tilde{z}_i X_i - \bar{\theta}(\tilde{z})\|^2, \quad \text{where} \quad \bar{\theta}(\tilde{z}) = \frac{1}{n} \sum_{i=1}^n \tilde{z}_i X_i = \mathbf{X}^T \tilde{z}/n.$$

So, in this context, Kmeans algorithm amounts to minimize

$$\widehat{z}_{Kmeans} \in \underset{\tilde{z} \in \{-1,+1\}^n}{\text{argmin}} \sum_{i=1}^n \|\tilde{z}_i X_i - \mathbf{X}^T \tilde{z}/n\|^2,$$

and then, to partition $\{1, \ldots, n\}$ according to the sign of the entries of $\widehat{z}_{Kmeans}$.

The corresponding Lloyd algorithm iterates the updates for $t = 1, 2, \ldots$

$$\widehat{z}^{(t+1)} \in \underset{\tilde{z} \in \{-1,+1\}^n}{\text{argmin}} \sum_{i=1}^n \|\tilde{z}_i X_i - \mathbf{X}^T \widehat{z}^{(t)}/n\|^2.$$

As $\|\tilde{z}_i X_i\|^2 = \|X_i\|^2$ and $\|\mathbf{X}^T \widehat{z}^{(t)}\|^2$ do not depend on $\tilde{z}$, expanding the squares, we get that

$$\begin{aligned}
\widehat{z}^{(t+1)} &\in \underset{\tilde{z} \in \{-1,+1\}^n}{\operatorname{argmax}} \frac{1}{n} \sum_{i=1}^{n} \langle \tilde{z}_i X_i, \mathbf{X}^T \widehat{z}^{(t)} \rangle \\
&= \underset{\tilde{z} \in \{-1,+1\}^n}{\operatorname{argmax}} \frac{1}{n} \tilde{z}^T \mathbf{X}\mathbf{X}^T \widehat{z}^{(t)} \\
&= \operatorname{sign}\left(\frac{1}{n} \mathbf{X}\mathbf{X}^T \widehat{z}^{(t)} \right).
\end{aligned}$$

Keeping in mind that Lloyd algorithm needs to be debiased, even in our case where $\Gamma = p\sigma^2 I_n$, we analyze below the debiased Lloyd algorithm started from some $\widehat{z}^{(0)}$ and with updates

$$\widehat{z}^{(t+1)} = \operatorname{sign}(\widehat{S}\, \widehat{z}^{(t)}), \quad \text{with} \quad \widehat{S} = \frac{1}{n}\left(\mathbf{X}\mathbf{X}^T - \widehat{\Gamma} \right), \tag{12.29}$$

where $\widehat{\Gamma}$ is an estimator of Γ, for example (12.15).
Next theorem provides an upper bound on the misclassification error of the debiased Lloyd algorithm (12.29).

Theorem 12.14 *Let us define the events*

$$\Omega_0 = \left\{ \operatorname{recov}(\widehat{z}^{(0)}) \le u_0 \right\} \quad \text{and} \quad \Omega_\Gamma = \left\{ \frac{1}{n} |\widehat{\Gamma} - \Gamma|_\infty \le \frac{\|\theta\|^2}{32} \right\}.$$

There exists a constant $C \in \mathbb{R}^+$, such that under the assumptions, $u_0 \le 1/32$ and $s^2 \ge C$, the Lloyd algorithm (12.29) fulfills at the iteration $t^* = \lceil \log_2 \left(u_0 \exp\left(\left(\frac{1}{162} \wedge \frac{n}{2C} \right) s^2 \right) \right) \rceil$

$$\mathbb{E}\left[\operatorname{recov}(\widehat{z}^{(t^*)}) \right] \le 16 \exp\left(- \left(\frac{1}{162} \wedge \frac{n}{2C} \right) s^2 \right) + \mathbb{P}[\Omega_0^c] + \mathbb{P}[\Omega_\Gamma^c], \tag{12.30}$$

with s^2 defined by (12.19), page 271.

Before proving this theorem, let us specify the above result for the choices (12.20) for the initialization $\widehat{z}^{(0)}$ and (12.15) for the debiasing $\widehat{\Gamma}$.

Corollary 12.15 *There exists $C \in \mathbb{R}^+$, such that for $s^2 \ge C$, the Lloyd algorithm (12.29) initialized with $\widehat{z}^{(0)}$ given by (12.20) and debiased with (12.15), fulfills at the iteration $t^* = \lceil \log_2 \left(\exp\left(\left(\frac{1}{162} \wedge \frac{n}{2C} \right) s^2 \right) / 32 \right) \rceil$*

$$\mathbb{E}\left[\operatorname{recov}(\widehat{z}^{(t^*)}) \right] \le 21 \exp\left(- \left(\frac{1}{162} \wedge \frac{n}{2C} \right) s^2 \right). \tag{12.31}$$

The proof of Corollary 12.15 is deferred after the proof of Theorem 12.14.

By initializing Lloyd algorithm with spectral clustering, we have obtained a recovery bound decreasing exponentially fast with s^2, as in the supervised learning case Section 12.7.1, page 270. So, in this setting, when there is enough separation between the means in order to be able to classify correctly at least a fixed fraction of the data points, clustering is not significantly harder than supervised classification as in Section 12.7.1. This feature also holds for a higher number K of groups, as long as $s^2 \geq c'K$ for some numerical constant c'.

The result in Corollary 12.15 is about the mean proportion of mismatches between $\widehat{G}^{(t^*)}$ and G^*. We might be interested by some other quantities, such as the probability of exact recovery of G^*. Since $\text{recov}(\widehat{z}^{(t^*)})$ takes value in $\{1/n, 2/n, \ldots, n/n\}$, we can recover a bound on the probability $\mathbb{P}\left[\widehat{G}^{(t^*)} \neq G^*\right]$ from (12.31)

$$\frac{1}{n}\mathbb{P}\left[\widehat{G}^{(t^*)} \neq G^*\right] \leq \mathbb{E}\left[\text{recov}(\widehat{z}^{(t^*)})\right] \leq 21 \exp\left(-\left(\frac{1}{162} \wedge \frac{n}{2C}\right)s^2\right).$$

Hence, when s^2 grows with n faster than $c\log(n)$ for some $c \geq 164$, the probability of exact recovery $\mathbb{P}\left[\widehat{G}^{(t^*)} = G^*\right]$ tends to one. While the constant 164 is not sharp at all, we can easily understand why a growth proportional to $\log(n)$ is needed for exact recovery from the supervised case described in Section 12.7.1, page 270. In a thought experiment where all the labels, but one, are revealed, the probability to misclassify the unknown label from the remaining data decays like $\exp(-cs^2)$ for some $c > 0$. Since we want to correctly classify the n data points with large probability, we need to classify each data point with a probability smaller than $1/n$. This can only be achieved if $s^2 \geq c^{-1}\log(n)$.

Proof of Theorem 12.14. We can decompose $\widehat{S}$ as $\widehat{S} = \frac{1}{n}\|\theta\|^2 zz^T + W$, with

$$W = \frac{1}{n}\left(EE^T - \Gamma + E\theta z^T + z\theta^T E^T + \Gamma - \widehat{\Gamma}\right).$$

The core of the proof is the following lemma proving a geometric decrease of the error.

Lemma 12.16 Geometric convergence.

Let us define

$$\mu = 2\sqrt{2}|W|_{\text{op}} + 4\|\theta\|^2 u_0 \quad \textit{and} \quad \mathscr{E}(\mu) = \sum_{i=1}^{n} \mathbf{1}_{z_i \widehat{S}_i^T z < \mu}.$$

Then, on the event $\Omega(\mu, u_0) = \{\mathscr{E}(\mu) \leq nu_0\} \cap \Omega_0$, we have for all $t \in \mathbb{N}$

$$\text{recov}(\widehat{z}^{(t)}) \leq 2^{-t}u_0 + \frac{1}{n}\mathscr{E}(\mu)\sum_{j=0}^{t-1} 2^{-j} \leq 2u_0. \tag{12.32}$$

Proof of Lemma 12.16. Bound (12.32) holds for $t=0$ on the event $\Omega(\mu,u_0)$. Let us prove by induction that it holds for all $t\in\mathbb{N}$. Assume that (12.32) holds up to iteration $t-1$ and let us check that it holds at iteration t.
With no loss of generality, we can assume that $\text{recov}(\widehat{z}^{(t-1)})=n^{-1}|\widehat{z}^{(t-1)}-z|_0$. Writing $\widehat{S}_i$ for the i-th row of $\widehat{S}$, we get from the definition (12.29) of $\widehat{z}^{(t)}$, that $\widehat{z}_i^{(t)}\neq z_i$ if and only if z_i and $\widehat{S}_i^T\widehat{z}^{(t-1)}$ are of a different sign. From the decomposition $z_i\widehat{S}_i^T\widehat{z}^{(t-1)}=z_i\widehat{S}_i^Tz+z_i\widehat{S}_i^T(\widehat{z}^{(t-1)}-z)$, we obtain

$$\mathbf{1}_{z_i\widehat{S}_i^T\widehat{z}^{(t-1)}<0}\leq\mathbf{1}_{z_i\widehat{S}_i^Tz<\mu}+\mathbf{1}_{z_i\widehat{S}_i^T(\widehat{z}^{(t-1)}-z)<-\mu}.$$

So, as

$$\begin{aligned}z_i\widehat{S}_i^T(\widehat{z}^{(t-1)}-z)&=\frac{\|\theta\|^2}{n}z^T(\widehat{z}^{(t-1)}-z)+z_iW_i^T(\widehat{z}^{(t-1)}-z)\\&=\frac{-2\|\theta\|^2}{n}|\widehat{z}^{(t-1)}-z|_0+z_iW_i^T(\widehat{z}^{(t-1)}-z),\end{aligned}$$

combining the above inequality with the inequalities $\mathbf{1}_{x>y}\leq(x/y_+)^2$ and $\sum_i(W_i^Tx)^2=\|Wx\|^2$, we get

$$\begin{aligned}|\widehat{z}^{(t)}-z|_0&\leq\sum_{i=1}^n\mathbf{1}_{z_i\widehat{S}_i^Tz<\mu}+\sum_{i=1}^n\mathbf{1}_{\left\{-z_iW_i^T(\widehat{z}^{(t-1)}-z)>\mu-\frac{2\|\theta\|^2}{n}|\widehat{z}^{(t-1)}-z|_0\right\}}\\&\leq\mathscr{E}(\mu)+\sum_{i=1}^n\left(\frac{W_i^T(\widehat{z}^{(t-1)}-z)}{\left(\mu-\frac{2\|\theta\|^2}{n}|\widehat{z}^{(t-1)}-z|_0\right)_+}\right)^2\\&\leq\mathscr{E}(\mu)+\left(\frac{|W|_{\text{op}}\|\widehat{z}^{(t-1)}-z\|}{\left(\mu-\frac{2\|\theta\|^2}{n}|\widehat{z}^{(t-1)}-z|_0\right)_+}\right)^2.\end{aligned}$$

By induction hypothesis, we have $|\widehat{z}^{(t-1)}-z|_0\leq 2nu_0$; so, by definition of μ, we have

$$\mu-\frac{2\|\theta\|^2}{n}|\widehat{z}^{(t-1)}-z|_0\geq 2\sqrt{2}|W|_{\text{op}}.$$

As $\|\widehat{z}^{(t-1)}-z\|^2=4|\widehat{z}^{(t-1)}-z|_0$, we conclude that, on $\Omega(\mu,u_0)$,

$$\begin{aligned}\frac{1}{n}|\widehat{z}^{(t)}-z|_0&\leq\frac{1}{n}\mathscr{E}(\mu)+\frac{1}{8n}\|\widehat{z}^{(t-1)}-z\|^2\\&\leq\frac{1}{n}\mathscr{E}(\mu)+\frac{1}{2n}|\widehat{z}^{(t-1)}-z|_0\\&\leq 2^{-t}u_0+\frac{1}{n}\mathscr{E}(\mu)\sum_{j=0}^{t-1}2^{-j}\leq 2u_0.\end{aligned}$$

Hence (12.32) holds at iteration t, and the proof of Lemma 12.16 is complete. □

To complete the proof of Theorem 12.14, we need to handle the event $(\Omega_0 \cap \Omega_\Gamma)^c$ and to bound the expectation of $\mathscr{E}(\mu)$ on $\Omega_0 \cap \Omega_\Gamma$.

Since $\text{recov}(\widehat{z}^{(t)}) \leq 1$, we have

$$\begin{aligned}\text{recov}(\widehat{z}^{(t)}) &\leq \text{recov}(\widehat{z}^{(t)})\mathbf{1}_{\Omega_0 \cap \Omega_\Gamma} + \mathbf{1}_{\Omega_0^c} + \mathbf{1}_{\Omega_\Gamma^c} \\ &\leq \left(2^{-t}u_0 + \frac{2}{n}\mathscr{E}(\mu) + \mathbf{1}_{\mathscr{E}(\mu) > nu_0}\right)\mathbf{1}_{\Omega_0 \cap \Omega_\Gamma} + \mathbf{1}_{\Omega_0^c} + \mathbf{1}_{\Omega_\Gamma^c}.\end{aligned}$$

The random variable $\mathscr{E}(\mu)$ is a sum of Bernoulli random variables. If these variables were independent, we could use a large deviation bound, as in Exercise 12.9.7. As the Bernoulli variables are not independent, we use instead the simple bound $\mathbf{1}_{\mathscr{E}(\mu) > nu_0} \leq (nu_0)^{-1}\mathscr{E}(\mu)$ to get

$$\mathbb{E}\left[\text{recov}(\widehat{z}^{(t)})\right] \leq 2^{-t}u_0 + \left(2 + \frac{1}{u_0}\right)\frac{1}{n}\mathbb{E}\left[\mathscr{E}(\mu)\mathbf{1}_{\Omega_0 \cap \Omega_\Gamma}\right] + \mathbb{P}\left[\Omega_0^c\right] + \mathbb{P}\left[\Omega_\Gamma^c\right]. \quad (12.33)$$

It remains to upper-bound the expectation $\mathbb{E}\left[\mathscr{E}(\mu)\mathbf{1}_{\Omega_0 \cap \Omega_\Gamma}\right]$.

Lemma 12.17 Probabilistic control.

There exists a constant $C > 0$ such that, for $s^2 \geq C$ and $n \geq 2$,

$$\frac{1}{n}\mathbb{E}\left[\mathscr{E}(\mu)\mathbf{1}_{\Omega_0 \cap \Omega_\Gamma}\right] \leq 5\exp\left(-\left(\frac{1}{81} \wedge \frac{n}{C}\right)s^2\right).$$

Before proving this bound, let us explain how Theorem 12.14 follows from Lemma 12.17.

By definition of t^*, we always have

$$2^{-t^*}u_0 \leq \exp\left(-\left(\frac{1}{162} \wedge \frac{n}{2C}\right)s^2\right).$$

When $u_0 \leq \exp\left(-\left(\frac{1}{162} \wedge \frac{n}{2C}\right)s^2\right)$, we have $t^* = 0$ and Bound (12.30) holds.

Let us focus now on the case $u_0 > \exp\left(-\left(\frac{1}{162} \wedge \frac{n}{2C}\right)s^2\right)$. Then, combining (12.33) with Lemma 12.17, we obtain

$$\begin{aligned}\mathbb{E}\left[\text{recov}(\widehat{z}^{(t^*)})\right] \leq &\exp\left(-\left(\frac{1}{162} \wedge \frac{n}{2C}\right)s^2\right) \\ &+ 5\left(2 + \exp\left(\left(\frac{1}{162} \wedge \frac{n}{2C}\right)s^2\right)\right)\exp\left(-\left(\frac{1}{81} \wedge \frac{n}{C}\right)s^2\right) \\ &+ \mathbb{P}\left[\Omega_0^c\right] + \mathbb{P}\left[\Omega_\Gamma^c\right].\end{aligned}$$

Bound (12.30) follows.

It remains to prove Lemma 12.17.

Proof of Lemma 12.17. Let us first observe that since $u_0 \leq 1/32$ and $\mu =$

$2\sqrt{2}|W|_{\text{op}} + 4u_0\|\theta\|^2$, splitting apart the cases $2\sqrt{2}|W|_{\text{op}} \le \|\theta\|^2/8$ and $2\sqrt{2}|W|_{\text{op}} > \|\theta\|^2/8$, we obtain

$$\begin{aligned}\frac{1}{n}\mathbb{E}\left[\mathscr{E}(\mu)\mathbf{1}_{\Omega_0\cap\Omega_\Gamma}\right] &= \mathbb{P}\left[\left\{z_i\widehat{S}_i^T z < \mu\right\}\cap\Omega_0\cap\Omega_\Gamma\right] \\ &\le \mathbb{P}\left[\left\{z_i\widehat{S}_i^T z < \frac{\|\theta\|^2}{4}\right\}\cap\Omega_\Gamma\right] + \mathbb{P}\left[\left\{|W|_{\text{op}} > \frac{\|\theta\|^2}{16\sqrt{2}}\right\}\cap\Omega_\Gamma\right].\end{aligned} \tag{12.34}$$

1- Bounding the probability $\mathbb{P}\left[\left\{|W|_{\text{op}} > \frac{\|\theta\|^2}{16\sqrt{2}}\right\}\cap\Omega_\Gamma\right]$.

From Proposition 12.9 with $D = \widehat{\Gamma}/n$, we have for any $L > 0$

$$\mathbb{P}\left[\left\{|W|_{\text{op}} > 4\sigma^2\sqrt{\frac{p}{n}(6+2L)} + (48+16L)\sigma^2 + 2\|\theta\|\sigma(1+\sqrt{2L}) + \frac{\|\theta\|^2}{32}\right\}\cap\Omega_\Gamma\right] \le 2e^{-nL}.$$

Let us set $L = s^2/C$. Since $s^2 = \|\theta\|^4/(\|\theta\|^2\sigma^2 + p\sigma^4/n)$, when $s^2 \ge C$, we have both $p\sigma^4/n \le \|\theta\|^4/C$ and $\sigma^2 \le \|\theta\|^2/C$. So using that $ps^2/n \le \|\theta\|^4/\sigma^4$ and $s^2 \le \|\theta\|^2/\sigma^2$, we get

$$\begin{aligned}&4\sigma^2\sqrt{\frac{p}{n}(6+2L)} + (48+16L)\sigma^2 + 2\|\theta\|\sigma(1+\sqrt{2L}) \\ &\le \|\theta\|^2\left(4\sqrt{\frac{8}{C}} + \frac{64}{C} + \frac{2(1+\sqrt{2})}{\sqrt{C}}\right) \\ &\le \frac{\|\theta\|^2}{32}\left(\sqrt{2}-1\right),\end{aligned}$$

for C large enough, for example $C = e^{15}$. Hence, for this choice of C, we have

$$\mathbb{P}\left[\left\{|W|_{\text{op}} > \frac{\|\theta\|^2}{16\sqrt{2}}\right\}\cap\Omega_\Gamma\right] \le 2\exp(-ns^2/C). \tag{12.35}$$

2- Bounding the probability $\mathbb{P}\left[\left\{z_i\widehat{S}_i^T z < \frac{\|\theta\|^2}{4}\right\}\cap\Omega_\Gamma\right]$.

The bound on $\mathbb{P}\left[\left\{z_i\widehat{S}_i^T z < \frac{\|\theta\|^2}{4}\right\}\cap\Omega_\Gamma\right]$ is very similar to the bound in Section 12.7.1 for the supervised setting. As

$$\widehat{S}_{ij} = \frac{1}{n}\langle X_i, X_j\rangle - \frac{1}{n}\widehat{\Gamma}_{ii} = \frac{1}{n}\langle z_i\theta + E_i, z_j\theta + E_j\rangle - \frac{1}{n}\widehat{\Gamma}_{ii},$$

defining $\varepsilon_i = z_i E_i$, we have

$$
\begin{aligned}
z_i \widehat{S}_i^T z &= \left\langle \theta + \varepsilon_i, \theta + \frac{1}{n}\sum_{j=1}^{n} \varepsilon_j \right\rangle - \frac{1}{n}\widehat{\Gamma}_{ii} \\
&\leq \|\theta\|^2 + \left\langle \theta, \left(1+\frac{1}{n}\right)\varepsilon_i + \frac{1}{n}\sum_{j:j\neq i}\varepsilon_j \right\rangle + \frac{1}{n}\left\langle \varepsilon_i, \sum_{j:j\neq i}\varepsilon_j \right\rangle + \frac{1}{n}\|\varepsilon_i\|^2 - \widehat{\Gamma}_{ii} \\
&\leq \|\theta\|^2 - \|\theta\|\sigma\sqrt{1+3/n}Z - \frac{\sqrt{n-1}}{n}\sigma^2 Q + \frac{1}{n}(\|\varepsilon_i\|^2 - \Gamma_{ii}) + \frac{1}{n}(\Gamma_{ii} - \widehat{\Gamma}_{ii}),
\end{aligned}
$$

with the Z a standard $\mathcal{N}(0,1)$ Gaussian random variable, and $Q = \langle \varepsilon, \varepsilon' \rangle$ the scalar product between two independent standard $\mathcal{N}(0, I_p)$ Gaussian random variables. Since $1 - 1/4 - 1/32 \geq 2/3$, we have

$$
\begin{aligned}
&\mathbb{P}\left[\left\{z_i \widehat{S}_i^T z < \frac{\|\theta\|^2}{4}\right\} \cap \Omega_\Gamma\right] \\
&\leq \mathbb{P}\left[\|\theta\|\sigma\sqrt{1+3/n}Z + \frac{\sqrt{n-1}}{n}\sigma^2 Q - \frac{1}{n}(\|\varepsilon_i\|^2 - \Gamma_{ii}) > \frac{2\|\theta\|^2}{3}\right] \\
&\leq \mathbb{P}\left[\|\theta\|\sigma Z > \frac{2\|\theta\|^2}{9}\right] + \mathbb{P}\left[\sqrt{\frac{1}{n}}\sigma^2 Q > \frac{2\|\theta\|^2}{9}\right] + \mathbb{P}\left[\frac{1}{n}(\Gamma_{ii} - \|\varepsilon_i\|^2) > \frac{2\|\theta\|^2}{9}\right].
\end{aligned}
$$

The Hanson-Wright inequalities (B.5) with $S = -I_p$ and (B.6) with $A = I_p$ (page 303) and Lemma B.4 (page 298) give us, for $n \geq 2$,

$$
\begin{aligned}
&\mathbb{P}\left[\left\{z_i \widehat{S}_i^T z < \frac{\|\theta\|^2}{4}\right\} \cap \Omega_\Gamma\right] \\
&\leq \exp\left(-\frac{2\|\theta\|^2}{81\sigma^2}\right) + \exp\left(-\frac{1}{4}\left(\frac{2\sqrt{n}\|\theta\|^2}{9\sigma^2} \wedge \left(\frac{2\sqrt{n}\|\theta\|^2}{9\sqrt{p}\sigma^2}\right)^2\right)\right) \\
&\quad + \exp\left(-\frac{1}{8}\left(\frac{2n\|\theta\|^2}{9\sigma^2} \wedge \left(\frac{2n\|\theta\|^2}{9\sqrt{p}\sigma^2}\right)^2\right)\right) \\
&\leq 3\exp\left(-\left(\frac{\|\theta\|^2}{42\sigma^2}\right) \wedge \left(\frac{n\|\theta\|^4}{81p\sigma^4}\right)\right) \leq e^{-s^2/81}.
\end{aligned}
$$

Combining this last bound with Bounds (12.34) and (12.35) complete the proof of Lemma 12.17. □

We now turn to the proof of Corollary 12.15.

Proof of Corollary 12.15.
According to (12.30), all we need is to prove the following lemma.

Lemma 12.18

There exists $C \in \mathbb{R}^+$, such that for $s^2 \geq C$ we have

$$\mathbb{P}[\Omega_0^c] + \mathbb{P}[\Omega_\Gamma^c] \leq 5e^{-ns^2/C}.$$

Let us prove this lemma. We first bound from above $\mathbb{P}[\Omega_\Gamma^c]$. According to Lemma 12.4, Estimator (12.15) fulfills with probability at least $1 - 3e^{-ns^2/C}$

$$\frac{1}{n}|\widehat{\Gamma} - \Gamma|_\infty \leq 3\sqrt{\frac{4p\log(n)\sigma^4}{n^2} + \frac{2p\sigma^4 s^2}{nC}} \bigvee \left(\frac{20\log(n)\sigma^2}{n} + \frac{10s^2\sigma^2}{C}\right).$$

Since $s^2 = \|\theta\|^4/(\|\theta\|^2\sigma^2 + p\sigma^4/n)$, when $s^2 \geq C$, we have both $p\sigma^4/n \leq \|\theta\|^4/C$ and $\sigma^2 \leq \|\theta\|^2/C$. So using that $ps^2/n \leq \|\theta\|^4/\sigma^4$ and $s^2 \leq \|\theta\|^2/\sigma^2$, we get

$$\frac{1}{n}|\widehat{\Gamma} - \Gamma|_\infty \leq \|\theta\|^2 \left(3\sqrt{\frac{2+4\log(n)/n}{C}} \bigvee \left(\frac{20\log(n)/n + 10}{C}\right)\right).$$

The right-hand side of the above inequality is smaller than $\|\theta\|^2/32$ for C large enough, hence $\mathbb{P}[\Omega_\Gamma^c] \leq 3e^{-ns^2/C}$.
Let us now bound from above $\mathbb{P}\left[\Omega_0^c\right]$ for $u_0 = 1/32$. Inequality (12.23) with $D = p\sigma^2 I_n/n$ ensures that with probability at least $1 - 2e^{-ns^2/C}$, we have

$$\begin{aligned}
|\frac{1}{n}\mathbf{X}\mathbf{X}^T - \frac{\|\theta\|^2}{n}zz^T - \frac{p\sigma^2}{n}I_n|_{\text{op}} &\leq 4\sigma^2\sqrt{\frac{p}{n}}\sqrt{6 + \frac{2s^2}{C}} + 48\sigma^2 + \frac{16s^2\sigma^2}{C} \\
&\quad + 2\|\theta\|\sigma\left(1 + \sqrt{\frac{2s^2}{C}}\right) \\
&\leq \|\theta\|^2\left(4\sqrt{\frac{8}{C}} + \frac{64}{C} + \frac{2(1+\sqrt{2})}{\sqrt{C}}\right),
\end{aligned} \tag{12.36}$$

where we used again the majorations $p\sigma^4/n \leq \|\theta\|^4/C$, $\sigma^2 \leq \|\theta\|^2/C$, $ps^2/n \leq \|\theta\|^4/\sigma^4$ and $s^2 \leq \|\theta\|^2/\sigma^2$ for the last inequality. The right-hand side of (12.36) is smaller than $(32 \times 8)^{-1}$ for C large enough. Hence, according to (12.22) with $\lambda = -p\sigma^2/n$, the spectral clustering algorithm $\widehat{z}^{(0)}$ given by (12.20) and debiased with (12.15) fulfills $\mathbb{P}\left[\Omega_0^c\right] \leq 2e^{-ns^2/C}$.
The proof of Corollary 12.15 is complete. □

12.8 Discussion and References

12.8.1 Take-Home Message

Clustering algorithms are very useful when dealing with data gathering some unknown subpopulations with different statistical behaviors. They provide a partition of the dataset, which, when the differences are strong enough, matches the underlying partition into subpopulations. This identification of the subpopulations can be of interest on its own, for scientific understanding of complex systems. It is also an important preliminary step for further analyzes, taking into account the heterogeneity in the data set.

There is a wide zoology of clustering algorithms. Some of them are based on some proximity/separation notions, such as hierarchical clustering algorithms. Some others are derived from probabilistic models. In such models, the MLE is usually computationally intractable. Again, some convex relaxations have been proposed, with strong theoretical garanties. Yet, while being convex, the minimization problems are difficult to solve efficiently, and current state-of-the-art optimization algorithms have a prohibitive computational time when sample sizes exceed a few thousands. Spectral algorithms offer an interesting alternative, coupled or not, with a local refinement, such as Lloyd algorithm.

12.8.2 References

Kmeans criterion has been introduced in [115, 146]. The convex relaxation (12.14) has been proposed by [128], and its statistical properties have been investigated in [40, 41, 57, 83, 122, 137]. Lloyd algorithm has been introduced by [113], and the analysis presented in this chapter is taken from [114, 124]. The spectral algorithm has been investigated in a long series of papers, we refer to [1, 157] for some strong results on exact and partial recovery.

12.9 Exercises

12.9.1 Exact Recovery with Hierarchical Clustering

In this exercise, we provide some conditions ensuring that hierarchical clustering exactly recovers the hidden partition in the setting (12.17) of Section 12.7.2 (page 269). We denote by $\mathscr{W} = \{(i,j) : z_i = z_j,\ i<j\}$ the set of pairs of points within the same cluster $G_1 = \{i : z_i = -1\}$ or $G_2 = \{i : z_i = 1\}$, and by $\mathscr{B} = \{(i,j) : z_i \neq z_j,\ i<j\}$ the set of pairs of points between the two clusters. The concentration bounds below are based on the concentration bounds of Exercise1.6.6 for the square norm of standard Gaussian random variables.

1. What is the value of $\mathbb{E}\left[\|X_i - X_j\|^2\right]$ for $(i,j) \in \mathscr{W}$? and for $(i,j) \in \mathscr{B}$?
2. Prove that

$$\mathbb{P}\left(\max_{(i,j)\in\mathscr{W}} \|X_i - X_j\|^2 \geq 2p\sigma^2 + 12\sigma^2\left(\sqrt{p\log(n)}) + \log(n)\right)\right) \leq \frac{1}{2n}.$$

3. Similarly, prove that

$$\mathbb{P}\left(\min_{(i,j)\in\mathscr{B}} \|X_i - X_j\|^2 \leq 2p\sigma^2 + 4\|\theta\|^2 - 8\sigma^2\sqrt{p\log(n)} - 16\sigma\|\theta\|\sqrt{\log(n)}\right) \leq \frac{1}{2n}.$$

4. Conclude that, when $\|\theta\|^2 \geq 10\sigma^2\left(\sqrt{p\log(n)} + 3\log(n)\right)$, the hierarchical clustering algorithm with Euclidean distance and single or complete linkage recovers the clusters G_1 and G_2 with probability at least $1 - 1/n$.

12.9.2 Bias of Kmeans

We consider the Gaussian cluster model (12.2). We assume that we have $K = 3$ groups of size s (with s even),

$$\theta_1 = (1,0,0)^T, \quad \theta_2 = (0,1,0)^T, \quad \theta_3 = \left(0, 1-\tau, \sqrt{1-(1-\tau)^2}\right)^T,$$

with $\tau > 0$, and

$$\mathrm{Tr}(\Lambda_1) = \gamma_+, \quad \mathrm{Tr}(\Lambda_2) = \mathrm{Tr}(\Lambda_3) = \gamma_-.$$

1. Check that $\|\theta_2 - \theta_3\|^2 = 2\tau$.
2. Compute $\mathbb{E}[\mathrm{Crit}_K(G^*)]$.
3. Let us define G' obtained by splitting G_1^* into two groups G_1', G_2' of equal size $s/2$ and by merging G_2^* and G_3^* into a single group G_3' of size $2s$. Check that
$$\mathbb{E}[\mathrm{Crit}_K(G')] = s(\gamma_+ + 2\gamma_- + \tau) - (2\gamma_+ + \gamma_-).$$
4. Check that we have $\mathbb{E}[\mathrm{Crit}_K(G^*)] < \mathbb{E}[\mathrm{Crit}_K(G')]$ only when $\|\theta_2 - \theta_3\|^2 > 2\left(\frac{\gamma_+ - \gamma_-}{s}\right)$.

12.9.3 Debiasing Kmeans

We consider again the Gaussian cluster model (12.2), on page 255. We assume also that all the clusters have a cardinality not smaller than 3. The risk bound in Lemma 12.2 (page 260) for the estimator $\widehat{\Gamma}^{(1)}$ of $\widetilde{\Gamma}$ still depends on $\|\Theta\|_{2\infty}$. In order to get rid of this dependence, we consider the more complex estimator

$$\widehat{\Gamma}^{(2)}_{ii} = \langle X_i - X_{j_1}, X_i - X_{j_2} \rangle$$

where j_1, j_2 are the two indices $j \in \{1,\ldots,n\} \setminus \{i\}$ for which

$$V_i(j) := \max_{a \neq b \notin \{i,j\}} \left\langle X_i - X_j, \frac{X_a - X_b}{\|X_a - X_b\|} \right\rangle$$

takes the two smallest values.

The goal of this exercise is to prove that there exists a numerical constant $c > 0$, such that, with probability larger than $1 - 5/n$, we have

$$|\Gamma^{(2)} - \widetilde{\Gamma}|_\infty \leq c\left(|\Lambda|_{\mathrm{op}} \log(n) + \sqrt{|\Lambda|_{\mathrm{op}} |\Lambda|_* \log(n)}\right) \tag{12.37}$$

where $|\Lambda|_{\mathrm{op}}$ and $\|\Lambda\|_F$ have been defined in (12.9) on page 259, and $|\Lambda|_* =$

$\max_k \mathrm{Tr}(\Lambda_k)$. We observe, in particular, that this bound does not depend on $\|\Theta\|_{2\infty}$ anymore.

A) Deterministic Bound
It is convenient to introduce the three random quantities

$$Z_1 = \max_{a=1,\ldots,n,\ k\neq\ell} \left\langle \frac{\theta_k - \theta_\ell}{\|\theta_k - \theta_\ell\|}, E_a \right\rangle^2, \quad Z_2 = \max_{i\neq j} |\langle E_i, E_j\rangle|, \quad Z_3 = \max_{i\neq j} \|E_i - E_j\|.$$

Let k and ℓ_1, ℓ_2 be such that $i \in G_k^*$, and $j_w \in G_{\ell_w}^*$ for $w = 1,2$.

1. For $c \in G_k^* \setminus \{i\}$ and $d \in G_{\ell_w}^* \setminus \{j_w\}$, prove the inequalities

$$\frac{1}{2}\|\theta_k - \theta_{\ell_w}\|^2 - 8Z_1 - 4Z_2 \leq |\langle X_i - X_{j_w}, X_c - X_d\rangle| \leq V_i(j_w)\left(\|\theta_k - \theta_{\ell_w}\| + Z_3\right)$$

2. From the previous question, prove that

$$\|\theta_k - \theta_{\ell_w}\|^2 \leq 4\left(V_i(j_w)^2 + Z_3 V_i(j_w) + 8Z_1 + 4Z_2\right).$$

3. Setting $V_i = V_i(j_1) \vee V_i(j_2)$, prove that

$$|\widehat{\Gamma}_{ii}^{(2)} - \widetilde{\Gamma}_{ii}| \leq \|\theta_k - \theta_{\ell_1}\|^2 + \|\theta_k - \theta_{\ell_2}\|^2 + Z_1 + 3Z_2 \leq 8(V_i^2 + Z_3 V_i) + 65Z_1 + 35Z_2.$$

B) Stochastic Controls

1. Let $i_1', i_2' \in G_k \setminus \{i\}$. For $w = 1,2$ and for $a \neq b \notin \{i, i_w'\}$, what is the distribution of the random variable $\left\langle E_i - E_{i_w'}, \frac{X_a - X_b}{\|X_a - X_b\|} \right\rangle$ conditionally on X_a, X_b?
2. Prove that with probability at least $1 - 2/n^2$, we have for some constant $c_V > 0$

$$V_i = V_i(j_1) \vee V_i(j_2) \leq V_i(i_1') \vee V_i(i_2') \leq c_V \sqrt{|\Lambda|_{\mathrm{op}} \log(n)}.$$

3. Prove that with probability at least $1 - 1/n$, we have for some constant $c_1 > 0$

$$Z_1 \leq c_1 |\Lambda|_{\mathrm{op}} \log(n).$$

4. With Hanson-Wright inequality (B.5) on page 303, prove that with probability at least $1 - 1/n$, we have for some constant $c_3 > 0$

$$Z_3 \leq c_3 \left(\sqrt{\|\Lambda\|_F^2 \log(n)} \vee (|\Lambda|_{\mathrm{op}} \log(n)) \vee |\Lambda|_*\right)^{1/2} \leq c_3 \sqrt{(|\Lambda|_{\mathrm{op}} \log(n)) \vee |\Lambda|_*}.$$

5. Combining the deterministic analysis of part (A), with the stochastic controls above and (12.12) on page 261, conclude the proof of (12.37).

12.9.4 Bayes Classifier for the Supervised Case

In this exercise, we consider the Gaussian supervised classification problem with two balanced classes, with identical covariances $\sigma^2 I_p$ and opposite means $\theta_{-1} = -\theta_1$ uniformly distributed on the Euclidean sphere $\partial B(0, \Delta/2)$ in $\mathbb{R}^p$.
We denote by $\mathscr{L} = (X_a, Z_a)_{a=1,\ldots,n}$ the learning sample distributed as follows. The labels $Z_1, \ldots, Z_n$ are i.i.d. with uniform distribution on $\{-1, 1\}$, a random vector $\theta \in \mathbb{R}^p$ is sampled uniformly over the sphere $\partial B(0, \Delta/2)$ independently of $Z_1, \ldots, Z_n$, and, conditionally on $Z_1, \ldots, Z_n, \theta$, the X_a are independent Gaussian random variables with mean $Z_a \theta$ and covariance $\sigma^2 I_p$.
The classifier minimizing the misclassification probability $\mathbb{P}\left[Z_{\text{new}} \neq \widehat{h}(X_{\text{new}})\right]$ over all the $\sigma(\mathscr{L})$-measurable classifiers $\widehat{h}$ is the Bayes classifier given by

$$\widehat{h}(x) = \text{sign}\big(\mathbb{P}[Z = 1 | X = x, \mathscr{L}] - \mathbb{P}[Z = -1 | X = x, \mathscr{L}]\big).$$

Let us compute the Bayes classifier in this setting.

1. Prove that

$$\mathbb{P}[Z = \delta | X = x, \mathscr{L}, \theta] = \mathbb{P}[Z = \delta | X = x, \theta] = \frac{e^{-0.5\|\delta x - \theta\|^2/\sigma^2}}{e^{-0.5\|x+\theta\|^2/\sigma^2} + e^{-0.5\|x-\theta\|^2/\sigma^2}}$$

   and

$$d\mathbb{P}[\theta | X = x, \mathscr{L}] \propto \left(e^{-0.5\|x+\theta\|^2/\sigma^2} + e^{-0.5\|x-\theta\|^2/\sigma^2}\right) e^{-0.5\sum_a \|Z_a X_a - \theta\|^2/\sigma^2}.$$

2. Denoting by γ the uniform distribution on $\partial B(0, \Delta/2)$, check that

$$\mathbb{P}[Z = \delta | X = x, \mathscr{L}]$$
$$= \frac{\int_{\partial B(0,\Delta/2)} e^{-\langle \delta x + \sum_a Z_a X_a, \theta \rangle/\sigma^2} d\gamma(\theta)}{\int_{\partial B(0,\Delta/2)} e^{-\langle x + \sum_a Z_a X_a, \theta' \rangle/\sigma^2} d\gamma(\theta') + \int_{\partial B(0,\Delta/2)} e^{-\langle -x + \sum_a Z_a X_a, \theta' \rangle/\sigma^2} d\gamma(\theta')}.$$

3. Check that $F(v) = \int_{\partial B(0,\Delta/2)} e^{\langle v, \theta \rangle} d\gamma(\theta)$ depends only on $\|v\|$ and is monotone increasing with $\|v\|$ and hence

$$\mathbb{P}[Z = 1 | X = x, \mathscr{L}] > \mathbb{P}[Z = -1 | X = x, \mathscr{L}] \iff \langle x, \sum_a Z_a X_a \rangle > 0,$$

   and

$$\widehat{h}(x) = \text{sign}\left(\left\langle \frac{1}{n} \sum_{i=1}^n Z_i X_i, x \right\rangle\right).$$

12.9.5 Cardinality of an ε-Net

In this exercise, we prove Lemma 12.12 (page 274). Let us define $\mathscr{N}_\varepsilon$ as follows. Start from any $x_1 \in \partial B_{\mathbb{R}^n}(0, 1)$, and for $k = 2, 3, \ldots$ choose recursively any $x_k \in \partial B_{\mathbb{R}^n}(0, 1)$ such that $x_k \notin \cup_{j=1,\ldots,k-1} B_{\mathbb{R}^n}(x_j, \varepsilon)$. When no such x_k remains, stop and define $\mathscr{N}_\varepsilon = \{x_1, x_2, \ldots\}$.

1. Why is the cardinality of $\mathcal{N}_\varepsilon$ finite?
2. Observe that $\mathcal{N}_\varepsilon$ is an ε-net of $\partial B_{\mathbb{R}^n}(0,1)$ and that $\|x-y\| > \varepsilon$ for any $x,y \in \mathcal{N}_\varepsilon$, with $x \neq y$.
3. Check that
   i) the balls $\{B_{\mathbb{R}^n}(x,\varepsilon/2) : x \in \mathcal{N}_\varepsilon\}$ are disjoint; and
   ii) $\bigcup_{x\in\mathcal{N}_\varepsilon} B_{\mathbb{R}^n}(x,\varepsilon/2) \subset B_{\mathbb{R}^n}(0,1+\varepsilon/2)$.
4. Conclude by comparing the volume of the balls $B_{\mathbb{R}^n}(x,\varepsilon/2)$ and $B_{\mathbb{R}^n}(0,1+\varepsilon/2)$.

12.9.6 Operator Norm of a Random Matrix

Let $E \in \mathbb{R}^{n\times p}$ be a matrix with i.i.d. entries distributed according to a Gaussian $\mathcal{N}(0,\sigma^2)$ distribution. For $q \leq n$, let $P \in \mathbb{R}^{n\times n}$ be an orthogonal projector onto a linear span $S \subset \mathbb{R}^n$ of dimension q. In this exercise, we derive from Lemma 12.10 some bounds on $|E|_{\text{op}}$ and $|PE|_{\text{op}}$, both in probability and in expectation.

1. Prove the sequence of inequalities

$$\begin{aligned}|E|_{\text{op}}^2 &\leq |p\sigma^2 I_n|_{\text{op}} + |EE^T - p\sigma^2 I_n|_{\text{op}} \\ &\leq \sigma^2\left(\sqrt{p}+7\sqrt{n+\xi}\right)^2,\end{aligned}$$

   where ξ is an exponential random variable with parameter 1.
2. Conclude that for any $L > 0$

$$\mathbb{E}\left[|E|_{\text{op}}\right] \leq \sigma\left(\sqrt{p}+7\sqrt{n+1}\right) \quad \text{and} \quad \mathbb{P}\left[|E|_{\text{op}} \geq \sigma\left(\sqrt{p}+7\sqrt{n+L}\right)\right] \leq e^{-L}.$$

3. Let $u_1,\ldots,u_q \in \mathbb{R}^n$ be an orthonormal basis of S and set $U = [u_1,\ldots,u_q] \in \mathbb{R}^{n\times q}$ and $W = U^T E \in \mathbb{R}^{q\times p}$. Check that the columns of W are independent with $\mathcal{N}(0,\sigma^2 I_q)$ distribution.
4. Check that $P = UU^T$ and $|PE|_{\text{op}} = |W|_{\text{op}}$.
5. Conclude that for any $L > 0$

$$\mathbb{E}\left[|PE|_{\text{op}}\right] \leq \sigma\left(\sqrt{p}+7\sqrt{q+1}\right) \text{ and } \mathbb{P}\left[|PE|_{\text{op}} \geq \sigma\left(\sqrt{p}+7\sqrt{q+L}\right)\right] \leq e^{-L}.$$

As explained on page 163, combining Lemma 8.3 and the Gaussian concentration inequality (Theorem B.7), we can get tighter constants in the bounds

$$\mathbb{E}\left[|PE|_{\text{op}}\right] \leq \sigma(\sqrt{p}+\sqrt{q}) \quad \text{and} \quad \mathbb{P}\left[|PE|_{\text{op}} \geq \sigma\left(\sqrt{p}+\sqrt{q}+\sqrt{2L}\right)\right] \leq e^{-L}.$$

12.9.7 Large Deviation for the Binomial Distribution

Let $X = \sum_{i=1}^n Y_i$ be the sum of n i.i.d. Bernoulli random variables with parameter q. Let $p \in [0,1]$, with $p > q$.

1. With Markov inequality (Lemma B.1, page 297), prove that for any $\lambda > 0$,

$$\mathbb{P}(X > pn) \leq e^{-\lambda pn} \left[e^{\lambda} q + 1 - q\right]^n.$$

2. Check that the above upper-bound is minimal for $e^{\lambda} = \frac{p(1-q)}{q(1-p)}$, so that

$$\mathbb{P}(X > pn) \leq e^{-n\,\mathrm{kl}(p,q)} \quad \text{with} \quad \mathrm{kl}(p,q) = p \log\left(\frac{p}{q}\right) - (1-p)\log\left(\frac{1-q}{1-p}\right).$$

3. For $p = \alpha q$, check that

$$\begin{aligned}\mathrm{kl}(\alpha q, q) &= \frac{\alpha q}{2}\log(\alpha) + \left[\frac{\alpha q}{2}\log(\alpha) - (1-\alpha q)\log\left(1 + \frac{q(\alpha-1)}{1-\alpha q}\right)\right] \\ &\geq \frac{\alpha q}{2}\log(\alpha) + q\left(\frac{\alpha}{2}\log(\alpha) - \alpha + 1\right).\end{aligned}$$

4. Conclude that for $\alpha \geq 5$, we have

$$\mathbb{P}(X > \alpha q n) \leq \exp\left(-\frac{\alpha q}{2}\log(\alpha)\right).$$

12.9.8 Sterling Numbers of the Second Kind

Let us denote by $S(n,K)$ the number of partitions of $\{1,\ldots,n\}$ into K (non-empty) clusters.

1. What is the value of $S(n,1)$? of $S(n,n)$?
2. With a combinatorial argument, prove the recursion formula

$$S(n,k) = kS(n-1,k) + S(n-1,k-1), \quad \text{for} \quad 2 \leq k \leq n-1.$$

3. Prove by induction that

$$S(n,k) = \frac{1}{k!}\sum_{j=0}^{k}(-1)^j C_k^j (k-j)^n,$$

   with $C_k^j = k!/(j!(k-j)!)$ the binomial coefficient.
4. With the recursion formula, prove the simple lower bound

$$S(n,k) \geq k^{n-k}.$$

The numbers $S(n,k)$ are called the Sterling numbers of the second kind. The total number $B_n = \sum_{k=1}^{n} S(n,k)$ of possible partitions of n elements (without constraints on the number of groups) are called the Bell numbers.

For a fixed k, we observe that $S(n,k)$ grows exponentially fast with n, and for $k = n/\log(n)$ the growth is even super-exponential: For any $0 < c < 1$, we have $S(n,k) \geq \exp(cn\log(n))$ for n large enough. In particular, the Bell number B_n grows super-exponentially fast with n.

12.9.9 Gaussian Mixture Model and EM Algorithm

We consider in this exercise the Gaussian mixture model, which can be described as follows. Let π be a probability distribution on $\{1,\ldots,K\}$. Let $Z_1,\ldots,Z_n$ be n independent random variables taking values in $\{1,\ldots,K\}$, with distribution π. Hence $\mathbb{P}(Z_i = k) = \pi_k$. Define the partition $G^* = \{G_1^*,\ldots,G_K^*\}$ of $\{1,\ldots,n\}$ by setting $G_k^* = \{i : Z_i = k\}$ for $k = 1,\ldots,K$. Then, conditionally on G^*, assume that the data points $X_1,\ldots,X_n$ follow the model (12.2) with $\Lambda_k = \sigma_k^2 I_p$. Compared to the cluster model described on page 255, we have then added a layer of modeling, by assuming that the partition G^* is issued from a sampling process governed by the distribution π.

The Expectation-Maximization (EM) algorithm is classically associated to the mixture models. We introduce and quickly analyze this algorithm in this exercise.

A) EM Algorithm

Let us set $w = (\theta_1,\ldots,\theta_K,\sigma_1^2,\ldots,\sigma_K^2,\pi) \in \mathscr{W} = \mathbb{R}^{pK} \times \mathbb{R}_+^K \times P_K$, where P_K is the simplex $\left\{x \in [0,1]^K : |x|_1 = 1\right\}$. Let also $g_{\theta,\sigma}$ denote the probability density function of the $\mathcal{N}(\theta,\sigma^2 I_p)$ Gaussian distribution.

1. Check that the likelihood of w relative to the observations $\mathbf{X} = \{X_1,\ldots,X_n\}$ is given by

$$L_{\mathbf{X}}(w) = \prod_{i=1}^n \sum_{k=1}^K \pi_k g_{\theta_k,\sigma_k^2}(X_i). \tag{12.38}$$

The log-likelihood of w relative to the observations $\mathbf{X}$ is given by

$$\log(L_{\mathbf{X}}(w)) = \sum_{i=1}^n \log\left(\sum_{k=1}^K \pi_k g_{\theta_k,\sigma_k^2}(X_i)\right),$$

which is not easily amenable to maximization.

Let us assume for a moment that the hidden variables Z are observed.

2. Prove that the likelihood of w relative to the observations $\mathbf{X}$ and $\mathbf{Z} = \{Z_1,\ldots,Z_n\}$ is given by

$$L_{\mathbf{X},\mathbf{Z}}(w) = \prod_{k=1}^K \prod_{i:Z_i=k} \pi_k g_{\theta_k,\sigma_k^2}(X_i).$$

This likelihood $L_{\mathbf{X},\mathbf{Z}}(w)$ is usually called complete likelihood, as it corresponds to the complete observation of the generating process. We observe that the log-likelihood of w relative to the observations $\mathbf{X}$ and $\mathbf{Z}$

$$\log(L_{\mathbf{X},\mathbf{Z}}(w)) = \sum_{i=1}^n \sum_{k=1}^K \mathbf{1}_{Z_i=k}\left(\log(\pi_k) + \log(g_{\theta_k,\sigma_k^2}(X_i))\right)$$

is easily amenable to optimization. Can we take profit of this nice feature for the case where $\mathbf{Z}$ is not observed?

The first recipe of the EM algorithm is to replace the maximization of $\log(L_{\mathbf{X}}(w))$ by an approximate maximization of $\mathbb{E}_w\left[\log(L_{\mathbf{X},\mathbf{Z}}(w))|\mathbf{X}\right]$, where $\mathbb{P}_w$ is the mixture distribution with parameter w. Let us compute this quantity.

3. Check that

$$\mathbb{E}_w\left[\log(L_{\mathbf{X},\mathbf{Z}}(w))|\mathbf{X}\right] = \sum_{i=1}^{n}\sum_{k=1}^{K}\mathbb{P}_w\left(Z_i = k|\mathbf{X}\right)\left(\log(\pi_k) + \log(g_{\theta_k,\sigma_k^2}(X_i))\right).$$

If the conditional probabilities $\mathbb{P}_w(Z_i = k|\mathbf{X})$ were not depending on w, the maximization of $\mathbb{E}_w[\log(L_{\mathbf{X},\mathbf{Z}}(w))|\mathbf{X}]$ would be easy to compute. Yet, the dependence of $\mathbb{P}_w(Z_i = k|\mathbf{X})$ on w makes the optimization hard. The second recipe of the EM algorithm is to iterate sequentially the maximization

$$w^{t+1} \in \underset{w \in \mathscr{W}}{\operatorname{argmax}}\, \mathbb{E}_{w^t}\left[\log(L_{\mathbf{X},\mathbf{Z}}(w))|\mathbf{X}\right], \tag{12.39}$$

starting from some randomly chosen w^0. As we will see, the maximization in w of $\mathbb{E}_{w^t}[\log(L_{\mathbf{X},\mathbf{Z}}(w))|\mathbf{X}]$ can be easily performed. Before proceeding to this maximization, notice that the iterates (12.39) proceed in two steps: first, compute the conditional expectation $\mathbb{E}_{w^t}[\log(L_{\mathbf{X},\mathbf{Z}}(w))|\mathbf{X}]$, and then maximize it. Hence the name Expectation-Maximization (EM) algorithm.

4. Let us set $p_t(i,k) = \mathbb{P}_{w^t}(Z_i = k|\mathbf{X})$ and $N_k = \sum_{i=1}^{n} p_t(i,k)$. Check that the update w^{t+1} is given by

$$\theta_k^{t+1} = \frac{1}{N_k}\sum_{i=1}^{n} p_t(i,k)X_i,$$

$$(\sigma_k^2)^{t+1} = \frac{1}{pN_k}\sum_{i=1}^{n} p_t(i,k)\|X_i - \theta_k^{t+1}\|^2,$$

$$\pi_k^{t+1} = \frac{N_k}{n},$$

for $k = 1,\ldots,K$.

Notice that the updates are very similar to those of Kmeans, except that, for computing the centers $\theta_k^{(t+1)}$, the hard assignments $\mathbf{1}_{i\in G_k^{(t)}}$ are replaced by the estimated conditional probabilities $p_t(i,k) = \mathbb{P}_{w^t}(Z_i = k|\mathbf{X})$. Hence, the EM algorithm is less greedy than Kmeans.

The EM algorithm does not directly provide a clustering of the data point $X_1,\ldots,X_n$, yet, a partition $\widehat{G}^t$ can be derived from w^t by applying the Bayes rule

$$\widehat{G}_k^t = \left\{i : p_t(i,k) = \max_{k'=1,\ldots,K} p_t(i,k')\right\},$$

with ties broken randomly.

B) The Likelihood is Non-Decreasing Along the EM Path

The EM updates are easy to compute, but do they make sense? In this part, we will show that the log-likelihood $\log(L_{\mathbf{X}}(w^t))$ is non-decreasing at each iteration.

We introduce the two following notations. First, we set

$$Q_v(w) = \mathbb{E}_v\left[\log(L_{\mathbf{X},\mathbf{Z}}(w))|\mathbf{X}\right], \quad \text{so that} \quad w^{t+1} \in \underset{w\in\mathscr{W}}{\operatorname{argmin}}\, Q_{w^t}(w).$$

Second, we set

$$p_w(z|\mathbf{X}) := \mathbb{P}_w(Z_1 = z_1, \ldots, Z_n = z_n|\mathbf{X}), \quad \text{for} \quad z \in \{1,\ldots,K\}^n.$$

1. Check that
$$\log(L_{\mathbf{X},\mathbf{Z}}(w)) = \log(L_{\mathbf{X}}(w)) + \log(p_w(\mathbf{Z}|\mathbf{X})),$$
and
$$Q_{w^t}(w) = \log(L_{\mathbf{X}}(w)) + \mathbb{E}_{w^t}\left[\log(p_w(\mathbf{Z}|\mathbf{X}))|\mathbf{X}\right].$$
2. Writing $\mathscr{K}(\mathbb{P},\mathbb{Q}) = \int \log(d\mathbb{P}/d\mathbb{Q})\, d\mathbb{P}$ for the Kullback–Leibler divergence between $\mathbb{P}$ and $\mathbb{Q}$, with $\mathbb{P} \ll \mathbb{Q}$, check that
$$Q_{w^t}(w) - Q_{w^t}(w^t) = \log(L_{\mathbf{X}}(w)) - \log(L_{\mathbf{X}}(w^t)) - \mathscr{K}\left(p_{w^t}(\cdot|\mathbf{X}), p_w(\cdot|\mathbf{X})\right).$$
3. Prove with Jensen inequality that $\mathscr{K}(\mathbb{P},\mathbb{Q}) \geq 0$ for any $\mathbb{P} \ll \mathbb{Q}$.
4. Conclude that $\log(L_{\mathbf{X}}(w^{t+1})) \geq \log(L_{\mathbf{X}}(w^t))$, i.e., the log-likelihood is non-decreasing at each iteration.

Appendix A

Gaussian Distribution

A.1 Gaussian Random Vectors

A random vector $Y \in \mathbb{R}^d$ is distributed according to the $\mathcal{N}(m,\Sigma)$ Gaussian distribution, with $m \in \mathbb{R}^d$ and $\Sigma \in \mathcal{S}_d^+$ (the set of all $d \times d$ symmetric positive semi-definite matrix), when

$$\mathbb{E}\left[e^{\mathrm{i}\langle\lambda,Y\rangle}\right] = \exp\left(\mathrm{i}\langle\lambda,m\rangle - \frac{1}{2}\lambda^T\Sigma\lambda\right), \quad \text{for all } \lambda \in \mathbb{R}^d. \tag{A.1}$$

When matrix Σ is non-singular (i.e., positive definite), the $\mathcal{N}(m,\Sigma)$ Gaussian distribution has a density with respect to the Lebesgue measure on $\mathbb{R}^d$ given by

$$\frac{1}{(2\pi)^{d/2}\det(\Sigma)^{1/2}} \exp\left(-\frac{1}{2}(y-m)^T\Sigma^{-1}(y-m)\right).$$

Affine transformations of Gaussian distribution are still Gaussian.

Lemma A.1 Affine transformation
Let $Y \in \mathbb{R}^d$ be a random vector with $\mathcal{N}(m,\Sigma)$ Gaussian distribution. Then for any $A \in \mathbb{R}^{n\times d}$ and $b \in \mathbb{R}^n$,

$$AY + b \sim \mathcal{N}(Am+b, A\Sigma A^T).$$

In particular, for $a \in \mathbb{R}^d$,

$$\langle a,Y\rangle \sim \mathcal{N}(\langle m,a\rangle, a^T\Sigma a).$$

Proof. The first identity is obtained by computing the characteristic function of $AY + b$

$$\mathbb{E}\left[e^{\mathrm{i}\langle\lambda,AY+b\rangle}\right] = \mathbb{E}\left[e^{\mathrm{i}\langle A^T\lambda,Y\rangle+\mathrm{i}\langle\lambda,b\rangle}\right] = \exp\left(\mathrm{i}\langle A^T\lambda,m\rangle - \frac{1}{2}(A^T\lambda)^T\Sigma A^T\lambda\right)e^{\mathrm{i}\langle\lambda,b\rangle}$$
$$= \exp\left(\mathrm{i}\langle\lambda,Am+b\rangle - \frac{1}{2}\lambda^T A\Sigma A^T\lambda\right).$$

The second identity is obtained with $A = a^T$ and $b = 0$. □

Lemma A.2 Orthogonal projections onto subspaces
Let $Y \in \mathbb{R}^d$ be a random vector with $\mathcal{N}(m,\Sigma)$ Gaussian distribution, and let S and V be two linear spans of $\mathbb{R}^d$ orthogonal with respect to the scalar product induced by Σ. Then the variables $\text{Proj}_S Y$ and $\text{Proj}_V Y$ are independent and follow, respectively, the $\mathcal{N}(\text{Proj}_S m, \text{Proj}_S \Sigma \text{Proj}_S)$ and $\mathcal{N}(\text{Proj}_V m, \text{Proj}_V \Sigma \text{Proj}_V)$ Gaussian distribution.

Proof. Since the projection matrices Proj_S and Proj_V are symmetric, we obtain that the joint characteristic function of $\text{Proj}_S Y$ and $\text{Proj}_V Y$ is

$$\begin{aligned}
\mathbb{E}\left[e^{\mathrm{i}\langle\lambda,\text{Proj}_S Y\rangle+\mathrm{i}\langle\gamma,\text{Proj}_V Y\rangle}\right] &= \mathbb{E}\left[e^{\mathrm{i}\langle\text{Proj}_S\lambda+\text{Proj}_V\gamma,Y\rangle}\right] \\
&= \exp\left(\mathrm{i}\langle\text{Proj}_S\lambda+\text{Proj}_V\gamma,m\rangle-\frac{1}{2}(\text{Proj}_S\lambda+\text{Proj}_V\gamma)^T\Sigma(\text{Proj}_S\lambda+\text{Proj}_V\gamma)\right) \\
&= \exp\left(\mathrm{i}\langle\lambda,\text{Proj}_S m\rangle-\frac{1}{2}\lambda^T\text{Proj}_S\Sigma\text{Proj}_S\lambda\right) \\
&\quad\times\exp\left(\mathrm{i}\langle\gamma,\text{Proj}_V m\rangle-\frac{1}{2}\gamma^T\text{Proj}_V\Sigma\text{Proj}_V\gamma\right) \\
&= \mathbb{E}\left[e^{\mathrm{i}\langle\lambda,\text{Proj}_S Y\rangle}\right]\mathbb{E}\left[e^{\mathrm{i}\langle\gamma,\text{Proj}_V Y\rangle}\right].
\end{aligned}$$

We conclude with Lemma A.1. □

A.2 Chi-Square Distribution

Let $Y \in \mathbb{R}^n$ be a random vector with $\mathcal{N}(0,I_n)$ Gaussian distribution. The χ^2 distribution with n degrees of freedom, corresponds to the distribution of $\|Y\|^2$. In particular, the mean of a $\chi^2(n)$ distribution is

$$\mathbb{E}\left[\|Y\|^2\right] = \sum_{i=1}^{n}\mathbb{E}\left[Y_i^2\right] = n.$$

Lemma A.3 Norms of projections
Let $Y \in \mathbb{R}^n$ be a random vector with $\mathcal{N}(0,I_n)$ Gaussian distribution, and let S be a linear subspace of $\mathbb{R}^n$ with dimension d. Then, the variable $\text{Proj}_S Y$ follows the $\mathcal{N}(0,\text{Proj}_S)$ Gaussian distribution and the square-norm $\|\text{Proj}_S Y\|^2$ follows a χ^2-distribution of degree d.

In particular, $\mathbb{E}\left[\|\text{Proj}_S Y\|^2\right] = \dim(S)$.

Proof. The projection Proj_S is symmetric, so $\text{Proj}_S\text{Proj}_S^T = \text{Proj}_S$ and $\text{Proj}_S Y$ follows a $\mathcal{N}(0,\text{Proj}_S)$ Gaussian distribution according to Lemma A.1.

Let $u_1,\dots,u_d$ be an orthonormal basis of S and set $U = [u_1,\dots,u_d]$. Since $U^TU = I_d$, the vector U^TY follows a $\mathcal{N}(0, I_d)$-distribution and

$$\|\mathrm{Proj}_S Y\|^2 = \sum_{k=1}^{d} (u_k^T Y)^2 = \|U^T Y\|^2$$

follows a χ^2 distribution of degree d. □

A.3 Gaussian Conditioning

We provide in this section a few useful results on Gaussian conditioning.

Lemma A.4
We consider two sets $A = \{1, \ldots, k\}$ and $B = \{1, \ldots, p\} \setminus A$, and a Gaussian random vector $X = \begin{bmatrix} X_A \\ X_B \end{bmatrix} \in \mathbb{R}^p$ with $\mathcal{N}(0, \Sigma)$ distribution. We assume that Σ is non-singular and write $K = \begin{bmatrix} K_{AA} & K_{AB} \\ K_{BA} & K_{BB} \end{bmatrix}$ for its inverse.

In the next formulas, K_{AA}^{-1} will refer to the inverse $(K_{AA})^{-1}$ of K_{AA} (and not to $(K^{-1})_{AA} = \Sigma_{AA}$).

Then, the conditional distribution of X_A given X_B is the Gaussian $\mathcal{N}\left(-K_{AA}^{-1} K_{AB} X_B, K_{AA}^{-1}\right)$ distribution. In others words, we have the decomposition

$$X_A = -K_{AA}^{-1} K_{AB} X_B + \varepsilon_A, \quad \text{where } \varepsilon_A \sim \mathcal{N}\left(0, K_{AA}^{-1}\right) \text{ is independent of } X_B. \tag{A.2}$$

Proof. We write $g(x_A, x_B)$, respectively, $g(x_A|x_B)$ and $g(x_B)$, for the density of the distribution of X, respectively, of X_A given $X_B = x_B$ and X_B. We have

$$\begin{aligned} g(x_A|x_B) &= g(x_A, x_B)/g(x_B) \\ &= \frac{1}{(2\pi)^{k/2}} \exp\left(-\frac{1}{2} x_A^T K_{AA} x_A - x_A^T K_{AB} x_B - \frac{1}{2} x_B^T \left(K_{BB} - \Sigma_{BB}^{-1}\right) x_B\right), \end{aligned}$$

with Σ_{BB} the covariance matrix of X_B. Since $\Sigma_{BB}^{-1} = K_{BB} - K_{BA} K_{AA}^{-1} K_{AB}$, we have

$$g(x_A|x_B) = \frac{1}{(2\pi)^{k/2}} \exp\left(-\frac{1}{2}(x_A + K_{AA}^{-1} K_{AB} x_B)^T K_{AA} (x_A + K_{AA}^{-1} K_{AB} x_B)\right).$$

We recognize the density of the Gaussian $\mathcal{N}\left(-K_{AA}^{-1} K_{AB} x_B, K_{AA}^{-1}\right)$ distribution. □

Corollary A.5 *For any $a \in \{1, \ldots, p\}$, we have*

$$X_a = -\sum_{b: b \neq a} \frac{K_{ab}}{K_{aa}} X_b + \varepsilon_a, \quad \text{where } \varepsilon_a \sim \mathcal{N}(0, K_{aa}^{-1}) \text{ is independent of } \{X_b : b \neq a\}. \tag{A.3}$$

Proof. We apply the previous lemma with $A = \{a\}$ and $B = A^c$. □

Finally, we derive from (A.2) the following simple formula for the conditional correlation of X_a and X_b given $\{X_c : c \neq a,b\}$, which is defined by

$$\mathrm{cor}(X_a, X_b | X_c : c \neq a,b) = \frac{\mathrm{cov}(X_a, X_b | X_c : c \neq a,b)}{\sqrt{\mathrm{var}(X_a | X_c : c \neq a,b)\, \mathrm{var}(X_b | X_c : c \neq a,b)}}.$$

Corollary A.6 *For any $a, b \in \{1,\ldots,p\}$, we have*

$$\mathrm{cor}(X_a, X_b | X_c : c \neq a,b) = \frac{-K_{ab}}{\sqrt{K_{aa}K_{bb}}}. \tag{A.4}$$

Proof. The previous lemma with $A = \{a,b\}$ and $B = A^c$ gives

$$\mathrm{cov}(X_A | X_B) = \begin{pmatrix} K_{aa} & K_{ab} \\ K_{ab} & K_{bb} \end{pmatrix}^{-1} = \frac{1}{K_{aa}K_{bb} - K_{ab}^2} \begin{pmatrix} K_{bb} & -K_{ab} \\ -K_{ab} & K_{aa} \end{pmatrix}.$$

Plugging this formula in the definition of the conditional correlation, we obtain Formula (A.4). □

Appendix B

Probabilistic Inequalities

B.1 Basic Inequalities

Markov inequality plays a central role in the control of the fluctuations of random variables.

Lemma B.1 Markov inequality

For any non-decreasing positive function $\varphi : \mathbb{R} \to \mathbb{R}^+$ and any real-valued random variable X, we have

$$\mathbb{P}(X \geq t) \leq \frac{1}{\varphi(t)} \mathbb{E}[\varphi(X)] \quad \text{for all} \quad t \in \mathbb{R}.$$

In particular, for any $\lambda > 0$, we have

$$\mathbb{P}(X \geq t) \leq e^{-\lambda t} \mathbb{E}\left[e^{\lambda X}\right] \quad \text{for all} \quad t \in \mathbb{R}.$$

Proof. Since φ is positive and non-decreasing, we have

$$\mathbb{P}(X \geq t) \leq \mathbb{E}\left[\frac{\varphi(X)}{\varphi(t)} \mathbf{1}_{X \geq t}\right] \leq \frac{1}{\varphi(t)} \mathbb{E}[\varphi(X)].$$

□

A consequence of Markov inequality is the Chernoff bound on the deviation of X from its expectation.

Lemma B.2 Chernoff bound

Let X be a real random variable, with finite expectation. For any $\lambda \geq 0$, we define $\Lambda(\lambda) = \log \mathbb{E}[\exp(\lambda(X - \mathbb{E}[X]))]$, with the convention $\log(+\infty) = +\infty$. We write

$$\Lambda^*(t) = \max_{\lambda \geq 0} \{\lambda t - \Lambda(\lambda)\},$$

for the Legendre transform of Λ. Then, we have

$$\mathbb{P}[X \geq \mathbb{E}[X] + t] \leq \exp(-\Lambda^*(t)).$$

Proof. The proof technique is more important than the result itself. From Markov inequality, we have for any $\lambda \geq 0$

$$\mathbb{P}[X \geq \mathbb{E}[X]+t] \leq \exp(-\lambda t + \Lambda(\lambda)).$$

Since the left-hand side does not depend on $\lambda \geq 0$, we get the result by taking the maximum over $\lambda \geq 0$. □

Jensen inequality is another important inequality that controls expectations.

Lemma B.3 Jensen inequality

For any convex function $\varphi : \mathbb{R}^d \to \mathbb{R}$ and any random variable X in $\mathbb{R}^d$, such that $\varphi(X)$ is integrable, we have

$$\varphi(\mathbb{E}[X]) \leq \mathbb{E}[\varphi(X)].$$

Proof. Let us denote by $\mathscr{L}_\varphi$ the set of affine functions from $\mathbb{R}^d$ to $\mathbb{R}$, such that $L(x) \leq \varphi(x)$ for all $x \in \mathbb{R}^d$. Since

$$\varphi(x) = \sup_{L \in \mathscr{L}_\varphi} L(x),$$

the linearity of the expectation gives

$$\mathbb{E}[\varphi(X)] = \mathbb{E}\left[\sup_{L \in \mathscr{L}_\varphi} L(X)\right] \geq \sup_{L \in \mathscr{L}_\varphi} \mathbb{E}[L(X)] = \sup_{L \in \mathscr{L}_\varphi} L(\mathbb{E}[X]) = \varphi(\mathbb{E}[X]).$$

□

Lemma B.4 Tail of the Gaussian distribution

Let Z be a standard Gaussian random variable. For any $x \geq 0$, we have

$$\mathbb{P}(|Z| \geq x) \leq e^{-x^2/2}.$$

Proof. The function

$$\varphi(x) = e^{-x^2/2} - \mathbb{P}(|Z| \geq x) = e^{-x^2/2} - \sqrt{\frac{2}{\pi}} \int_x^\infty e^{-t^2/2}\, dt, \quad x \geq 0,$$

takes value 0 at $x = 0$ and its derivative $\varphi'(x) = \left(\sqrt{2/\pi} - x\right) e^{-x^2/2}$ is positive for $x \leq \sqrt{2/\pi}$, so it is positive on $[0, \sqrt{2/\pi}]$. Furthermore, for $x \geq \sqrt{2/\pi}$ we have

$$\sqrt{\frac{2}{\pi}} \int_x^\infty e^{-t^2/2}\, dt \leq \int_x^\infty t e^{-t^2/2}\, dt = e^{-x^2/2},$$

so φ is positive on $\mathbb{R}^+$. □

B.2 Concentration Inequalities

Concentration inequalities provide bounds on the fluctuation of functions of independent random variables around their means. They are central tools for designing and analyzing statistical procedures. We refer to the books by Ledoux [105] and by Boucheron, Lugosi, and Massart [38] for detailed accounts on this topic.

B.2.1 McDiarmid Inequality

McDiarmid concentration inequality [119] is adapted to the setting where the variables are bounded, as in supervised classification.

Theorem B.5 McDiarmid inequality

Let $\mathscr{X}$ be some measurable set and $F : \mathscr{X}^n \to \mathbb{R}$ be a measurable function, such that there exists $\delta_1, \ldots, \delta_n$, fulfilling

$$\left|F(x_1,\ldots,x_i',\ldots,x_n) - F(x_1,\ldots,x_i,\ldots,x_n)\right| \leq \delta_i, \quad \textit{for all} \quad x_1,\ldots,x_n,x_i' \in \mathscr{X},$$

for all $i = 1,\ldots,n$. Then, for any $t > 0$ and any independent random variables $X_1,\ldots,X_n$, with values in $\mathscr{X}$, we have

$$\mathbb{P}\big(F(X_1,\ldots,X_n) > \mathbb{E}\left[F(X_1,\ldots,X_n)\right] + t\big) \leq \exp\left(-\frac{2t^2}{\delta_1^2 + \ldots + \delta_n^2}\right).$$

In other words, under the assumptions of Theorem B.5, there exists a random variable ξ with exponential distribution of parameter 1, such that

$$F(X_1,\ldots,X_n) \leq \mathbb{E}\left[F(X_1,\ldots,X_n)\right] + \sqrt{\frac{\delta_1^2 + \ldots + \delta_n^2}{2}\,\xi}\,.$$

We give here the original proof, combining Markov inequality with a martingale argument due to Azuma [12]. We refer to Boucheron, Lugosi, and Massart [38], Chapter 6, for a more conceptual proof based on the entropy method.

Proof. Let us denote by $\mathscr{F}_k$ the σ-field $\sigma(X_1,\ldots,X_k)$ with $\mathscr{F}_0 = \{\emptyset, \Omega\}$. For simplicity, we write in the following F for $F(X_1,\ldots,X_n)$, and we define for $k = 1,\ldots,n$

$$\Delta_k = \mathbb{E}\left[F|\mathscr{F}_k\right] - \mathbb{E}\left[F|\mathscr{F}_{k-1}\right].$$

Let us fix some $\lambda, t > 0$. The principle of the proof is to start from the Markov inequality

$$\begin{aligned} \mathbb{P}(F > \mathbb{E}[F] + t) &\leq e^{-\lambda t}\,\mathbb{E}\left[e^{\lambda(F - \mathbb{E}[F])}\right] \\ &= e^{-\lambda t}\,\mathbb{E}\left[\prod_{k=1}^{n} e^{\lambda \Delta_k}\right], \end{aligned} \tag{B.1}$$

and then apply repeatedly the following lemma.

Lemma B.6

For any $\lambda > 0$ and $k \in \{1,\dots,n\}$, we have

$$\mathbb{E}\left[e^{\lambda \Delta_k} \middle| \mathscr{F}_{k-1}\right] \le e^{\lambda^2 \delta_k^2/8}.$$

Proof of the lemma. Let us define the function F_k by $F_k(X_1,\dots,X_k) = \mathbb{E}[F|\mathscr{F}_k]$ and the variables S_k and I_k by

$$\begin{aligned} S_k &= \sup_{x\in\mathscr{X}} F_k(X_1,\dots,X_{k-1},x) - \mathbb{E}[F|\mathscr{F}_{k-1}] \\ \text{and} \quad I_k &= \inf_{x\in\mathscr{X}} F_k(X_1,\dots,X_{k-1},x) - \mathbb{E}[F|\mathscr{F}_{k-1}]. \end{aligned}$$

We have almost surely $I_k \le \Delta_k \le S_k$ and $0 \le S_k - I_k \le \delta_k$. The convexity of $x \to e^{\lambda x}$ ensures that

$$e^{\lambda \Delta_k} \le \frac{\Delta_k - I_k}{S_k - I_k} e^{\lambda S_k} + \frac{S_k - \Delta_k}{S_k - I_k} e^{\lambda I_k}.$$

Since I_k, S_k are $\mathscr{F}_{k-1}$-measurable and $\mathbb{E}[\Delta_k|\mathscr{F}_{k-1}] = 0$, we obtain

$$\begin{aligned} \mathbb{E}\left[e^{\lambda\Delta_k}\middle|\mathscr{F}_{k-1}\right] &\le \mathbb{E}\left[\frac{\Delta_k - I_k}{S_k - I_k} e^{\lambda S_k} + \frac{S_k - \Delta_k}{S_k - I_k} e^{\lambda I_k}\middle| \mathscr{F}_{k-1}\right] \\ &= \frac{-I_k}{S_k - I_k} e^{\lambda S_k} + \frac{S_k}{S_k - I_k} e^{\lambda I_k} = e^{\phi(\lambda(S_k - I_k))}, \end{aligned}$$

with

$$\phi(x) = \frac{I_k x}{S_k - I_k} + \log\left(\frac{S_k - I_k e^x}{S_k - I_k}\right).$$

Since $0 \le S_k - I_k \le \delta_k$, we only have to check that $\phi(x) \le x^2/8$ for all $x \ge 0$, in order to conclude the proof of the lemma. Straightforward computations give

$$\phi(0) = 0, \quad \phi'(0) = 0, \quad \text{and} \quad \phi''(x) = \frac{-S_k I_k e^x}{(S_k - I_k e^x)^2} \le \frac{1}{4},$$

where the last inequality follows from $4ab \le (a+b)^2$. Taylor inequality then ensures that

$$\phi(x) \le \frac{x^2}{2} \sup_{t\in[0,x]} \phi''(t) \le \frac{x^2}{8}, \quad \text{for all } x \ge 0,$$

which conclude the proof of Lemma B.6. □

Let us now prove the McDiarmid inequality. Applying repeatedly Lemma B.6, we obtain

$$\begin{aligned} \mathbb{E}\left[\prod_{k=1}^{n} e^{\lambda\Delta_k}\right] &= \mathbb{E}\left[\prod_{k=1}^{n-1} e^{\lambda\Delta_k} \mathbb{E}\left[e^{\lambda\Delta_n}|\mathscr{F}_{n-1}\right]\right] \\ &\le e^{\lambda^2\delta_n^2/8} \mathbb{E}\left[\prod_{k=1}^{n-1} e^{\lambda\Delta_k}\right] \le \ldots \le e^{\lambda^2 \sum_{k=1}^{n} \delta_k^2/8}. \end{aligned}$$

Combining this inequality with (B.1), we get

$$\mathbb{P}(F > \mathbb{E}[F] + t) \leq e^{-\lambda t} e^{\lambda^2 \sum_{k=1}^n \delta_k^2/8}.$$

For $\lambda = 4t \left(\sum_{k=1}^n \delta_k^2\right)^{-1}$, it gives the McDiarmid inequality. □

B.2.2 Gaussian Concentration Inequality

Lipschitz functions of Gaussian random variables also fulfill a good concentration around their mean.

Theorem B.7 *Assume that* $F : \mathbb{R}^d \to \mathbb{R}$ *is 1-Lipschitz and* Z *has a Gaussian* $\mathcal{N}(0, \sigma^2 I_d)$ *distribution.*

Then, there exists a variable ξ *with exponential distribution of parameter 1, such that*

$$F(Z) \leq \mathbb{E}[F(Z)] + \sigma\sqrt{2\xi}. \tag{B.2}$$

A striking point with the Gaussian concentration is that the size of the fluctuation $\sigma\sqrt{2\xi}$ does not depend on the dimension d. We give here a short proof based on Itô calculus due to Ibragimov, Sudakov, and Tsirel'son [94]. We refer to Exercise 1.6.7, on page 25, for a simple proof of a less tight version of (B.2) and to Ledoux [105] for some more classical and elementary proofs and further references.

Proof. We can assume in the following that $Z \sim \mathcal{N}(0, I_d)$, since $F(Z) = \sigma\widetilde{F}(Z/\sigma)$, where $\widetilde{F}(x) = \sigma^{-1}F(\sigma x)$ is 1-Lipschitz and Z/σ has a $\mathcal{N}(0, I_d)$ distribution. Let $(W_t)_{t\geq 0}$ be a standard Brownian motion in $\mathbb{R}^d$ and set for $x \in \mathbb{R}^d$ and $t \in [0,1]$

$$G(t,x) = \mathbb{E}\left[F(x + W_{1-t})\right].$$

The function G is continuous on $[0,1] \times \mathbb{R}^d$, differentiable in t, and infinitely differentiable in x for any $t \in [0,1[$. Furthermore, since the infinitesimal generator of the standard Brownian motion is $\frac{1}{2}\Delta$, we have for all $(t,x) \in]0,1[\times\mathbb{R}^d$

$$\frac{\partial}{\partial t}G(t,x) = -\frac{1}{2}\Delta_x G(t,x), \tag{B.3}$$

where Δ_x represents the Laplacian in the variable x. Let $(B_t)_{t\geq 0}$ be another standard Brownian motion in $\mathbb{R}^d$, independent of W. Itô's formula (see Revuz–Yor [131], Chapter 4, Theorem 3.3) gives

$$\begin{aligned}
\underbrace{G(1,B_1)}_{=F(B_1)} &= G(0,0) + \int_0^1 \frac{\partial}{\partial t}G(s,B_s)\,ds + \int_0^1 \nabla_x G(s,B_s)\cdot dB_s + \frac{1}{2}\int_0^1 \Delta_x G(s,B_s)\,ds \\
&= \underbrace{G(0,0)}_{=\mathbb{E}[F(W_1)]} + \int_0^1 \nabla_x G(s,B_s)\cdot dB_s,
\end{aligned}$$

where the last equality follows from Equation (B.3). The stochastic integral $\int_0^t \nabla_x G(s, B_s) \cdot dB_s$ defines a continuous martingale starting from 0, so according to Dubins–Schwarz's representation of continuous martingales ([131], Chapter 5, Theorem 1.6), there exists a 1-dimensional standard Brownian motion β, such that

$$\int_0^1 \nabla_x G(s, B_s) \cdot dB_s = \beta_{T_1}, \quad \text{where} \quad T_1 = \int_0^1 \|\nabla_x G(s, B_s)\|^2 ds.$$

Since F is 1-Lipschitz, the function $x \to G(t,x)$ is 1-Lipschitz for all $t \in [0,1]$, and therefore $\|\nabla_x G(t,x)\| \leq 1$ for all $(t,x) \in [0,1[\times \mathbb{R}^d$. As a consequence, we have that $T_1 \leq 1$ a.s. and

$$\left| \int_0^1 \nabla_x G(s, B_s) \cdot dB_s \right| \leq \sup_{t \in [0,1]} \beta_t \quad \text{a.s.}$$

Finally, the random variable $\sup_{t\in[0,1]} \beta_t$ has the same distribution as the absolute value of a standard Gaussian random variable ([131], Chapter 3, Theorem 3.7), so

$$\begin{aligned} \mathbb{P}(F(B_1) \geq \mathbb{E}[F(W_1)] + x) &\leq \mathbb{P}\left(\sup_{t\in[0,1]} \beta_t \geq x \right) \\ &\leq 2\int_x^{+\infty} e^{-t^2/2} \frac{dt}{\sqrt{2\pi}} \leq e^{-x^2/2}. \end{aligned}$$

We conclude by noticing that B_1 and W_1 have a Gaussian $\mathcal{N}(0, I_d)$ distribution. □

We note that replacing F by $-F$, we also have $F(Z) \geq \mathbb{E}[F(Z)] - \sigma\sqrt{2\xi'}$ for some ξ' with exponential distribution of parameter 1.

Remark 1. A typical example of use of the Gaussian concentration inequality is with the norm $\|Z\|$ of a $\mathcal{N}(0, \sigma^2 I_d)$ Gaussian random variable. Since the norm is 1-Lipschitz, and since we have $\mathbb{E}[\|Z\|] \leq \sqrt{\mathbb{E}[\|Z\|^2]} = \sigma\sqrt{d}$, the norm $\|Z\|$ fulfills the inequality

$$\|Z\| \leq \sigma\sqrt{d} + \sigma\sqrt{2\xi},$$

where ξ is an exponential random variable of parameter 1.

Remark 2. Another consequence of the Gaussian concentration is that the variance of the norm $\|Z\|$ of a $\mathcal{N}(0, \sigma^2 I_d)$ Gaussian random variable can be bounded independently of the dimension d. Actually, there exists two standard exponential random variables ξ and ξ', such that

$$\mathbb{E}[\|Z\|] - \sigma\sqrt{2\xi'} \leq \|Z\| \leq \mathbb{E}[\|Z\|] + \sigma\sqrt{2\xi}.$$

Accordingly, we have the upper bound

$$\begin{aligned} \text{var}[\|Z\|] &= \mathbb{E}\left[(\|Z\| - \mathbb{E}[\|Z\|])_+^2 \right] + \mathbb{E}\left[(\mathbb{E}[\|Z\|] - \|Z\|)_+^2 \right] \\ &\leq 2\mathbb{E}[\xi]\sigma^2 + 2\mathbb{E}[\xi']\sigma^2 = 4\sigma^2. \end{aligned}$$

As a consequence, we have the following bounds on the expectation of $\|Z\|$

$$(d-4)\,\sigma^2 \leq \mathbb{E}\left[\|Z\|^2\right] - \text{var}[\|Z\|] = \mathbb{E}\left[\|Z\|\right]^2 \leq \mathbb{E}\left[\|Z\|^2\right] = d\,\sigma^2. \tag{B.4}$$

B.2.3 Concentration of Quadratic Forms of Gaussian Vectors

The next lemma gathers two simple versions of Hanson-Wright inequality for concentration of quadratic forms of Gaussian vectors.

Theorem B.8 Concentration of Quadratic forms of Gaussian vectors
Symmetric forms. *Let ε be a standard Gaussian random variable $\mathcal{N}(0,I_p)$ in $\mathbb{R}^p$ and S be a real symmetric $p \times p$ matrix. Then, we have for any $L \geq 0$*

$$\mathbb{P}\left[\varepsilon^T S\varepsilon - \operatorname{Tr}(S) > \sqrt{8\|S\|_F^2 L} \vee (8|S|_{\mathrm{op}} L)\right] \leq e^{-L}. \tag{B.5}$$

Cross-products. *Let $\varepsilon, \varepsilon'$ be two independent standard Gaussian random variables $\mathcal{N}(0,I_p)$ in $\mathbb{R}^p$ and A be* any *real $p \times p$ matrix. Then, we have for any $L \geq 0$*

$$\mathbb{P}\left[\varepsilon^T A\varepsilon' > \sqrt{4\|A\|_F^2 L} \vee (4|A|_{\mathrm{op}} L)\right] \leq e^{-L}. \tag{B.6}$$

Proof of Theorem B.8.
Symmetric forms. The proof is based on the Chernoff argument, page 297. Next lemma provides an upper bound on the Laplace transform of a square Gaussian random variable.

Lemma B.9

Let Z be a $\mathcal{N}(0,1)$ standard Gaussian random variable. Then, for any $|s| \leq 1/4$, we have

$$\mathbb{E}\left[\exp(s(Z^2-1))\right] \leq e^{2s^2}.$$

Proof of Lemma B.9. Since $-\log(1-x) \leq x + x^2$ for $|x| \leq 1/2$, we have

$$\mathbb{E}\left[\exp(s(Z^2-1))\right] = \frac{e^{-s}}{(1-2s)^{1/2}} \leq e^{2s^2}.$$

The proof of Lemma B.9 is complete. □

Since S is symmetric we can diagonalize it, $S = \sum_{k=1}^p \lambda_k v_k v_k^T$ and

$$\varepsilon^T S\varepsilon = \sum_{k=1}^p \lambda_k (v_k^T \varepsilon)^2.$$

Since the eigenvectors $\{v_1, \ldots, v_p\}$ form an orthonormal basis of $\mathbb{R}^p$, the matrix $V = [v_1, \ldots, v_p]$ fulfills $V^T V = I$. Hence, $Z = V^T \varepsilon$ follows a $\mathcal{N}(0,I)$ distribution, which means that the random variables $Z_k = v_k^T \varepsilon$, for $k = 1, \ldots, p$ are i.i.d. $\mathcal{N}(0,1)$-random variables.
Applying Markov inequality (Lemma B.1, page 297), we get for $t \geq 0$ and $|s| \leq$

$(4|S|_{op})^{-1}$

$$\begin{aligned}\mathbb{P}\left[\varepsilon^T S\varepsilon - \mathrm{Tr}(S) > t\right] &\leq e^{-st}\,\mathbb{E}\left[e^{s(\varepsilon^T S\varepsilon - \mathrm{Tr}(S))}\right] \\ &\leq e^{-st}\prod_{k=1}^{p}\mathbb{E}\left[\exp\left(s\lambda_k(Z_k^2-1)\right)\right] \\ &\leq \exp\left(-st+2s^2\sum_{k=1}^{p}\lambda_k^2\right) = \exp\left(-st+2\|S\|_F^2 s^2\right)\end{aligned}$$

The minimum of $s \to -st + 2\|S\|_F^2 s^2$ over $|s| \leq (4|S|_{op})^{-1}$ is achieved for $s = \frac{1}{4}(t/\|S\|_F^2) \wedge (1/|S|_{op})$ and hence

$$\begin{aligned}\min_{|s|\leq(4|S|_{op})^{-1}}\left(-st+2\|S\|_F^2 s^2\right) &= -\frac{t^2}{8\|S\|_F^2}\mathbf{1}_{\{t\leq\|S\|_F^2/|S|_{op}\}} + \left(\frac{\|S\|_F^2}{8|S|_{op}^2} - \frac{t}{4|S|_{op}}\right)\mathbf{1}_{\{t>\|S\|_F^2/|S|_{op}\}} \\ &\leq -\frac{1}{8}\left(\frac{t^2}{\|S\|_F^2}\wedge\frac{t}{|S|_{op}}\right).\end{aligned}$$

The bound (B.5) follows.

Cross-products. The trick for (B.6) is to notice that

$$\varepsilon^T A\varepsilon' = \begin{bmatrix}\varepsilon \\ \varepsilon'\end{bmatrix}^T S \begin{bmatrix}\varepsilon \\ \varepsilon'\end{bmatrix}, \quad \text{with} \quad S = \frac{1}{2}\begin{bmatrix}0 & A \\ A^T & 0\end{bmatrix}.$$

Since S is symmetric, we can apply (B.5). The conclusion follows by noticing that $\mathrm{Tr}(S) = 0$, $|S|_{op} = |A|_{op}/2$ and $\|S\|_F^2 = \|A\|_F^2/2$. □

Remark. When S is symmetric positive semi-definite, we have $\varepsilon^T S\varepsilon = \|S^{1/2}\varepsilon\|^2$. Since

$$\left|\|S^{1/2}y\| - \|S^{1/2}x\|\right| \leq \|S^{1/2}(y-x)\| \leq |S|_{op}^{1/2}\|y-x\|,$$

the application $x \to \|S^{1/2}x\|$ is $|S|_{op}^{1/2}$-Lipschitz, and we can use the Gaussian concentration inequality in order to bound from above $\varepsilon^T S\varepsilon - \mathrm{Tr}(S)$. Yet, as explained below, the bound obtained is less tight than (B.5) when the spectrum of S is not flat.

Indeed, since

$$\mathbb{E}\left[\|S^{1/2}\varepsilon\|\right]^2 \leq \mathbb{E}\left[\|S^{1/2}\varepsilon\|^2\right] = \mathrm{Tr}(S),$$

the Gaussian concentration inequality (Theorem B.7, page 301) ensures that for $L > 0$

$$\mathbb{P}\left[\|S^{1/2}\varepsilon\| \geq \sqrt{\mathrm{Tr}(S)} + \sqrt{2|S|_{op}L}\right] \leq e^{-L}.$$

It then follows the concentration bound

$$\mathbb{P}\left[\varepsilon^T S\varepsilon - \mathrm{Tr}(S) \geq 2\sqrt{2|S|_{op}\mathrm{Tr}(S)L} + 2|S|_{op}L\right] \leq e^{-L}.$$

This bound is similar to (B.5), except that $\|S\|_F^2$ has been replaced by $|S|_{op}\text{Tr}(S)$. When the spectrum of S is not flat and the dimension p is large, this last quantity $|S|_{op}\text{Tr}(S) = \sum_{k=1}^p (\lambda_1 \lambda_k)$ can be much larger than $\|S\|_F^2 = \sum_{k=1}^p \lambda_k^2$. For example, if $\lambda_k = 1/k$, the quantity $|S|_{op}\text{Tr}(S)$ diverges to infinity when p goes to infinity, while $\|S\|_F^2$ remains bounded.

B.3 Symmetrization and Contraction Lemmas

B.3.1 Symmetrization Lemma

The symmetrization lemma is a simple bound, very useful for bounding the expectation of suprema of empirical processes, as, for example, in the proof of Theorem 11.1 in Chapter 11.

Theorem B.10 Symmetrization lemma

Let $\mathcal{Z}$ be a measurable set, and $\mathcal{F}$ be a set of integrable functions $f : \mathcal{Z} \to \mathbb{R}$. Let $Z_1, \ldots, Z_n$ be i.i.d. random variables in $\mathcal{Z}$, and $\sigma_1, \ldots, \sigma_n$ be i.i.d. random variables independent of $Z_1, \ldots, Z_n$, with uniform distribution on $\{-1, +1\}$. Then, we have

$$\mathbb{E}\left[\sup_{f\in\mathcal{F}}\left|\frac{1}{n}\sum_{i=1}^n (f(Z_i) - \mathbb{E}[f(Z_i)])\right|\right] \leq 2\mathbb{E}\mathbb{E}_\sigma\left[\sup_{f\in\mathcal{F}}\left|\frac{1}{n}\sum_{i=1}^n \sigma_i f(Z_i)\right|\right], \tag{B.7}$$

where $\mathbb{E}$ refers to the expectation with respect to $Z_1, \ldots, Z_n$, and $\mathbb{E}_\sigma$ refers to the expectation with respect to $\sigma_1, \ldots, \sigma_n$.

Proof. Let $(\widetilde{Z}_i)_{i=1,\ldots,n}$ be an independent copy of $(Z_i)_{i=1,\ldots,n}$. We first observe that

$$\frac{1}{n}\sum_{i=1}^n \mathbb{E}[f(Z_i)] = \widetilde{\mathbb{E}}\left[\frac{1}{n}\sum_{i=1}^n f(\widetilde{Z}_i)\right],$$

where $\widetilde{\mathbb{E}}$ refers to the expectation with respect to the random variables $(\widetilde{Z}_i)_{i=1,\ldots,n}$. According to Jensen inequality, we have

$$\begin{aligned}
\mathbb{E}\left[\sup_{f\in\mathcal{F}}\left|\frac{1}{n}\sum_{i=1}^n (f(Z_i) - \mathbb{E}[f(Z_i)])\right|\right] &= \mathbb{E}\left[\sup_{f\in\mathcal{F}}\left|\frac{1}{n}\sum_{i=1}^n f(Z_i) - \widetilde{\mathbb{E}}\left[\frac{1}{n}\sum_{i=1}^n f(\widetilde{Z}_i)\right]\right|\right] \\
&\leq \mathbb{E}\left[\sup_{f\in\mathcal{F}}\widetilde{\mathbb{E}}\left[\left|\frac{1}{n}\sum_{i=1}^n f(Z_i) - \frac{1}{n}\sum_{i=1}^n f(\widetilde{Z}_i)\right|\right]\right] \\
&\leq \mathbb{E}\widetilde{\mathbb{E}}\left[\sup_{f\in\mathcal{F}}\left|\frac{1}{n}\sum_{i=1}^n \left(f(Z_i) - f(\widetilde{Z}_i)\right)\right|\right].
\end{aligned}$$

By symmetry, we notice that $\left(\sigma_i(f(Z_i) - f(\widetilde{Z}_i))\right)_{i=1,\ldots,n}$ has the same distribution as

$\left(f(Z_i)-f(\widetilde{Z}_i)\right)_{i=1,\ldots,n}$, so the previous bound and the triangular inequality give

$$
\begin{aligned}
&\mathbb{E}\left[\sup_{f\in\mathscr{F}}\left|\frac{1}{n}\sum_{i=1}^{n}(f(Z_i)-\mathbb{E}[f(Z_i)])\right|\right]\\
&\leq \mathbb{E}\widetilde{\mathbb{E}}\mathbb{E}_\sigma\left[\sup_{f\in\mathscr{F}}\left|\frac{1}{n}\sum_{i=1}^{n}\sigma_i\left(f(Z_i)-f(\widetilde{Z}_i)\right)\right|\right]\\
&\leq \mathbb{E}\mathbb{E}_\sigma\left[\sup_{f\in\mathscr{F}}\left|\frac{1}{n}\sum_{i=1}^{n}\sigma_i f(Z_i)\right|\right]+\widetilde{\mathbb{E}}\mathbb{E}_\sigma\left[\sup_{f\in\mathscr{F}}\left|\frac{1}{n}\sum_{i=1}^{n}\sigma_i f(\widetilde{Z}_i)\right|\right]\\
&\leq 2\mathbb{E}\mathbb{E}_\sigma\left[\sup_{f\in\mathscr{F}}\left|\frac{1}{n}\sum_{i=1}^{n}\sigma_i f(Z_i)\right|\right].
\end{aligned}
$$

The proof of Theorem B.10 is complete. □

B.3.2 Contraction Principle

The contraction principle of Ledoux and Talagrand (Theorem 4.12 in [106]) is a useful tool for analyzing empirical processes, as, for example, in the proof of Theorem 11.10 in Chapter 11.

Theorem B.11 Contraction principle

Let $\mathscr{Z}$ be a bounded subset of $\mathbb{R}^n$, and $\varphi:\mathbb{R}\to\mathbb{R}$ be an α-Lipschitz function fulfilling $\varphi(0)=0$. For $\sigma_1,\ldots,\sigma_n$ i.i.d. random variables with distribution $\mathbb{P}_\sigma(\sigma_i=1)=\mathbb{P}_\sigma(\sigma_i=-1)=1/2$, we have

$$\mathbb{E}_\sigma\left[\sup_{z\in\mathscr{Z}}\left|\sum_{i=1}^{n}\sigma_i\varphi(z_i)\right|\right]\leq 2\alpha\,\mathbb{E}_\sigma\left[\sup_{z\in\mathscr{Z}}\left|\sum_{i=1}^{n}\sigma_i z_i\right|\right]. \tag{B.8}$$

Proof. We will actually prove the following stronger result: For any function $g:\mathbb{R}^n\to\mathbb{R}$ and any integer $d\leq n$,

$$\mathbb{E}_\sigma\left[\sup_{z\in\mathscr{Z}}\left(g(z)+\sum_{i=1}^{d}\sigma_i\varphi(z_i)\right)_+\right]\leq\alpha\,\mathbb{E}_\sigma\left[\sup_{z\in\mathscr{Z}}\left(g(z)+\sum_{i=1}^{d}\sigma_i z_i\right)_+\right]. \tag{B.9}$$

Let us first check that (B.8) follows from (B.9). Since $|x|=(x)_++(-x)_+$, we have

$$\mathbb{E}_\sigma\left[\sup_{z\in\mathscr{Z}}\left|\sum_{i=1}^{n}\sigma_i\varphi(z_i)\right|\right]\leq\mathbb{E}_\sigma\left[\sup_{z\in\mathscr{Z}}\left(\sum_{i=1}^{n}\sigma_i\varphi(z_i)\right)_+\right]+\mathbb{E}_\sigma\left[\sup_{z\in\mathscr{Z}}\left(-\sum_{i=1}^{n}\sigma_i\varphi(z_i)\right)_+\right].$$

The random variable $-\sigma$ has the same distribution as σ, so applying (B.9) with $g=0$

and $d = n$, we obtain

$$\begin{aligned}\mathbb{E}_\sigma\left[\sup_{z\in\mathscr{Z}}\left|\sum_{i=1}^n \sigma_i\varphi(z_i)\right|\right] &\leq 2\alpha\,\mathbb{E}_\sigma\left[\sup_{z\in\mathscr{Z}}\left(\sum_{i=1}^n \sigma_i z_i\right)_+\right] \\ &\leq 2\alpha\,\mathbb{E}_\sigma\left[\sup_{z\in\mathscr{Z}}\left|\sum_{i=1}^n \sigma_i z_i\right|\right],\end{aligned}$$

which gives (B.8).

To conclude the proof of Theorem B.11, it remains to prove (B.9). Replacing φ by φ/α, we can assume that φ is 1-Lipschitz. The following technical lemma is the key of the proof.

Lemma B.12

Let $\mathscr{Z}$ be a bounded subset of $\mathbb{R}^n$, and consider $\varphi : \mathbb{R} \to \mathbb{R}$ a 1-Lipschitz function fulfilling $\varphi(0) = 0$. Then for any function $g : \mathbb{R}^n \to \mathbb{R}$, we have

$$\sup_{z,z'\in\mathscr{Z}}\left\{(g(z)+\varphi(z_1))_+ + (g(z')-\varphi(z'_1))_+\right\} \leq \sup_{z,z'\in\mathscr{Z}}\left\{(g(z)+z_1)_+ + (g(z')-z'_1)_+\right\}$$

where z_1 denotes the first coordinate of z.

Proof of Lemma B.12. For any $z, z' \in \mathscr{Z}$, we have for any $\varphi : \mathbb{R} \to \mathbb{R}$ and $g : \mathbb{R}^n \to \mathbb{R}$

$$\begin{aligned}&(g(z)+\varphi(z_1))_+ + (g(z')-\varphi(z'_1))_+ \\ &\quad = \max\left\{g(z)+\varphi(z_1)+g(z')-\varphi(z'_1), g(z)+\varphi(z_1), g(z')-\varphi(z'_1), 0\right\}. \qquad \text{(B.10)}\end{aligned}$$

Let us bound each term in the right-hand side maximum.

1. Let us bound the first term. Since φ is 1-Lipschitz, we have

$$\begin{aligned}g(z)+\varphi(z_1)+g(z')-\varphi(z'_1) &\leq g(z)+g(z')+|z_1-z'_1| \\ &\leq \sup_{z,z'\in\mathscr{Z}}\left\{g(z)+g(z')+|z_1-z'_1|\right\}.\end{aligned}$$

   Due to the symmetry in z and z', we have

$$\sup_{z,z'\in\mathscr{Z}}\left\{g(z)+g(z')+|z_1-z'_1|\right\} = \sup_{z,z'\in\mathscr{Z}}\left\{g(z)+g(z')+z_1-z'_1\right\},$$

   and therefore

$$\begin{aligned}g(z)+\varphi(z_1)+g(z')-\varphi(z'_1) &\leq \sup_{z,z'\in\mathscr{Z}}\left\{g(z)+g(z')+z_1-z'_1\right\} \\ &\leq \sup_{z,z'\in\mathscr{Z}}\left\{(g(z)+z_1)_+ + (g(z')-z'_1)_+\right\}. \qquad \text{(B.11)}\end{aligned}$$

2. Let us bound the two other terms in the right-hand side of (B.10). Since φ is

1-Lipschitz and $\varphi(0) = 0$, we have $|\varphi(z_1)| \leq |z_1|$ for all $z_1 \in \mathbb{R}$. This inequality gives

$$\begin{aligned} g(z) \pm \varphi(z_1) &\leq g(z) + |z_1| \\ &\leq (g(z) + z_1)_+ + (g(z) - z_1)_+ \\ &\leq \sup_{z,z' \in \mathscr{Z}} \left\{ (g(z) + z_1)_+ + (g(z') - z'_1)_+ \right\}. \end{aligned} \tag{B.12}$$

Combining (B.10) with (B.11) and (B.12) completes the proof of Lemma B.12. □

We now prove (B.9) by induction on d.

- For $d = 1$, we have

$$\begin{aligned} \mathbb{E}_{\sigma_1}\left[\sup_{z \in \mathscr{Z}} (g(z) + \sigma_1 \varphi(z_1))_+\right] &= \frac{1}{2} \sup_{z \in \mathscr{Z}} (g(z) + \varphi(z_1))_+ + \frac{1}{2} \sup_{z' \in \mathscr{Z}} \left(g(z') - \varphi(z'_1)\right)_+ \\ &= \frac{1}{2} \sup_{z,z' \in \mathscr{Z}} \left\{ (g(z) + \varphi(z_1))_+ + \left(g(z') - \varphi(z'_1)\right)_+ \right\}. \end{aligned}$$

Lemma B.12 ensures that

$$\begin{aligned} \mathbb{E}_{\sigma_1}\left[\sup_{z \in \mathscr{Z}} (g(z) + \sigma_1 \varphi(z_1))_+\right] &\leq \frac{1}{2} \sup_{z,z' \in \mathscr{Z}} \left\{ (g(z) + z_1)_+ + (g(z') - z'_1)_+ \right\} \\ &= \mathbb{E}_{\sigma_1}\left[\sup_{z \in \mathscr{Z}} (g(z) + \sigma_1 z_1)_+\right], \end{aligned}$$

which proves (B.9) for $d = 1$.

- Assume now that Inequality (B.9) has been proved up to $d = k - 1$, with $k \geq 2$. Below, we will denote by $\mathbb{E}_{\sigma_k}$ the expectation under the variable σ_k and by $\mathbb{E}_{\sigma_{-k}}$ the expectation under the variables $\sigma_1, \ldots, \sigma_{k-1}$. Applying successively Inequality (B.9) with $d = k - 1$ and $d = 1$, we obtain

$$\begin{aligned} &\mathbb{E}_{\sigma}\left[\sup_{z \in \mathscr{Z}} \left(g(z) + \sum_{i=1}^{k} \sigma_i \varphi(z_i)\right)_+\right] \\ &= \mathbb{E}_{\sigma_k}\left[\mathbb{E}_{\sigma_{-k}}\left[\sup_{z \in \mathscr{Z}} \left((g(z) + \sigma_k \varphi(z_k)) + \sum_{i=1}^{k-1} \sigma_i \varphi(z_i)\right)_+\right]\right] \quad \text{(Fubini)} \\ &\leq \mathbb{E}_{\sigma_k}\left[\mathbb{E}_{\sigma_{-k}}\left[\sup_{z \in \mathscr{Z}} \left((g(z) + \sigma_k \varphi(z_k)) + \sum_{i=1}^{k-1} \sigma_i z_i\right)_+\right]\right] \quad \text{(Ineq. (B.9) with } d = k-1) \\ &= \mathbb{E}_{\sigma_{-k}}\left[\mathbb{E}_{\sigma_k}\left[\sup_{z \in \mathscr{Z}} \left(\left(g(z) + \sum_{i=1}^{k-1} \sigma_i z_i\right) + \sigma_k \varphi(z_k)\right)_+\right]\right] \quad \text{(Fubini)} \\ &\leq \mathbb{E}_{\sigma_{-k}}\left[\mathbb{E}_{\sigma_k}\left[\sup_{z \in \mathscr{Z}} \left(g(z) + \sum_{i=1}^{k} \sigma_i z_i\right)_+\right]\right] \quad \text{(Inequality (B.9) with } d = 1), \end{aligned}$$

which gives (B.9) for $d = k$. By induction, this completes the proof of (B.9) for any $d \leq n$. □

B.4 Birgé's Inequality

Birgé's inequality [30] is very useful for deriving minimax lower bounds, as, for example, in Theorem 3.5 and Exercise 3.6.2. We state here a simple version of the inequality, and we refer to Corollary 2.18 in Massart's lecture notes [118] for a stronger result.

In the following, we denote by

$$KL(\mathbb{P},\mathbb{Q}) = \begin{cases} \int \log \frac{d\mathbb{P}}{d\mathbb{Q}}\, d\mathbb{P} & \text{when } \mathbb{P} \ll \mathbb{Q} \\ +\infty & \text{else} \end{cases}$$

the Kullback–Leibler divergence between $\mathbb{P}$ and $\mathbb{Q}$.

Theorem B.13 Birgé's Inequality
Let us consider a family $(A_i)_{i=1,\ldots,N}$ of disjointed events, and a collection $(\mathbb{P}_i)_{i=1,\ldots,N}$ of probability measures. Then, we have

$$\min_{i=1,\ldots,N} \mathbb{P}_i(A_i) \leq \frac{2e}{2e+1} \bigvee \frac{\max_{i\neq j} KL(\mathbb{P}_i,\mathbb{P}_j)}{\log(N)}. \qquad \text{(B.13)}$$

Proof. We can assume that $\mathbb{P}_i \ll \mathbb{P}_j$ for all $i \neq j$, otherwise $\max_{i\neq j} KL(\mathbb{P}_i,\mathbb{P}_j) = +\infty$ and (B.13) is obvious. The proof is mainly based on the following simple inequality.

Lemma B.14

Let X be a bounded random variable and $\mathbb{P}_1, \mathbb{P}_2$ be two probability measures, such that $\mathbb{P}_1 \ll \mathbb{P}_2$ and $\mathbb{P}_2 \ll \mathbb{P}_1$. Then, we have

$$\mathbb{E}_2[X] - \log \mathbb{E}_1\left[e^X\right] \leq KL(\mathbb{P}_2,\mathbb{P}_1), \qquad \text{(B.14)}$$

where $\mathbb{E}_i$ denotes the expectation with respect to $\mathbb{P}_i$.

Proof of the lemma. Since $-\log$ is convex, Jensen inequality ensures that

$$\begin{aligned} -\log \mathbb{E}_1\left[e^X\right] &= -\log\left(\int e^X \frac{d\mathbb{P}_1}{d\mathbb{P}_2}\, d\mathbb{P}_2\right) \\ &\leq \int -\log\left(e^X \frac{d\mathbb{P}_1}{d\mathbb{P}_2}\right) d\mathbb{P}_2 = -\mathbb{E}_2[X] + KL(\mathbb{P}_2,\mathbb{P}_1), \end{aligned}$$

which concludes the proof of Lemma B.14. □

We set $m = \min_{i=1,\ldots,N} \mathbb{P}_i(A_i)$, and for $i \leq N-1$, we define $X = \mathbf{1}_{A_i} \log(m/q)$ with $q = (1-m)/(N-1)$. Inequality (B.14) gives

$$\mathbb{P}_i(A_i)\log(m/q) - \log \mathbb{E}_N\left[\left(\frac{m}{q}\right)^{\mathbf{1}_{A_i}}\right] \leq KL(\mathbb{P}_i,\mathbb{P}_N).$$

Computing the expectation, we find

$$\log \mathbb{E}_N \left[\left(\frac{m}{q} \right)^{\mathbf{1}_{A_i}} \right] = \log \left(\mathbb{P}_N(A_i) \left(\frac{m}{q} - 1 \right) + 1 \right)$$
$$\leq \mathbb{P}_N(A_i) \left(\frac{m}{q} - 1 \right) \leq \frac{m}{q} \mathbb{P}_N(A_i).$$

We have $\mathbb{P}_i(A_i) \geq m$, so averaging over $i \in \{1, \ldots, N-1\}$ gives

$$m \log(m/q) - \frac{m}{q(N-1)} \sum_{i=1}^{N-1} \mathbb{P}_N(A_i) \leq \overline{KL} = \frac{1}{N-1} \sum_{i=1}^{N-1} KL(\mathbb{P}_i, \mathbb{P}_N).$$

Since

$$\sum_{i=1}^{N-1} \mathbb{P}_N(A_i) \leq 1 - \mathbb{P}_N(A_N) \leq 1 - m \quad \text{and} \quad q = \frac{1-m}{N-1},$$

we finally obtain

$$m \log \left(\frac{m(N-1)}{e(1-m)} \right) \leq \overline{KL}.$$

To conclude the proof of (B.13), we simply check that for $m \geq 2e/(2e+1)$, we have

$$\log \left(\frac{m(N-1)}{e(1-m)} \right) \geq \log(2(N-1)) \geq \log(N),$$

and that $\overline{KL} \leq \max_{i \neq j} KL(\mathbb{P}_i, \mathbb{P}_j)$. □

Appendix C

Linear Algebra

C.1 Singular Value Decomposition

The Singular Value Decomposition (SVD) is a matrix decomposition that is very useful in many fields of applied mathematics. In the following, we will use that, for any $n \times p$ matrix A, the matrices $A^T A$ and AA^T are symmetric positive semi-definite.

Theorem C.1 Singular value decomposition
Any $n \times p$ matrix A of rank r can be decomposed as

$$A = \sum_{j=1}^{r} \sigma_j u_j v_j^T, \tag{C.1}$$

where

- $r = \text{rank}(A)$,
- $\sigma_1 \geq \ldots \geq \sigma_r > 0$,
- $\{\sigma_1^2, \ldots, \sigma_r^2\}$ *are the nonzero eigenvalues of $A^T A$ (which are also the nonzero eigenvalues of AA^T), and*
- $\{u_1, \ldots, u_r\}$ *and* $\{v_1, \ldots, v_r\}$ *are two orthonormal families of $\mathbb{R}^n$ and $\mathbb{R}^p$, such that*

$$AA^T u_j = \sigma_j^2 u_j \quad \textit{and} \quad A^T A v_j = \sigma_j^2 v_j.$$

The values $\sigma_1, \ldots, \sigma_r$ are called the singular values of A. The vectors $\{u_1, \ldots, u_r\}$ and $\{v_1, \ldots, v_r\}$ are said to be left-singular vectors and right-singular vectors, respectively.

Proof. Let us prove that such a decomposition exists. Since AA^T is positive semi-definite, we have a spectral decomposition

$$AA^T = \sum_{j=1}^{r} \lambda_j u_j u_j^T,$$

with $\lambda_1 \geq \ldots \geq \lambda_r > 0$ and $\{u_1, \ldots, u_r\}$ an orthonormal family of $\mathbb{R}^n$. Let us define $v_1, \ldots, v_r$ by $v_j = \lambda_j^{-1/2} A^T u_j$ for $j = 1, \ldots, r$. We have

$$\|v_j\|^2 = \lambda_j^{-1} u_j^T AA^T u_j = u_j^T u_j = 1,$$

and

$$A^T A v_j = \lambda_j^{-1/2} A^T (AA^T) u_j = \lambda_j^{1/2} A^T u_j = \lambda_j v_j,$$

so $\{v_1, \ldots, v_r\}$ is an orthonormal family of eigenvectors of $A^T A$. Setting $\sigma_j = \lambda_j^{1/2}$, we obtain

$$\begin{aligned}\sum_{j=1}^{r} \sigma_j u_j v_j^T &= \sum_{j=1}^{r} \lambda_j^{1/2} \lambda_j^{-1/2} u_j u_j^T A \\ &= \left(\sum_{j=1}^{r} u_j u_j^T\right) A.\end{aligned}$$

We notice that $\sum_{j=1}^{r} u_j u_j^T$ is the projection onto the range of AA^T. To conclude, we recall that $\mathbb{R}^p = \ker(A) \bigoplus \text{range}(A^T)$ is the orthogonal sum of $\ker(A)$ and $\text{range}(A^T)$, so the range of A and the range of AA^T coincide and

$$\sum_{j=1}^{r} \sigma_j u_j v_j^T = \left(\sum_{j=1}^{r} u_j u_j^T\right) A = \text{Proj}_{\text{range}(A)}\, A = A.$$

The proof of Lemma C.1 is complete. □

C.2 Moore–Penrose Pseudo-Inverse

The Moore–Penrose pseudo-inverse A^+ of a matrix A generalizes the notion of inverse for singular matrices. It is a matrix such that $AA^+y = y$ for all y in the range of A and $A^+Ax = x$ for all x in the range of A^+. Furthermore, the matrices AA^+ and A^+A are symmetric. When A is non-singular, we have the identity $A^+ = A^{-1}$. We first describe A^+ for diagonal matrices, then for symmetric matrices, and finally for arbitrary matrices.

Diagonal matrices

The Moore–Penrose pseudo-inverse of a diagonal matrix D is a diagonal matrix D^+, with diagonal entries $[D^+]_{jj} = 1/D_{jj}$ when $D_{jj} \neq 0$ and $[D^+]_{jj} = 0$ otherwise.

Symmetric matrices

Write $A = UDU^T$ for a spectral decomposition of A with D diagonal and U unitary[1]. The Moore–Penrose pseudo-inverse of A is given by $A^+ = UD^+U^T$.

Arbitrary matrices

Write $A = \sum_{j=1}^{r} \sigma_j(A) u_j v_j^T$ for a singular value decomposition of A with $r = \text{rank}(A)$. The Moore–Penrose pseudo-inverse of A is given by

$$A^+ = \sum_{j=1}^{r} \sigma_j(A)^{-1} v_j u_j^T. \tag{C.2}$$

---

[1] U unitary if $U^T U = UU^T = I$.

We notice that

$$A^+A = \sum_{j=1}^{r} v_j v_j^T = \mathrm{Proj}_{\mathrm{range}(A^T)} \quad \text{and} \quad AA^+ = \sum_{j=1}^{r} u_j u_j^T = \mathrm{Proj}_{\mathrm{range}(A)}. \tag{C.3}$$

In particular, $AA^+ = A^+A = I$ when A is non-singular.

C.3 Matrix Norms

In the following, we denote by $\sigma_1(A) \geq \sigma_2(A) \geq \ldots$ the singular values of A. Several interesting norms are related to the singular values.

Frobenius norm
The standard scalar product on matrices is $\langle A, B\rangle_F = \sum_{i,j} A_{i,j} B_{i,j}$. It induces the Frobenius norm

$$\|A\|_F^2 = \sum_{i,j} A_{i,j}^2 = \mathrm{Tr}(A^T A) = \sum_k \sigma_k(A)^2.$$

The last equality follows from the fact that the $\sigma_k(A)^2$ are the eigenvalues of $A^T A$.

Operator norm
The $\ell^2 \to \ell^2$ operator norm is defined by

$$|A|_{\mathrm{op}} = \sup_{\|x\| \leq 1} \|Ax\| = \sigma_1(A).$$

Let us prove this last equality. We have $Ax = \sum_k \sigma_k(A) u_k v_k^T x$, so $\|Ax\|^2 = \sum_k \sigma_k^2(A) \langle v_k, x\rangle^2 \leq \sigma_1(A)^2 \|x\|^2$, with equality for $x = v_1$.

Nuclear norm
The nuclear norm is defined by

$$|A|_* = \sum_{k=1}^{r} \sigma_k(A).$$

The three following inequalities are very useful.

Lemma C.2 *We have*

1. $|A|_* \leq \sqrt{\mathrm{rank}(A)}\, \|A\|_F$,
2. $\langle A, B\rangle_F \leq |A|_* |B|_{\mathrm{op}}$,
3. $\|AB\|_F \leq |A|_{\mathrm{op}} \|B\|_F$.

Proof. The first inequality is simply Cauchy–Schwartz inequality. For the second inequality, we start from

$$\langle A, B\rangle_F = \sum_k \sigma_k(A) \langle u_k v_k^T, B\rangle_F = \sum_k \sigma_k(A) \langle u_k, B v_k\rangle$$

and notice that $\langle u_k, Bv_k\rangle \leq \|Bv_k\| \leq |B|_{op}$ since $\|u_k\| = \|v_k\| = 1$. The inequality

$$\langle A, B\rangle_F \leq \sum_k \sigma_k(A)|B|_{op}$$

then follows. Let us turn to the third inequality. We denote by B_j the j-th column of B. We observe that $\|B\|_F^2 = \sum_j \|B_j\|^2$, so

$$\|AB\|_F^2 = \sum_j \|(AB)_j\|^2 = \sum_j \|AB_j\|^2 \leq \sum_j |A|_{op}^2 \|B_j\|^2 = |A|_{op}^2 \|B\|_F^2.$$

The proof of Lemma C.2 is complete. □

C.4 Matrix Analysis

C.4.1 Characterization of the Singular Values

Next result is a geometric characterization of the singular values.

Theorem C.3 Max–Min / Min–Max formula
For any $n \times p$ matrix A and $k \leq \min(n, p)$, we have

$$\sigma_k(A) = \max_{S:\dim(S)=k} \min_{x\in S\setminus\{0\}} \frac{\|Ax\|}{\|x\|}, \tag{C.4}$$

where the maximum is taken over all the linear spans $S \subset \mathbb{R}^p$ with dimension k. Symmetrically, we have

$$\sigma_k(A) = \min_{S:codim(S)=k-1} \max_{x\in S\setminus\{0\}} \frac{\|Ax\|}{\|x\|}, \tag{C.5}$$

where the minimum is taken over all the linear spans $S \subset \mathbb{R}^p$ with codimension $k-1$.

Proof. We start from the singular value decomposition $A = \sum_{j=1}^r \sigma_j(A) u_j v_j^T$ and we consider $\{v_{r+1}, \ldots, v_p\}$, such that $\{v_1, \ldots, v_p\}$ is an orthonormal basis of $\mathbb{R}^p$. We define $S_k = \text{span}\{v_1, \ldots, v_k\}$ and $W_k = \text{span}\{v_k, \ldots, v_p\}$. For any linear span $S \subset \mathbb{R}^p$ with dimension k, we have $\dim(S) + \dim(W_k) = p + 1$, so $S \cap W_k \neq \{0\}$. For any nonzero $x \in S \cap W_k$ we have

$$\frac{\|Ax\|^2}{\|x\|^2} = \frac{\sum_{j=k}^r \sigma_j(A)^2 \langle v_j, x\rangle^2}{\sum_{j=k}^p \langle v_j, x\rangle^2} \leq \sigma_k(A)^2,$$

so

$$\max_{S:\dim(S)=k} \min_{x\in S\setminus\{0\}} \frac{\|Ax\|}{\|x\|} \leq \sigma_k(A).$$

Conversely, for all $x \in S_k \setminus \{0\}$, we have

$$\frac{\|Ax\|^2}{\|x\|^2} = \frac{\sum_{j=1}^k \sigma_j(A)^2 \langle v_j, x\rangle^2}{\sum_{j=1}^k \langle v_j, x\rangle^2} \geq \sigma_k(A)^2,$$

with equality for $x = v_k$. As a consequence,

$$\max_{S:\dim(S)=k} \min_{x \in S \setminus \{0\}} \frac{\|Ax\|}{\|x\|} = \sigma_k(A),$$

with equality for $S = S_k$, which proves (C.4). The min–max formula (C.5) is proved similarly. □

Corollary C.4 *For an $n \times p$ matrix A and $k \leq \min(n,p)$, we have for any $P \in \mathbb{R}^{n\times n}$ with $|P|_{\text{op}} \leq 1$*

$$\sigma_k(PA) \leq \sigma_k(A) \quad \text{and} \quad \|PA\|_F \leq \|A\|_F. \tag{C.6}$$

Similarly, we have for any $P \in \mathbb{R}^{p\times p}$ with $|P|_{\text{op}} \leq 1$

$$\sigma_k(AP) \leq \sigma_k(A) \quad \text{and} \quad \|AP\|_F \leq \|A\|_F. \tag{C.7}$$

Proof. Since $|P|_{\text{op}} \leq 1$, we have $\|PAx\| \leq \|Ax\|$. The inequality (C.6) then follows from (C.4). Furthermore, we have $\sigma_k(AP) = \sigma_k(P^T A^T) \leq \sigma_k(A^T) = \sigma_k(A)$, which gives (C.7). □

C.4.2 Best Low-Rank Approximation

The next theorem characterizes the "projection" on the set of matrices of rank r. It also provides an improvement of the Cauchy–Schwartz inequality $\langle A,B\rangle_F \leq \|A\|_F\|B\|_F$ in terms of the Ky–Fan $(2,q)$-norm

$$\|A\|_{(2,q)}^2 = \sum_{k=1}^{q} \sigma_k(A)^2, \tag{C.8}$$

with $q = \text{rank}(A) \wedge \text{rank}(B)$. We observe that $\|A\|_{(2,q)} \leq \|A\|_F$, with strict inequality if $q < \text{rank}(A)$.

Theorem C.5 *For any matrices $A, B \in \mathbb{R}^{n\times p}$, we set $q = \text{rank}(A) \wedge \text{rank}(B)$. We then have*

$$\langle A,B\rangle_F \leq \|A\|_{(2,q)}\, \|B\|_{(2,q)},$$

where the Ky–Fan $(2,q)$-norm $\|A\|_{(2,q)}$ is defined in (C.8).

As a consequence, for $A = \sum_{k=1}^{r} \sigma_k(A) u_k v_k^T$ and $q < r$, we have

$$\min_{B:\text{rank}(B)\leq q} \|A-B\|_F^2 = \sum_{k=q+1}^{r} \sigma_k(A)^2.$$

In addition, the minimum is achieved for

$$B = \sum_{k=1}^{q} \sigma_k(A) u_k v_k^T.$$

Proof. We can assume, e.g., that the rank of B is not larger than the rank of A. Let us denote by q the rank of B and P_B the projection on the range of B. We have

$$\langle A,B\rangle_F = \langle P_BA,B\rangle_F \leq \|P_BA\|_F\|B\|_F.$$

The rank of P_BA is at most q and previous corollary ensures that $\sigma_k(P_BA) \leq \sigma_k(A)$, since $|P_B|_{\text{op}} \leq 1$. So,

$$\|P_BA\|_F^2 = \sum_{k=1}^{q} \sigma_k(P_BA)^2 \leq \sum_{k=1}^{q} \sigma_k(A)^2 = \|A\|_{(2,q)}^2.$$

Since $q = \text{rank}(B)$, we have $\|B\|_F = \|B\|_{(2,q)}$, and the first part of the theorem is proved.

According to the first part of the theorem, for any matrix B of rank q, we have

$$\|A-B\|_F^2 = \|A\|_F^2 - 2\langle A,B\rangle_F + \|B\|_F^2 \geq \|A\|_F^2 - 2\|A\|_{(2,q)}\|B\|_F + \|B\|_F^2.$$

The right-hand side is minimum for $\|B\|_F = \|A\|_{(2,q)}$, so

$$\|A-B\|_F^2 \geq \|A\|_F^2 - \|A\|_{(2,q)}^2 = \sum_{k=q+1}^{r} \sigma_k(A)^2.$$

Finally, we observe that this lower bound is achieved for $B = \sum_{k=1}^{q} \sigma_k(A)u_kv_k^T$. □

C.5 Perturbation Bounds

In statistics and in machine learning, it is useful to relate the SVD of the observed matrix $B = A+E$ to the SVD of the signal matrix A.

C.5.1 Weyl Inequality

Weyl inequality states that the singular values are 1-Lipschitz with respect to the operator norm.

Theorem C.6 Weyl inequality
For two $n\times p$ matrices A and B, we have for any $k \leq \min(n,p)$

$$|\sigma_k(A) - \sigma_k(B)| \leq \sigma_1(A-B) = |A-B|_{\text{op}}.$$

□

Proof. For any $x \in \mathbb{R}^p \setminus \{0\}$, we have

$$\frac{\|Ax\|}{\|x\|} \leq \frac{\|Bx\|}{\|x\|} + \frac{\|(A-B)x\|}{\|x\|} \leq \frac{\|Bx\|}{\|x\|} + \sigma_1(A-B).$$

The inequality follows by applying the Max–Min formula (C.3). □

C.5.2 Eigenspaces Localization

Let $A, B \in \mathbb{R}^{n\times n}$ be two symmetric matrices and let $A = \sum_k \lambda_k u_k u_k^T$ and $B = \sum_k \rho_k v_k v_k^T$ be their eigenvalue decomposition with $\lambda_1 \geq \lambda_2 \geq \cdots$ and $\rho_1 \geq \rho_2 \geq \cdots$. We want to compare the eigenspaces span$\{u_1,\ldots,u_r\}$ and span$\{v_1,\ldots,v_r\}$, spanned by the r leading eigenvectors of A and B.
A first idea could be to compare the two matrices $U_r = [u_1,\ldots,u_r]$ and $V_r = [v_1,\ldots,v_r]$. Yet, there exist some orthogonal transformation R, such that span$\{Ru_1,\ldots,Ru_r\}$ = span$\{u_1,\ldots,u_r\}$, but $RU_r \neq U_r$, so a directed comparison of U_r and V_r is not suited. Instead, we will compare

$$U_rU_r^T = \sum_{k=1}^{r} u_k u_k^T \quad \text{and} \quad V_rV_r^T = \sum_{k=1}^{r} v_k v_k^T,$$

which are the orthogonal projectors in $\mathbb{R}^n$ onto span$\{u_1,\ldots,u_r\}$ and span$\{v_1,\ldots,v_r\}$, respectively. The next proposition relates the Frobenius distance between $U_rU_r^T$ and $V_rV_r^T$ to the Frobenius norm of $U_{-r}^T V_r$.

Proposition C.7 *Let $U_{-r} = [u_{r+1},\ldots,u_n]$ and $V_{-r} = [v_{r+1},\ldots,v_n]$. Then, we have*

$$\|U_rU_r^T - V_rV_r^T\|_F^2 = 2\|V_{-r}^T U_r\|_F^2 = 2\|U_{-r}^T V_r\|_F^2.$$

Proof of Proposition C.7. We first expand the squares and use that the Frobenius norm of a projector is equal to its rank

$$\begin{aligned}\|U_rU_r^T - V_rV_r^T\|_F^2 &= \|U_rU_r^T\|_F^2 + \|V_rV_r^T\|_F^2 - 2\langle U_rU_r^T, V_rV_r^T\rangle_F \\ &= 2r - 2\mathrm{Tr}(\mathrm{U_r^T V_r V_r^T U_r}).\end{aligned}$$

Then, since span$\{v_{r+1},\ldots,v_n\}$ is the orthogonal complement of span$\{v_1,\ldots,v_r\}$, we have $V_rV_r^T = I_n - V_{-r}V_{-r}^T$. So, as $U_r^T U_r = I_r$

$$\begin{aligned}\|U_rU_r^T - V_rV_r^T\|_F^2 &= 2r - 2\mathrm{Tr}(\mathrm{I_r - U_r^T V_{-r} V_{-r}^T U_r}) \\ &= 2\mathrm{Tr}(\mathrm{U_r^T V_{-r} V_{-r}^T U_r}) = 2\|\mathrm{V_{-r}^T U_r}\|_\mathrm{F}^2.\end{aligned}$$

The second equality of Lemma C.7 follows by symmetry. □

A classical inequality to bound the norm $\|U_{-r}^T V_r\|_F^2$ is the Davis-Kahan perturbation bound.

Theorem C.8 Davis-Kahan perturbation bound.
Let $A, B \in \mathbb{R}^{n\times n}$ be two symmetric matrices and let $A = \sum_k \lambda_k u_k u_k^T$ and $B = \sum_k \rho_k v_k v_k^T$ be their eigenvalue decomposition with $\lambda_1 \geq \cdots \geq \lambda_n$ and $\rho_1 \geq \cdots \geq \rho_n$. Let $U_r = [u_1, \ldots, u_r]$, $U_{-r} = [u_{r+1}, \ldots, u_n]$ and similarly $V_r = [v_1, \ldots, v_r]$, $V_{-r} = [v_{r+1}, \ldots, v_n]$. Then, we have

$$\|U_{-r}^T V_r\|_F \leq \frac{(\sqrt{r}|A-B|_{\mathrm{op}}) \wedge \|A-B\|_F}{(\rho_r - \lambda_{r+1}) \vee (\lambda_r - \rho_{r+1})} \tag{C.9}$$

$$\leq 2\,\frac{(\sqrt{r}|A-B|_{\mathrm{op}}) \wedge \|A-B\|_F}{\lambda_r - \lambda_{r+1}}. \tag{C.10}$$

In many cases, we only wish to compare the two leading eigenvectors of A and B, which corresponds to the case $r = 1$.

Corollary C.9 Comparing leading eigenvectors.

$$\sqrt{1 - \langle u_1, v_1\rangle^2} \leq \frac{2\inf_{\lambda\in\mathbb{R}} |A + \lambda I - B|_{\mathrm{op}}}{\lambda_1 - \lambda_2}. \tag{C.11}$$

Proof of Corollary C.9.
We first observe that

$$\|U_{-1}^T v_1\|^2 = v_1^T U_{-1} U_{-1}^T v_1 = v_1^T (I - u_1 u_1^T) v_1 = 1 - (u_1^T v_1)^2.$$

In addition, the eigenvectors of A and $A + \lambda I$ are the same, while the eigenvalues are all translated by λ, preserving the eigengap $\lambda_1 - \lambda_2$ between the two largest eigenvalues. So we have for any $\lambda \in \mathbb{R}$, the Inequality (C.10) applied to $A + \lambda I$ and B gives

$$\sqrt{1 - \langle u_1, v_1\rangle^2} \leq \frac{2|A + \lambda I - B|_{\mathrm{op}}}{\lambda_1 - \lambda_2}.$$

The proof of Corollary C.9 is complete. □

Proof of Theorem C.8.
We first observe that the Bound (C.10) directly follows from (C.9) and the inequalities

$$\begin{aligned}\lambda_r - \lambda_{r+1} &= \lambda_r - \rho_{r+1} - (\rho_r - \rho_{r+1}) + \rho_r - \lambda_{r+1} \\ &\leq (\lambda_r - \rho_{r+1}) + (\rho_r - \lambda_{r+1}) \leq 2((\rho_r - \lambda_{r+1}) \vee (\lambda_r - \rho_{r+1})).\end{aligned}$$

Let us prove (C.9). As a starting point, we notice that either $\rho_r > \lambda_{r+1}$ or $\lambda_r \geq \rho_{r+1}$, so

$$(\rho_r - \lambda_{r+1}) \vee (\lambda_r - \rho_{r+1}) = (\rho_r - \lambda_{r+1})_+ \vee (\lambda_r - \rho_{r+1})_+.$$

In addition, we observe from Lemma C.7 that the roles of A and B are symmetric. Hence, we only need to prove

$$\|U_{-r}^T V_r\|_F \leq \frac{(\sqrt{r}|A-B|_{\mathrm{op}}) \wedge \|A-B\|_F}{(\rho_r - \lambda_{r+1})_+}. \tag{C.12}$$

When $\rho_r \leq \lambda_{r+1}$ the right-hand side is infinite, so we only need to focus on the case where $\rho_r > \lambda_{r+1}$.
We have the decomposition

$$\|U_{-r}^T V_r\|_F^2 = \sum_{k=1}^{r} \|U_{-r}^T v_k\|^2, \tag{C.13}$$

so we will start by bounding the square norms $\|U_{-r}^T v_k\|^2$. Since

$$A = \sum_{k=1}^{n} \lambda_k u_k u_k^T = U_r \operatorname{diag}(\lambda_1, \ldots, \lambda_r) U_r^T + U_{-r} \operatorname{diag}(\lambda_{r+1}, \ldots, \lambda_n) U_{-r}^T,$$

we have $U_{-r}^T A = \operatorname{diag}(\lambda_{r+1}, \ldots, \lambda_n) U_{-r}^T$. Hence, with $Bv_k = \rho_k v_k$, we have for $k = 1, \ldots, r$

$$\begin{aligned} \rho_k U_{-r}^T v_k &= U_{-r}^T B v_k = U_{-r}^T (A + B - A) v_k \\ &= \operatorname{diag}(\lambda_{r+1}, \ldots, \lambda_n) U_{-r}^T v_k + U_{-r}^T (B - A) v_k. \end{aligned}$$

Hence

$$U_{-r}^T v_k = \operatorname{diag}(\rho_k - \lambda_{r+1}, \ldots, \rho_k - \lambda_n)^{-1} U_{-r}^T (B - A) v_k,$$

and then, since $\rho_k \geq \rho_r > \lambda_{r+1}$,

$$\begin{aligned} \|U_{-r}^T v_k\|^2 &\leq |\operatorname{diag}(\rho_k - \lambda_{r+1}, \ldots, \rho_k - \lambda_n)^{-1} U_{-r}^T|_{\text{op}}^2 \, \|(B - A) v_k\|^2 \\ &\leq \frac{\|(B - A) v_k\|^2}{(\rho_k - \lambda_{r+1})^2} \leq \frac{\|(B - A) v_k\|^2}{(\rho_r - \lambda_{r+1})^2}, \end{aligned}$$

for $k = 1, \ldots, r$. To conclude, we observe that

$$\sum_{k=1}^{r} \|(B - A) v_k\|^2 \leq |B - A|_{\text{op}}^2 \sum_{k=1}^{r} \|v_k\|^2 = r|B - A|_{\text{op}},$$

since $\|v_k\| = 1$. So, with (C.13) we get

$$\|U_{-r}^T V_r\|_F^2 \leq \frac{r|A - B|_{\text{op}}^2}{(\rho_r - \lambda_{r+1})^2}. \tag{C.14}$$

In addition, since $I_n = V_r V_r^T + V_{-r} V_{-r}^T$, we have

$$\|A - B\|_F^2 = \langle (A - B)(V_r V_r^T + V_{-r} V_{-r}^T), A - B \rangle_F = \|(A - B) V_r\|_F^2 + \|(A - B) V_{-r}\|_F^2,$$

so

$$\begin{aligned} \sum_{k=1}^{r} \|(B - A) v_k\|^2 &= \|(B - A) V_r\|_F^2 \\ &\leq \|(B - A) V_r\|_F^2 + \|(B - A) V_{-r}\|_F^2 = \|B - A\|_F^2. \end{aligned}$$

Combining this bound with (C.13), we get

$$\|U_{-r}^T V_r\|_F^2 \leq \frac{\|A - B\|_F^2}{(\rho_r - \lambda_{r+1})^2}. \tag{C.15}$$

Combining (C.14) and (C.15), we get (C.12), completing the proof of Theorem C.8. □

We refer to Horn and Johnson [93] for more involved results on matrix analysis.

Appendix D

Subdifferentials of Convex Functions

D.1 Subdifferentials and Subgradients

A function $F : \mathbb{R}^n \to \mathbb{R}$ is convex if $F(\lambda x + (1-\lambda)y) \leq \lambda F(x) + (1-\lambda)F(y)$ for all $x, y \in \mathbb{R}^n$ and $\lambda \in [0,1]$. An equivalent definition is that the epigraph $\{(x,y),\ x \in \mathbb{R}^n,\ y \in [F(x), +\infty[\}$ is a convex subset of $\mathbb{R}^{n+1}$.

Lemma D.1 *When the function $F : \mathbb{R}^n \to \mathbb{R}$ is convex and differentiable, we have*

$$F(y) \geq F(x) + \langle \nabla F(x), y - x \rangle, \quad \text{for all } x, y \in \mathbb{R}^n.$$

Proof. Let $x, h \in \mathbb{R}^n$, and define $f : \mathbb{R} \to \mathbb{R}$ by $f(t) = F(x + th)$. Since F is differentiable, so is f and $f'(t) = \langle \nabla F(x + th), h \rangle$. By Taylor's expansion, we have for some $t^* \in [0,1]$

$$F(x+h) - F(x) = \langle \nabla F(x + t^* h), h \rangle = f'(t^*).$$

Since

$$f(\lambda t + (1-\lambda)s) = F(\lambda(x+th) + (1-\lambda)(x+sh)) \leq \lambda f(t) + (1-\lambda) f(s),$$

the function f is convex, so

$$F(x+h) - F(x) = f'(t^*) \geq f'(0) = \langle \nabla F(x), h \rangle.$$

We conclude by setting $h = y - x$. □

We define the subdifferential ∂F of a convex function $F : \mathbb{R}^n \to \mathbb{R}$ by

$$\partial F(x) = \{ w \in \mathbb{R}^n : \ F(y) \geq F(x) + \langle w, y - x \rangle \ \text{ for all } y \in \mathbb{R}^n \}. \qquad \text{(D.1)}$$

A vector $w \in \partial F(x)$ is called a subgradient of F in x.

Lemma D.2

1. *$F : \mathbb{R}^n \to \mathbb{R}$ is convex if and only if the set $\partial F(x)$ is non-empty for all $x \in \mathbb{R}^n$.*
2. *When F is convex and differentiable in x, $\partial F(x) = \{\nabla F(x)\}$.*

Proof.
1. Assume that $\partial F(x)$ is non-empty for all $x \in \mathbb{R}^n$. For any $x, y \in \mathbb{R}^n$ and $\lambda \in [0,1]$, there exists $w \in \partial F(\lambda x + (1-\lambda)y)$. By definition of the subdifferential, we have $F(y) \geq F(\lambda x + (1-\lambda)y) + \langle w, \lambda(y-x)\rangle$ and $F(x) \geq F(\lambda x + (1-\lambda)y) + \langle w, (1-\lambda)(x-y)\rangle$. Multiplying the first inequality by $(1-\lambda)$ and the second by λ, we obtain by summing the results

$$(1-\lambda)F(y) + \lambda F(x) \geq F(\lambda x + (1-\lambda)y),$$

so F is convex.

Conversely, if F is convex, then its epigraph is convex in $\mathbb{R}^{n+1}$, so there exists a supporting hyperplane H_x separating the epigraph from any (x,z) with $z < F(x)$. Since F is finite on $\mathbb{R}^n$, this hyperplane is not vertical, so there exists $u \in \mathbb{R}^n$ and $a \in \mathbb{R}$, such that $H_x = \{(\alpha, \beta) : \langle u, \alpha\rangle + \beta = a\}$ and such that any (α, β) in the epigraph of F fulfills $\langle u, \alpha\rangle + \beta \geq a$. Since $(x, F(x)) \in H_x$, we have $a = \langle u, x\rangle + F(x)$. For any $y \in \mathbb{R}^n$, the couple $(y, F(y))$ belongs to the epigraph of F, and therefore

$$\langle u, y\rangle + F(y) \geq a = \langle u, x\rangle + F(x).$$

This ensures that $-u \in \partial F(x)$.

2. Let w be a subgradient of F. When F is differentiable, Taylor's formula gives for any $x, h \in \mathbb{R}^n$ and $t > 0$

$$F(x \pm th) - F(x) = \pm t\langle \nabla F(x), h\rangle + o(t) \geq \pm t\langle w, h\rangle.$$

Letting t go to zero, this enforces $\langle \nabla F(x), h\rangle = \langle w, h\rangle$ for all $h \in \mathbb{R}^n$, so $w = \nabla F(x)$.
□

More generally, when $F : \mathscr{D} \to \mathbb{R}$ is convex on a convex domain $\mathscr{D}$ of $\mathbb{R}^n$, the subdifferential $\partial F(x)$ is non-empty for all x in the interior of $\mathscr{D}$.

It is well-known that the derivative f' of a smooth convex function $f : \mathbb{R} \to \mathbb{R}$ is increasing. The next lemma shows that this result remains valid for subdifferentials.

Lemma D.3 Monotonicity
The subdifferential of a convex function F is monotone increasing:

$$\langle w_x - w_y, x - y\rangle \geq 0, \quad \textit{for all } w_x \in \partial F(x) \textit{ and } w_y \in \partial F(y). \tag{D.2}$$

Proof. By definition, we have $F(y) \geq F(x) + \langle w_x, y - x\rangle$ and $F(x) \geq F(y) + \langle w_y, x - y\rangle$. Summing these two inequalities gives $\langle w_x - w_y, x - y\rangle \geq 0$. □

Finally, the minimum of a convex function can be easily characterized in terms of its subdifferential.

Lemma D.4 First-order optimality condition
For any convex function $F : \mathbb{R}^n \to \mathbb{R}$, we have

$$x_* \in \underset{x\in\mathbb{R}^n}{\operatorname{argmin}} F(x) \iff 0 \in \partial F(x_*). \tag{D.3}$$

Proof. Both conditions are equivalent to $F(y) \geq F(x_*) + \langle 0, y - x_* \rangle$ for all $y \in \mathbb{R}^n$. □

D.2 Examples of Subdifferentials

As examples, we compute the subdifferential of several common norms. For $x \in \mathbb{R}$, we set $\operatorname{sign}(x) = \mathbf{1}_{x>0} - \mathbf{1}_{x\leq 0}$.

Lemma D.5 Subdifferential of ℓ^1 and ℓ^∞ norms

1. *For $x \in \mathbb{R}^n$, let us set $J(x) = \{j : x_j \neq 0\}$. We have*

$$\partial |x|_1 = \{w \in \mathbb{R}^n : \ w_j = \operatorname{sign}(x_j) \text{ for } j \in J(x), \ w_j \in [-1,1] \text{ for } j \notin J(x)\}.$$

2. *Let us set $J_* = \{j : \ |x_j| = |x|_\infty\}$ and write $\mathscr{P}(J_*)$ for the set of probabilities on J_*. We have for $x \neq 0$*

$$\partial |x|_\infty = \left\{ w \in \mathbb{R}^n : \begin{array}{l} w_j = 0 \ \text{ for } j \notin J_* \\ w_j = \lambda_j \, sign(x_j) \ \text{ for } j \in J_* \text{ with } \lambda \in \mathscr{P}(J_*) \end{array} \right\}.$$

Proof. For $p \in [0, +\infty]$ and q, such that $1/p + 1/q = 1$, Hölder's inequality ensures that $|x|_p = \sup\{\langle \phi, x \rangle : \ |\phi|_q \leq 1\}$. To prove Lemma D.5, all we need is to check that

$$\partial |x|_p = \{\phi \in \mathbb{R}^n : \langle \phi, x \rangle = |x|_p \text{ and } |\phi|_q \leq 1\}.$$

i) Consider ϕ_x, such that $\langle \phi_x, x \rangle = |x|_p$ and $|\phi_x|_q \leq 1$. Then, we have for any $y \in \mathbb{R}^n$

$$|y|_p \geq \langle \phi_x, y \rangle = |x|_p + \langle \phi_x, y - x \rangle,$$

and therefore $\phi_x \in \partial |x|_p$.
ii) Conversely, let us consider $w \in \partial |x|_p$. For $y = 0$ and $y = 2x$, Equation (D.1) gives

$$0 \geq |x|_p - \langle w, x \rangle \quad \text{and} \quad 2|x|_p \geq |x|_p + \langle w, x \rangle,$$

from which we get $|x|_p = \langle w, x \rangle$. Furthermore, we have $|w|_q = \langle w, \phi_w \rangle$ for some $\phi_w \in \mathbb{R}^n$ fulfilling $|\phi_w|_p \leq 1$. The triangular inequality and (D.1) give

$$|x|_p + |\phi_w|_p \geq |x + \phi_w|_p \geq |x|_p + \langle w, \phi_w \rangle,$$

which finally ensures that $|w|_q = \langle w, \phi_w \rangle \leq |\phi_w|_p \leq 1$. The proof of Lemma D.5 is complete. □

The next lemma characterizes the subdifferential of the nuclear norm. We refer to Appendix C for the definition of the nuclear norm $|\cdot|_*$, the operator norm $|\cdot|_{op}$, and the singular value decomposition (SVD).

Lemma D.6 Subdifferential of the nuclear norm

Let us consider a matrix B with rank r and singular value decomposition $B = \sum_{k=1}^r \sigma_k u_k v_k^T$. We write P_u for the orthogonal projector onto span$\{u_1,\ldots,u_r\}$ *and P_v for the orthogonal projector onto* span$\{v_1,\ldots,v_r\}$. *We also set $P_u^\perp = I - P_u$ and $P_v^\perp = I - P_v$. Then, we have*

$$\partial|B|_* = \left\{ \sum_{k=1}^r u_k v_k^T + P_u^\perp W P_v^\perp : |W|_{op} \leq 1 \right\}. \tag{D.4}$$

Proof. With the same reasoning as in the proof of Lemma D.5, we have

$$\partial|B|_* = \left\{ Z : \langle Z, B\rangle_F = |B|_* \text{ and } |Z|_{op} \leq 1 \right\}.$$

All we need is then to check that this set coincides with (D.4).

i) First, we observe that any matrix $Z = \sum_{k=1}^r u_k v_k^T + P_u^\perp W P_v^\perp$ with $|W|_{op} \leq 1$ fulfills $|Z|_{op} \leq 1$ and $\langle Z, B\rangle_F = |B|_*$. Therefore, such a matrix Z is in the subdifferential of $|B|_*$.

ii) Conversely, let Z be a matrix fulfilling $\langle Z, B\rangle_F = |B|_*$ and $|Z|_{op} \leq 1$. Since

$$\sum_{k=1}^r \sigma_k = \langle Z, B\rangle_F = \sum_{k=1}^r \sigma_k \langle Z, u_k v_k^T\rangle_F = \sum_{k=1}^r \sigma_k \langle Z v_k, u_k\rangle_F$$

and $\langle Z v_k, u_k\rangle_F \leq \|Z v_k\| \leq |Z|_{op} \leq 1$, we then have $\langle Z v_k, u_k\rangle_F = \|Z v_k\| = 1$. Since $\|u_k\| = 1$, this enforces $Z v_k = u_k$. Since $\langle Z v_k, u_k\rangle_F = \langle v_k, Z^T u_k\rangle_F$, we have for the same reasons $Z^T u_k = v_k$. In particular, $u_1,\ldots,u_r$ are eigenvectors of ZZ^T associated to the eigenvalue 1, and $v_1,\ldots,v_r$ are eigenvectors of $Z^T Z$ also associated to the eigenvalue 1. As a consequence, an SVD of Z is given by

$$Z = \sum_{k=1}^r u_k v_k^T + \sum_{k=r+1}^{\text{rank}(Z)} \widetilde{\sigma}_k \widetilde{u}_k \widetilde{v}_k^T,$$

where $\widetilde{u}_k$ is orthogonal to $u_1,\ldots,u_r$ and $\widetilde{v}_k$ is orthogonal to $v_1,\ldots,v_r$. Furthermore, we have $\widetilde{\sigma}_k \leq 1$ since $|Z|_{op} \leq 1$. In particular, we can write

$$\sum_{k=r+1}^{\text{rank}(Z)} \widetilde{\sigma}_k \widetilde{u}_k \widetilde{v}_k^T = P_u^\perp W P_v^\perp$$

for some matrix W, fulfilling $|W|_{op} \leq 1$. The derivation of (D.4) is complete. □

Appendix E

Reproducing Kernel Hilbert Spaces

Reproducing Kernel Hilbert Spaces (RKHS) are some functional Hilbert spaces, where the smoothness of a function is driven by its norm. RKHS also fulfill a special "reproducing property" that is crucial in practice, since it allows efficient numerical computations as in Proposition 11.9 in Chapter 11.

A function $k : \mathcal{X} \times \mathcal{X} \to \mathbb{R}$ is said to be a positive definite kernel if it is symmetric ($k(x,y) = k(y,x)$ for all $x,y \in \mathcal{X}$), and if for any $N \in \mathbb{N}$, $x_1,\ldots,x_N \in \mathcal{X}$ and $a_1,\ldots,a_N \in \mathbb{R}$ we have

$$\sum_{i,j=1}^{N} a_i a_j k(x_i,x_j) \geq 0. \tag{E.1}$$

Examples of positive definite kernels in $\mathcal{X} = \mathbb{R}^d$:

- linear kernel: $k(x,y) = \langle x,y \rangle$
- Gaussian kernel: $k(x,y) = e^{-\|x-y\|^2/2\sigma^2}$
- histogram kernel ($d = 1$): $k(x,y) = \min(x,y)$
- exponential kernel: $k(x,y) = e^{-\|x-y\|/\sigma}$.

We can associate to a positive definite kernel k a special Hilbert space $\mathcal{F} \subset \mathbb{R}^{\mathcal{X}}$ called Reproducing Kernel Hilbert Space associated to k. In the following, the notation $k(x,.)$ refers to the map $y \to k(x,y)$.

Proposition E.1 Reproducing Kernel Hilbert Space (RKHS)

To any positive definite kernel k on $\mathcal{X}$, we can associate a (unique) Hilbert space $\mathcal{F} \subset \mathbb{R}^{\mathcal{X}}$ fulfilling

1. $k(x,.) \in \mathcal{F}$ *for all* $x \in \mathcal{X}$
2. **reproducing property:**

$$f(x) = \langle f, k(x,.) \rangle_{\mathcal{F}} \quad \text{for all } x \in \mathcal{X} \text{ and } f \in \mathcal{F}. \tag{E.2}$$

The space $\mathcal{F}$ is called the Reproducing Kernel Hilbert Space associated to k.

Proof. From the first property, if the Hilbert space $\mathcal{F}$ exists, it must include the linear space $\mathscr{F}_0$ spanned by the family $\{k(x,.) : x \in \mathscr{X}\}$

$$\mathscr{F}_0 = \left\{ f : \mathscr{X} \to \mathbb{R} : f(x) = \sum_{i=1}^{N} a_i k(x_i, x),\ N \in \mathbb{N},\ x_1, \ldots, x_N \in \mathscr{X},\ a_1, \ldots, a_N \in \mathbb{R} \right\}.$$

Furthermore, from the reproducing property, if $\mathscr{F}$ exists, we must have $\langle k(x,.), k(y,.)\rangle_{\mathscr{F}} = k(x,y)$ and

$$\left\langle \sum_{i=1}^{N} a_i k(x_i,.), \sum_{j=1}^{M} b_j k(y_j,.) \right\rangle_{\mathscr{F}} = \sum_{i=1}^{N} \sum_{j=1}^{M} a_i b_j k(x_i, y_j).$$

Accordingly, we define for any $f = \sum_{i=1}^{N} a_i k(x_i,.)$ and $g = \sum_{j=1}^{M} b_j k(y_j,.)$ in $\mathscr{F}_0$

$$\langle f, g\rangle_{\mathscr{F}_0} := \sum_{i=1}^{N} \sum_{j=1}^{M} a_i b_j k(x_i, y_j) = \sum_{i=1}^{N} a_i g(x_i) = \sum_{j=1}^{M} b_j f(y_j),$$

where the last two equalities ensures that $\langle f, g\rangle_{\mathscr{F}_0}$ does not depend on the choice of the expansion of f and g, so $\langle f, g\rangle_{\mathscr{F}_0}$ is well-defined. The application $(f,g) \to \langle f, g\rangle_{\mathscr{F}_0}$ is bilinear, symmetric, positive (according to (E.1)), and we have the reproducing property

$$f(x) = \langle f, k(x,.)\rangle_{\mathscr{F}_0} \quad \text{for all } x \in \mathscr{X} \quad \text{and } f \in \mathscr{F}_0. \tag{E.3}$$

The Cauchy–Schwartz inequality $\langle f, g\rangle_{\mathscr{F}_0} \leq \|f\|_{\mathscr{F}_0} \|g\|_{\mathscr{F}_0}$ and the reproducing formula (E.3) give

$$|f(x)| \leq \sqrt{k(x,x)}\ \|f\|_{\mathscr{F}_0}. \tag{E.4}$$

As a consequence $\|f\|_{\mathscr{F}_0} = 0$ implies $f = 0$ so $\langle f, g\rangle_{\mathscr{F}_0}$ is a scalar product on $\mathscr{F}_0$. Therefore $\mathscr{F}_0$ is a pre-Hilbert space fulfilling the reproducing property. We obtain $\mathscr{F}$ by completing $\mathscr{F}_0$. □

Remark 1. Let us consider two sequences $(x_i) \in \mathscr{X}^{\mathbb{N}}$ and $(a_i) \in \mathbb{R}^{\mathbb{N}}$ fulfilling $\sum_{i,j \geq 1} a_i a_j k(x_i, x_j) < +\infty$. According to (E.4), for any $M < N$ and $x \in \mathscr{X}$, we have

$$\left| \sum_{i=M+1}^{N} a_i k(x_i, x) \right| \leq \sqrt{k(x,x)} \sum_{i,j=M+1}^{N} a_i a_j k(x_i, x_j).$$

When $\sum_{i,j \geq 1} a_i a_j k(x_i, x_j)$ is finite, the right-hand side goes to 0 when M, N goes to infinity, so the partial series $\sum_{i=1}^{N} a_i k(x_i, x)$ is Cauchy and it converges when $N \to \infty$. We can therefore define the space

$$\mathscr{F}_0' = \left\{ f : \mathscr{X} \to \mathbb{R} : f(x) = \sum_{i=1}^{\infty} a_i k(x_i, x), \right.$$
$$\left. (x_i) \in \mathscr{X}^{\mathbb{N}},\ (a_i) \in \mathbb{R}^{\mathbb{N}},\ \sum_{i,j \geq 1} a_i a_j k(x_i, x_j) < +\infty \right\}$$

and the bilinear form

$$\langle f,g\rangle_{\mathscr{F}_0'} := \sum_{i,j=1}^{\infty} a_i b_j k(x_i,y_j) = \sum_{i=1}^{\infty} a_i g(x_i) = \sum_{j=1}^{\infty} b_j f(y_j)$$

for $f = \sum_{i=1}^{\infty} a_i k(x_i,.)$ and $g = \sum_{j=1}^{\infty} b_j k(y_j,.)$ in $\mathscr{F}_0'$. Exactly as above, the application $(f,g) \to \langle f,g\rangle_{\mathscr{F}_0'}$ is a scalar product fulfilling the reproduction property

$$f(x) = \langle f, k(x,.)\rangle_{\mathscr{F}_0'} \quad \text{for all } x \in \mathscr{X} \text{ and } f \in \mathscr{F}_0'.$$

In addition, the partial sums $f_N = \sum_{i=1}^{N} a_i k(x_i,.)$ relative to a function $f = \sum_{i=1}^{\infty} a_i k(x_i,.) \in \mathscr{F}_0'$ are Cauchy, since

$$\|f_M - f_N\|_{\mathscr{F}_0}^2 = \sum_{i,j=N+1}^{M} a_i a_j k(x_i,x_j) \overset{N,M\to\infty}{\to} 0,$$

and they converge to f. As a consequence, $\mathscr{F}_0'$ is included in the completion $\mathscr{F}$ of $\mathscr{F}_0$ and the scalar product $\langle .,.\rangle_{\mathscr{F}}$ restricted to $\mathscr{F}_0'$ coincides with $\langle .,.\rangle_{\mathscr{F}_0'}$.

Remark 2. The norm of a function f in an RKHS $\mathscr{F}$ is strongly linked to its smoothness. This appears clearly in the inequality

$$|f(x) - f(x')| = |\langle f, k(x,.) - k(x',.)\rangle_{\mathscr{F}}| \leq \|f\|_{\mathscr{F}} \, \|k(x,.) - k(x',.)\|_{\mathscr{F}}. \qquad \text{(E.5)}$$

Let us illustrate this point by describing the RKHS associated to the histogram and Gaussian kernels.

Example 1: RKHS associated to the histogram kernel.
The Sobolev space

$$\mathscr{F} = \left\{ f \in C([0,1],\mathbb{R}) : f \text{ is a.e. differentiable, with } f' \in L^2([0,1]) \text{ and } f(0) = 0 \right\}$$

endowed with the scalar product $\langle f,g\rangle_{\mathscr{F}} = \int_0^1 f'g'$ is an RKHS with reproducing kernel $k(x,y) = \min(x,y)$ on $[0,1]$. Actually, $k(x,.) \in \mathscr{F}$ for all $x \in [0,1]$ and

$$f(x) = \int_0^1 f'(y) \mathbf{1}_{y \leq x} \, dy = \langle f, k(x,.)\rangle_{\mathscr{F}}, \quad \text{for all } f \in \mathscr{F} \text{ and } x \in [0,1].$$

In this case the norm $\|f\|_{\mathscr{F}}$ corresponds simply to the L^2-norm of the derivative of f. The smaller is this norm, the smoother is f.

Example 2: RKHS associated to the Gaussian kernel.
Let us write $\mathbf{F}[f]$ for the Fourier transform in $\mathbb{R}^d$ with normalization

$$\mathbf{F}[f](\omega) = \frac{1}{(2\pi)^{d/2}} \int_{\mathbb{R}^d} f(t) e^{-\mathrm{i}\langle \omega,t\rangle}, \quad \text{for } f \in L^1(\mathbb{R}^d) \cap L^2(\mathbb{R}^d) \text{ and } \omega \in \mathbb{R}^d.$$

For any $\sigma > 0$, the functional space

$$\mathscr{F}_\sigma = \left\{ f \in C_0(\mathbb{R}^d) \cap L^1(\mathbb{R}^d) \text{ such that } \int_{\mathbb{R}^d} \left|\mathbf{F}[f](\omega)\right|^2 e^{\sigma|\omega|^2/2}\, d\omega < +\infty \right\},$$

endowed with the scalar product

$$\langle f, g \rangle_{\mathscr{F}_\sigma} = (2\pi\sigma^2)^{-d/2} \int_{\mathbb{R}^d} \overline{\mathbf{F}[f](\omega)} \mathbf{F}[g](\omega) e^{\sigma|\omega|^2/2}\, d\omega,$$

is an RKHS associated with the Gaussian kernel $k(x,y) = \exp(-\|y-x\|^2/2\sigma^2)$. Actually, for all $x \in \mathbb{R}^d$ the function $k(x,.)$ belongs to $\mathscr{F}_\sigma$, and straightforward computations give

$$\langle k(x,.), f \rangle_{\mathscr{F}_\sigma} = \mathbf{F}^{-1}\left[\mathbf{F}[f]\right](x) = f(x) \quad \text{for all } f \in \mathscr{F} \text{ and all } x \in \mathbb{R}^d.$$

The space $\mathscr{F}_\sigma$ gathers very regular functions, and the norm $\|f\|_{\mathscr{F}_\sigma}$ directly controls the smoothness of f. We note that when σ increases, the space $\mathscr{F}_\sigma$ shrinks and contains smoother and smoother functions.

We refer to Aronszajn [10] and Schölkopf and Smola [140] for more details on RKHS.

Notations

$$
\begin{aligned}
(x)_+ &= \max(0,x) \\
x \vee y &= \max(x,y) \\
x \wedge y &= \min(x,y) \\
\|\beta\| &= \sqrt{\sum_j \beta_j^2} \\
\langle x,y \rangle &= x^T y = \sum_j x_j y_j \\
|\beta|_1 &= \sum_j |\beta_j| \\
|\beta|_\infty &= \max_j |\beta_j| \quad \text{or} \quad \max_{i,j} |\beta_{ij}| \text{ if } \beta \text{ is a matrix} \\
\text{supp}(\beta) &= \{j : \beta_j \neq 0\} \\
|\beta|_0 &= \text{card}(\text{supp}(\beta)) \\
\beta_S &= [\beta_j]_{j \in S} \\
|S| &= \text{card}(S) \\
I_n &= \text{identity matrix on } \mathbb{R}^n \\
\partial_i F &= \text{partial derivative of } F \text{ according to variable } i \\
\text{div}(F) &= \sum_i \partial_i F \quad \text{(divergence of } F\text{)} \\
\nabla F &= \text{gradient of } F \\
\partial F(x) &= \text{subdifferential of } F \text{ at point } x \\
\sigma_j(A) &= j\text{-th largest singular value of } A \\
|A|_{\text{op}} &= \sigma_1(A) = \sup_{\|x\| \leq 1} \|Ax\| \\
|A|_* &= \sum_j \sigma_j(A) \\
\|A\|_F &= \sqrt{\sum_{ij} A_{ij}^2} = \sqrt{\sum_j \sigma_j(A)^2} \\
\|A\|_{(2,q)} &= \sqrt{\sum_{k=1}^q \sigma_k(A)^2}
\end{aligned}
$$

$$
\begin{aligned}
|A|_{1,\infty} &= \max_j \sum_i |A_{ij}| \\
A_{j:} &= j\text{-th row of matrix } A \\
A_j &= j\text{-th column of matrix } A \\
\mathrm{Proj}_S &= \text{orthogonal projector onto the linear span } S \\
C_p^d &= \frac{p!}{d!(n-d)!} \\
\mathrm{sign}(x) &= \mathbf{1}_{x>0} - \mathbf{1}_{x\leq 0} \\
x^T &= \text{transpose of vector or matrix } x \\
\underset{\beta\in\mathscr{C}}{\mathrm{argmin}}\, F(\beta) &= \text{set of the minimizers in } \mathscr{C} \text{ of } F \\
\mathscr{P}(E) &= \text{set gathering all the subsets of } E \\
E = F \bigoplus G &= \text{decomposition } E = F + G \text{ with } F \text{ orthogonal to } G \\
]a,b] &= \{x \in \mathbb{R} : a < x \leq b\} \\
\mathrm{i} &= \text{imaginary unit} \\
\mathrm{var}(X) &= \mathbb{E}\left[(X - \mathbb{E}[X])^2\right] \\
\mathrm{sdev}(X) &= \sqrt{\mathrm{var}(X)}
\end{aligned}
$$

Bibliography

[1] E. Abbe, J. Fan, and K. Wang. An ℓ_p theory of pca and spectral clustering. *preprint arXiv:2006.14062*, 2020.

[2] H. Akaike. Information theory and an extension of the maximum likelihood principle. In *Second International Symposium on Information Theory (Tsahkadsor, 1971)*, pages 267–281. Akadémiai Kiadó, Budapest, 1973.

[3] T. Anderson. *An Introduction to Multivariate Statistical Analysis*. Wiley, New York, NY, second edition, 1984.

[4] A. Antoniadis. Comments on: ℓ_1-penalization for mixture regression models. *TEST*, 19(2):257–258, 2010.

[5] S. Arlot. Model selection by resampling penalization. *Electron. J. Stat.*, 3:557–624, 2009.

[6] S. Arlot and F. Bach. Data-driven calibration of linear estimators with minimal penalties. In Y. Bengio, D. Schuurmans, J. Lafferty, C. K. I. Williams, and A. Culotta, editors, *Advances in Neural Information Processing Systems 22*, pages 46–54. 2009.

[7] S. Arlot and A. Célisse. A survey of cross-validation procedures for model selection. *Stat. Surv.*, 4:40–79, 2010.

[8] S. Arlot and A. Célisse. Segmentation of the mean of heteroscedastic data via cross-validation. *Stat. Comput.*, 21(4):613–632, 2011.

[9] S. Arlot and P. Massart. Data-driven calibration of penalties for least-squares regression. *J. Mach. Learn. Res.*, 10:245–279, 2010.

[10] N. Aronszajn. Theory of reproducing kernels. *Trans. Amer. Math. Soc.*, 68:337–404, 1950.

[11] P. Awasthi, M. Charikar, R. Krishnaswamy, and A. K. Sinop. The hardness of approximation of Euclidean K-means. *arXiv preprint arXiv:1502.03316*, 2015.

[12] K. Azuma. Weighted sums of certain dependent random variables. *Tohoku Mathematical Journal*, 19(3):357–367, 1967.

[13] F. Bach. Consistency of trace norm minimization. *J. Mach. Learn. Res.*, 9:1019–1048, 2008.

[14] F. Bach, R. Jenatton, J. Mairal, and G. Obozinski. Structured sparsity through convex optimization. *Statistical Science*, 27(4):450–468, 2012.

[15] O. Banerjee, L. El Ghaoui, and A. d'Aspremont. Model selection through

sparse maximum likelihood estimation for multivariate Gaussian or binary data. *J. Mach. Learn. Res.*, 9:485–516, 2008.

[16] Y. Baraud. Estimator selection with respect to Hellinger-type risks. *Probab. Theory Related Fields*, 151(1-2):353–401, 2011.

[17] Y. Baraud. About the lower bounds for the multiple testing problem. *preprint arXiv:1807.05410*, 2018.

[18] Y. Baraud, L. Birgé, and M. Sart. A new method for estimation and model selection: ρ-estimation. *Invent. math.*, 2017:425–517, 2017.

[19] Y. Baraud, C. Giraud, and S. Huet. Gaussian model selection with an unknown variance. *Ann. Statist.*, 37(2):630–672, 2009.

[20] Y. Baraud, C. Giraud, and S. Huet. Estimator selection in the gaussian setting. *Annales de l'Institut Henri Poincaré, Probabilités et Statistiques*, 50(3):1092–1119, 2014.

[21] A. Barron. *Are Bayes rules consistent in information? in Open problems in communication and computation. Cover, T.M. and Gopinath, B.* Springer-Verlag, 1987.

[22] A. Barron, L. Birgé, and P. Massart. Risk bounds for model selection via penalization. *Probab. Theory Related Fields*, 113(3):301–413, 1999.

[23] J.-P. Baudry, C. Maugis, and B. Michel. Slope heuristics: Overview and implementation. *Statist. Comput.*, 22(2):455–470, 2012.

[24] A. Beck and M. Teboulle. A fast iterative shrinkage-thresholding algorithm for linear inverse problems. *SIAM J. Img. Sci.*, 2(1):183–202, 2009.

[25] A. Belloni, V. Chernozhukov, and L. Wang. Square-root lasso: Pivotal recovery of sparse signals via conic programming. *Biometrika*, 98(4):791–806, 2011.

[26] Y. Benjamini and Y. Hochberg. Controlling the false discovery rate: a practical and powerful approach to multiple testing. *JRSS B*, 57(1):289–300, 1995.

[27] Y. Benjamini and D. Yekutieli. The control of the false discovery rate in multiple testing under dependency. *Annals of Statistics*, 29:1165–1188, 2001.

[28] Q. Berthet and P. Rigollet. Optimal detection of sparse principal components in high dimension. *The Annals of Statistics*, 41(4):1780–1815, 2013.

[29] P. Bickel, Y. Ritov, and A. Tsybakov. Simultaneous analysis of lasso and Dantzig selector. *Ann. Statist.*, 37(4):1705–1732, 2009.

[30] L. Birgé. A new lower bound for multiple hypothesis testing. *IEEE Trans. Inform. Theory*, 51:1611–1615, 2005.

[31] L. Birgé. Model selection via testing: an alternative to (penalized) maximum likelihood estimators. *Ann. Inst. H. Poincaré Probab. Statist.*, 42(3):273–325, 2006.

[32] L. Birgé and P. Massart. Gaussian model selection. *J. Eur. Math. Soc. (JEMS)*, 3(3):203–268, 2001.

[33] L. Birgé and P. Massart. Minimal penalties for Gaussian model selection. *Probab. Theory Related Fields*, 138(1-2):33–73, 2007.

[34] G. Blanchard and E. Roquain. Two simple sufficient conditions for FDR control. *Electron. J. Stat.*, 2:963–992, 2008.

[35] T. Blumensath and M. E. Davies. Iterative hard thresholding for compressed sensing. *Applied and Computational Harmonic Analysis*, 27(3):265 - 274, 2009.

[36] M. Bogdan, E. van den Berg, C. Sabatti, W. Su, and E. J. Candès. Slope - adaptive variable selection via convex optimization. *Ann. Appl. Stat.*, 9(3):1103–1140, 2015.

[37] S. Boucheron, O. Bousquet, and G. Lugosi. Theory of classification: some recent advances. *ESAIM Probability & Statistics*, 9:323–375, 2005.

[38] S. Boucheron, G. Lugosi, and P. Massart. *Concentration Inequalities*. Oxford University Press, 2013.

[39] S. Bubeck. Theory of convex optimization for machine learning. *arXiv:1405.4980*, 2014.

[40] F. Bunea, C. Giraud, X. Luo, M. Royer, and N. Verzelen. Model assisted variable clustering: Minimax-optimal recovery and algorithms. *The Annals of Statistics*, 48(1), 2020.

[41] F. Bunea, C. Giraud, M. Royer, and N. Verzelen. PECOK: a convex optimization approach to variable clustering. *arXiv preprint arXiv:1606.05100*, 2016.

[42] F. Bunea, J. Lederer, and Y. She. The square root group lasso: theoretical properties and fast algorithms. *IEEE Transactions on Information Theory*, 30(2):1313–1325, 2014.

[43] F. Bunea, Y. She, and M. Wegkamp. Optimal selection of reduced rank estimators of high-dimensional matrices. *Ann. Stat.*, 39(2):1282–1309, 2011.

[44] F. Bunea, Y. She, and M. Wegkamp. Joint variable and rank selection for parsimonious estimation of high-dimensional matrices. *The annals of statistics*, 40(5):2359–2388, 2012.

[45] F. Bunea, A. Tsybakov, and M. Wegkamp. Sparsity oracle inequalities for the lasso. *Electronic Journal of Statistics*, 1:169–194, 2007.

[46] T. Cai, W. Liu, and X. Luo. A constrained ℓ^1 minimization approach to sparse precision matrix estimation. *Journal of the American Statistical Association*, 106(494):594–607, 2011.

[47] E. Candès and T. Tao. The Dantzig selector: statistical estimation when p is much larger than n. *Ann. Statist.*, 35(6):2313–2351, 2007.

[48] E. J. Candès, J. Romberg, and T. Tao. Robust uncertainty principles: exact signal reconstruction from highly incomplete frequency information. *IEEE Transactions on Information Theory*, 52(2):489–509, 2006.

[49] R. Castelo and A. Roverato. A robust procedure for Gaussian graphical model

search from microarray data with p larger than n. *J. Mach. Learn. Res.*, 7:2621–2650, 2006.

[50] O. Catoni. A mixture approach to universal model selection. Technical report, Ecole Normale Superieure, 1997.

[51] O. Catoni. Statistical learning theory and stochastic optimization. In *Lectures on Probability Theory and Statistics, École d'Été de Probabilités de Saint-Flour XXXI — 2001*, volume 1851 of *Lecture Notes in Mathematics*, pages 1–269, New York, 2004. Springer.

[52] D. Ćevid, P. Bühlmann, and N. Meinshausen. Spectral deconfounding via perturbed sparse linear models. *Journal of Machine Learning Research*, 21(232):1–41, 2020.

[53] G. Chagny. Penalization versus goldenshluger–lepski strategies in warped bases regression. *ESAIM: Probability and Statistics*, 17:328–358, 2013.

[54] V. Chandrasekaran, P. Parrilo, and A. Willsky. Latent variable graphical model selection via convex optimization. *The Annals of Statistics*, 40(4):1935–1967, 2012.

[55] S. Chatterjee. A new perspective on least squares under convex constraint. *Ann. Statist.*, 42(6):2340–2381, 2014.

[56] S. Chen, D. Donoho, and M. Saunders. Atomic decomposition by basis pursuit. *SIAM J. Sci. Comput.*, 20(1):33–61, 1998.

[57] X. Chen and Y. Yang. Cutoff for exact recovery of gaussian mixture models. *preprint arXiv:2001.01194*, 2020.

[58] S. Clémençon, G. Lugosi, and N. Vayatis. Ranking and empirical risk minimization of u-statistics. *The Annals of Statistics*, 36:844–874, 2008.

[59] A. Dalalyan and A. Tsybakov. Aggregation by exponential weighting, sharp oracle inequalities and sparsity. *Machine Learning*, 72(1-2):39– 61, 2008.

[60] A. S. Dalalyan, E. Grappin, and Q. Paris. On the exponentially weighted aggregate with the laplace prior. *Ann. Statist.*, 46(5):2452–2478, 2018.

[61] K. Davidson and S. Szarek. Local operator theory, random matrices and Banach spaces. In *Handbook of the geometry of Banach spaces, Vol. I*, pages 317–366. North-Holland, Amsterdam, 2001.

[62] L. Devroye, L. Györfi, and G. Lugosi. *A Probabilistic Theory of Pattern Recognition*. Springer, Berlin, 1996.

[63] L. Devroye and T. Wagner. The L_1 convergence of kernel density estimates. *Ann. Statist.*, 7(5):1136–1139, 1979.

[64] T. Dickhaus. *Simultaneous Statistical Inference*. Springer, Berlin, 2014.

[65] D. Donoho. High-dimensional data analysis: The curses and blessings of dimensionality, 2000.

[66] D. Donoho, I. Johnstone, and A. Montanari. Accurate prediction of phase transitions in compressed sensing via a connection to minimax denoising. *Information Theory, IEEE Transactions on*, 59(6):3396–3433, 2013.

[67] D. L. Donoho. Compressed sensing. *IEEE Transactions on Information Theory*, 52(4):1289–1306, 2006.

[68] S. Dudoit, J. Fridlyand, and T. Speed. Comparison of discrimination methods for the classification of tumors using gene expression data. *JASA*, 97(457):77–87, 2002.

[69] S. Dudoit and M. van der Laan. *Multiple Testing Procedures with Applications to Genomics*. Springer Series in Statistics, 2008.

[70] B. Efron, T. Hastie, I. Johnstone, and R. Tibshirani. Least angle regression. *Ann. Statist.*, 32(2):407–499, 2004.

[71] R. M. Fano and D. Hawkins. Transmission of information: A statistical theory of communications. *American Journal of Physics*, 29(11):793–794, 1961.

[72] J. Friedman, T. Hastie, H. Hofling, and R. Tibshirani. Pathwise coordinate optimization. *Ann. Appl. Statist.*, (1):302–332, 2007.

[73] J. Friedman, T. Hastie, and R. Tibshirani. Sparse inverse covariance estimation with the graphical Lasso. *Biostatistics*, 9(3):432–441, 2008.

[74] J. Friedman, T. Hastie, and R. Tibshirani. A note on the group lasso and a sparse group lasso, 2010. arXiv:1001.0736.

[75] S. Geisser. The predictive sample reuse method with applications. *J. Amer. Statist. Assoc.*, 70:320–328, 1975.

[76] S. Gerchinovitz. Sparsity regret bounds for individual sequences in online linear regression. *JMLR Workshop and Conference Proceedings*, 19 (COLT 2011 Proceedings):377–396, 2011.

[77] S. Gerchinovitz, P. Ménard, and G. Stoltz. Fano's inequality for random variables. *Statist. Sci.*, 35(2):178–201, 2020.

[78] C. Giraud. Estimation of Gaussian graphs by model selection. *Electron. J. Stat.*, 2:542–563, 2008.

[79] C. Giraud. Mixing least-squares estimators when the variance is unknown. *Bernoulli*, 14(4):1089–1107, 2008.

[80] C. Giraud. Low rank multivariate regression. *Electron. J. Stat.*, 5:775–799, 2011.

[81] C. Giraud, S. Huet, and N. Verzelen. Graph selection with GGMselect. *Stat. Appl. Genet. Mol. Biol.*, 11(3):1–50, 2012.

[82] C. Giraud, S. Huet, and N. Verzelen. High-dimensional regression with unknown variance. *Stat. Sci.*, 27(4):500–518, 2012.

[83] C. Giraud and N. Verzelen. Partial recovery bounds for clustering with the relaxed k-means. *Mathematical Statistics and Learning*, pages 317–374, 2018.

[84] J. Goeman and A. Solari. Multiple hypothesis testing in genomics. *Statistics in Medicine*, 33(11):1946–1978, 2014.

[85] A. Goldenshluger and O. Lepski. Bandwidth selection in kernel density estimation: oracle inequalities and adaptive minimax optimality. *Ann. Statist.*,

39(3):1608–1632, 2011.

[86] T. Golub, D. Slonim, P. Tamayo, C. Huard, M. Gaasenbeek, J. Mesirov, H. Coller, M. Loh, J. Downing, M. Caligiuri, and C. Bloomfield. Molecular classification of cancer: class discovery and class prediction by gene expression monitoring. *Science*, 286:531–537, 1999.

[87] A. Guntuboyina. Lower bounds for the minimax risk using f-divergences, and applications. *IEEE Transactions on Information Theory*, 57(4):2386–2399, 2011.

[88] W. Guo and M. Rao. On control of the false discovery rate under no assumption of dependency. *J. Statist. Plann. Inference*, 138(10):3176–3188, 2008.

[89] J. Hammersley and P. Clifford. Markov fields on finite graphs and lattices. Technical report, 1971.

[90] R. Z. Has'minskii. A lower bound on the risks of non-parametric estimates of densities in the uniform metric. *Theory of Probability & Its Applications*, 23(4):794–798, 1979.

[91] T. Hastie, R. Tibshirani, and J. Friedman. *The elements of statistical learning*. Springer Series in Statistics. Springer, New York, second edition, 2009.

[92] H. Hazimeh, R. Mazumder, and A. Saab. Sparse regression at scale: Branch-and-bound rooted in first-order optimization. *preprint arXiv:2004.06152*, 2020.

[93] R. Horn and C. Johnson. *Matrix analysis. 2nd ed.* Cambridge: Cambridge University Press xviii, 2013.

[94] I. Ibragimov, V. Sudakov, and B. Tsirel'son. *Norms of Gaussian sample functions*, volume 550 of *Lecture Notes in Mathematics*. Springer, 1976.

[95] P. Jain, A. Tewari, and P. Kar. On iterative hard thresholding methods for high-dimensional m-estimation. In Z. Ghahramani, M. Welling, C. Cortes, N. D. Lawrence, and K. Q. Weinberger, editors, *Advances in Neural Information Processing Systems 27*, pages 685–693. Curran Associates, Inc., 2014.

[96] J. Jin. Impossibility of successful classification when useful features are rare and weak. *Proceedings of the National Academy of Sciences*, 106(22):8859–8864, 2009.

[97] M. Kalisch and P. Bühlmann. Estimating high-dimensional directed acyclic graphs with the pc-algorithm. *J. Mach. Learn. Res.*, 8:613–636, 2007.

[98] M. Kalisch and P. Bühlmann. Robustification of the pc-algorithm for directed acyclic graphs. *J. Comput. Graph. Statist.*, 17(4):773–789, 2008.

[99] V. Koltchinskii, K. Lounici, and A. Tsybakov. Nuclear norm penalization and optimal rates for noisy low rank matrix completion. *Annals of Statistics*, 39(5):2302–2329, 2011.

[100] J. Lafferty, H. Liu, and L. Wasserman. Sparse nonparametric graphical models. *Statistical Science*, 27:519–537, 2012.

[101] S. Lauritzen. *Graphical Models*. Oxford University Press, 1996.

[102] E. Lebarbier. Detecting multiple change-points in the mean of gaussian process by model selection. *Signal Processing*, 85:717–736, 2005.

[103] L. LeCam. Convergence of estimates under dimensionality restrictions. *Ann. Statist.*, 1(1):38–53, 1973.

[104] Y. LeCun, L. Bottou, Y. Bengio, and P. Haffner. Gradient-based learning applied to document recognition. In *Proceedings of the IEEE*, volume 86, pages 2278–2324, 1998.

[105] M. Ledoux. *The concentration of measure phenomenon*. Mathematical surveys and monographs. American Mathematical Society, Providence (R.I.), 2001.

[106] M. Ledoux and M. Talagrand. *Probability in Banach Spaces*. Springer, 1991.

[107] O. Lepski. Asymptotically minimax adaptive estimation. i. upper bounds. optimally adaptive estimates. *Teor. Veroyatnost i Primenen*, 36:645–659, 1991.

[108] M. Lerasle and S. Arlot. Why v=5 is enough in v-fold cross-validation. arXiv:1210.5830, 2012.

[109] G. Leung and A. Barron. Information theory and mixing least-squares regressions. *IEEE Trans. Inform. Theory*, 52(8):3396–3410, 2006.

[110] K.-C. Li. Asymptotic optimality for C_p, C_L, cross-validation and generalized cross-validation: discrete index set. *Ann. Statist.*, 15(3):958–975, 1987.

[111] H. Liu and R. Foygel Barber. Between hard and soft thresholding: optimal iterative thresholding algorithms. *Information and Inference: A Journal of the IMA*, 9:899–933, 2020.

[112] H. Liu, F. Han, M. Yuan, J. Lafferty, and L. Wasserman. High-dimensional semiparametric gaussian copula graphical models. *The Annals of Statistics*, 40(4):2293–2326, 2012.

[113] S. Lloyd. Least Squares Quantization in PCM. *IEEE Trans. Inf. Theor.*, 28(2):129–137, 1982.

[114] Y. Lu and H. H. Zhou. Statistical and Computational Guarantees of Lloyd's Algorithm and its Variants. *preprint ArXiv:1612.02099*, 2016.

[115] J. MacQueen. Some methods for classification and analysis of multivariate observations. Proc. 5th Berkeley Symp. Math. Stat. Probab., Univ. Calif. 1965/66, 1, 281-297, 1967.

[116] C. Mallows. Some comments on c_p. *Technometrics*, 15:661–675, 1973.

[117] V. Marčenko and L. Pastur. Distribution of eigenvalues for some sets of random matrices. *Mathematics of the USSR-Sbornik*, 1(4):457–483, 2007.

[118] P. Massart. *Concentration inequalities and model selection*, volume 1896 of *Lecture Notes in Mathematics*. Springer, Berlin, 2007. Lectures from the 33rd Summer School on Probability Theory held in Saint-Flour, July 6–23, 2003.

[119] C. McDiarmid. On the method of bounded differences. *Surveys in Combinatorics*, 141:148–188, 1989.

[120] N. Meinshausen and P. Bühlmann. High-dimensional graphs and variable selection with the lasso. *Ann. Statist.*, 34(3):1436–1462, 2006.

[121] N. Meinshausen, G. Rocha, and B. Yu. Discussion: a tale of three cousins: Lasso, L2boosting and Dantzig. *Annals of Statistics*, 35(6):2373–2384, 2007.

[122] D. G. Mixon, S. Villar, and R. Ward. Clustering subgaussian mixtures by semidefinite programming. *Information and Inference: A Journal of the IMA*, 6(4):389–415, 2017.

[123] F. Mosteller and J. Tukey. Data analysis, including statistics. In G. Lindsey and E. Aronson, editors, *Handbook of Social Psychology, Vol. 2*. Addison-Wesley, 1968.

[124] M. Ndaoud. Sharp optimal recovery in the Two Component Gaussian Mixture Model. *arXiv:1812.08078*, 2018.

[125] M. Ndaoud. Scaled minimax optimality in high-dimensional linear regression: A non-convex algorithmic regularization approach. *preprint arXiv:2008.12236*, 2020.

[126] Y. Nesterov. A method of solving a convex programming problem with convergence rate $O(1/\sqrt{k})$. *Soviet Mathematics Doklady*, 27:372–376, 1983.

[127] J. Norris. *Markov chains*. Cambridge series in statistical and probabilistic mathematics. Cambridge University Press, 1998.

[128] J. Peng and Y. Wei. Approximating K-means-type clustering via semi-definite programming. *SIAM J. on Optimization*, 18(1):186–205, 2007.

[129] G. Pisier. Probabilistic methods in the geometry of banach spaces. In G. Letta and M. Pratelli, editors, *Probability and Analysis*, pages 167–241. Springer Berlin Heidelberg, 1986.

[130] P. Ravikumar, M. Wainwright, G. Raskutti, and B. Yu. High-dimensional covariance estimation by minimizing, ℓ^1-penalized log-determinant divergence. *Electronic Journal of Statistics*, 5:935–980, 2011.

[131] D. Revuz and M. Yor. *Continuous Martingales and Brownian Motion*. Grundlehren der mathematischen Wissenschaften. Springer-Verlag, 3rd edition, 1999.

[132] P. Rigollet and A. Tsybakov. Exponential screening and optimal rates of sparse estimation. *Annals of Statistics*, 39(2):731–771, 2011.

[133] P. Rigollet and A. Tsybakov. Sparse estimation by exponential weighting. *Statistical Science*, 27(4):558–575, 2012.

[134] C. Robert and G. Casella. *Monte Carlo Statistical Methods*. Springer-Verlag, 2nd edition, 2004.

[135] C. Robert and G. Casella. *Méthodes de Monte-Carlo avec R*. Springer Paris, 2011.

[136] E. Roquain. Type I error rate control for testing many hypotheses: a survey with proofs. *J. SFdS*, 152(2):3–38, 2011.

[137] M. Royer. Adaptive clustering through semidefinite programming. In

I. Guyon, U. V. Luxburg, S. Bengio, H. Wallach, R. Fergus, S. Vishwanathan, and R. Garnett, editors, *Advances in Neural Information Processing Systems*, volume 30. Curran Associates, Inc., 2017.

[138] V. Sabarly, O. Bouvet, J. Glodt, O. Clermont, D. Skurnik, L. Diancourt, D. de Vienne, E. Denamur, and C. Dillmann. The decoupling between genetic structure and metabolic phenotypes in escherichia coli leads to continuous phenotypic diversity. *Journal of Evolutionary Biology*, 24:1559–1571, 2011.

[139] A. Sanchez-Perez. Time series prediction via aggregation : an oracle bound including numerical cost. In *Modeling and Stochastic Learning for Forecasting in High Dimension*, Lecture Notes in Statistics. Springer, 2014.

[140] B. Schölkopf and A. Smola. *Learning with kernels: support vector machines, regularization, optimization, and beyond.* MIT Press, Cambridge, MA, USA, 2001.

[141] G. Schwarz. Estimating the dimension of a model. *Ann. Statist.*, 6(2):461–464, 1978.

[142] J. Shao. Linear model selection by cross-validation. *J. Amer. Statist. Assoc.*, 88(422):486–494, 1993.

[143] D. Slepian. The one-sided barrier problem for gaussian noise. *Bell System Technical Journal*, 41(2):463–501, 1962.

[144] P. Spirtes, C. Glymour, and R. Scheines. *Causation, prediction, and search.* Adaptive Computation and Machine Learning. MIT Press, Cambridge, MA, second edition, 2000.

[145] N. Städler, P. Bühlmann, and S. van de Geer. ℓ_1-penalization for mixture regression models. *TEST*, 19(2):209–256, 2010.

[146] H. Steinhaus. Sur la division des corps matériels en parties. *Bull. Acad. Pol. Sci., Cl. III*, 4:801–804, 1957.

[147] M. Stone. Cross-validatory choice and assessment of statistical predictions. *J. Roy. Statist. Soc. Ser. B*, 36:111–147, 1974.

[148] W. Su, M. Bogdan, and E. Candès. False discoveries occur early on the lasso path. *Ann. Statist.*, 45(5):2133–2150, 2017.

[149] T. Sun and C.-H. Zhang. Scaled sparse linear regression. *Biometrika*, 99(4):879–898, 2012.

[150] R. Tibshirani. Regression shrinkage and selection via the lasso. *J. Roy. Statist. Soc. Ser. B*, 58(1):267–288, 1996.

[151] R. Tibshirani, M. Saunders, S. Rosset, J. Zhu, and K. Knight. Sparsity and smoothness via the fused lasso. *J. R. Stat. Soc. Ser. B Stat. Methodol.*, 67(1):91–108, 2005.

[152] L. Tierney and J. Kadane. Accurate approximations for posterior moments and marginal densities. *J. Am. Stat. Assoc.*, 81:82–86, 1986.

[153] A. B. Tsybakov. *Introduction to Nonparametric Estimation.* Springer Pub-

lishing Company, Incorporated, 1st edition, 2008.

[154] S. Van de Geer. High-dimensional generalized linear models and the lasso. *The Annals of Statistics*, 36(2):614–645, 2008.

[155] V. Vapnik and A. Chervonenkis. On the uniform convergence of relative frequencies of events to their probabilities. *Theory of Probability and its Applications*, 16(2):264–280, 1971.

[156] V. Vapnik and A. Chervonenkis. *Theory of Pattern Recognition.* Nauka, Moscow, 1974.

[157] S. Vempala and G. Wang. A spectral algorithm for learning mixture models. *Journal of Computer and System Sciences*, 68(4):841–860, 2004. (Special Issue on FOCS 2002).

[158] R. Vershynin. Introduction to the non-asymptotic analysis of random matrices. arXiv:1011.3027, 2011.

[159] R. Vershynin. *High-Dimensional Probability: An Introduction with Applications in Data Science.* Cambridge Series in Statistical and Probabilistic Mathematics. Cambridge University Press, 2018.

[160] N. Verzelen. Minimax risks for sparse regressions: Ultra-high-dimensional phenomenons. *Electron. J. Stat.*, 6:38–90, 2012.

[161] F. Villers, B. Schaeffer, C. Bertin, and S. Huet. Assessing the validity domains of graphical Gaussian models in order to infer relationships among components of complex biological systems. *Statistical Applications in Genetics and Molecular Biology*, 7, 2008.

[162] M. Wainwright. Information-theoretic limits on sparsity recovery in the high-dimensional and noisy setting. *IEEE Trans. Inform. Theory*, 55(12):5728–5741, 2009.

[163] M. J. Wainwright. *High-dimensional statistics: a non-asymptotic viewpoint.* Cambridge series in statistical and probabilistic mathematics. Cambridge University Press, 2019.

[164] A. Wille and P. Bühlmann. Low-order conditional independence graphs for inferring genetic networks. *Stat. Appl. Genet. Mol. Biol.*, 5:Art. 1, 34 pp. (electronic), 2006.

[165] L. Xue and H. Zou. Regularized rank-based estimation of high-dimensional nonparanormal graphical models. *The Annals of Statistics*, 40(5):2541–2571, 2012.

[166] M. Yu, V. Gupta, and M. Kolar. Recovery of simultaneous low rank and two-way sparse coefficient matrices, a nonconvex approach. *Electron. J. Statist.*, 14(1):413–457, 2020.

[167] M. Yuan and Y. Lin. Model selection and estimation in regression with grouped variables. *J. R. Stat. Soc. Ser. B Stat. Methodol.*, 68(1):49–67, 2006.

[168] Y. Yuhong and A. Barron. Information-theoretic determination of minimax rates of convergence. *Ann. Statist.*, 27(5):1564–1599, 1999.

[169] T. Zhang. Adaptive forward-backward greedy algorithm for learning sparse representations. *IEEE Trans. Inform. Theory*, 57(7):4689–4708, 2011.

[170] Y. Zhang, M. Wainwright, and M. Jordan. Lower bounds on the performance of polynomial-time algorithms for sparse linear regression. *JMLR: Workshop and Conference Proceedings*, 35:1–28, 2014.

[171] H. Zou. The adaptive lasso and its oracle properties. *Journal of the American Statistical Association*, 101:1418–1429, 2006.

[172] H. Zou and T. Hastie. Regularization and variable selection via the elastic net. *J. R. Stat. Soc. Ser. B Stat. Methodol.*, 67(2):301–320, 2005.

Index

adaptive-Lasso, **103**
AIC, 33, 40, 45, 47, 51

Bayes classifier, **220**, 221, 223, 236, 247
Benjamini–Hochberg procedure, **213**, 216
Benjamini–Yekutieli procedure, **211**, 216
BIC, **45**, 47
Birgé's inequality, **309**
block coordinate descent, 98, 103, 106, 113, 116, 189, 195
Bonferroni correction, 207, 216
boosting, 235, 244

chi-square distribution, 38, 144, 155, **294**
classifier, 219
compatibility constant, 95, 98, 111, 145, 146, 148
compress sensing, 120
concentration inequality, 17, 24, 25, 39, 150, 156, 164, 166, 171, 175, 178, 192, 197, 199, 200, 226, 238, **299**, **301**
conditional correlation, 186, 296
conditional independence, **180**
contraction principle, 238, **306**
Convex constrained estimators, 53
convex function, 24, 89, 91, 104, 113, 115, 150, 172, 187, 189, 233, 237, 241, 298, 321
cross-validation, **141**, 142, 154, 245

Dantzig selector, **113**, 196, 199
Data processing inequality, 71
Davis-Kahan perturbation bound, 318
Dendrogram, 255
dictionary, 223, 231
dynamic programming, 51

Elastic Net, 116
EM algorithm, 290
empirical covariance matrix, 12, 185, 186, 193, 196, 221
empirical risk classification, 233
empirical risk minimization, 223

factorization formula, **182**, 184, 194
Fano's lemma, 57, 73
FDR: False Discovery Rate, **207**, 209, 211, 213, 216
first order optimality condition, **323**
FISTA: Fast Iterative Shrinkage Thresholding Algorithm, **100**
forward algorithm, 157
forward–backward algorithm, 43
Frobenius norm, 23, **160**, 161, 198, 250, **313**
fused–Lasso estimator, **107**

Gaussian concentration inequality, 25, 301
Gaussian copula graphical models, 198
Gibbs distribution, 76, 81, 83, 85
Goldenshluger–Lepski method, 52
graphical models
    directed acyclic models, **181**, 194
    Gaussian, **185**, 195, 196
    non-directed models, **183**
graphical-Lasso estimator, **187**, 196
group sparse, 32, 48, 84, 103, 112, 145, 167, 175, 189, 195
group-Lasso estimator, 103, **104**, 105, 106, 112, 145, 168, 189, 195

Hanson-Wright inequality, 303
hard thresholding, 47
hidden variables, 192, 195

Hierarchical clustering, **253**, 284

Iterative Group Thresholding, 131, 136
Iterative Hard Thresholding, 124

Jensen inequality, 227, 239, **298**, 305, 309

Kendall's tau, **199**
kernel trick, 244
Kmeans, 252, 257, 285
Kullback–Leibler divergence, 56, 76, 83, **292**, 309
Ky–Fan norm, **160**, 163, 166, 170, 178, **315**

LARS algorithm, 101
Lasso estimator, **91**, 95, 98, 102, 108, 109, 141, 144, 147, 155, 172, 187, 188
LDA: Linear Discriminant Analysis, 15, **220**, 241, 246, 270
linear regression, 7, 13, 23, 28–30, 48, 84, 144, 146, 157, 159, 185, 188
LINselect, **143**, 144, 155
Lloyd algorithm, 264, 276
logistic regression, **222**
loss function, 234

Mahalanobis distance, **247**
margin hyperplane, **241**
Markov inequality, 9, **297**, 299
matrix
- coordinate-sparse, 187, 188, 195, 196
- low rank, 161, 165, 168, 172, 177, 195, 196
- row sparse, 167, 168

Max–Min formula, **314**, 316
McDiarmid inequality, 199, 226, 238, **299**
Metropolis–Hastings sampling, **79**, 80, 84
minimax, 48, 67, 69, 164, 309
Minimax risk, 56
minimax risk, 64
misclassification probability, 223
model, **31**, 75, 91, 223
model selection estimator, **35**, 46, 91, 94, 97, 168, 223
monotonicity (subgradient), 95, 173, **322**
Moore–Penrose pseudo-inverse, 94, 115, 161, 174, **312**
moral graph, 183, 184, 194
multiple testing, 11, 186, 204
multivariate regression, 159, 188

nuclear norm, **172**, 175, **313**, 324

operator norm, **160**, 163, 166, 168, **313**, 316, 324
oracle, **31**, 37, 44, 76, 95, 152, 165, 169, 172, 231

p-value, 204, 206, 211, 214, 217
PCA: Principal Component Analysis, 14, **22**, 42, 243, 249
penalty, **35**, 47, 144, 155, 165, 169, 188, 189, 195, 232, 235
precision matrix, 185, 187
Proximal method, 124

R package
- flare, 113, 198
- gglasso, 106
- GGMselect, 194
- glasso, 187
- glmnet, 99
- huge, 193
- kernalb, 241
- lars, 101
- LINselect, 144
- multtest, 214
- scalreg, 150
- SGL, 107

random matrix, 13, 163
rank selection, 165, 177
ranking, 246
representation formula, **235**, 240, 249
reproducing property, 235, 239, 243, **325**, 326

restricted eigenvalue, 145, 189
restricted isometry, 64, 200
reversible Markov chain, 80
ridge regression, **115**
RKHS: Reproducing Kernel Hilbert Space, **235**, 240, 243, 248, 249, **325**

scaled Lasso, **151**
separating hyperplane, 221, 222, 241
shattering coefficient, 224, 225, 228
shrinkage, 94, 102
singular value, 160, 162, 172, 176, 200, **311**, 313, 314, 316, 329
singular vector, **311**
Slope estimator, 118
slope heuristic, 153
sparse regression, 28, 30, 44, 46, 63, 67, 81, 82, 84, 90, 103, 106, 144, 146, 167, 188, 196
sparse–group Lasso estimator, **106**
Spectral clustering, 267, 271
square-root Lasso estimator, **147**, 151, 155
Stein formula, **76**, 77
Sterling numbers of second kind, 289
subdifferential, **89**, 90, 92, 104, 173, **321**
subgradient, **89**, 90, 95, 173, **321**
supervised classification, 219
SVD: Singular Value Decomposition, 22, 113, 115, 160, 161, 163, 172, 176, **311**, 314, 324
SVM: Support Vector Machine, **240**, 241, 243, 248
symmetrization lemma, 226, **305**

VC dimension, **228**, 229–231, 247, 248

weak learner, **235**, 245
Weyl inequality, 177, **316**
witness approach, 109
WPRD: Weak Positive Regression Dependency, 211, 213, 217

Printed in the United States
by Baker & Taylor Publisher Services